SUPERCONTINUUM GENERATION IN OPTICAL FIBERS

The optical fiber based supercontinuum source has recently become a significant scientific and commercial success, with applications ranging from frequency comb production to advanced medical imaging. This unique book explains the theory of fiber supercontinuum broadening, describes the diverse operational regimes and indicates principal areas of applications, making it an indispensable guide for researchers and graduate students.

With contributions from major figures and groups who have pioneered research in this field, the book describes the historical development of the subject, provides a background to the associated nonlinear optical processes, treats the generation mechanisms from continuous wave to femtosecond pulse pump regimes and highlights several important applications. A full discussion of numerical methods and comprehensive computer code are also provided, enabling readers to confidently predict and model supercontinuum generation characteristics under realistic conditions.

J. M. DUDLEY is a Professor in the Département d'Optique P. M. Duffieux at the Université de Franche-Comté and CNRS research institute FEMTO-ST in Besançon, France. His research has spanned both experimental and theoretical nonlinear optics and photonics.

J. R. TAYLOR is a Professor in the Physics Department at Imperial College London. He has made contributions in various aspects of laser research, photonics, ultrafast processes, optical fibers and nonlinear optics.

SUPERCONTINUUM GENERATION IN OPTICAL FIBERS

Edited by

J. M. DUDLEY
Université de Franche-Comté

J. R. TAYLOR
Imperial College of Science, Technology and Medicine, London

CAMBRIDGE
UNIVERSITY PRESS

CAMBRIDGE
UNIVERSITY PRESS

Shaftesbury Road, Cambridge CB2 8EA, United Kingdom

One Liberty Plaza, 20th Floor, New York, NY 10006, USA

477 Williamstown Road, Port Melbourne, VIC 3207, Australia

314–321, 3rd Floor, Plot 3, Splendor Forum, Jasola District Centre, New Delhi – 110025, India

103 Penang Road, #05–06/07, Visioncrest Commercial, Singapore 238467

Cambridge University Press is part of Cambridge University Press & Assessment, a department of the University of Cambridge.

We share the University's mission to contribute to society through the pursuit of education, learning and research at the highest international levels of excellence.

www.cambridge.org
Information on this title: www.cambridge.org/9780521514804

First published 2010

A catalogue record for this publication is available from the British Library

ISBN 978-0-521-51480-4 Hardback

Contents

Contributors

S. Afshar V.
Centre of Expertise in Photonics
School of Chemistry & Physics
University of Adelaide
Adelaide SA 5005
Australia

O. Bang
DTU Fotonik
Department of Photonics Engineering
Technical University of Denmark,
Ørsteds Plads 343
DK-2800 Kgs. Lyngby
Denmark

S. Coen
Department of Physics
University of Auckland
Private Bag 92019
Auckland
New Zealand

S. T. Cundiff
JILA
National Institute of Standards and Technology and University of Colorado
440 UCB
Boulder, CO 80309-0440
USA

J. M. Dudley
Laboratoire d'Optique P.M. Duffieux
Université de Franche-Comté
Institut FEMTO CNRS UMR 6174
Besançon
France

C. Dunsby
Physics Department and
Department of Histopathology,
Division of Investigative Science
Imperial College London
South Kensington Campus
London SW7 2AZ
United Kingdom

B. J. Eggleton
Centre for Ultrahigh-bandwidth
Devices for Optical Systems
(CUDOS)
School of Physics
University of Sydney
NSW 2006
Australia

P. M. W. French
Photonics Group
Physics Department
Imperial College London
South Kensington Campus
London SW7 2AZ
United Kingdom

M. H. Frosz
DTU Fotonik
Department of Photonics Engineering
Technical University of Denmark,
Ørsteds Plads 343
DK-2800 Kgs. Lyngby
Denmark

G. Genty
Optics Laboratory
Department of Physics
Tampere University of Technology
P.O. Box 692
FIN-33101 Tampere
Finland

A. V. Gorbach
Centre for Photonics and Photonic Materials
Department of Physics
University of Bath
Claverton Down
Bath BA2 7AY
United Kingdom

J. C. Knight
Centre for Photonics and
Photonic Materials
Department of Physics
University of Bath
Claverton Down
Bath BA2 7AY
United Kingdom

M. Lamont
Centre for Ultrahigh-bandwidth
Devices for Optical Systems
(CUDOS)
School of Physics
University of Sydney
NSW 2006
Australia

B. Luther Davies
Centre for Ultrahigh-bandwidth
Devices for Optical Systems
(CUDOS)
Laser Physics Centre
Australian National University
Canberra
ACT 0200
Australia

T. M. Monro
Centre of Expertise in Photonics
School of Chemistry & Physics
University of Adelaide
Adelaide SA 5005
Australia

P. M. Moselund
DTU Fotonik
Department of Photonics Engineering
Technical University of Denmark,
Ørsteds Plads 343
DK-2800 Kgs. Lyngby
Denmark

S. G. Murdoch
Department of Physics
University of Auckland
Private Bag 92019
Auckland
New Zealand

J. H. V. Price
Optoelectronics Research Centre
University of Southampton
Southampton SO17 1BJ
United Kingdom

J. K. Ranka
Lincoln Laboratory
Massachusetts Institute of Technology
244 Wood Street
Lexington, MA 02420
USA

P. D. Rasmussen
DTU Fotonik
Department of Photonics Engineering
Technical University of Denmark,
Ørsteds Plads 343
DK-2800 Kgs. Lyngby
Denmark

D. J. Richardson
Optoelectronics Research Centre
University of Southampton
Southampton SO17 1BJ
United Kingdom

D. V. Skryabin
Centre for Photonics and
Photonic Materials
Department of Physics
University of Bath
Claverton Down
Bath BA2 7AY
United Kingdom

J. R. Taylor
Femtosecond Optics Group
Photonics, Blackett Laboratory
Imperial College London
Prince Consort Road
London SW7 2BW
United Kingdom

C. L. Thomsen
NKT Photonics A/S
Blokken 84
DK-3460 Birkerød
Denmark

J. C. Travers
Femtosecond Optics Group
Photonics, Blackett Laboratory
Imperial College London
Prince Consort Road
London SW7 2BW
United Kingdom

M. D. Turner
Centre of Expertise in Photonics
School of Chemistry & Physics
University of Adelaide
Adelaide SA 5005
Australia

F. Vanholsbeeck
Department of Physics
University of Auckland
Private Bag 92019
Auckland
New Zealand

W. J. Wadsworth
Centre for Photonics and
Photonic Materials
Department of Physics
University of Bath
Claverton Down
Bath BA2 7AY
United Kingdom

D. I. Yeom
Division of Energy Systems Research
Ajou University
San5 Wonchun-dong, Yeongtong-gu
Suwon, 443-749
Korea

A. M. Zheltikov
Physics Department
International Laser Center
M. V. Lomonosov Moscow State University
Vorob'evy gory, 119992 Moscow
Russian Federation

Preface

Spectral broadening and the generation of new frequency components is an inherent feature of nonlinear optics, and has been studied in both bulk media and optical fiber waveguides since the 1960s. However, it was not until the early 1970s that the mechanism was widely applied to provide an extended "white-light" source for time resolved spectroscopy, which was later coined "a supercontinuum" by the Alfano group. Subsequent developments in the late 1970s in low-loss optical fibers with conventional structures for telecommunications led to the introduction of fiber as an ideal platform for supercontinuum generation. At the same time, the development of optical soliton physics throughout the late 1980s and early 1990s laid the theoretical foundation and established all the experimental mechanisms required for the production of this versatile source. Despite this progress, however, extensive laboratory deployment remained inhibited by unwieldy pump sources and unreliable system integration.

The advent of photonic crystal fiber in the late 1990s, together with developments in efficient high power and short pulse fiber lasers, fuelled a revolution in the generation of ultrabroadband high brightness optical spectra through the process of supercontinuum generation. Experiments using photonic crystal fiber in 1999–2000 attracted widespread interest and excitement because of the combination of high power, high coherence and the possibility to generate spectra spanning more than an octave. Moreover, the design freedom of photonic crystal fiber allowed supercontinuum generation to be optimized to the wider range of available pump sources, and experiments reported broadband spectra covering the complete window of transmission of silica based fiber using input pulses with durations ranging from several nanoseconds to several tens of femtoseconds, as well as high power continuous wave sources. Supercontinuum generation in PCF was rapidly applied to a range of fields including optical coherence tomography, spectroscopy, and optical frequency metrology and, indeed, this latter result was explicitly cited in the award of the 2005 Nobel Prize in physics.

These results have since led to a huge research effort studying nonlinear spectral broadening in PCF, and have also renewed interest in similar nonlinear phenomena in standard optical fiber. Recent results have provided new insight into the spectral broadening mechanisms, tailored supercontinuum properties to specific applications, and extended supercontinuum generation into new fibers and waveguides using engineered dispersion profiles and/or non-silica materials. Improvements in numerical modeling techniques have also led to remarkable agreement between theoretical prediction and experimental realization.

This progress has of course been well documented in the archival literature, but researchers are now facing the problem that there is no single resource that explains the physics of fiber supercontinuum generation, describes the important properties of new fibers and waveguides, and outlines the features of supercontinuum generation relevant to specific applications. Our aim with this book is to address this problem explicitly through a series of invited papers written by experts familiar with all aspects of this field: the fundamentals and recent developments in supercontinuum generation physics, the different possibilities raised by the availability of new fibers and materials; and the diverse applications where supercontinuum sources can be used.

The book begins with two chapters describing the historical development of the field preceding a concise introduction to nonlinear fiber optics and the numerical modeling of supercontinuum generation. This is followed by a chapter providing an overview of the fiber supercontinuum generation processes under a wide range of conditions. These first four introductory chapters are aimed to ensure that the book is self-contained and accessible to advanced undergraduates and beginning doctoral students requiring a broad introduction to the field. The most significant technical content of the book appears in the subsequent chapters where various aspects of fiber waveguide properties and fiber supercontinuum processes are described in detail by researchers who have been responsible for seminal contributions to the field.

At this point it is perhaps appropriate to add a short word about citing the work in this book. Books and monographs sometimes develop the tendency to become general references that are cited in lieu of the original literature. Whilst this can sometimes be useful, it can also sometimes be detrimental in hiding the contributions of primary journal papers and the original authors. As a solution to this problem, we wish to suggest that readers please take due care that they do not forget to cite the primary literature where appropriate. When material is described both in the primary literature and in this book, there is of course the possibility to cite both.

In closing, we wish to say that we have been very fortunate in being able to include chapters from pioneering and leading research groups from across the world, and we are very grateful to all contributors for their agreement, their effort and their patience. We hope that this book and the excellent contributions that we have been lucky enough to solicit from our colleagues will allow professionals to develop their research even further, and students to enter this field more effectively.

1

Introduction and history

J. R. Taylor

With the invention of the laser (Maiman 1960), rapid technological development of Q-switching (McClung and Hellwarth 1962) and mode locking techniques (Mocker and Collins 1965, DeMaria et al. 1966) allowed the achievement of the shortest, controllable, man-made pulse durations, and, consequently, for even modest pulse energies, unprecedented optical peak powers were achievable with ever-decreasing pulse durations, establishing a trend which continues to the present day. The enormous optical field strengths generated at the focal point of a pulsed laser ensured that the corresponding electronic polarization response of a transparent medium was nonlinear, in that higher order terms of the expansion describing the polarization needed to be considered despite the then insignificance of the magnitude of the second and third order susceptibilities and as a consequence ushered in the era of nonlinear optics. The first nonlinear optical process to be reported was second harmonic generation (Franken et al. 1961), which although observable, is of little importance in relation to the subject matter of this book, supercontinuum generation in optical fibres. However, this was followed by reports of frequency mixing (Bass et al. 1962) and parametric generation (Giordmaine and Miller 1965, Akhmanov et al. 1965). Essential for supercontinuum generation are the processes that result from the third order nonlinear term (Maker and Terhune 1965). In addition to third harmonic generation (New and Ward 1967), again extensively observed but of little importance in supercontinuum generation, these third order processes include the optical Kerr effect or intensity dependent refractive index (Maker et al. 1964), self-focusing (Askaryan 1962, Shen and Shaham 1965), four-wave mixing (Carman et al. 1966), stimulated Brillouin scattering (Chiao et al. 1964) and stimulated Raman scattering (Woodbury and Ng 1962, Eckhardt et al. 1962), all theoretically proposed and experimentally characterized within a few years of the development of the laser and clearly illustrating the richness of the field in those early days.

Supercontinuum Generation in Optical Fibers, ed. by J. M. Dudley and J. R. Taylor. Published by Cambridge University Press.

Researchers were well aware of the processes leading to self-focusing instabilities and spectral broadening in early laser driven systems (Brewer and Lifsitz 1966), with these causing damage to laser rods and primarily looked upon as deleterious effects rather than as a resource. However, as early as 1964, Jones and Stoicheff utilized a nominal "continuum" generated via anti-Stokes scattering in liquid to probe the Raman absorption spectra of other organic species in an effective nanosecond time scale transient absorption experiment. Although the continuum utilized was only a few nanometres wide, it did illustrate the principle of nonlinear spectrally broadened sources applied to spectroscopic measurement. Of course, this was not a new technique; Kirchoff and Bunsen (1860) in their systematic investigations of line reversal in the alkali and alkali earth elements in the nineteenth century had utilized a continuum or "white light" source, however, all measurements were time integrated. Significant spectral broadening of Q-switched ruby lasers in self-focused filaments in carbon disulphide cells was also later reported (Ueda and Shimoda 1967, Brewer 1967) and based on experimental observation, Shimizu (1967) theoretically demonstrated that the spectral broadening and observed interference was due to self phase modulation arising from the intensity dependent refractive index.

In 1969, Alfano and Shapiro undertook a series of measurements to characterize self phase modulation in crystals and glasses using picosecond pulse excitation from a frequency doubled Nd: glass laser (Alfano and Shapiro, 1970a). However, it should be noted that the role of self phase modulation in glass leading to spectral broadening and a linear frequency chirp had been identified by Treacy (1968), who had used a pair of diffraction gratings to directly compress to sub-picosecond durations the 10 nm, 4 ps chirped pulses from a passively mode locked Nd:glass laser. Despite the earlier results reporting spectral broadening in a variety of liquid, crystal and glass samples, the first report of "supercontinuum generation" is widely recognized as Alfano and Shapiro (1970b), recording spectral coverage from 400 nm to 700 nm, a "white light" source, in a borosilicate glass sample pumped by GW picosecond pulses from a frequency doubled Nd: glass laser. Alfano and Shapiro immediately recognized the importance of this unique source in transient absorption measurements, subsequently deploying it in undertaking the first spectroscopic measurements in the picosecond domain of Raman absorption spectra (Alfano and Shapiro 1970c). Throughout the 1970s and 1980s the technique of focusing amplified picosecond and femtosecond pulses (Shank et al. 1979, Knox et al. 1984), primarily from dye laser sources, into liquid filled cells or jets generated white light continua, with self phase modulation identified as the major contributing effect (Fork et al. 1983), that were extensively used in time resolved spectroscopy. It is interesting to note that over the first two decades of research the phenomenon was most commonly referred to as frequency broadening, anomalous frequency broadening or white light generation. A simple reference to any bibliography search engine reveals that the first use of "supercontinuum" to describe the process was in 1980 by Gersten et al. of the Alfano group.

Time resolved spectroscopy remained the principal application of the various sources. However, the technology remained very much in the basic research laboratories primarily because of the quite extensive nature of the experimental configurations. The physics, technology and applications of these first generation supercontinuum sources are best reviewed in Alfano's seminal text *The Supercontinuum Laser Source* (1989).

Driven by the potential application in telecommunications, the development of low loss, single mode optical fibre in the 1970s provided the platform for a new field of study – nonlinear fibre optics. The advantage of fibre over bulk is very clear, despite the exceedingly low nonlinear coefficient of silica, simply by considering the many orders of magnitude improvement ($\sim 10^7$–10^8) in interaction length achieved through propagation over the loss length of a single mode fibre compared to the achievable confocal interaction length of lens coupling to a bulk medium. The interaction in a single mode fibre also allowed more control over the nonlinear process by eliminating the problems of self-focusing and filamentation that were often necessary to observe nonlinearity in bulk media but which also led to irreproducibility of results and quite frequently damage.

Stimulated Raman scattering was the first nonlinear effect reported using the enhancement offered by a carbon disulphide liquid-filled hollow core fibre (Ippen 1970), a concept that once again has come into vogue with the availability of air core photonic band gap fibre. A similar experimental configuration was also used to make the first observation of self phase modulation in an optical fibre (Ippen et al. 1974). With the availability of conventional low loss fibres, however, all the principal nonlinear effects that had previously been observed in bulk materials were rapidly characterized and reported, but, and importantly, at much lower power levels. These included stimulated Raman scattering (Stolen at al. 1972), stimulated Brillouin scattering (Ippen and Stolen 1972), the optical Kerr effect (Stolen and Ashkin 1973), four-wave mixing (Stolen et al. 1974) and self phase modulation (Stolen and Lin 1978), all of which can play important roles in supercontinuum generation in fibres.

A key nonlinear process and a vital component in supercontinuum generation was proposed by Hasegawa and Tappert in 1973 arising through the balance of self phase modulation and anomalous dispersion. Optical soliton generation had to wait a further seven years before it was unambiguously demonstrated and characterized in a series of classic experiments by Mollenauer (Mollenauer et al. 1980, 1983; Mollenauer and Gordon 2006). The long delay between theoretical prediction and experimental realization was a result of the technological challenges involved in developing an appropriate source of transform-limited picosecond pulses in the anomalous dispersion regime, i.e. at wavelengths greater than 1.27 μm for conventional silica-based fibres. The early experiments relied on launching pulses with power and transform limited spectral characteristics to match the fundamental soliton requirements of the particular fibre. It is known, however, that a pulse with any reasonable shape will evolve into a soliton (Hasegawa and Kodama 1981). In such a case, the energy not required to establish the soliton appears as a dispersive wave and this again is an important process in the supercontinuum generation process. It is interesting to note that in the early reports of supercontinuum generation utilizing single pass cascaded Raman generation and prior to the first reports of optical soliton pulse realization, solitons most certainly were generated as evidenced by the continua in the spectral region above 1.3 μm (Cohen and Lin 1978, Lin et al. 1978). Pulses generated in what is now designated the soliton Raman continuum were detector limited and if autocorrelations traces had been taken, the signature of sub-picosecond soliton generation, without doubt, would have been recorded.

Another nonlinear process closely related to soliton generation, resulting from the interplay of the intensity dependent refractive index and anomalous dispersion is modulational instability, which was first proposed in 1980 by Hasegawa and Brinkman and is an important effect in the initiation of supercontinuum generation particularly under cw or long pulse pumping in the region of low dispersion. Many nonlinear systems exhibit such an instability leading to modulation of the steady state, for example in plasmas (Hasegawa 1970) or in fluids (Benjamin and Feir 1967) and was first observed in optical fibre by Tai et al. (1986a), where the generated picosecond modulations appeared on an effective cw background of 100 ps pump pulses. In conventional fibre, the unavailability of adequately powered cw sources at suitable wavelengths above 1.3 μm inhibited observation with true cw excitation until reported by Itoh et al. in 1989 using 1.319 μm from a cw Nd: YAG laser in 5 km of a silica fibre with a fluoride doped depressed cladding.

The modulational instability process can be envisaged as a four-wave mixing process phase matched through self phase modulation, where exponential growth of the Stokes and anti-Stokes sidebands takes place at the expense of two photons from the pump. Modulation instability is, most commonly, self-starting from noise at the frequency separation of the maximum of the gain (Hasegawa and Brinkman 1980). However, it is possible to initiate the process by seeding with an additional signal at a frequency separation from the pump lying within the gain window. This process of induced modulational instability was initially proposed by Hasegawa in 1984 and was experimentally verified by Tai et al. (1986b). This mechanism introduces a control to the modulational instability process that allows manipulation and enhancement of the supercontinuum generation process. As will be discussed below for long pulse and cw pumping, supercontinuum generation processes are dominated by soliton Raman effects that proceed via modulational instability (Gouveia-Neto et al. 1989a) and by seeding the modulational instability process via Raman amplification of the sidebands, enhanced spectral coverage of the continuum associated with cleaner pulses and reduced pedestal components is achieved (Gouveia-Neto et al. 1988a).

Cross phase modulation, which is inherent in the Raman generation process (Gersten et al. 1980, Schadt and Jaskorzynska 1987) can also be used to induce modulational instability on weak signals in the anomalously dispersive regime and is particularly effective when the signal is group velocity matched with a pump in the normal dispersive regime (Gouveia-Neto et al. 1988b).

For supercontinuum generation in the anomalously dispersive regime, in addition to soliton effects, the Raman process contributes in several ways to the formation. Vysloukh and Serkin first proposed the use of the stimulated Raman process for soliton generation (Vysloukh and Serkin 1983, 1984) and the technique was first experimentally demonstrated by Dianov et al. (1985). As well as demonstrating a soliton Raman continuum, for the first time, this latter paper also describes for the first time the important mechanism of Raman self-interaction of the generated femtosecond solitons to account for the continuous extension of the Stokes continuum with propagation or increased pump power. This mechanism was later rediscovered and renamed the soliton self-frequency shift (Mitschke and Mollenauer 1986, Gordon 1986). Throughout the 1980s and early 1990s the Dianov group undertook an immense catalogue of work, both experimental and theoretical (Serkin 1987a,b, Grudinin et al. 1987, Golovchenko et al. 1987a, 1987b) investigating the generation, propagation, stability and decay of femtosecond soliton structures in fibres. This definitive work laid a foundation for the understanding of the mechanisms contributing to supercontinuum generation in fibres. However, much of this seminal work has quite often been overlooked or perhaps translations of the original Soviet texts were unavailable to researchers. As these are too numerous to list, reference should be made to those listed above and the review text *Nonlinear Effects in Optical Fibres* (Dianov et al. 1989a).

The decay of high order solitons launched in the region of minimum dispersion was primarily investigated as a route for extreme pulse compression (Mollenauer et al. 1980, Grudinin et al. 1987, Tai and Tomita 1986a, 1986b, Gouveia-Neto et al. 1987a, 1988c, Beaud et al. 1987), although naturally the process was also accompanied by substantial spectral broadening. Perturbations to high order solitons in the region of minimum dispersion caused by the effects of higher order dispersion (Vysloukh 1983, Wai et al. 1986) lead to instability and soliton fragmentation into its numerous constituent fundamental solitons. In fact, high order solitons are extremely susceptible to any external perturbation, such as from Raman gain

(Tai et al., 1988) or other self-effects, rapidly decaying into their various coloured solitons (Golovchenko et al. 1985, 1987a, 1987b), which in recent years has been renamed soliton fission.

Launched around the region of the minimum dispersion, Wai et al. in 1987 theoretically predicted that solitons would emerge from pulses of any arbitrary shape and amplitude. It was also shown that with increased amplitude at launch, the solitons would frequency down shift with increasing amplitude and that the central frequency of the dispersive wave component would correspondingly increase. These theoretical predictions were experimentally verified by Gouveia-Neto et al. (1988d). Beaud et al. (1987) had earlier investigated the decay of high order solitons launched near the zero dispersion and experimentally characterized the trapping of dispersive waves by femtosecond solitons. As the solitons experienced the soliton self-frequency shift on propagating over increasing fibre length, the trapped dispersive wave correspondingly shifted to shorter wavelengths. The authors clearly identified the mechanism and demonstrated the essential group velocity matching to maintain the process. This too is an essential ingredient in the short wavelength extension of both pulsed and cw pumped supercontinuum generation.

Following the characterization of the basic nonlinear processes in fibre as described above, Lin and Stolen reported the first continuum generation in fibre in 1976. Pumped by various nanosecond pulsed dye lasers, the continua extend from 392 nm to 685 nm, depending on pump wavelength, but were typically 100 nm to 200 nm broad. The high Raman gain coefficient in the visible accompanied by self phase modulation, spectrally broadened the cascaded Raman orders into a continuum and the application of such versatile pulsed sources to spectroscopy was clearly identified by the authors. Extension of the technique into the infrared followed, using Q-switched and Q-switched and mode locked Nd:YAG lasers as the pump (Cohen and Lin 1978, Lin et al. 1978) generating the familiar, distinct, cascaded Raman orders in the normal dispersion regime, and a soliton Raman continuum in the region of anomalous dispersion of the fibres which were hundreds of metres in length. In these early systems utilizing pump wavelengths around 1 μm and dispersion zero wavelengths greater than 1.3 μm, four-wave mixing processes were inefficiently phase matched and wavelength generation below the pump was orders of magnitude less intense than the Stokes Raman dominated contributions. However, by utilizing higher order mode propagation to phase matching, short wavelength generation enhancement was possible (Stolen 1975, Sasaki and Ohmori 1983).

The role of sum frequency generation between the pump radiation and long wavelength generated Stokes radiation was also recognized (Fuji et al. 1980) in the contribution to the short wavelength in the blue/green components of a supercontinuum, generated in a few metres of fibre, that extended from 300 nm to 2100 nm pumped by the 100 kW pulses from a Q-switched and mode locked Nd:YAG laser in a conventional early experimental configuration.

The importance of pumping in the region of low anomalous dispersion was also demonstrated (Washio et al. 1980) through the use of a Q-switched and mode locked Nd:YAG laser operating at 1.34 μm, such that the generated supercontinua exhibited the smooth spectral profile of a soliton Raman continuum instead of the more recognizable signature of discrete cascaded lines obtained in the normal dispersion regime. However, the mechanism for the smooth spectrum was not alluded to since soliton generation was in its infancy and it would also be several years before soliton Raman effects would be theoretically proposed and demonstrated (Vysloukh and Serkin 1983, Dianov et al. 1985). The efficiency of the Raman generated continuum and reduction of the required pump power with increasing fibre length was noted together with the highly efficient four-wave mixing in the region of the zero dispersion.

Early investigations were also carried out on continuum enhancement in multimode fibre in the region of low dispersion using dual pump wavelengths, a technique that would later be examined particularly in relation to seeded modulational instability in single mode fibre. It was shown (Nakazawa and Tokuda 1983) that the presence of the 1.34 μm component of a Nd:YAG laser gave rise to four-wave mixing with the 1.32 μm component pump which spectrally enhanced the generated supercontinuum compared to the discrete cascaded Raman orders obtained when pumping solely with the latter wavelength. The dual pump technique also led to a reduction in pump power for supercontinuum generation.

The early 1980s saw improved understanding of the processes contributing to supercontinuum generation and it was realized that self phase modulation alone could not account for the extent of the generated spectral broadening at a given pump power. Grigoryants et al. (1982) using nanosecond pumping of a multimode fibre demonstrated that it was essential to consider four-wave mixing as an important contribution to the continuum generation process, in addition to the role of sub-nanosecond pulse spiking within the envelope of their pump pulse. They also observed that a better understanding of supercontinuum generation would be gained by considering stochastic spontaneous oscillations and while the effects of noise on supercontinuum generation and soliton evolution dynamics have been considered in the intervening twenty-five years, it is still an important field of study even to the time of writing (Solli et al. 2007, Dudley et al. 2008).

By the mid-1980s research volume on supercontinuum generation was declining. However, as mentioned above, interest was still maintained in spectral broadening in fibres primarily as a result of extreme pulse compression through soliton effects. It had been shown (Mollenauer et al. 1980, 1983) that the breathing of high order solitons led to pulse compression and for solitons of order N, an optimum compression ratio of $4.1N$ was achievable. For low soliton orders and relatively long, picosecond input pulse durations the breathing solitons recovered their input durations following compression. However, for input pulses of high soliton order, extreme pulse compression led to femtosecond pulse durations (Tai and Tomita 1986b) in optimized fibre lengths. This technique was used to produce pulses of 18 femtoseconds, four optical cycles at 1.32 μm, with an associated self phase modulated dominated spectrum extending more than 200 nm, from 1200 nm to 1400 nm (Gouveia-Neto et al. 1988c). On propagation beyond the optimum compression length, perturbations, primarily arising from higher order effects such as self Raman interaction or higher order dispersion (Golovchenko et al. 1987a, Kodama and Hasegawa 1987) led to fragmentation of the high order soliton into its constituent or various "coloured" solitons which was later termed soliton "fission" (Hermann et al. 2002). This soliton compression, fragmentation and spectral shifting is an important mechanism contributing to the long wavelength extension of supercontinuum generation.

Following the theoretical prediction and experimental realization of broad band continua based on soliton Raman effects (Vysloukh and Serkin 1983, 1984, Dianov et al. 1985) the latter part of the decade saw extensive investigations of this mechanism for broad band generation using a variety of pump laser sources and pump pulse durations (Gouveia-Neto et al. 1987b, Grudinin et al. 1987, Vodop'yanov et al. 1987, Islam et al. 1989a). Many of these systems were pumped by pulses with durations in excess of 100 picoseconds and also laid the foundations for high average power supercontinuum generation with average output powers in some cases in the watt regime. Where the pump wavelength was in the region of normal dispersion the generated spectra exhibited the classic cascaded Raman orders up to the region of anomalous dispersion where soliton effects dominated and a continuum was formed. Pumping in the region of anomalous dispersion using a Nd:YAG laser operating at 1.32 μm, Gouveia-Neto et al. 1987b identified that modulational instability initiated the process and that the generated spectrum contained many fundamental solitons formed randomly in time from the Raman amplification of noise structures, which also gave rise to the characteristic smooth spectral profile. Multisoliton collisions in the presence of Raman gain were also shown to play a very important role in the wavelength extension of the continuum (Islam et al. 1989b). The spectral and temporal evolution of the continuum from modulational instability was characterized (Gouveia-Neto et al. 1989a) and it was also demonstrated (Gouveia-Neto et al. 1988a) that by Raman amplification of a seeded modulational instability signal the continuum could be generated at substantially lower overall pump power levels and the autocorrelations of the temporal signature demonstrated shorter pulses and lower pedestals indicative of fewer yet more powerful solitons within the continuum.

The soliton Raman continuum provided a source of widely tunable ultrashort pulses and the technique of spectral selection to provide such sources was patented as early as 1988 (Taylor et al.), however, since these Raman schemes are based on the evolution from noise signatures they are not appropriate for sources where low temporal jitter is essential (Keller et al. 1989). By the early 1990s, however, supercontinuum generation for source applications in wavelength division multiplexed systems was intensively investigated. Initially demonstrated by Morioka et al. (1993), a mode locked Nd:YLF laser producing 7.6 ps pulses was used to generate a relatively narrow supercontinuum extending from 1224 nm to 1394 nm in a 450 m long polarization maintaining fibre and from which 100 wavelength channels on a 1.9 nm spacing were selected using a periodic birefringent filter. The demands placed upon the spectral extent of the source for this application are somewhat less than usual, lying within the second and third telecommunications windows, however, it is important that the spectrum remains relatively flat over the required wavelength of operation and that the noise induced jitter of each channel is substantially less than the time window of the receiver. The role of amplified spontaneous emission or noise perturbing the amplitude and carrier frequency of a fundamental soliton, giving rise to jitter, has been well documented (Gordon and Haus 1986) as well as the elegant, yet simple, technique of spectral filtering, the sliding-guiding filter (Mollenauer et al. 1992), to minimize or negate the effect of jitter. Similarly, the Nakazawa group at NTT laboratories actively investigated the role of modulational instability or noise on high order solitons and demonstrated that spectral filtering in the region of the maximum modulational instability gave rise to improved supercontinuum stability (Kubota et al. 1999). The supercontinuum source was simplified through the use of an all-fibre amplified Er fibre laser and the nearly penalty-free transmission of 6.3 Gbit/s pulse trains was demonstrated (Morioka et al. 1994). The technique was developed extensively throughout the latter part of the decade with up to 1021 channels (Collings et al. 2000) being spectrally selected from the moderately broad continua and total transmission rates of 1 Tbit/s (10 channels at 100 Gbit/s) over 40 km (Morioka et al. 1996), while by employing distributed Raman amplification a 10 GHz repetition rate supercontinuum with an average power in excess of 1 W was generated (Lewis et al. 1998). To date, however, although successfully demonstrated in the laboratory the technique has not been implemented in the field.

Various techniques were used in order to develop spectrally flat continua with low temporal jitter for the wavelength division multiplexed application, although the demand on spectral coverage was quite modest since application was effectively limited to the Er gain bandwidth. Tapered fibres were employed (Mori et al. 1997, Okuno et al. 1998) where the fibres exhibited a dispersion flattened profile and the group velocity dispersion decreased along the length of the fibre, anomalous at input and normal at the output. The flattened third order dispersion inhibited break up of the short pulse solitons, while the adiabatic nature of the dispersion decreasing profile gave rise to soliton pulse compression, with associated spectral broadening and these highly compressed, high peak power pulses generate a self phase modulated dominated spectral broadening in the normal dispersion regime. On propagation through the region of nominal zero dispersion, the generation of very high order solitons leads to spectral and temporal instability. The enhancement of supercontinuum generation in a tapered, dispersion-decreasing fibre had been initially reported by Lou et al. (1997). However, the concept of the dispersion-decreasing fibre had been first proposed by Tajima (1987) to compensate for fibre loss and allow enhanced soliton transmission distances. Its application in the adiabatic compression of solitons was first proposed by Dianov et al. (1989b) and theoretically expanded and experimentally verified by Mamyshev et al. (1991). With extensive development of the technique for soliton pulse compression, particularly from sinusoidal beat signals in the early 1990s (Chernikov et al. 1992, 1993), for high bit rate generation, it was shown that the dispersion-decreasing fibre could be effectively broken up into its individual nonlinear and dispersive elements (Chernikov et al. 1994a) and controlled in a comb-like dispersion profiled multi-element fibre or more simply as a step-like dispersion-decreasing profiled fibre consisting of several segments (Chernikov et al. 1994b). The step-like dispersion profile was later reintroduced (Travers et al. 2005a) for the control of the phase matching conditions required for enhanced short wavelength generation for supercontinuum in photonic crystal fibres.

Increased stability in supercontinuum generation for WDM application was also achieved simply by pumping in the normal dispersion regime (Takushima et al. 1998), hence avoiding problems with soliton instability, or with solitons but by using short fibre lengths (Nowak et al. 1999) and consequently high powers, such that the high order solitons rapidly compress in the optimized length, and hence spectrally expand (Tai and Tomita 1986a, 1986b, Gouveia-Neto et al.1988c), while the spectrum is extracted before instability and soliton fission into various coloured solitons that would take place with further fibre propagation. Nowak et al. also employed a two-step dispersion-decreasing profiled fibre configuration. Through using tapers with relatively short (~30 mm) and long waists (~100 mm) manufactured through heating and stretching in a hydrogen flame, Birks et al. (2000) generated spectra that extended over two octaves from 370 nm to 1545 nm pumped by an unamplified 100 fs pulse length Ti:sapphire laser, with average power around 400 mW.

By the end of the 1990s the processes contributing to supercontinuum generation in fibres were well established and improved theoretical modelling (Gross and Manassah 1992) of these allowed similarly improved agreement between predicted performance and experimental observation. Although the principal thrust of supercontinuum research in the decade was directed towards spectral slicing of modest continua for WDM application, other potential applications areas such as optical coherence tomography, metrology and remote sensing were placing different demands on the spectral coverage. In addition, the compact, high power fibre laser was becoming a mature and commercial technology and the integration of these with the nonlinear fibre was to lay the foundation of a versatile source.

The improving average output powers from fibre lasers allowed the first supercontinuum source based upon cw pumping, which employed a 1 W Yb:Er fibre laser, exciting a 2.3 km long 18% Ge doped silica fibre in a ring configuration, with the approximately 400 nm broad spectrum dominated by soliton Raman effects (Persephonis et al. 1996). Using diode pumping (Chernikov et al. 1997) a novel self Q-switched ytterbium fibre laser supercontinuum source was developed, producing an average power up to 1 W, passively Q-switched using Rayleigh and Brillouin backscatter, to provide pulses as short as 2 ns at selectable repetition rates in the range 1–20 kHz, with pulse energies up to 50 μJ and peak power in excess of 10 kW. Supercontinuum generation took place in this fully fibre integrated configuration that was dominated by cascaded Raman generation from the 1.06 μm pump and a soliton Raman continuum up to 2.3 μm, spectrally reminiscent of the early supercontinua (Lin et al.1978). Four-wave mixing extended to the short wavelength side of the pump and second and third harmonic components of the pump and dominant Raman lines were observed, with effectively the complete transmission window of silica covered by this unique source. In a compact package of around 500 cm^3, the source, with spectral shaping, was attractive for applications in optical coherence tomography. However, by the end of the decade, research interest in supercontinuum had significantly waned, but all that was about to change with the introduction of photonic crystal fibre, pioneered by researchers at the University of Bath (Knight et al. 1996, Knight 2003, Russell 2006).

For a given pump wavelength the generated supercontinuum is critically dependent on the nonlinearity and the dispersion. For conventional silica-based fibre geometries the minimum dispersion zero is achieved at 1.27 μm. With the most commonly available pump sources being either solid state lasers based on Ti^{3+} or Nd^{3+} or fibre lasers based on Yb^{3+}, it was impossible for the pump to directly initiate or take advantage of the solitonic effects described above for efficient supercontinuum generation. With photonic crystal fibre, however, it was shown that by adjustment of the pitch and diameter of the photonic crystal cladding around the core, control of the dispersion was possible with practical fibres exhibiting dispersion zeros well into the visible (Mogilevtsev et al. 1998, Knight et al. 2000), while the ability to manufacture these fibres with single transverse mode operation throughout the spectral range of supercontinuum generation (Birks et al. 1997) improved output beam characteristics compared to conventional fibre-based systems. In addition, the smaller effective mode field diameters of the pcf led to enhanced nonlinearity (Broderick et al. 1999). The culmination of this exquisite and highly innovative technological progress was Ranka's report of supercontinuum generation from 400 nm to 1600 nm pumped in the anomalous dispersion regime of a 75 cm long pcf by the 100 fs 800 pJ pulses from a Ti:sapphire laser (Ranka et al. 2000). Although this was neither the broadest nor the most powerful supercontinuum ever generated, the result provided the impetus to initiate a revival of nonlinear fibre optics and in particular supercontinuum generation, the fruits of which over the past decade are the subject matter of this book. Consequently, to minimize duplication, in this historical introductory chapter less emphasis is placed on the developments with pcf as the key references are described in the accompanying chapters. As recognized by Ranka, soliton formation and self-interaction, as in conventional fibre, played a vital role in the supercontinuum generation process. Soliton effects in pcfs were widely characterized with particular relevance to their role in the supercontinuum (Wadsworth et al. 2000, Price et al. 2002) and although no new physical concepts resulted, this vitally important technological step demonstrated the versatility of tailoring the fibre parameters to suit the available pump sources.

The generation process was subsequently characterized under various pump conditions extending over the picosecond and nanosecond regimes (Provino et al. 2001, Coen et al. 2001, 2002, Champert et al. 2002a, 2002b) and the first truly cw pumped supercontinuum generation was reported (Popov et al. 2002) using a 10 W cw Yb and also Yb:Er fibre lasers to pump both pcf and highly nonlinear conventional fibre structures generating soliton Raman dominated supercontinua, with the latter 1560 nm pumped system exhibiting a 2 W average power continuum extending from 1250 nm to 2000 nm with a less than 5 dB spectral intensity variation over the complete continuum.

Simultaneously, research continued deploying the more highly developed pulsed, femtosecond Er fibre laser based pumps and highly nonlinear fibres for supercontinuum generation. In a beautifully executed experimental measurement using 100 fs pulses at 1560 nm in various short lengths of polarization maintaining highly nonlinear fibre characterized supercontinuum evolution and using cross correlation frequency resolved gating, Nishizawa and Goto (2001) mapped the spectro-temporal evolution of the continuum, clearly showing evidence of soliton trapping of dispersive wave components, with supercontinua extending from 1250 nm to 1950 nm. This effect, which had been predicted and observed earlier by Beaud et al. (1987) and experimentally investigated by Gouveia-Neto et al. (1988d), was unequivocally characterized by Nishizawa and Goto (2002a, 2002b) showing the trapping of dispersive waves in the normal dispersion regime that were continuously pulled to shorter wavelengths by the trapping solitons as they experienced soliton self frequency shift.

The concept of the short step-like dispersion decreasing fibre spans was reintroduced (Nicholson et al. 2003), configuring an all-fibre system pumped by a 188 fs, 50 mW average power mode locked erbium fibre laser. The power scaling of this was later increased to 400 mW (Nicholson et al. 2004a) using a chirped pulse amplification scheme to provide 34 fs pulses that were used to generate a stable octave-spanning supercontinuum proposed for metrology applications. Other amplified passively mode locked erbium fibre laser schemes (Hori et al. 2004, Takayanagi et al. 2005) together with short lengths of polarization maintaining highly nonlinear fibre and also in hybrid arrangements allowed the generation of spectrally flat low noise supercontinua as well as continua that extended from 1000 nm to 2750 nm.

The high stability of supercontinua generated by femtosecond pulse, mode locked Er fibre lasers in short lengths of highly nonlinear fibre and the excellent coherence characteristics of these sources made them highly attractive for applications such as frequency combs in metrology (Washburn et al. 2004). Extensive development of this has been undertaken, which has been well reviewed by Ye and Cundiff (2005).

For increased spectral power density in the supercontinua, higher average power pump schemes based on picosecond and femtosecond solid state Nd doped lasers bulk optically coupled to photonic crystal fibres were initially deployed (Seefeldt et al. 2003, Schreiber et al. 2003) with up to 5 W average power being achieved 500 nm to 1800 nm. For full fibre integration, however, the Yb based fibre laser, which over the past decade has experienced extensive and technologically sophisticated development, is the pump source of choice. In the first demonstration of a high average power fully fibre integrated supercontinuum source (Rulkov et al. 2004a, 2005), an all normally dispersive Yb fibre laser passively mode locked using polarization rotation was amplified and coupled to a 65 m long pcf with a zero dispersion at 1040 nm, allowing mW/nm spectral power density to be achieved over the spectral range 525 nm to 1800 nm. This simple compact configuration was later investigated by others (Rusu et al. 2005) examining the temporal characteristics of the spectral components and it also forms the basis of commercial supercontinuum units.

It was noted that in pumping with picosecond pulses around 1060 nm in the region of the dispersion zero of the pcf that the generated supercontinua did not extend much below 525 nm. This was recognized as being due to the limitations on the four-wave mixing process impaired by loss to the long wavelengths primarily through guiding and scatter losses which limited the maximum extent to about 2200 nm. Consequently, phase matched four-wave mixing of this upper wavelength with the pump around 1060 nm limited the short wavelength edge to around 550 nm. This was overcome through using a simple two-step dispersion decreasing arrangement configured from pcfs. The lower dispersion zero of the second fibre permitted phase matching of supercontinuum components generated in the first fibre to extend the output continuum to below 450 nm (Travers et al. 2005a). The process of the dispersion decreasing pcf for short wavelength extension was further exploited in optimized pcf structures manufactured directly at the pulling tower and exhibiting a continuously decreasing optimized dispersion profile (Kudlinski et al. 2006). Operation down to 330 nm was demonstrated and spectral power densities in excess of 5 mW/nm achieved throughout the visible. Stone and Knight (2008) have also shown that through the simple expedient of modifying the fibre structure of a fixed zero dispersion wavelength pcf it is possible to achieve spectral coverage 400 nm to 2450 nm pumped by the 600 ps pulses from a microchip laser operating at 1064 nm. It has also been shown that by irradiating highly nonlinear fibres with uv light, the uv induced refractive index changes are sufficient to shift the dispersion zero wavelength by up to 100 nm and that as a consequence the short wavelength extent of the generated supercontinuum can be shifted by up to 200 nm (Nicholson et al. 2004b, Westbrook et al. 2005).

In addition to four-wave mixing, the trapping of dispersive waves and the evolution of these dispersive waves to shorter wavelengths group velocity matched to the self frequency shifting solitons has been revisited many times to extend and, with improved fibre design, to enhance the spectral coverage of the supercontinuum and which with improved theoretical modelling has led to remarkable agreement between theory and experiment (Tartara et al. 2003, Cristiani et al. 2004, Genty et al. 2005, Efimov et al. 2005, Gorbach et al. 2007a, 2007b). The relative contributions of solitonic or parametric processes to the continuum are clearly dispersion dependent and in addition to the dispersion managed schemes described above the role of pcfs with, for example, two zero-dispersion wavelengths have also been investigated with the contributions of high order soliton fission and self phase modulation controlled (Hilligsøe et al. 2004, Genty et al. 2004, Frosz et al. 2005).

Another interesting and unique technique for the spectral control of supercontinuum in the visible utilized two-wavelength pumping (Champert et al. 2004). A frequency doubled Q-switched 600 ps pulsed microchip laser provided pumps at 532 nm and 1064 nm incident on a pcf with a zero-dispersion wavelength around 870 nm. When pumped at 532 nm in the normal dispersion regime, the 1.5 kW pump generated a conventional cascade of up to seven discrete Raman orders extending to 660 nm. When the 1064 nm fundamental and 532 nm pumps were coincidental and with sufficient power in the fundamental, the supercontinuum in the visible extended from 375 nm to 700 nm and was smooth and flat with no evidence of discrete Raman lines. The authors did not offer a complete explanation of the processes involved, but it is likely that the high power fundamental generated a soliton continuum above 870 nm and that the extended Raman orders generated by the 532 nm pump acted as seeds for dispersive waves along with the dispersive waves generated by the high order soliton pump which with four-wave mixing processes led to the smooth profile of the supercontinuum in the visible.

As interest in commercial picosecond and femtosecond laser pumped supercontinuum sources escalated so too did the average powers reported (Liu et al. 2005) but not necessarily with compact pump or system configurations. Using highly nonlinear fibres the simplest technique to increase the average power in the continuum is to utilize high average power cw fibre lasers integrated with longer fibre lengths to achieve the power-length factors necessary for efficient nonlinear processes to be observed. The application of high average power fibre lasers for supercontinuum generation was based upon the much earlier technique deploying long pulse picosecond pump sources, where a soliton Raman continuum was initiated via modulational instability (Gouveia-Neto et al. 1987b, 1988a, 1989a) and cw pump schemes had been successfully modelled by Golovchenko et al. (1991). The first reported high power fibre laser pumped cw supercontinua utilized both pcf and highly nonlinear conventional fibres pumped in the region of low anomalous dispersion (Popov et al. 2002). Modulational instability initiates the process leading to a rapid evolution of single solitons which experience Raman gain and as a consequence adiabatically temporally compress, experience collisions and self-Raman interaction and the continuum extends to the long wavelength region with increasing power and fibre length. Pumped at 1065 nm by a cw Yb fibre laser an integrated 100 m length of 2.3 μm core pcf generated a 3.8 W average power supercontinuum that extended from the pump up to 1380 nm where the strong loss associated with the absorption of water at the air hole interface prohibited further evolution (Avdokhin et al. 2003).

The importance of operation in the region of anomalous dispersion was demonstrated through the excitation of a conventional highly nonlinear fibre with a zero dispersion at 1457 nm using a high power tunable fibre Raman laser. When pumped in the region of low normal dispersion, evolution of the continuum occurred only from the first Stokes component of the pump that lay in the region of anomalous dispersion, as had been much earlier observed for long picosecond pulse pumping (Gouveia-Neto and Taylor 1988). When the pump was tuned to be in the anomalously dispersive regime, evolution was via modulational instability of the pump (Rulkov et al. 2004b). The use of a high power fibre Raman laser provides wavelength flexibility of the pump, optimized with reference to the fibre's zero dispersion (González-Herráez et al. 2003, Abeeluck et al. 2004) and permitted spectrally flat supercontinuum generation 1480 nm to 2050 nm with an average power of approximately 10 W (Rulkov et al. 2004b).

In early pcf structures, water loss severely limited the spectral extent and power scaling of one-micron cw pumped sources. Technological advances in the fibre manufacture stages allowed nearly an order of magnitude reduction in the peak absorption coefficient and permitted extension of cw pumped continua beyond 1550 nm (Travers et al. 2005b). Dominated by soliton Raman generation, cw pumped supercontinuum sources are characterized by wavelengths longer than the pump. Significant extension to the short wavelength range has been reported (Popov et al. 2002). Abeeluck and Headley (2005) have also made similar observations, erroneously attributing this to the dispersive waves associated with high order soliton fission. However, in the cw pumped regime only single solitons adiabatically evolve from the amplified noise background and high order solitons are not involved in the process. Spectral extension beyond the water loss resonance around 1380 nm is possible in cw pumped systems by using shorter fibre lengths but these correspondingly require increased pump powers. In a 20 m fibre with double zero dispersion wavelengths at 810 nm and 1730 nm a 29 W average power continuum was generated (Cumberland et al. 2008a). The spectrum exhibited evidence of dispersive wave generation beyond the upper zero dispersion wavelength but no radiation below the lower zero dispersion wavelength. In order to generate radiation below the dispersion zero of the pump it is vital for the pump to be in the region of anomalous dispersion but close to the dispersion zero such that ultrashort pulse solitons evolving from the pump extend into the region of normal dispersion. In addition, four-wave mixing between the pump and anti-Stokes components of modulational instability can give rise to short wavelength generation in the normal dispersion regime (Cumberland et al. 2008b). Consequently these two processes can lead to significant short wavelength generation in cw pumped configurations. The critical dependence of the location of the dispersion zero relative to the pump was demonstrated in an experiment characterizing several fibres under remarkably high average power excitation deploying an industrial scale single-mode cw Yb fibre laser launching up to 170 W into the samples. Spectral extension down to 600 nm was observed, with spectral power densities in excess of 3 mW/nm in the visible and more than 100 mW/nm in the infrared. Soliton dispersive wave trapping was reported to be the major contribution to the short wavelength extension (Travers et al. 2008).

The role of system noise is very important in supercontinuum generation. Early studies established that solitons could evolve from adequately powered noise bursts (Gouveia-Neto and Taylor 1989) or from the amplification of noise (Gouveia-Neto et al. 1989b). In cw pumped systems despite the evolution from modulational instability and amplitude or phase perturbations on the input signal, supercontinuum noise performance (DeMatos et al. 2004) is adequate for application to optical coherence tomography and seeded master oscillator power amplifier configurations have been successfully deployed. System seeding has also been demonstrated using low coherence diodes (Abeeluck and Headley 2004). Seeding of the modulational instability signal in addition provides a route to the control of the evolving supercontinuum and a means to control the spectral extent, the number of solitons and pulse quality within the continuum (Gouveia-Neto et al. 1988a). With ever improving models of the influence of noise both from pump and the system (Kobtsev and Smirnov 2005, Mussot et al. 2004, 2007, Vanholsbeeck et al. 2005, Frosz et al. 2006, see also Chapter 8 in this book) experiment and theory have reached remarkable agreement and should lead to enhanced supercontinua with superior and controlled temporal and spectral characteristics. Such control has recently been directed towards so-called "rogue waves" (Solli et al. 2007) which are really just a result of the stochastic nature of the noise seeding process. By seeding weak signals into the modulational instability sidebands of the pump signal, much in the same manner as Gouveia-Neto et al. (1988a), enhanced soliton evolution generation leading to long wavelength spectral extension is observed with improved coherence (Solli et al. 2008). Modulation of the input pulse at THz rates has also been theoretically shown to significantly increase the rate of generation of the "rogue" or long wavelength solitonic spectral components of the supercontinuum (Dudley et al. 2008).

In addition to the high level of improvements in technology, both in pump lasers and in optical fibre, the sophistication and efficiency of numerical modelling has significantly contributed to recent developments in supercontinuum generation and has led to a complete understanding of all the contributing nonlinear processes. Clearly all the contributing mechanisms were well established by the end of the 1980s and the introduction of pcf and highly nonlinear fibres did not lead to any new physical processes. However, present day agreement between experimentally generated supercontinua and theoretically predicted temporal and spectral performance has never been better in all regimes of pumping from cw to femtosecond. One key area where significant advance has been made is in the modelling of the effects of noise and its effects on jitter, spectral structure and the effect on coherence. Developments in modelling have been excellently reviewed by Dudley et al. (2006) and Genty et al. (2007) in addition to the content of several chapters contributing to this book.

All the above reports of supercontinuum characterization have effectively focused upon generation in silica-based fibres, primarily limiting generation to the window of transmission between 300 nm and 2200 nm. Over the past decade, however, numerous studies have highlighted alternative glasses with higher nonlinearities, and alternative transmission windows to provide alternative wavelength operation regions. In addition, the advent of hollow core photonic bandgap fibre (Cregan et al. 1999) has allowed these devices to be filled with both liquid and gases for nonlinear generation.

Supercontinuum generation in high pressure gas cells was initially investigated by Corkum et al. (1986). In gas filled hollow core microstructured fibres, research has primarily centred upon studies of stimulated Raman scattering (Benabid 2006) and where the strong filtering effect of the photonic bandgap has been innovatively applied to selectively permit operation on the much weaker rotational bands while suppressing the more commonly observed vibrational components (Benabid et al. 2004). An advantage of liquid filling is that materials with exceedingly high nonlinear coefficients can be used, going back to the original idea of Ippen (1970), consequently shorter interaction lengths are required and potentially more stable supercontinua generated. With optimal design (Zhang et al. 2006) it has been theoretically shown that a 10 cm long carbon disulphide filled hollow core fibre pumped around 1550 nm by 1 kW 200 fs pulses would generate a continuum extending from around 1100 nm to 2100 nm. Experimentally Bozolan et al. (2008) demonstrated a 500 nm wide continuum in the 5 cm long water filled core of a hcf pumped by 940 kW pulses from a 60 fs optical parametric oscillator. To date, however, neither gas nor liquid filled hcf continuum sources can compete with the conventional all-fibre approaches described previously either in the simplicity of the technology or in the spectral coverage achieved.

Considerably greater potential to extend the capabilities of supercontinuum generation potential is afforded by the use of non-silica glasses both to extend the wavelength range of operation, in particular to the near and mid-infrared (Omenetto et al. 2006), and to reduce peak power demands or fibre lengths through the use of highly nonlinear materials. Mid-infrared supercontinuum generation has been reviewed in microstructured optical fibres manufactured from a variety of glasses (Price et al. 2007). The ability to simplify the manufacture of microstructured fibres using extrusion techniques, for example with tellurite (Kumar et al. 2003), is particularly attractive, and rather impressive supercontinua, extending over 4000 nm (700 nm to 5000 nm) have been generated in only 8 mm of fibre pumped by 19 kW 100 fs pulses at 1550 nm (Domachuk et al. 2008).

Chalcogenide holey fibres (Monro et al. 2000) with their broad infrared transmission window and high nonlinearity are also highly applicable to mid-infrared supercontinuum generation. Pumped by the 100 fs 2500 nm pulses from an optical parametric amplifier system, supercontinuum generation in the spectral range 2000 nm to 3600 nm was observed in As-Se and As-S pcf of about 1 m length (Sanghera et al. 2009). Tapered chalcogenide nanowires have also been used (Yeom et al. 2008) to produce an effective nonlinearity, which was nearly five orders of magnitude greater than a standard silica fibre, to allow low threshold continuum generation. This technique of nonlinear enhancement in sub-micron tapered fibres both in pcf and standard fibres had been previously successfully demonstrated with conventional silica technology for supercontinuum generation (Leon-Saval et al. 2004, Foster and Gaeta 2004, Gattas et al. 2006).

Fluoride fibres also permit low loss transmission in the mid-infrared. An initial demonstration using a relatively low average pump power Er fibre laser delivering

900 fs 1.5 μJ pulses at 20 kHz repetition rate were used to generate a continuum in a two-fibre cascade configuration of a standard silica fibre followed by a fluoride fibre, to produce a 5 mW supercontinuum extending to beyond 3000 nm (Hagen et al. 2006). This technique was further refined (Xia et al. 2007) and the same group has recently reported an impressive average power in excess of 10 W in a supercontinuum extending from around 1000 nm to beyond 4000 nm (Xia et al. 2009).

Although supercontinuum sources have been widely available since 1970 and the physics describing their operation has been theoretically predicted and identified since the end of the 1980s, it really is quite remarkable that it has only been the last decade that has seen explosive growth in optical fibre-based supercontinuum source research and development. Although not essential for the process itself, the photonic crystal fibre most certainly has been the driving force behind this and of course the technological development of reliable, compact and efficient pump sources, such as the short pulse and/or high power fibre laser, to create a powerful synergy. Completely fibre integrated supercontinuum sources are now widely commercially available based upon several differing pumping schemes and these devices now underpin laboratory and clinical application. The work described in this chapter hopefully gives some indication of how this extensive catalogue of nonlinear optical processes in fibres has led to this development. It represents a personal view and I am certain that along the way I have unintentionally omitted some of the key contributions to this remarkable technology, for which I apologize and hope that due reference will be given by colleagues in the following chapters.

References

Abeeluck, A.K. and Headley, C. (2004). Supercontinuum growth in a highly nonlinear fiber with a low-coherence semiconductor laser diode, *App. Phys. Lett.* **85**, 4863–4865.

Abeeluck, A.K. and Headley, C. (2005). Continuous-wave pumping in the anomalous and normal dispersion regimes of nonlinear fibers for supercontinuum generation, *Opt. Lett.* **30**, 61–63.

Abeeluck, A.K., Headley, C. and Jørgensen, C.G. (2004). High power supercontinuum generation in highly nonlinear, dispersion shifted fibers by use of a continuous-wave Raman fiber laser, *Opt. Lett.* **29**, 2163–2165.

Akhmanov, S.A., Kovrigin, A.I., Piskarskas, A.S., Fadeev, V.V. and Khoklov, R.V. (1965). Observation of parametric amplification in the optical range, *JETP Lett.* **2**, 191–194.

Alfano, R.R. (1989). *The Supercontinuum Laser Source*, Springer, New York, ISBN 0387969462.

Alfano, R.R. and Shapiro, S.L. (1970a). Observation of self-phase modulation and small-scale filaments in crystals and glasses, *Phys. Rev. Lett.* **24**, 592–594.

Alfano, R.R. and Shapiro, S.L. (1970b). Emission in the region 4000 to 7000 A via four-photon coupling in glass, *Phys. Rev. Lett.* **24**, 584–587.

Alfano, R.R. and Shapiro, S.L. (1970c). Picosecond spectroscopy using the inverse Raman effect, *Chem. Phys. Lett.* **8**, 631–633.

Askaryan, G.A. (1962). Effects of the gradient of a strong electromagnetic beam on electrons and atoms, *JETP*, **15**, 1088–1090.

Avdokhin A.V., Popov, S.V. and Taylor, J.R. (2003). Continuous-wave, high-power, Raman continuum generation in holey fibers, *Opt. Lett.* **28**, 1353–1355.

Bass, M., Franken, P.A., Hill, A.E., Peters, C.W. and Weinreich, G. (1962). Optical mixing, *Phys. Rev. Lett.* **8**, 18.

Beaud, P., Hodel, W., Zysset, B. and Weber, H.P. (1987). Ultrashort pulse propagation, pulse breakup and fundamental soliton formation in a single-mode optical fiber, *IEEE J. Quantum Elect.* **QE23**, 1938–1946.

Benabid, F. (2006). Hollow-core photonic bandgap fibre: new light guidance for new science and technology, *Phil. Trans. R. Soc. A* **364**, 3439–3462.

Benabid, F., Bouwmans, G., Knight, J.C. and Russell, P. St.J. (2004). Ultrahigh efficiency laser wavelength conversion in a gas-filled hollow core photonic crystal fiber by pure stimulated rotational Raman scattering in molecular hydrogen, *Phys. Rev. Lett.* **93**, 123903.

Benjamin, T.B. and Feir, J.E. (1967). The disintegration of wave trains on deep water Part 1. Theory, *J. Fluid Mech.* **27**, 417–430.

Birks, T.A., Knight, J.C. and Russell, P. St.J. (1997). Endlessly single-mode photonic crystal fiber, *Opt. Lett.* **22**, 961–963.

Birks, T.A., Wadsworth, W.J. and Russell, P.St.J. (2000). Supercontinuum generation in tapered fibers, *Opt. Lett.* **25**, 1415–1417.

Bozolan, A., deMatos, C.J.S., Cordeiro, C.M.B., dos Santos, E.M. and Travers, J.C. (2008). Supercontinuum generation in a water-core photonic crystal fiber, *Opt. Exp.* **16**, 9671–9676.

Brewer, R.G. (1967). Frequency shifts in self-focused light, *Phys. Rev. Lett.* **19**, 8–10.

Brewer, R.G. and Lifsitz, J.R. (1966). Narrow optical waveguides and instabilities induced in liquids, *Phys. Lett.* **23**, 79–81.

Broderick, N.G.R., Monro, T.M., Bennett, P.J. and Richardson, D.J. (1999). Nonlinearity in holey optical fibers: measurement and future opportunities, *Opt. Lett.* **24**, 1395–1397.

Carman, R.L, Chiao, R.Y. and Kelly, P.L. (1966). Observation of degenerate stimulated four-photon interaction and four-wave parametric amplification, *Phys. Rev. Lett.* **17**, 1281–1283.

Champert, P.A., Popov, S.V. and Taylor, J.R. (2002a). Generation of multiwatt, broadband continua in holey fibers, *Opt. Lett.* **27**, 122–124.

Champert, P.A., Popov, S.V., Solodyankin, M.A. and Taylor, J.R. (2002b). Multiwatt average power continua generation in holey fibers pumped by kilowatt peak power seeded ytterbium fiber amplifier, *App. Phys. Lett.* **81**, 2157–2159.

Champert, P.A., Couderc, V., Leproux, P., Février, S., Tombelaine, V., Labonté, L., Roy, P. and Froehly, C. (2004). White-light supercontinuum generation in normally dispersive optical fiber using original multi-wavelength pumping system, *Opt. Exp.* **12**, 4366–4371.

Chernikov, S.V., Taylor, J.R., Mamyshev, P.V. and Dianov, E.M. (1992). Generation of soliton pulse train in optical fibre using two cw singlemode diode lasers, *Elect. Lett.* **28**, 931–932.

Chernikov, S.V., Richardson, D.J., Laming, R.I., Dianov, E.M. and Payne, D.N. (1993). 70 Gbits/s fibre based source of fundamental solitons at 1550 nm, *Elect. Lett.* **28**, 1210–1212.

Chernikov, S.V., Taylor, J.R. and Kashyap, R. (1994a). Comblike dispersion-profiled fiber for soliton pulse train generation, *Opt. Lett.* **19**, 539–541.

Chernikov, S.V., Taylor, J.R. and Kashyap, R. (1994b). Experimental demonstration of step-like dispersion profiling in optical fibre for soliton pulse generation and compression, *Elect. Lett.* **30**, 433–435.

Chernikov, S.V., Zhu, Y., Taylor, J.R. and Gapontsev, V.P. (1997). Supercontinuum self-Q switched ytterbium fiber laser, *Opt. Lett.* **22**, 298–300.

Chiao, R.Y., Townes, C.H. and Stoicheff, B.P. (1964). Stimulated Brillouin scattering and coherent generation of intense hypersonic waves, *Phys. Rev. Lett.* **12**, 592–595.

Coen, S., Chau, A.H.L., Leonhardt, R., Harvey, J.D., Knight, J.C., Wadsworth, W.J. and Russell, P. St.J. (2001). White-light supercontinuum generation with 60-ps pump pulses in a photonic crystal fiber, *Opt. Lett.* **26**, 1356–1358.

Coen, S., Chau, A.H.L., Leonhardt, R., Harvey, J.D., Knight, J.C., Wadsworth, W.J. and Russell, P. St.J. (2002). Supercontinuum generation by stimulated Raman scattering and parametric four-wave mixing in photonic crystal fibers, *J. Opt. Soc. Am. B* **19**, 758–764.

Cohen, L.G. and Lin, C. (1978). A universal fiber-optic (UFO) measurement system based upon a near-ir fiber Raman laser, *IEEE J. Quantum Elect.* **QE14**, 855–859.

Collings, B.C., Mitchell, M.L., Boivin, L. and Knox, W.H. (2000). A 1021 channel WDM system, *IEEE Phot. Tech. Lett.* **12**, 906–908.

Corkum, P.B., Rolland, C. and Srinivasan-Rao, T. (1986). Supercontinuum generation in gases, *Phys. Rev. Lett.* **57**, 2268–2271.

Cregan, R.F., Mangan, B.J., Knight, J.C., Birks, T.A., Russell, P.St.J., Roberts, P.J. and Allan, D.C. (1999), Single mode photonic band gap guidance of light in air, *Science* **285**, 1537–1539.

Cristiani, I., Tediosi, R., Tartara, L. and Degiorgio, V. (2004). Dispersive wave generation by solitons in microstructured optical fibers, *Opt. Exp.* **12**, 124–135.

Cumberland, B.A., Travers, J.C., Popov, S.V. and Taylor, J.R. (2008a) 29 W high power cw supercontinuum source, *Opt. Exp.* **16**, 5954–5962.

Cumberland, B.A., Travers, J.C., Popov, S.V. and Taylor, J.R. (2008b). Towards visible cw-pumped supercontinua, *Opt. Lett.* **33**, 2122–2124.

DeMaria, A.J., Ferrar, C.M. and Danielsonn, G.E. (1966). Mode locking of a Nd^{3+} doped glass laser, *App. Phys. Lett.* **8**, 22–24.

DeMatos, C.J.S., Popov, S.V. and Taylor, J.R. (2004). Temporal and noise characteristics of continuous-wave pumped continuum generation in holey fibers around 1300 nm, *App. Phys. Lett.* **85**, 2706–2708.

Dianov, E.M., Karasik, A.Ya, Mamyshev, P.V., Prokhorov, A.M., Serkin, V.N., Stelmakh, M.F. and Fomichev, A.A. (1985). Stimulated-Raman conversion of multisoliton pulses in quartz optical fibers, *JETP Lett.* **41**, 294–297.

Dianov, E.M., Mamyshev, P.V., Prokhorov, A.M. and Serkin, V.N. (1989a) *Nonlinear Effects in Optical Fibres*, Harwood Academic Publishers, ISBN 3-7186-4889-X.

Dianov, E.M., Mamyshev, P.V., Prokhorov, A.M. and Chernikov, S.V. (1989b). Generation of a train of fundamental solitons at a high repetition rate in optical fibers, *Opt. Lett* **14**, 1008–1010.

Domachuk, P., Wolchover, N.A., Cronin-Golomb, M., Wang, A., George, A.K., Cordeiro, C.M.B., Knight, J.C. and Omenetto, F.G. (2008). Over 4000 nm bandwidth of mid-IR supercontinuum generation in sub-centimeter segments of highly nonlinear tellurite PCFs, *Opt. Exp.* **16**, 7161–7168.

Dudley, J.M., Genty, G. and Coen, S. (2006). Supercontinuum generation in photonic crystal fiber, *Rev. Mod. Phys.* **78**, 1135–1184.

Dudley, J.M., Genty, G. and Eggleton, B.J. (2008). Harnessing and control of optical rogue waves in supercontinuum generation, *Opt. Express* **16**, 3644–3651.

Eckhardt, G., Hellwarth, R.W., McClung, F.J., Schwarz, S.E., Weiner, D. and Woodbury, E.J. (1962). Stimulated Raman scattering from organic liquids, *Phys. Rev. Lett.* **9**, 455–457.

Efimov, A., Yulin, A.V., Skyrabin, D.V., Knight, J.C., Joly, N., Omenetto, F.G., Taylor, A.J. and Russell P. St.J. (2005). Interaction of an optical soliton with a dispersive wave, *Phys. Rev. Lett.* **95**, 213902.

Fork, R.L., Shank, C.V., Hirlimann, C., Yen, R. and Tomlinson, W.J. (1983). Femtosecond white-light continuum pulses, *Opt. Lett.* **8**, 1–3.

Foster, M.A. and Gaeta, A.L. (2004). Ultra-low threshold supercontinuum generation in sub-wavelength waveguides, *Opt. Exp.* **12**, 3137–3143.

Franken, P.A., Hill, A.E., Peters, C.W. and Weinreich, G. (1961). Generation of optical harmonics, *Phys. Rev. Lett.* **7**, 118–119.

Frosz, M.H., Falk, P. and Bang, O. (2005). The role of the second zero-dispersion wavelength in generation of supercontinua and bright-bright soliton pairs across the zero-dispersion wavelength, *Opt. Exp.* **13**, 6181–6192.

Frosz, M.H., Bang, O. and Bjarklev, A. (2006). Soliton collision and Raman gain regimes in continuous-wave pumped supercontinuum generation, *Opt. Exp.* **14**, 9391–9407.

Fuji, Y., Kawasaki, B.S., Hill, K.O. and Johnson, D.C. (1980). Sum-frequency light generation in optical fibers, *Opt. Lett.* **5**, 48–50.

Gattas, R.R., Svacha, G.T., Tong, L. and Mazur, E. (2006). Supercontinuum generation in submicrometer diameter silica fibers, *Opt. Exp.* **14**, 9408–9414.

Genty, G., Lehtonen, M., Ludvigsen, H. and Kaivola, M. (2004). Enhanced bandwidth of supercontinuum generated in microstructured fibers, *Opt. Exp.* **12**, 3471–3480.

Genty, G., Lehtonen, M. and Ludvigsen, H. (2005). Route to broadband blue-light generation in microstructured fibers, *Opt. Lett.* **30**, 756–758.

Genty, G., Coen, S. and Dudley, J.M. (2007). Fiber supercontinuum sources, *J. Opt. Soc. Am. B* **24**, 1771–1785.

Gersten, J., Alfano, R.R. and Belic, M. (1980). Combined stimulated Raman scattering and continuum self-phase modulation, *Phys. Rev. A* **21**, 1222–1224.

Giordmaine, J.A. and Miller, R.C. (1965). Tunable coherent parametric oscillation in $LiNbO_3$ at optical frequencies, *Phys. Rev. Lett.* **14**, 973–976.

Golovchenko, E.A., Dianov, E.M., Prokhorov, A.M. and Serkin, V.N. (1985). Decay of optical solitons, *JETP Lett.* **42**, 74–77.

Golovchenko, E.A., Dianov, E.M., Pilipetski, A.N., Prokhorov, A.M. and Serkin, V.N. (1987a). Self-effect and maximum contraction of optical femtosecond wave packets in a nonlinear dispersive medium, *JETP Lett.* 45, 91–95.

Golovchenko, E.A., Dianov, E.M., Karasik, A.Ya, Mamyshev, P.V., Pilipetskii, A.N. and Prokhorov, A.M. (1987b). Stimulated-Raman self-scattering of laser pulses, *Sov. J. Quant. Elect.* **19**, 391–392.

Golovchenko, E.A., Mamyshev, P.V., Pilipetskii, A.N. and Dianov, E.M. (1991). Numerical analysis of the Raman spectrum evolution and soliton pulse generation in single-mode fibers, *J. Opt. Soc. Am. B* **8**, 1626–1632.

González-Herráez, M., Martin-López, S., Corredera, P., Hernanz, M.L. and Horche, P.R. (2003). Supercontinuum generation using a continuous-wave Raman fiber laser, *Opt. Commun.* **226**, 323–328.

Gorbach, A.V. and Skyrabin, D.V. (2007a). Theory of radiation trapping by accelerating solitons in optical fibers, *Phys. Rev. A* **76**, 053803.

Gorbach, A.V. and Skyrabin, D.V. (2007b). Light trapping in gravity-like potentials and expansion of supercontinuum spectra in photonic-crystal fibers, *Nature Phot.* **1**, 653–657.

Gordon, J.P. (1986). Theory of the soliton self-frequency shift, *Opt. Lett.* **11**, 662–664.

Gordon, J.P. and Haus, H.A. (1986). Random walk of coherently amplified solitons in optical fiber transmission, *Opt. Lett.* **11**, 665–667.

Gouveia-Neto, A.S. and Taylor, J.R. (1988). Soliton-Raman generation with pump radiation in normal dispersion regime, *Elect. Lett.* **24**, 1544–1546.

Gouveia-Neto, A.S. and Taylor, J.R. (1989). Soliton evolution from noise bursts, *Elect. Lett.* **25**, 736–737.

Gouveia-Neto, A.S., Gomes, A.S.L. and Taylor, J.R. (1987a). High order soliton pulse compression and splitting at 1.32 μm in a single-mode optical fiber, *IEEE J. Quantum Elect.* **QE23**, 1193–1198.

Gouveia-Neto, A.S., Gomes, A.S.L. and Taylor, J.R. (1987b). High-efficiency single-pass solitonlike compression of Raman radiation in an optical fiber around 1.4 μm, *Opt. Lett.* **12**, 1035–1037.

Gouveia-Neto, A.S., Faldon, M.E. and Taylor, J.R. (1988a). Raman amplification of modulational instability and solitary-wave formation, *Opt. Lett.* **13**, 1029–1031.

Gouveia-Neto, A.S., Faldon, M.E., Sombra, A.B., Wigley, P.G.J. and Taylor, J.R. (1988b). Subpicosecond-pulse generation through cross-phase-modulation-induced modulational instability in optical fibers, *Opt. Lett.* **13**, 901–903.

Gouveia-Neto, A.S., Gomes, A.S.L. and Taylor, J.R. (1988c). Pulses of four optical cycles from an optimized optical fibre/grating pair/soliton pulse compressor at 1.32 μm, *J. Mod. Opt.* **35**, 7–10.

Gouveia-Neto, A.S., Faldon, M.E. and Taylor, J.R. (1988d). Temporal and spectral evolution of femtosecond solitons in the region of the zero group velocity dispersion of a single mode optical fibre, *Optics Commun.* **69**, 173–176.

Gouveia-Neto, A.S., Faldon, M.E. and Taylor, J.R. (1989a). Spectral and temporal study of the evolution from modulational instability to solitary wave, *Optics Commun.* **69**, 325–328.

Gouveia-Neto, A.S., Wigley, P.G.J. and Taylor, J.R. (1989b). Soliton generation through Raman amplification of noise bursts, *Opt. Lett.* **14**, 1122–1124.

Gross, B. and Manassah, J.T. (1992). Supercontinuum in the anomalous group velocity dispersion region, *J. Opt. Soc.* B **9**, 1813–1815.

Grigoryants, E.E., Smirnov, V.I. and Chamorovskii, Yu.K. (1982). Generation of wide-band optical continuum in fiber waveguides, *Sov. J. Quantum Elect.* **12**, 841–847.

Grudinin, A.B., Dianov, E.M., Korobkin, D.V., Prokhorov, A.M., Serkin, V.N. and Kaidarov, D.V. (1987). Stimulated-Raman-scattering excitation of 18-fs pulses in 1.6-μm region during pumping of a single-mode optical fiber by the beam from a Nd:YAG laser ($\lambda = 1.064\mu$m), *JETP Lett.* **45**, 260–263.

Hagen, C.L., Walewski, J.W. and Sanders, S.T. (2006). Generation of a continuum extending to the midinfrared by pumping ZBLAN fiber with an ultrafast 1550 nm source, *IEEE Phot. Tech. Lett.* **18**, 91–93.

Hasegawa, A. (1970). Observation of self-trapping instability of a plasma cyclotron wave in a computer experiment, *Phys. Rev. Lett.* **24**, 1165–1168.

Hasegawa, A. (1984). Generation of a train of soliton pulses by induced modulational instability in optical fibers, *Opt. Lett.* **9**, 288–290.

Hasegawa, A. and Brinkman, F. (1980). Tunable coherent IR and FIR sources utilizing modulational instability, *IEEE J. Quantum Elect.* **QE16**, 694–699.

Hasegawa, A. and Kodama, Y. (1981). Signal transmission by optical solitons in monomode fiber, *Proc IEEE* **69**, 1145–1150.

Hasegawa, A. and Tappert, F. (1973). Transmission of stationary nonlinear optical pulses in dispersive dielectric fibers 1. Anomalous dispersion, *App. Phys. Lett.* **23**, 142–144.

Hermann, J., Griebner, U., Zhavoronkov, N., Husakou, A., Nickel, D., Knight, J.C., Wadsworth, W.J., Russell, P.St.J. and Korn, G. (2002). Experimental evidence for supercontinuum generation by fusion of higher-order solitons in photonic fibers, *Phys. Rev. Lett.* **88**, 173901.

Hilligsøe, K.M., Andersen, T.V., Paulsen, H.N., Nielsen, C.K., Mølmer, K., Keiding, S., Kristiansen, R., Hansen, K.P. and Larsen, J.J. (2004). Supercontinuum generation in a photonic crystal fiber with two zero dispersion wavelengths, *Opt. Exp.* **12**, 1045–1054.

Hori, T., Takayanagi, J., Nishizawa, N. and Goto, T. (2004). Flatly broadened, wideband and low noise supercontinuum generation in highly nonlinear hybrid fiber, *Opt. Exp.* **12**, 317–324.

Ippen, E.P. (1970). Low-power quasi-cw Raman oscillator, *App. Phys. Lett.* **16**, 303–305.

Ippen, E.P. and Stolen, R.H. (1972). Stimulated Brillouin scattering in optical fibers, *App. Phys. Lett.* **21**, 539–541.

Ippen, E.P., Shank, C.V. and Gustavson, T.K. (1974). Self-phase modulation of picosecond pulses in optical fibers, *App. Phys. Lett.* **24**, 190–192.

Islam, M.N., Sucha, G., Bar-Joseph, I., Wegener, M., Gordon, J.P. and Chemla, D.S. (1989a). Broad bandwidths from frequency-shifting solitons in fibers, *Opt. Lett.* **14**, 370–372.

Islam, M.N., Sucha, G., Bar-Joseph, I., Wegener, M., Gordon, J.P. and Chemla, D.S. (1989b). Femtosecond distributed soliton spectrum in fibers, *J. Opt. Soc. Am. B* **6**, 1149–1158.

Itoh, H., Davis, G.M. and Sudo, S. (1989). Continuous-wave-pumped modulational instability in an optical fiber, *Optics Lett.* **14**, 1368–1370.

Jones, W.J. and Stoicheff, B.P. (1964). Inverse Raman spectra: induced absorption at optical frequencies, *Phys. Rev. Lett.* **13**, 657–659.

Keller, U., Li, K.D., Rodwell, M. and Bloom, D.M. (1989). Noise characterization of femtosecond fiber Raman soliton lasers, *IEEE J. Quant Elect.* **QE25**, 280–288.

Kirchoff, G. and Bunsen, R. (1860). Chemical analysis by spectrum-observations, *Phil. Mag.* **22**, 89–109.

Knight, J.C. (2003). Photonic crystal fibres, *Nature* **424**, 847–851.

Knight, J.C., Birks, T.A., Russell, P.St.J. and Atkin, D.M. (1996). All-silica single-mode optical fiber with photonic crystal cladding, *Opt. Lett.* **21**, 1547–1549.

Knight, J.C., Arriaga, J., Birks, T.A., Ortigosa-Blanch, A., Wadsworth, W.J. and Russell, P.St.J. (2000). Anomalous dispersion in photonic crystal fiber, *IEEE Photon. Tech. Lett.* **12**, 807–809.

Knox, W.H., Downer, M.C., Fork, R.L. and Shank, C.V. (1984). Amplified femtosecond optical pulses and continuum generation at 5-kHz repetition rate, *Opt. Lett.* **9**, 552–554.

Kobtsev, S.M. and Smirnov, S.V. (2005). Modelling of high-power supercontinuum generation in highly nonlinear, dispersion shifted fibers at CW pump, *Opt. Exp.* **13**, 6912–6918.

Kodama, Y. and Hasegawa, A. (1987). Nonlinear pulse propagation in a monomode dielectric waveguide, *IEEE J. Quant. Elect.* **QE23**, 510–524.

Kubota, H., Tamura, K.R. and Nakazawa, M. (1999). Analyses of coherence-maintained ultrashort optical pulse trains and supercontinuum generation in the presence of soliton-amplified spontaneous-emission interaction, *J. Opt. Soc. Am. B* **16**, 2223–2232.

Kudlinski, A., George, A.K., Knight, J.C., Travers, J.C., Rulkov, A.B., Popov, S.V. and Taylor, J.R. (2006). Zero-dispersion wavelength decreasing photonic crystal fibers for ultraviolet-extended supercontinuum generation, *Opt. Exp.* **14**, 5715–5722.

Kumar, V.V.R.K., George, A.K., Knight, J.C. and Russell, P.St.J. (2003). Tellurite photonic crystal fiber, *Opt. Exp.* **11**, 2641–2645.

Leon-Saval, S.G., Birks, T.A., Wadsworth, W.J. and Russell, P.St.J. (2004). Supercontinuum generation in submicron fibre waveguides, *Opt. Exp.* **12**, 2864–2869.

Lewis, S.A.E., Chernikov, S.V. and Taylor, J.R. (1998). Ultra-broad-bandwidth spectral continuum generation in fibre Raman amplifier, *Elect. Lett.* **34,** 2267–2268.

Lin, C. and Stolen, R.H. (1976). New nanosecond continuum for excited-state spectroscopy, *App. Phys. Lett.* **28**, 216–218.

Lin, C., Nguyen, V.T. and French, W.G. (1978). Wideband near i.r. continuum (0.7–2.1μm) generated in low-loss optical fibres, *Elect. Lett.* **14**, 822–833.

Liu, A., Norsen, M.A. and Mead, R.D. (2005). 7.2 W supercontinuum generation in photonic crystal fibers pumped by a nanosecond fiber laser, *SPIE Proc.* **5709**, 206–213.

Lou, J.W., Xia, T.J., Boyraz, O., Shi, C.-X., Nowak, G.A. and Islam, M.N. (1997). Broader and flatter supercontinuum spectra in dispersion-tailored fibers, Paper TuH6, Technical Digest Optical Fibre Communications, 32–34.

Maiman, T.H. (1960). Stimulated optical radiation in ruby, *Nature* **187**, 493–494.

Maker, P.D. and Terhune, R.W. (1965). Study of optical effects due to an induced polarization third order in the electric field strength, *Phys. Rev. A* **137**, 810–818.

Maker, P.D., Terhune, R.W. and Savage, C.M. (1964). Intensity-dependent changes in the refractive index of liquids, *Phys. Rev. Lett.* **12**, 507–509.

Mamyshev, P.V., Chernikov, S.V. and Dianov, E.M. (1991). Generation of fundamental soliton trains for high-bit-rate optical fiber communication lines, *IEEE J. Quant. Elect.* **QE 27**, 2347–2355.

McClung, F.J. and Hellwarth, R.W. (1962). Giant optical pulsations from ruby, *J. App. Phys.* **33**, 828–829.

Mitschke, F.M. and Mollenauer, L.F. (1986). Discovery of the soliton self-frequency shift, *Opt. Lett.* **11**, 659–661.

Mocker, H.W. and Collins, R.J. (1965). Mode competition and self-locking effects in a Q-switched ruby laser, *App. Phys. Lett.* **7**, 270–273.

Mogilevtsev, D., Birks, T.A. and Russell, P.St.J. (1998). Group-velocity dispersion in photonic crystal fibers, *Opt. Lett.* **23**, 1662–1664.

Mollenauer, L.F. and Gordon, J.P. (2006). *Solitons in Optical Fibers*, Elsevier-Academic Press, ISBN 13: 978-0-12-504190-4.

Mollenauer, L.F., Stolen, R.H. and Gordon, J.P. (1980). Experimental observation of picosecond pulse narrowing and solitons in optical fibers, *Phys. Rev. Lett.* **45**, 1095–1098.

Mollenauer, L.F., Stolen, R.H., Gordon, J.P. and Tomlinson, W.J. (1983). Extreme picosecond pulse narrowing by means of soliton effect in single-mode optical fibers, *Opt. Lett.* **8**, 289–291.

Mollenauer, L.F., Gordon, J.P. and Evangelides, S.G. (1992). The sliding-frequency guiding filter: an improved form of soliton jitter control, *Opt. Lett.* **17**, 1575–1577.

Monro, T.M., West, Y.D., Hewak, D.W., Broderick, N.G.R. and Richardson, D.J. (2000). Chalcogenide holey fibres, *Elect. Lett.* **36**, 1998–2000.

Mori, K., Takara H., Kawanishi, T., Saruwatari, M. and Morioka, T. (1997). Flatly broadened supercontinuum spectrum generated in a dispersion decreasing fibre with convex dispersion profile, *Elect. Lett.* **33**, 1806–1808.

Morioka, T., Mori, K. and Saruwatari, M. (1993). More than 100 wavelength channel picosecond optical pulse generation from single laser source using supercontinuum in optical fibres, *Electron. Lett.* **29**, 862–864.

Morioka, T., Kawanishi, S., Mori, K. and Saruwatari, M. (1994). Nearly penalty-free, < 4ps supercontinuum Gbit/s pulse generation over 1535-1560 nm, *Elect. Lett.* **30**, 790–791.

Morioka, T., Takara, H., Kawanishi, S., Kamatani, O., Takiguchi, K., Uchiyama, K., Saruwatari, M., Takahashi, H., Yamada, M., Kanamori, T. and Ono, H. (1996). 1 Tbit/s (100Gbit/s x10 channel) OTDM/WDM transmission using a single supercontinuum WDM source, *Elect. Lett.* **32**, 906–907.

Mussot, A., Lantz, E., Maillotte, H. and Sylvestre, T. (2004). Spectral broadening of a partially coherent CW laser beam in single-mode optical fibers, *Opt. Exp.* **12**, 2838–2843.

Mussot, A., Beaugeois, M., Bouazaoui, M. and Sylvestre, T. (2007). Tailoring CW supercontinuum generation in microstructured fibers with two-zero dispersion wavelengths, *Opt. Exp.* **15**, 11553–11563.

Nakazawa, M. and Tokuda, M. (1983). Continuum spectrum generation in a multimode fiber using two pump beams at 1.3 μm wavelength region, *Jap. J. App. Phys.* **22**, L239–241.

New, G.H.C. and Ward, J.F. (1967). Optical third-harmonic generation in gases, *Phys. Rev. Lett.* **19**, 556–559.

Nicholson, J.W., Yan, M.F., Wisk, P., Fleming, J., DiMarcello, F., Monberg, E., Yablon, A., Jørgensen, C. and Veng, T. (2003). All-fiber, octave spanning supercontinuum, *Opt. Lett.* **28**, 643–645.

Nicholson, J.W., Yablon, A.D., Westbrook, P.S., Feder, K.S. and Yan, M.E. (2004a). High power, single mode, all-fiber source of femtosecond pulses at 1550 nm and its use in supercontinuum generation, *Opt. Exp.* **12**, 3025–3034.

Nicholson, J.W., Westbrook, P.S., Feder, K.S. and Vablon, A.D. (2004b). Supercontinuum generation in ultraviolet-irradiated fibers, *Opt. Lett.* **29**, 2363–2365.

Nishizawa, N. and Goto, T. (2001). Widely broadened supercontinuum generation using highly nonlinear dispersion shifted fibers and femtosecond fiber lasers, *Jap. J. App. Phys.* **40**, L365–367.

Nishizawa, N. and Goto, T. (2002a). Pulse trapping by ultrashort soliton pulses in optical fibers across zero-dispersion wavelength, *Opt. Lett.* **27**, 152–154.

Nishizawa, N. and Goto, T. (2002b). Characteristics of pulse trapping by use of ultrashort soliton pulses in optical fibers across the zero-dispersion wavelength, *Opt. Exp.* **10**, 1151–1159.

Nowak, G.A., Kim, J. and Islam, M.N. (1999). Stable supercontinuum generation in short lengths of conventional dispersion-shifted fiber, *App. Opt.* **38**, 7364–7369.

Okuno T., Onishi M. and Nishimura M. (1998). Generation of ultra-broad-band supercontinuum by dispersion-flattened and decreasing fiber, *IEEE Phot. Tech. Lett.* **10**, 72–74.

Omenetto, F.G., Wolchover, N.A., Wehner, M.R., Ross, M., Efimov, A., Taylor, A.J., Kumar, V.V.R.K., George, A.K., Knight, J.C., Joly, N.Y. and Russell, P.St.J. (2006). Spectrally smooth supercontinuum from 350nm to 3μm in sub-centimeter lengths of soft-glass photonic crystal fibers, *Opt. Exp.* **14**, 4928–4934.

Persephonis, P., Chernikov, S.V. and Taylor, J.R. (1996). Cascaded CW fibre Raman laser source 1.6–1.9 μm, *Elect. Lett.* **32**, 1486–1487.

Popov, S.V., Champert, P.A., Solodyankin, M.A. and Taylor, J.R. (2002). Seeded fibre amplifiers and multi-watt average power continuum generation in holey fibres, Paper WKK2, Proceedings OSA Annual Meeting 117.

Price, J.H.V., Belardi, W., Monro, T.M., Malinowski, A., Piper, A. and Richardson, D.J. (2002). Soliton transmission and supercontinuum generation in holey fiber, using a diode pumped Ytterbium fiber source, *Opt. Exp.* **10**, 382–387.

Price, J.H.V., Monro, T.M., Ebendorff-Heidepriem, H., Poletti, F., Horak, P., Finazzi, V., Leong, J.Y.Y., Petropoulos, P., Flanagan, J.C., Brambilla, G., Feng, X. and Richardson, D.J. (2007). Mid-IR supercontinuum from nonsilica microstructured optical fibers, *IEEE J. Sel. Top. Quant. Elect.* **13**, 738–749.

Provino, L., Dudley, J.M., Maillotte, H., Grossard, N., Windeler, R.S. and Eggleton, B.J. (2001). Compact broadband continuum source based on microchip laser pumped microstructured fibre, *Elect. Lett.* **37**, 558–560.

Ranka, J.K., Windeler, R.S. and Stentz, A.J. (2000). Visible continuum generation in air-silica microstructure optical fibers with anomalous dispersion at 800 nm, *Opt. Lett.* **25**, 25–27.

Rulkov, A.B., Getman, A.G., Vyatkin, M.Y., Popov, S.V., Gapontsev, V.P. and Taylor, J.R. (2004a). 525–1800 nm, watt-level, all-fibre picosecond source, Paper CDPC7, Conference on Lasers and Electro Optics, San Francisco.

Rulkov, A.B., Popov, S.V., Taylor, J.R. and Gapontsev, V.P. (2004b). 1.5–2.0 μm, multi-watt white-light generation in cw format in highly-nonlinear fibre, Paper TuA6, Advanced Solid State Photonics, Trends in Optics and Photonics Series **94**, 131–134.

Rulkov, A.B., Vyatkin, M.Y., Popov, S.V., Taylor, J.R. and Gapontsev, V.P. (2005). High brightness picosecond all-fiber generation in 525–1800 nm range with picosecond Yb pumping, *Opt. Exp.* **13**, 377–381.

Russell, P.St.J. (2006). Photonic-crystal fibers, *J. Lightwave Tech.* **24**, 4729–4749.

Rusu, M., Grudinin, A.B. and Okhotnikov, O.G. (2005). Slicing the supercontinuum radiation generated in photonic crystal fiber using an all-fiber chirped-pulse amplification scheme, *Opt. Exp.* **13**, 6390–6400.

Sanghera, J.S., Shaw, L.B. and Aggarwal, I.D. (2009). Chalcogenide glass-fiber-based mid-IR sources and applications, *IEEE J. Sel. Top. Quant. Elect.* **15**, 114–119.

Sasaki, Y. and Ohmori, Y. (1983). Sum-frequency wave generation in optical fibers, *J. Optical Commun.* **4**, 83–90.

Schadt, D. and Jaskorzynska, B. (1987). Frequency chirp and spectra due to self phase modulation and stimulated Raman scattering influenced by pulse walk-off in optical fibers, *J. Opt. Soc. Am. B* **4**, 856–862.

Schreiber, T., Limpert, J., Zellmer, H., Tünnermann, A. and Hansen, K.P. (2003). High average power supercontinuum generation in photonic crystal fibers, *Opt. Commun.* **228**, 71–78.

Seefeldt, M., Heuer, A. and Menzel, R. (2003). Compact white-light source with an average output power of 2.4 W and 900 nm spectral bandwidth, *Opt. Commun.* **216**, 199–202.

Serkin, V.N. (1987a). Self-compression and decay of femtosecond optical wave packets in fiber light guides, *Sov. Phys. Lebedev Inst. Rep.* **6**, 49–53.

Serkin, V.N. (1987b). Colored envelope solitons in optical fibers, *Sov. Tech. Phys. Lett.* **13**, 320–321.

Shank, C.V., Fork, R.L., Lehny, R.F. and Shah, J. (1979). Dynamics of photoexcited GaAs band edge absorption with subpicosecond resolution, *Phys. Rev. Lett.* **42**, 112–115.

Shen, Y.R. and Shaham, Y.J. (1965). Beam deterioration and stimulated Raman effect, *Phys. Rev. Lett.* **15**, 1008–1010.

Shimizu, F. (1967). Frequency broadening in liquids by a short light pulse, *Phys. Rev. Lett.* **19**, 1097–1100.

Solli, D.R., Ropers, C., Koonath, P. and Jalali, B. (2007). Optical rogue waves, *Nature* **450**, 1054–1058.

Solli, D.R., Ropers, C. and Jalali, B. (2008). Active control of rogue waves for stimulated supercontinuum generation, *Phys. Rev. Letts.* **101**, 233902.

Stolen, R.H. (1975). Phase-matched-stimulated four-photon mixing in silica fiber waveguides, *IEEE J. Quantum Elect.* **QE11**, 100–103.

Stolen, R.H. and Ashkin, A. (1973). Optical Kerr effect in glass waveguide, *App. Phys. Lett.* **22**, 294–296.

Stolen, R.H. and Lin, C. (1978). Self-phase-modulation in silica optical fibers, *Phys. Rev. A* **17**, 1448–1453.

Stolen, R.H., Ippen, E.P. and Tynes, A.R. (1972). Raman oscillation in glass optical waveguides, *App. Phys. Lett.* **20**, 62–64.

Stolen, R.H., Bjorkholm, J.E. and Ashkin, A. (1974). Phase-matched three-wave mixing in silica fiber optical waveguides, *App. Phys. Lett.* **24**, 308–310.

Stone, J.M. and Knight, J.C. (2008). Visibly "white" light generation in uniform photonic crystal fiber using a microchip laser, *Opt. Exp.* **16**, 2670–2675.

Tai, K. and Tomita, A. (1986a). 1100x optical fiber pulse compression using grating pair and soliton effect at 1.319 μm, *App. Phys. Lett.* **48**, 1033–1035.

Tai, K. and Tomita, A. (1986b). 50 x optical fiber pulse compression at 1.319 μm, *App. Phys. Lett.* **48**, 309–311.

Tai, K., Hasegawa, A. and Tomita, A. (1986a). Observation of modulational instability in optical fibers, *Phys. Rev. Lett.* **56**, 135–138.

Tai, K., Tomita, A., Jewell, J.L. and Hasegawa, A. (1986b). Generation of subpicosecond solitonlike optical pulses at 0.3 THz repetition rate by induced modulational instability, *App. Phys. Lett.* **49**, 236–238.

Tai, K., Hasegawa, A. and Bekki, N. (1988). Fission of optical solitons induced by stimulated Raman effect, *Opt. Lett.* **13**, 392–394.

Tajima, K., (1987). Compensation of soliton broadening in nonlinear optical fibers with loss, *Opt. Lett.* **12**, 54–56.

Takayanagi, J., Nishizawa, N., Nagai, H., Yoshida, M. and Goto, T. (2005). Generation of high-power femtosecond pulse and octave-spanning ultrabroad supercontinuum using all-fiber system, *IEEE Phot. Tech. Lett.* **17**, 37–39.

Takushima, Y., Futami, F. and Kikuchi, K. (1998). Generation of over 140-nm-wide supercontinuum from a normal dispersion fiber by using a mode-locked semiconductor laser source, *IEEE Phot. Tech. Lett.* **10**, 1560–1562.

Tartara, L., Cristiani, I. and Degiorgio, V. (2003). Blue light and infrared continuum generation by soliton fission in a microstructured fiber, *App. Phys. B* **77**, 307–311.

Taylor, J.R., Gomes, A.S.L. and Gouveia-Neto, A.S. (1988). Radiation pulse generation, European Patent 88302981.

Travers, J.C., Popov, S.V. and Taylor, J.R. (2005a). Extended blue supercontinuum generation in cascaded holey fibers, *Opt. Lett.* **30**, 3132–3134.

Travers, J.C., Kennedy, R.E., Popov, S.V., Taylor, J.R., Sabert, H. and Mangan, B. (2005b). Extended continuous-wave supercontinuum generation in a low-water-loss holey fiber, *Opt. Lett.* **30**, 1938–1940.

Travers, J.C., Rulkov, A.B., Cumberland, B.A., Popov, S.V. and Taylor, J.R. (2008). Visible supercontinuum generation in photonic crystal fibers with a 400 W continuous wave fiber laser, *Opt. Exp.* **16**, 14435–14447.

Treacy, E.B. (1968). Compression of picosecond light pulses, *Phys Lett.* A **28**, 34–35.

Ueda, Y. and Shimoda, K. (1967). Observation of stimulated Raman emission and stimulated Rayleigh-wing scattering from self-trapped filaments of a laser beam, *Jap. J. App. Phys.* **6**, 628–633.

Vanholsbeeck, F., Martin-Lopez, S., González-Herráez, M. and Coen, S. (2005). The role of pump incoherence in continuous-wave supercontinuum generation, *Opt. Exp.* **13**, 6615–6625.

Vodop'yanov, K.L., Grudinin, A.B., Dianov, E.M., Kulevskii, L.A., Prokhorov, A.M. and Khaidarov, D.V. (1987). Generation of pulses of 100–200 fsec duration by stimulated Raman scattering in a single-mode fiber waveguide at wavelengths 1.5–1.7 μm, *Sov. J. Quant. Elect.* **17**, 1311–1313.

Vysloukh, V.A. (1983). Propagation of pulses in optical fibers in the region of a dispersion minimum. Role of nonlinearity and higher-order dispersion, *Sov. J. Quantum Elect.* **13**, 1113–1114.

Vysloukh, V.A. and Serkin, V.N. (1983). Generation of high-energy solitons of stimulated Raman radiation in fiber light guides, *JETP Lett.* **38**, 199–202.

Vysloukh, V.A. and Serkin, V.N. (1984). Nonlinear transformation of solitons in fiber lightguides, *Bull. Acad. Sci. USSR, Phys. Ser.* **48**, 125–129.

Wadsworth, W.J., Knight, J.C., Ortigosa-Blanch, A., Arriaga, J., Silvestre, E. and Russell, P.St.J. (2000). Soliton effects in photonic crystal fibers at 600 nm, *Elect. Lett.* **36**, 53–55.

Wai, P.K.A., Menyuk, C.R., Lee, C. and Chen, H.H. (1986). Nonlinear pulse propagation in the neighbourhood of the zero-dispersion wavelength of monomode optical fibers, *Opt. Lett.* **11**, 464–466.

Wai, P.K.A., Menyuk, C.R., Lee, C. and Chen, H.H. (1987). Soliton at the zero-group-dispersion wavelength of a single mode fiber, *Opt. Lett.* **12**, 628–630.

Washburn, B.R., Diddams, S.A., Newbury, N.R., Nicholson, J.W., Yan, M.F. and Jørgensen, C. (2004). Phase-locked, erbium-fiber-laser-based frequency comb in the near infra red, *Opt. Lett.* **29**, 250–252.

Washio, K., Inoue, K. and Tanigawa, T. (1980). Efficient generation of near i.r. stimulated light scattering in optical fibres pumped in low-dispersion region at 1.3 μm, *Elect. Lett.* **16**, 331–333.

Westbrook, P.S., Nicholson, J.W., Feder, K.S. and Yablon, A.D. (2005). Improved supercontinuum generation through UV processing of highly nonlinear fibers, *J. Lightwave Tech.* **23**, 13–18.

Woodbury, E.J. and Ng, W.K. (1962). Ruby laser operation in the near IR, *Proc. IRE* **50**, 2367.

Xia, C., Kumar, M., Cheng, M.Y., Hegde, R.S., Islam, M.N., Galvanauskas, A., Winful, H.C. and Terry Jr, F.I. (2007). Power scalable mid-infrared supercontinuum generation in ZBLAN fluoride fibers with up to 1.3 watts time-averaged power, *Opt. Exp.* **15**, 865–871.

Xia, C., Islam, M.N., Terry Jr, F.L., Freeman, M.J. and Mauricio, J. (2009). 10.5 watts time-averaged power mid-infrared supercontinuum generation extending beyond 4 μm with direct pulse pattern modulation, *IEEE J. Sel. Top. Quant. Elect.* **15**, 422–434.

Ye, J. and Cundiff, S.T. (2005). *Femtosecond Optical Frequency Comb Technology*, Springer Science+Business Media, ISBN 0-387-23790-9.

Yeom, D.-I., Mägi, E.C., Lamont, M.R.E., Roelens, M.A.F., Fu, L. and Eggleton, B.E. (2008). Low-threshold supercontinuum generation in highly nonlinear chalcogenide nanowires, *Opt. Lett.* **33**, 660–662.

Zhang, R., Teipel, J. and Giessen, H. (2006), Theoretical design of a liquid-core photonic crystal fiber for supercontinuum generation, *Opt. Exp.* **14**, 6800–6812.

2

Supercontinuum generation in microstructure fibers – a historical note

J. K. Ranka

At the time of writing, it is just over 10 years since a seemingly simple research project at Bell Laboratories led to a new revolution in optical frequency metrology, when in 1999, in a packed CLEO postdeadline session, the first public presentation of supercontinuum generation in photonic crystal (or microstructure) fibers was given (Ranka et al. 1999).

This work had started a year earlier as a curiosity I had had as a post-doc at Bell Laboratories on the possibilities of extreme nonlinear interactions in small-core high-numerical aperture (NA) silica fibers that were fabricated with a transverse microstructure or photonic crystal fiber cladding. In all honesty, at the time I did not consider the possibility that the fiber structure would so substantially alter the dispersive and modal properties of the fiber so far from the zero dispersion wavelength of bulk silica. The success of this work is a tribute to the ability of a commercial research organization to understand and encourage basic research with a long-range outlook. Having a management team step back and allow a post-doc to work independently, with support, free from bureaucratic or technical interference, allowed the project to proceed unhindered.

After a survey of a number of different fibers that our optical fiber research group had fabricated and initial calculations of potential nonlinear effects that would occur with femtosecond duration pulses, a simple fiber design was chosen to start experimenting with. Because the necessary laser equipment, a mode-locked Ti:sapphire laser, was not available within the group, I simply arranged to use the equipment in a neighboring biophysics group. It was pure luck, but the laser system was tuned to the perfect wavelength for the fiber chosen, and as soon I coupled some light into the fiber, a white continuum was generated. I was completely astonished and first thought that the fiber was melting. After settling down and realizing the magnitude of the continuum result, the hard part started. Even coupling light into the webbing of the fiber produced light shifted to different colors. Over the next few months

Supercontinuum Generation in Optical Fibers, ed. by J. M. Dudley and J. R. Taylor. Published by Cambridge University Press.

I characterized the fiber and spectrum in detail, measuring the nonlinear index, continuum thresholds, bandwidths, spatial properties, pump wavelength dependence, and pulse duration. None of the conventional calculations or simulations could explain the effect. The fiber needed to be thought of from outside the box. The only possibility was that the waveguide substantially altered the dispersive properties of the fiber. After a few initial calculations, I spent the next few weeks characterizing the dispersion of the fiber, confirming that the zero-dispersion wavelength of the initial test fiber was 780 nm. At that point everything fell into place, how the continuum was generated and the mode properties were a natural outcome of having a small core, high-NA fiber (Ranka et al. 2000a, 2000b).

A collaboration with JILA at the University of Colorado started accidentally in September 1998 during a fiber measurement conference at NIST Boulder. Scott Diddams and John Hall gave me a tour of the lab and an overview of their efforts in optical frequency metrology. While they were unaware of the fiber work, the meeting provided an additional sense of direction on the importance of the continuum source. The role that this has played in the development of frequency metrology has been discussed elsewhere (Hall & Haensch 2005).

After the 1999 CLEO postdeadline presentation, there was enormous interest in the fiber work. As our group did not have the equipment necessary to take advantage of the fiber, a collaborative effort was established with several groups and organizations, including JILA. Later on, the fiber was made available to the wider optics community. At this point the history of the continuum work can be traced through the large number of publications. Interestingly, the original fiber chosen still exhibits one of the broadest continuum spectra demonstrated using Ti:sapphire pumping.

I left Bell Labs in 2000 at which point my direct work on the fiber ended. I consider myself to be fortunate to have worked on this project, from initial conception to fundamental understanding of the physics behind the phenomena, and the opportunity to present such high-impact work to the scientific community.

References

Hall, J.L. & Haensch, T.W. (2005) History of Optical Comb Development, in Ye, J. & Cundiff, S.T. (Eds), *Femtosecond Optical Frequency Comb Technology: Principle, Operation, and Application*, Springer, New York, 1–11.

Ranka, J.K., Windeler, R.S. & Stentz, A.J. (1999) Efficient visible continuum generation in air-silica microstructure optical fibers with anomalous dispersion at 800 nm. Conference on Lasers and Electro-Optics (CLEO), Baltimore, Postdeadline Paper CPD8.

Ranka, J.K., Windeler, R.S. & Stentz, A.J. (2000a) Visible continuum generation in air-silica microstructure optical fibers with anomalous dispersion at 800 nm. *Opt. Lett.* **25**, 25–27.

Ranka, J.K., Windeler, R.S. & Stentz, A.J. (2000b) Optical properties of high-delta air-silica microstructure optical fibers, *Opt. Lett.* **25**, 796–798.

3

Nonlinear fibre optics overview

J. C. Travers, M. H. Frosz and J. M. Dudley

3.1 Introduction

This chapter provides a succinct overview of the various nonlinear effects that can occur when a light field propagates in an optical fibre. Given that nonlinear fibre optics is a very mature research field, it has been covered in much detail by many previous reviews and monographs. The reader is particularly referred to Agrawal (2007) for a treatment that combines a review of both theory and experiments in a way that is simultaneously accessible and technically comprehensive. Other monographs that contain valuable material and references to the original literature include Taylor (2005) and Alfano (2006). In this treatment we provide only a brief introduction to the major concepts, placing particular emphasis on effects that play an important role in supercontinuum generation. Where appropriate, these effects will be discussed in more detail in other chapters. We do, however, treat the numerical modelling of nonlinear pulse propagation in more depth than is usually found in the literature.

3.2 Modelling nonlinear pulse propagation

The propagation of an electromagnetic wave or pulse depends on the medium in which it propagates. In vacuum a pulse can propagate unchanged. When propagating in a medium, however, an electromagnetic field interacts with the atoms, which generally means that the pulse experiences loss and dispersion. The latter effect occurs because the different wavelength components of the pulse travel at different velocities due to the wavelength dependence of the refractive index. In an optical waveguide, the total dispersion has an additional component due to the light confinement called waveguide dispersion, which cannot be neglected. In addition, if the field intensity is high enough, the medium also responds in a nonlinear way. Most notably, the refractive index becomes intensity dependent (the Kerr effect) and photons can interact with phonons (molecular vibrations) of the medium (the Raman effect).

Supercontinuum Generation in Optical Fibers, ed. by J. M. Dudley and J. R. Taylor. Published by Cambridge University Press.

The first step in modelling nonlinear pulse propagation is therefore to include the nonlinear polarisation of the medium in Maxwell's equations. From these one derives a second-order wave equation which is approximated to a first-order propagation equation for the pulse. A number of derivations of such first-order equations have been presented in the literature, and the most useful of the early papers are Blow and Wood (1989), Mamyshev and Chernikov (1990) and Francois (1991). More recent papers have clarified various aspects of the derivations, or examined the validity of the assumptions used in more detail (Brabec and Krausz, 1997; Karasawa et al., 2001; Ferrando et al., 2005; Laegsgaard, 2007). Of particular importance to emphasise is the fact that the use of an envelope-carrier decomposition in the derivation of a first-order propagation equation does not in itself imply that the envelope needs to be slowly varying. Rather, the key assumptions are that propagation is unidirectional and any backward propagating waves (and any nonlinear coupling between forward and backward propagating waves) can be neglected. These aspects have been discussed in detail in Genty et al. (2007) and Kinsler (2007).

In order to focus on the major spectral broadening processes, we consider only a scalar treatment here. Discussions of broadband coupled propagation equations suitable for describing polarisation effects have been given in Trillo and Wabnitz (1992), Coen et al. (2002) and Zhu and Brown (2004). Generalisations to consider vectorial contributions due to longitudinal field components are discussed in Chapter 11 of this book.

Nonlinear propagation equations have been derived in both the time- and frequency-domains. Time-domain formulations have often been preferred because of analytic similarity to the nonlinear Schrödinger equation (NLSE) about which there is a vast literature in many fields (see for example Sulem and Sulem, 1999). However, the frequency-domain formulation does show more directly the frequency-dependence of effects such as dispersion, loss and the effective mode area. Both approaches have been successfully used in the study of supercontinuum generation, and thus we present useful forms of both equations here.

We first define the electric field (linearly polarised along $\boldsymbol{x}$) as: $\boldsymbol{E}(\boldsymbol{r},t) = (1/2)\boldsymbol{x}\,\{E(x,y,z,t)\exp(-i\omega_0 t) + \text{c.c.}\}$. In the frequency domain, the Fourier transform of $E(x,y,z,t)$ is: $\tilde{E}(x,y,z,\omega) = F(x,y,\omega)\tilde{A}(z,\omega-\omega_0)\,\exp(i\beta_0 z)$, where $\tilde{A}(z,\omega)$ is the complex spectral envelope, while ω_0 is a chosen reference frequency (often the centre frequency of the input pulse) and β_0 the wavenumber at that frequency. $F(x,y,\omega)$ is the transverse modal distribution. The time-domain envelope is obtained from

$$A(z,t) = \mathcal{F}^{-1}\left\{\tilde{A}(z,\omega-\omega_0)\right\} = \frac{1}{2\pi}\int_{-\infty}^{\infty}\tilde{A}(z,\omega-\omega_0)\exp[-i(\omega-\omega_0)t]\,d\omega, \tag{3.1}$$

where the amplitude is normalised such that $|A(z,t)|^2$ gives the instantaneous power in watts[1] and $\mathcal{F}^{-1}$ denotes the inverse Fourier transform.

Using this notation, and implementing the change of variable $T = t - \beta_1 z$ to transform into a co-moving frame at the envelope group velocity β_1^{-1}, we obtain a time-domain generalised NLSE for the evolution of $A(z,T)$:

$$\frac{\partial A}{\partial z} + \frac{\alpha}{2}A - \sum_{k\geq 2} \frac{\mathrm{i}^{k+1}}{k!}\beta_k \frac{\partial^k A}{\partial T^k} = i\gamma \left(1 + i\tau_{\text{shock}} \frac{\partial}{\partial T}\right) \times \left(A(z,T) \int_{-\infty}^{\infty} R(T') \left|A(z,T-T')\right|^2 \mathrm{d}T'\right). \tag{3.2}$$

The left-hand side of this equation models linear propagation effects, with α being the linear power attenuation and the β_k the dispersion coefficients associated with the Taylor series expansion of the propagation constant $\beta(\omega)$ about ω_0. In fact, given $\beta(\omega)$, the dispersion operator can be applied directly and in an approximation-free manner in the frequency domain through multiplication of the complex spectral envelope $\tilde{A}(z,\omega)$ by the operator $\tilde{\beta}(\omega) = (\beta(\omega) - (\omega - \omega_0)\beta_1 - \beta_0)$, and this is to be recommended. The right-hand side models nonlinear effects:

$$\gamma = \frac{\omega_0 n_2(\omega_0)}{c A_{\text{eff}}(\omega_0)} \tag{3.3}$$

is the usual nonlinear coefficient, where $n_2(\omega_0)$ is the nonlinear refractive index and $A_{\text{eff}}(\omega_0)$ is the mode effective area, both evaluated at the reference frequency ω_0. When the frequency dependence of n_2 and A_{eff} is not explicitly included, as it is not in Eq. (3.2) but will be in the following propagation equations, it is particularly important to choose ω_0 as the centre (angular) frequency of the input pulse for the approximation of frequency-independence to be reasonable. As pulse components of the propagated field can shift far away from ω_0 during propagation, this approximation must be used with care, see e.g. Kibler et al. (2005). Note that the correct way to define the effective area can be subtle, and the reader is referred to Tzolov et al. (1995) and Laegsgaard (2007) for details.

[1] Since $F(x,y,\omega)$ is typically defined as non-dimensional, this normalisation actually requires the previously defined electric field $E(x,y,z,t)$ to be scaled by a factor of $\sqrt{\frac{1}{2}\epsilon_0 cn}$ to obtain the electric field in V/m, where ϵ_0 is the vacuum permittivity, c is the speed of light in vacuum and n is the refractive index.

$R(t)$ is the Raman response function usually modelled as (Blow and Wood, 1989; Agrawal, 2007)

$$\begin{aligned} R(t) &= (1-f_R)\delta(t) + f_R h_R(t) \\ &= (1-f_R)\delta(t) + f_R \frac{\tau_1^2+\tau_2^2}{\tau_1\tau_2^2}\exp(-t/\tau_2)\sin(t/\tau_1)\Theta(t) \end{aligned} \tag{3.4}$$

where $f_R = 0.18$ is the fractional contribution of the delayed Raman response, $\tau_1 =$ 12.2 fs, and $\tau_2 = 32$ fs. $\Theta(t)$ is the Heaviside step function and $\delta(t)$ is the Dirac delta function. There exist both a parallel and an orthogonally polarised delayed Raman response (Hellwarth, 1977), but the orthogonal component is generally negligible (Dinda et al., 1998) and therefore usually not considered, which is also the approach taken here. More accurate forms of the response function have also recently been investigated by Lin and Agrawal (2006). Alternatively, the measured imaginary part of the Raman response function, and the real part obtained using the Kramers–Kronig relations, can be directly included; although essentially the same results can be obtained with no loss of accuracy by using the simpler multi-vibrational-mode model of Hollenbeck and Cantrell (2002).

The time derivative term on the right-hand side models the dispersion of the nonlinearity. This is usually associated with effects such as self-steepening and optical shock formation, characterised by a timescale $\tau_{\text{shock}} = \tau_0 = 1/\omega_0$. In the context of fibre propagation, additional dispersion of the nonlinearity arises from the frequency dependence of the effective area, and τ_{shock} can be generalised to account for this in an approximate manner (Blow and Wood, 1989; Kibler et al., 2005).

A better approach to include the dispersion of the nonlinear response is to describe it directly in the frequency domain (Mamyshev and Chernikov, 1990; Laegsgaard, 2007). In this case, we can derive a generalised NLSE for the pulse evolution as:

$$\frac{\partial \tilde{A}'}{\partial z} = i\bar{\gamma}(\omega)\exp(-\hat{L}(\omega)z)\mathcal{F}\left\{\bar{A}(z,T)\int_{-\infty}^{\infty} R(T')|\bar{A}(z,T-T')|^2 \mathrm{d}T'\right\} \tag{3.5}$$

with the frequency dependent nonlinear coefficient defined as

$$\bar{\gamma}(\omega) = \frac{n_2 n_0 \omega}{c n_{\text{eff}}(\omega) A_{\text{eff}}^{1/4}(\omega)}, \tag{3.6}$$

and

$$\bar{A}(z,T) = \mathcal{F}^{-1}\left\{\frac{\tilde{A}(z,\omega)}{A_{\text{eff}}^{1/4}(\omega)}\right\}. \tag{3.7}$$

Note that defining $\bar{\gamma}(\omega)$ as in Eq. (3.6) means that the nonlinear coefficient has a different unit than the standard γ defined in Eq. (3.3); the two nonlinear coefficients can therefore not be directly compared. This inconsistency is resolved when we later rewrite the GNLSE in the form of Eq. (3.12).

Eq. (3.5) also makes use of the change of variables:

$$\tilde{A}'(z,\omega) = \tilde{A}(z,\omega)\exp(-\hat{L}(\omega)z). \tag{3.8}$$

$\hat{L}(\omega)$ is the linear operator, given by $\hat{L}(\omega) = i(\beta(\omega) - \beta(\omega_0) - \beta_1(\omega_0)[\omega - \omega_0]) - \alpha(\omega)/2$. In these equations n_{eff} is the frequency dependent effective index of the guided mode and n_0 is the linear refractive index used when determining n_2. The frequency dependence of the nonlinear coefficient, arising through the frequency variation of A_{eff} and n_{eff}, is now explicitly evident. Another advantage of this formulation is that Eq. (3.5) is an ordinary differential equation (ODE), because of the change of variables given in Eq. (3.8). This means that it can be directly integrated using standard ODE codes (an example is presented in Section 3.4). To see how this is related to our time dependent equation [Eq. (3.2)], and to obtain a form of the GNLSE with a definition of γ which can be directly compared with the standard definition [Eq. (3.3)], we insert Eqs. (3.7)–(3.8) into Eq. (3.5) to obtain

$$\begin{aligned}&\frac{\partial \tilde{C}}{\partial z} - i\tilde{\beta}(\omega)\tilde{C}(z,\omega)\\ &= i\frac{n_2 n_0 \omega}{c n_{\text{eff}}(\omega)\sqrt{A_{\text{eff}}(\omega)A_{\text{eff}}(\omega_0)}}\mathcal{F}\left\{C(z,t)\int_{-\infty}^{\infty} R(T')|C(z,T-T')|^2 \mathrm{d}T'\right\},\end{aligned} \tag{3.9}$$

where the substitution

$$\mathcal{F}\{C(z,t)\} = \tilde{C}(z,\omega) = \left[\frac{A_{\text{eff}}(\omega)}{A_{\text{eff}}(\omega_0)}\right]^{-1/4} \tilde{A}(z,\omega) \tag{3.10}$$

was also used. If we now define a frequency dependent nonlinear coefficient $\gamma(\omega)$ as

$$\gamma(\omega) = \frac{n_2 n_0 \omega_0}{c n_{\text{eff}}(\omega)\sqrt{A_{\text{eff}}(\omega)A_{\text{eff}}(\omega_0)}}, \tag{3.11}$$

then Eq. (3.9) can be written as

$$\begin{aligned}&\frac{\partial \tilde{C}}{\partial z} - i[\beta(\omega) - \beta(\omega_0) - \beta_1(\omega_0)[\omega - \omega_0]]\tilde{C}(z,\omega) + \frac{\alpha(\omega)}{2}\tilde{C}(z,\omega)\\ &= i\gamma(\omega)\left[1 + \frac{\omega - \omega_0}{\omega_0}\right]\mathcal{F}\left\{C(z,t)\int_{-\infty}^{\infty} R(T')\left|C(z,T-T')\right|^2 \mathrm{d}T'\right\}.\end{aligned} \tag{3.12}$$

This form of the GNLSE has the following advantages. First, the definition of $\gamma(\omega)$ reduces to the standard definition Eq. (3.3) when the frequency dependencies of n_{eff} and A_{eff} are neglected. Second, neglecting the same frequency dependencies, expanding $\beta(\omega)$ as a Taylor series and inverse-Fourier transforming Eq. (3.12) leads to the well-known time-domain equation Eq. (3.2).

Eqs. (3.2), (3.5) and (3.12) have been successfully used to model the dynamics of nonlinear pulse propagation and fibre supercontinuum generation over a wide parameter range (Dudley et al., 2006). As written above, no noise sources are explicitly included in these propagation equations, but both input pulse noise and forcing noise due to spontaneous Raman scattering can be added (Drummond and Corney, 2001). This allows numerical simulations based on these equations to be used to determine the statistical properties of the generated supercontinuum spectra. A detailed comparison between such stochastic simulations and measured RF noise properties of supercontinuum spectra was presented in Corwin et al. (2003).

We close this section with a brief discussion of the numerical techniques used to solve these equations. The workhorse of numerical methods in nonlinear fibre optics is the split-step Fourier method, which essentially involves alternating between the time- and frequency-domains in order to solve the nonlinear and linear terms in Eq. (3.2) or Eq. (3.12) in the most natural and efficient way possible. This technique is in the class of pseudo-spectral methods, which are commonly employed to deal with stiff differential equations containing diffractive or dispersive terms. Using this technique, the linear term is exactly accounted for in each step, as it is completely integrable, whereas the nonlinear term is usually integrated using Runge–Kutta techniques. More details can be found in Bandelow et al. (2003), Cristiani et al. (2004), Agrawal (2007) and Hult (2007).

Alternatively, Eq. (3.5) can be integrated directly using Runge–Kutta techniques because the stiff dispersive part has been removed due to the change of variables. This is a generalisation of the method described in Hult (2007). Section 3.4 contains an implementation of this method.

Other methods such as the direct integration of Eq. (3.12), without the frequency dependence of n_{eff} and A_{eff}, as described in Francois (1991), or finite difference methods, can be used, but are generally less efficient.

3.3 Physical mechanisms

Supercontinuum generation in optical fibre involves the interplay between nonlinear and linear effects that occur during the propagation of an optical field (pulsed or continuous wave) in a fibre waveguide. Although the complex dynamics of this interplay can sometimes make it difficult to identify contributing mechanisms in isolation, an understanding of how different effects act individually is nonetheless indispensable developing a physical understanding of the spectral broadening process.

3.3.1 Dispersion

Dispersion is a linear effect but one that plays a crucial role in influencing the character of nonlinear interactions in a fibre. As mentioned previously, dispersion arises because of the frequency variation of the effective index of the guided mode, and depends on both material and waveguide contributions. Although for standard fibres with cylindrical symmetry, the fibre dispersion characteristics can be determined analytically, the dispersion properties of photonic crystal fibres generally require numerical computation. For most of the results presented in this book, we consider only the chromatic dispersion of the fundamental guided mode. However, more generally, it is also necessary to consider additional dispersion properties due to polarisation mode dispersion in the case of a birefringent fibre or the higher order intermodal dispersion between transverse modes in a multimode fibre.

In the context of supercontinuum generation, it is important to note that dispersion influences both the phase velocity and group velocity of a signal propagating in a fibre. Phasematching criteria play a very important role in determining nonlinear frequency conversion processes, and in the general case, both linear and nonlinear contributions to overall phase mismatch must be considered. However, even under conditions where a particular process is phasematched, group velocity mismatch can play an important role in limiting the interaction length. On the other hand, the matching of group velocities at different wavelengths can also result in mutual trapping between widely separated spectral components (Nishizawa and Goto, 2002; Gorbach and Skryabin, 2007).

The group velocity dispersion (GVD) β_2 and higher order dispersion terms are defined in terms of the coefficients associated with the Taylor series expansion of the propagation constant. Two complementary definitions of GVD are used in the literature: β_2 (with units of $\mathrm{s^2\,m^{-1}}$) and $D = -(2\pi c/\lambda^2)\beta_2$ (with units of $\mathrm{ps\,nm^{-1}\,km^{-1}}$). The former definition is generally employed by the physics and ultrafast optics communities; the latter by the engineering community. The wavelength range where $\beta_2 > 0$ is referred to as the normal dispersion regime, whereas the wavelength range where $\beta_2 < 0$ is referred to as the anomalous dispersion regime. The wavelength where $\beta_2 = 0$ is referred to as the zero dispersion wavelength (ZDW) and, depending on the fibre structure, both standard fibres and PCF can exhibit multiple ZDWs (Mamyshev et al., 1993; Skryabin et al., 2003). In a purely linear regime of propagation, an initially-unchirped input pulse will temporally broaden in an identical way irrespective of the sign of β_2. However, in the presence of nonlinearity, the sign of the dispersion influences the propagation dynamics significantly. The detailed way in which the sign of the GVD at the pump wavelength used influences supercontinuum generation will be seen in some of the subsequent chapters of this book.

3.3.2 Self and cross-phase modulation

If we neglect dispersion and only consider the Kerr nonlinearity of the fibre then we can observe self and cross-phase modulation (SPM and XPM respectively). The time dependent intensity $I(t)$ of a pulse causes a corresponding modulation to the local refractive index in the fibre which can be expressed as $\Delta n = n_2 I(t)$. This in turn causes a time dependent phase delay to the same pulse (for SPM) or a co-propagating pulse (XPM). This amounts to applying a corresponding nonlinear chirp with the generation of new frequency components.

When considering SPM in isolation for a temporally symmetric unchirped input pulse, the spectral broadening is also symmetric and the time domain envelope of the pulse is unchanged. In general, however, the dispersion cannot be neglected and it is the interplay of the two which is important. This leads to two regimes of operation which have distinctly different behaviour. In one, SPM and normal dispersion interact to lead to simultaneous spectral and temporal broadening with the development of a linearised chirp across the central region of the pulse (Grischkowsky and Balant, 1982). Under some conditions, this evolution can lead to novel self-similar propagation dynamics (Dudley et al., 2007). The second broad range of interaction between SPM and dispersion occurs in the anomalous dispersion regime, where the resulting dynamics leads to the generation of solitons. The interaction dynamics of XPM in supercontinuum generation can be rather complex, as generally the two interacting pulses can be propagating in different regimes of dispersion. Efficient interaction generally only occurs between wavelength components where there is low group velocity mismatch. Aspects of these interactions will be considered in Chapter 9.

3.3.3 Solitons

In the anomalous dispersion regime of an optical fibre, the nonlinear chirp induced by SPM and the linear chirp from GVD (i.e. β_2 only) can combine in such a way as to yield either stable or periodic evolution of optical solitons. Solitons are analytic solutions to the NLSE Eq. (3.2) above without higher order dispersion, the shock terms or Raman, and their properties have been extensively studied, notably using the advanced mathematical technique of inverse scattering. This analysis yields different classes of soliton solutions corresponding to initially injected pulses of the form $A(0,T) = N\text{sech}(T/T_0)$. The soliton order of the input pulse N is determined by both pulse and fibre parameters through $N^2 = L_D/L_{NL}$. Here, $L_D = T_0^2/|\beta_2|$ and $L_{NL} = 1/(\gamma P_0)$ are the characteristic dispersive and nonlinear length scales respectively. Here P_0 is the peak power, T_0 a measure of pulse width and the other parameters are as above. The fundamental soliton is described by $N = 1$ and, for this case, both the spectral and temporal profiles remain unchanged during propagation.

This particular case can also be usefully understood using a simplified analysis based on chirp cancellation (Dudley et al., 2001). Integral values of $N \geq 2$ correspond to higher-order solitons that undergo periodic spectral and temporal evolution. Although the general form of the evolution can be very complex, the initial phase of propagation is always associated with simultaneous nonlinear spectral broadening and temporal compression due to the fibre anomalous GVD.

3.3.4 Soliton perturbation, Raman scattering, dispersive waves

Although ideal soliton propagation is observed only under extremely stringent conditions, the injection of high-power pulses in the anomalous dispersion regime of a fibre very often leads to propagation effects characteristic of what can be classed as "soliton dynamics". In the context of understanding supercontinuum generation, a number of such propagation effects are of particular interest.

Firstly, a high-power pulse injected in the anomalous dispersion regime of a fibre generally is associated with a soliton number $N \gg 1$. In this case (even if N is non-integral) the initial stage of evolution will be qualitatively similar to that of a higher order soliton in that there will be an initial phase of spectral broadening and temporal compression. However, higher-order solitons are intrinsically unstable, and in the presence of perturbation readily break up (or "decay") into a series of lower amplitude sub-pulses. This process has now come to be referred to as soliton fission. In the femtosecond regime, higher-order dispersion and Raman scattering are the two most significant effects that can perturb higher-order soliton evolution and induce pulse break-up.

The distance at which fission occurs usually corresponds to the point where the evolving input pulse soliton reaches its maximum bandwidth. Optimising the fibre length just shorter than this point is the basis of the higher-order soliton effect compression technique, and a number of empirical expressions for this characteristic distance have been obtained (Dianov et al., 1986; Chen and Kelley, 2002). We have found it particularly useful to refer to this distance as the "fission length", written in the simplified form: $L_{\text{fiss}} \sim L_{\text{D}}/N$.

Apart from soliton fission, there are two important effects crucial to the generation of additional bandwidth beyond the initial spectral broadening phase of the injected pulse. Firstly, a soliton propagating close to the ZDW can transfer a portion of its energy to the normal GVD regime to generate what is variously referred to as a "dispersive wave", "non-solitonic radiation" or "Cherenkov radiation". This effect was first observed numerically (Wai et al., 1986) and can be analysed in terms of a phasematching argument involving the soliton linear and nonlinear phase and the linear phase of a continuous wave at a different frequency (Akhmediev and Karlsson, 1995). In addition, even if perturbed by dispersive wave generation, solitons can, nonetheless, experience continuous displacement to longer wavelengths due to the soliton self-frequency shift, because the individual soliton bandwidths overlap the Raman gain. The dynamics of the frequency shift ν_R can be expressed simply as $d\nu_R/dz \propto |\beta_2|/T_0^4$ (Gordon, 1986), but dispersion of the nonlinear response can decelerate this frequency shift as the solitons shift to longer wavelengths (Voronin and Zheltikov, 2008). Both of these effects occur simultaneously, and their interaction can be subtle. Moreover, the effects of group velocity matching and cross-phase modulation between the soliton and dispersive wave radiation can induce trapping dynamics and lead to further increase in the bandwidth on both the long- and short-wavelength edges of the supercontinuum (Nishizawa and Goto, 2002; Genty et al., 2004; Gorbach and Skryabin, 2007).

In closing this section, it is worth making one additional comment. Our discussion of perturbed propagation dynamics following soliton fission may give the impression that such effects are important only in regimes where supercontinuum generation is induced by high-power femtosecond pulses. This is not the case, however, as even with longer pulses or quasi-CW radiation, modulation instability (discussed in the next section) can lead to the break up of a field envelope into a train of individual soliton pulses. These pulses can suffer the same dynamics of dispersive wave generation and Raman shifting as described above. In fact, the careful design of fibres to exploit these dynamics has led to the development of compact supercontinuum sources using nanosecond pulse sources (Stone and Knight, 2008), or even continuous-wave pump sources, as described elsewhere in this book.

3.3.5 Modulation instability and four-wave mixing

Four-wave mixing (FWM) between narrowband optical waves is one of the most fundamental processes in nonlinear optics (Stolen and Bjorkholm, 1982). Many of the key features of the process are seen in the case where the two pump waves have the same frequency, and the nonlinear interaction involves the conversion of the pump into a pair of sidebands that are frequency downshifted (Stokes) and upshifted (anti-Stokes) relative to the pump. Assuming an undepleted pump of power P_0, the growth of the sidebands is exponential with amplitude gain g (units of m^{-1}) given by $g = [(\gamma P_0)^2 - (\kappa/2)^2]^{1/2}$. For single-mode fibres, the phase mismatch $\kappa = 2\gamma P_0 + 2\sum_{m=1}^{\infty}(\beta_{2m}/(2m)!)\Omega^{2m}$ where Ω is the angular frequency shift from the pump and the β_{2m} terms represent the even dispersion coefficients of the fibre mode at the pump wavelength (corresponding to Taylor-expanding the propagation constant β around the pump wavelength).

In the absence of initial seeding, FWM corresponds to an instability of the propagating CW pump and the growth from noise of sidebands symmetric in frequency about the pump. The maximum growth rate is at frequencies satisfying phasematching where $\kappa = 0$, with maximum amplitude gain of $g_{\max} = \gamma P_0 = 1/L_{\mathrm{NL}}$. In the time domain, this leads to a temporal modulation, and the process is referred to as a modulation instability (MI). It is important to note that FWM and MI can be considered as frequency-domain and time-domain descriptions of the same underlying physics (Stolen et al., 1989), but the respective frequency-domain and time-domain pictures can be useful for explaining different aspects of the field evolution, and both terms are often used in the literature.

In the context of supercontinuum generation, FWM/MI processes dominate the broadening mechanisms in the case of input pulses of picosecond-nanosecond duration, or for the case of CW pumping. Efficient spectral broadening requires pumping close to the fibre ZDW, although it is important to note that the contribution of higher-order dispersion terms to the phasematching condition can yield significant gain windows with a pump in either the normal or the anomalous dispersion regime (Reeves et al., 2003; Harvey et al., 2003). A FWM analysis has also proven useful in describing the appearance of certain narrowband spectral features in the case of femtosecond pulse excitation due to the interactions of solitons and dispersive waves (Skryabin and Yulin, 2005).

3.4 The numerical solution of the GNLSE

In this section we provide implementation details for the numerical solution of the generalised nonlinear Schrödinger equation, based on the propagation equations described above. The code presented here is not production ready, in the sense that we have traded some efficiency for clarity and because the quality of the integration is not closely monitored so care needs to be exercised when assessing accuracy. However, we hope it will provide a practical starting point for investigating some of the physics described in this book. Updates and extensions to the code are available from www.scgbook.info.

Most of the discussion will apply to solving any of the propagation equations in this chapter, but the code we will give will solve

$$\frac{\partial \tilde{A}'}{\partial z} = i\frac{\gamma\omega}{\omega_0}\exp(-\hat{L}z)\mathcal{F}\left\{A(z,T)\int_{-\infty}^{\infty}R(T')|A(z,T-T')|^2\mathrm{d}T'\right\}, \tag{3.13}$$

which is obtained by ignoring the frequency dependence of n_2, n_{eff} and A_{eff} in Eq. (3.5). This equation integrates the field $A(z,T)$ in the frequency domain while making use of a change of variables to shift into the so-called interaction picture. This avoids the stiff dispersive part of the equation. The shift is given by Eq. (3.8). See Section 3.2 for a fuller discussion.

3.4.1 Grid selection

To numerically solve continuous problems on a computer it is usual to approximate the solution on a discrete grid of points. The span and sampling density of this grid determine how close the approximation is to the true solution. To solve Eq. (3.13) we represent the field on a temporal grid (and via the Fourier transform a corresponding frequency grid) and then propagate stepwise through z (fibre length).

The temporal grid must be selected under two constraints:

(i) The width of the grid must contain the maximum time delay of the field.
(ii) The sampling must be fast enough to contain the maximum frequency of the field.

Both of these conditions can be hard to predict before running some numerical experiments. The first can be estimated from the walk-off between the two most separated frequencies of the field over the fibre length. The full temporal width of the grid must accommodate the pump pulse duration and any delay due to group velocity dispersion which may arise. When simulating broadband supercontinua over long fibre lengths, the group delay of the extreme spectrum limits can be up to several hundred picoseconds.

The second condition is due to the Nyquist sampling theorem – one must choose a temporal grid spacing corresponding to twice our highest frequency. For supercontinuum generation the limit is usually on the short wavelength edge. We can write the minimum correctly included wavelength as:

$$\lambda_{\text{min}} = \frac{c}{\frac{1}{2\Delta t} + \frac{\omega_0}{2\pi}} = \frac{1}{\frac{1}{2c\Delta t} + \frac{1}{\lambda_0}} \tag{3.14}$$

where Δt is our temporal grid spacing and λ_0 is the reference wavelength corresponding to ω_0. For example, in the simulation discussed below, our short wavelength edge is ~500 nm while our reference wavelength is 835 nm. This requires $\Delta t < 2$ fs. Given a maximum delay of around 5 ps and therefore a total temporal width of 10 ps we need at least 5000 points. The largest wavelength included in the spectral window is given by

$$\lambda_{\text{max}} = \frac{1}{\frac{1}{\lambda_0} - \frac{1}{2c\Delta t}}. \tag{3.15}$$

From this it is seen that there is also a lower limit on Δt given by $\Delta t > \lambda_0/(2c)$, in order to avoid negative frequencies in the spectral window. For the case of a reference wavelength of 835 nm, this means that Δt must be larger than 1.39 fs.

It is usual to set the number of grid points to a power of 2, as the fast Fourier transform algorithm is particularly fast for this case, unless the grid size is extremely large (over 2^{20}).

Most methods used to propagate the solution of Eq. (3.13), or similar equations, along a fibre have to evaluate the full right-hand side (RHS) at each point of our z (length) grid. As this involves considerable computational effort it is common to employ some sort of adaptive step selection algorithm. At each step an estimate of the solution error is obtained and used to either increase or decrease the step size, in order to choose the smallest step which maintains a certain level of accuracy.

A second problem is that, especially for supercontinuum simulations, the scale of the problem can be hard to judge before running some test problems and therefore one should run a convergence test to check that important dynamics are correctly modelled. The use of an adaptive algorithm helps one to get to the correct scale more quickly.

The conventional split-step algorithm, or its adaption, as described in Agrawal (2007) and Hult (2007), do not provide any direct knowledge of the error of each step. In this case a step-doubling technique is used, where two fine steps and one longer equivalent step are taken from the same starting point, and their differences are compared to obtain an estimate of the error. Such a technique is described by Sinkin et al. (2003), and usually works very well.

In the case of Eq. (3.13), we can use standard ordinary differential equation (ODE) integrators, which have been extensively studied, and some of which have built-in error estimators and methods for adaption (Hairer et al., 1993; Brankin and Gladwell, 1994; Shampine and Reichelt, 1997).

It should be emphasised that none of these techniques guarantees a certain global error in the solution, but simply help one to converge more quickly. Independent convergence tests and monitoring of conserved quantities, such as photon number, should always be performed to ensure accuracy. A convergence test could, for example, consist of repeating the simulation with a tolerance parameter smaller by a factor of 10 and checking whether the output spectrum did not change significantly.

3.4.2 Implementation details

Compared to the nonlinear Schrödinger equation (for which some example code is given in Agrawal (2007)) there are two new difficulties to deal with when implementing a numerical code to solve the generalised equations. One is the shock term, which appears as a time derivative in Eq. (3.2). In a frequency domain formulation, as will be described here, this term is quite simply dealt with as a multiplication by $i\omega$, as is implicit in Eq. (3.13). In a time domain code it can be treated as a perturbation; details of this technique are given in Blow and Wood (1989) and Cristiani et al. (2004).

The second difficulty is the convolution integral between the nonlinear response function and the time domain field. All three formulations of this chapter require this to be solved. From the definition of $R(t)$ given by Eq. (3.4) we can write the convolution integral as:

$$\begin{aligned}
&\int_{-\infty}^{\infty} R(T')|A(z,T-T')|^2 \mathrm{d}T' \\
&= \int_{-\infty}^{\infty} [(1-f_\mathrm{R})\delta(T') + f_\mathrm{R} h_\mathrm{R}(T')]|A(z,T-T')|^2 \mathrm{d}T' \\
&= (1-f_\mathrm{R})|A(z,T)|^2 + f_\mathrm{R} \int_{-\infty}^{\infty} h_\mathrm{R}(T')|A(z,T-T')|^2 \mathrm{d}T'. \qquad (3.16)
\end{aligned}$$

The first term on the last line is simply the Kerr effect term from the plain NLSE. The second term is still a convolution integral which describes stimulated Raman scattering. To numerically compute it we can either use a simple recurrence relation – if $h_\mathrm{R}(t)$ is a simple analytic form – or we can make use of the convolution theorem. The latter is more commonly employed and allows one to solve for an arbitrary Raman response efficiently using fast Fourier transforms. For details consult a standard Fourier textbook such as Bracewell (2000); specifically for our case the integral in the last line of Eq. (3.16) can be evaluated by transforming $h_\mathrm{R}(T)$ and $|A(z,T)|^2$ into the frequency domain, multiplying, then transforming back. Explicitly we compute:

$$\int_{-\infty}^{\infty} h_\mathrm{R}(T')|A(z,T-T')|^2 \mathrm{d}T' = \Delta t \mathcal{F}^{-1}[\mathcal{F}(h_\mathrm{R}(T))\mathcal{F}(|A(z,T)|^2)]. \qquad (3.17)$$

3.4.3 A basic implementation

The code listings which we provide below are written in the Matlab language as this is popular for scientific computing and relatively clear. Note that the definitions of the FFT and IFFT in Matlab are the opposite to ours, defined in Eq. (3.1). Therefore, in the code listed below we use IFFT for the FFT and vice versa.

As noted above this code was written with space limitations and simplicity in mind and therefore suffers somewhat in efficiency and completeness. But it has been tested to reproduce a number of supercontinuum simulations in the literature and this book, and should serve as a good starting point. The line numbers indicated in the code listing are used for cross-referencing with the comments in the following text, aimed at explaining some of the details.

The following listing contains the code for a function which propagates an input field through a fibre using Eq. (3.13).

Listing 3.1: gnlse.m – an implementation of Eqs. (3.13), (3.16) and (3.17).

```
function [Z, AT, AW, W] = gnlse(T, A, w0, gamma, betas, ...
                          loss, fr, RT, flength, nsaves)
% Propagate an optical field using the generalised NLSE
% This code integrates Eqs. (3.13), (3.16) and (3.17).
% For usage see the exampe of test_Dudley.m (below)
% Written by J.C. Travers, M.H. Frosz and J.M. Dudley (2009)
% Please cite this chapter in any publication using this code.
% Updates to this code are available at www.scgbook.info
n = length(T); dT = T(2)-T(1); % grid parameters
V = 2*pi*(-n/2:n/2-1)'/(n*dT); % frequency grid
alpha = log(10.^(loss/10));    % attenuation coefficient
B = 0;
for i = 1:length(betas)        % Taylor expansion of betas
  B = B + betas(i)/factorial(i+1).*V.^(i+1);
end
L = 1i*B - alpha/2;            % linear operator
if abs(w0) > eps               % if w0>0 then include shock
    gamma = gamma/w0;
    W = V + w0;                % for shock W is true freq
else
    W = 1;                     % set W to 1 when no shock
end
RW = n*ifft(fftshift(RT.'));   % frequency domain Raman
L = fftshift(L); W = fftshift(W); % shift to fft space
% === define function to return the RHS of Eq. (3.13)
function R = rhs(z, AW)
  AT = fft(AW.*exp(L*z));         % time domain field
  IT = abs(AT).^2;                % time domain intensity
  if (length(RT) == 1) || (abs(fr) < eps) % no Raman case
    M = ifft(AT.*IT);             % response function
  else
    RS = dT*fr*fft(ifft(IT).*RW); % Raman convolution
    M = ifft(AT.*((1-fr).*IT + RS));% response function
  end
  R = 1i*gamma*W.*M.*exp(-L*z);   % full RHS of Eq. (3.13)
end
% === define function to print ODE integrator status
function status = report(z, y, flag) %
  status = 0;
  if isempty(flag)
    fprintf('%05.1f %% complete\n', z/flength*100);
  end
end
% === setup and run the ODE integrator
Z = linspace(0, flength, nsaves); % select output z points
% === set error control options
options = odeset('RelTol', 1e-5, 'AbsTol', 1e-12, ...
                 'NormControl', 'on', ...
                 'OutputFcn', @report);
[Z, AW] = ode45(@rhs, Z, ifft(A), options); % run integrator
```

```
51  % === process output of integrator
52  AT = zeros(size(AW(1,:)));
53  for i = 1:length(AW(:,1))
54    AW(i,:) = AW(i,:).*exp(L.'*Z(i)); % change variables
55    AT(i,:) = fft(AW(i,:));            % time domain output
56    AW(i,:) = fftshift(AW(i,:))./dT;  % scale
57  end
58  W = V + w0; % the absolute frequency grid
59  end
```

Details of the input parameters and an example of the usage of this function are given later. For now we introduce the variables as they occur.

Line 9 derives the temporal grid spacing and number of points from the input time grid provided in `T`, and line 10 uses these to compute the relative frequency grid. The next lines, 11–24, compute some of the propagation variables, including: the attenuation coefficient from the loss (line 11); the Taylor expansion to obtain $\beta(\omega)$ (lines 12–15); and the combined linear operator $\hat{L}$ (line 16). Lines 17–22 are a convenience allowing one to propagate without the shock term if `w0=0` on input. This allows one to test NLSE propagation such as for a soliton. Line 23 pre-computes the frequency domain Raman response for use in the convolution later on. Line 24 shift the operator grids to be aligned with the output of the fast Fourier transform, to avoid shifting the field in the inner loops of the computation.

Lines 26–36 define the main part of the computation, the right-hand side of Eq. (3.13). This nested function is passed to the ODE integrator later to integrate the field propagation through the fibre. Line 27 shifts the input field back from the interaction picture, using the variable change of Eq. (3.8) and then transforms it to the time domain. Line 28 computes the temporal intensity. The next lines include an efficiency saving if the input Raman response was defined as a scalar, in which case no Raman convolution is computed and line 30 computes just the Kerr response. Instead, if a full Raman response function was passed in as `RT`, then line 32 computes the Raman convolution as per Eq. (3.17) and line 33 computes the complete response function as in Eq. (3.16). These are then combined, along with a shift back into the interaction picture and multiplication by `gamma` to obtain the full right-hand side of Eq. (3.13) as the output on line 35.

The following section, lines 38–50, utilises the Matlab ODE suite (Shampine and Reichelt, 1997) to integrate the `rhs` function over the fibre. Similar sets of codes in other languages, such as RKSUITE in Fortran (Brankin and Gladwell, 1994) can be used for the same purpose. Line 45 determines the grid of z points to output the field on so that we can look at the evolution of it through the fibre. Lines 47 and 48 select the error control conditions we wish to use, these should be set with great care. Line 50 actually performs the integration.

Finally lines 52–56 shift the array of results from the integration back from the interaction picture (line 54), convert to the time domain (line 55) and scale the frequency domain results so that they correspond to amplitudes in units of [$W^{1/2}$] (line 56). Line 58 ensures `W` contains the absolute frequency grid for use in the calling code.

As an example of how to use this function, the listing below shows code that runs a supercontinuum simulation similar to that presented in Fig. 3 of Dudley et al. (2006). Most of this code should be relatively self-explanatory, dealing with the grid, input field, simulation and fibre parameters.

Listing 3.2: An example of how to call the gnlse function in Listing 3.1 for the conditions described in Dudley et al. (2006).

```
% Simulate supercontinuum generation for parameters similar
% to Fig.3 of Dudley et. al, RMP 78 1135 (2006)
% Written by J.C. Travers, M.H Frosz and J.M. Dudley (2009)
% Please cite this chapter in any publication using this code.
% Updates to this code are available at www.scgbook.info
n = 2^13;                       % number of grid points
twidth = 12.5;                  % width of time window [ps]
c = 299792458*1e9/1e12;         % speed of light [nm/ps]
wavelength = 835;               % reference wavelength [nm]
w0 = (2.0*pi*c)/wavelength; % reference frequency [2*pi*THz]
T = linspace(-twidth/2, twidth/2, n); % time grid
% === input pulse
power = 10000;                  % peak power of input [W]
t0 = 0.0284;                    % duration of input [ps]
A = sqrt(power)*sech(T/t0); % input field [W^(1/2)]
% === fibre parameters
flength = 0.15;                 % fibre length [m]
% betas = [beta2, beta3, ...] in units [ps^2/m, ps^3/m ...]
betas = [-11.830e-3, 8.1038e-5, -9.5205e-8, 2.0737e-10, ...
         -5.3943e-13, 1.3486e-15, -2.5495e-18, 3.0524e-21, ...
         -1.7140e-24];
gamma = 0.11;                   % nonlinear coefficient [1/W/m]
loss = 0;                       % loss [dB/m]
% === Raman response
fr = 0.18;                      % fractional Raman contribution
tau1 = 0.0122; tau2 = 0.032;
RT = (tau1^2+tau2^2)/tau1/tau2^2*exp(-T/tau2).*sin(T/tau1);
RT(T<0) = 0;              % heaviside step function
RT = RT/trapz(T,RT);      % normalise RT to unit integral
% === simulation parameters
nsaves = 200;       % number of length steps to save field at
% propagate field
[Z, AT, AW, W] = gnlse(T, A, w0, gamma, betas, loss, ...
                       fr, RT, flength, nsaves);
% === plot output
figure();
lIW = 10*log10(abs(AW).^2); % log scale spectral intensity
mlIW = max(max(lIW));           % max value, for scaling plot
WL = 2*pi*c./W; iis = (WL>400 & WL<1350); % wavelength grid
```

```
40  subplot(1,2,1);
41  pcolor(WL(iis), Z, lIW(:,iis)); % plot as pseudocolor map
42  caxis([mlIW-40.0, mlIW]);  xlim([400,1350]); shading interp;
43  xlabel('Wavelength / nm'); ylabel('Distance / m');
44
45  lIT = 10*log10(abs(AT).^2); % log scale temporal intensity
46  mlIT = max(max(lIT));       % max value, for scaling plot
47  subplot(1,2,2);
48  pcolor(T, Z, lIT);          % plot as pseudocolor map
49  caxis([mlIT-40.0, mlIT]);  xlim([-0.5,5]); shading interp;
50  xlabel('Delay / ps'); ylabel('Distance / m');
```

The definition of the Raman response function is given on lines 26–29. Any of the methods described in this chapter can be used. Here we implement Eq. (3.4). One point to note is that the response function should have unit integral, which is enforced on line 29.

The code for plotting the output on lines 37–50 is designed to reproduce Fig. 3 in Dudley et al. (2006). Note that the authors of that paper used a different definition of the Raman response and included the effect of the frequency dependent effective area, and hence some differences from Fig. 3 of that paper and the figures from this code will be apparent.

References

Agrawal G. P. (2007). *Nonlinear Fiber Optics*, 4th edn. San Diego: Academic Press.

Akhmediev N., Karlsson M. (1995). Cherenkov radiation emitted by solitons in optical fibers. *Phys. Rev. A* **51**, 2602–2607.

Alfano R. R., Ed. (2006). *The Supercontinuum Laser Source*. Springer, New York.

Bandelow U., Demircan A., Kesting M. (2003). Simulation of pulse propagation in nonlinear optical fibers. *WIAS Report, Weierstrass Institute for Applied Analysis and Stochastics* **23**.

Blow K. J., Wood D. (1989). Theoretical description of transient stimulated Raman scattering in optical fibers. *IEEE J. Quantum Electron. 25*, 12 (Dec.), 2665–2673.

Brabec T., Krausz F. (1997). Nonlinear optical pulse propagation in the single-cycle regime. *Phys. Rev. Lett. 78*, 17 (Apr), 3282–3285.

Bracewell R. (2000). *The Fourier transform and its applications*. McGraw-Hill.

Brankin R., Gladwell I. (1994). A Fortran 90 Version of RKSUITE: An ODE Initial Value Solver. *Annals of Numer. Math* **1**, 363–375.

Chen C.-M., Kelley P. L. (2002). Nonlinear pulse compression in optical fibers: scaling laws and numerical analysis. *J. Opt. Soc. Am. B* **19**, 1961–1967.

Coen S., Chau A., Leonhardt R. et al. (2002). Supercontinuum generation by stimulated Raman scattering and parametric four-wave mixing in photonic crystal fibers. *J. Opt. Soc. Am. B 19*, 4 (APR), 753–764.

Corwin K., Newbury N., Dudley J. et al. (2003). Fundamental noise limitations to supercontinuum generation in microstructure fiber. *Phys. Rev. Lett. 90*, 11 (MAR 21).

Cristiani I., Tediosi R., Tartara L., Degiorgio V. (2004). Dispersive wave generation by solitons in microstructured optical fibers. *Opt. Express 12*, 1, 124–135.

Dianov E. M., Nikonova Z. S., Prokhorov A. M., Serkin V. N. (1986). Optimal compression of multi-soliton pulses in optical fibers. *Pis'ma Zh. Tekh. Fiz.* **12**, 756–760 [Sov. Tech. Phys. Lett. **12**, 311–313 (1986)].

Dinda P. T., Millot G., Wabnitz S. (1998). Polarization switching and suppression of stimulated Raman scattering in birefringent optical fibers. *J. Opt. Soc. Am. B 15*, 5 (May), 1433–1441.

Drummond P. D., Corney J. F. (2001). Quantum noise in optical fibers. I. Stochastic equations. *J. Opt. Soc. Am. B 18*, 2 (Feb.), 139–152.

Dudley J. M., Finot C., Richardson D. J., Millot G. (2007). Self-similarity in ultrafast nonlinear optics. *Nat. Physics 3*, 9 (SEP), 597–603.

Dudley J. M., Genty G., Coen S. (2006). Supercontinuum generation in photonic crystal fiber. *Rev. Mod. Phys.* **78**, 1135–1184.

Dudley J. M., Peacock A. C., Millot G. (2001). The cancellation of nonlinear and dispersive phase components on the fundamental optical fiber soliton: a pedagogical note. *Opt. Commun.* **193**, 253–259.

Ferrando A., Zacarés M., Fernández de Córdoba P., Binosi D., Montero Á. (2005). Forward-backward equations for nonlinear propagation in axially invariant optical systems. *Phys. Rev. E 71*, 1, 16601.

Francois P. (1991). Nonlinear propagation of ultrashort pulses in optical fibers: total field formulation in the frequency domain. *J. Opt. Soc. Am. B 8*, 2, 276–293.

Genty G., Kinsler P., Kibler B., Dudley J. (2007). Nonlinear envelope equation modeling of sub-cycle dynamics and harmonic generation in nonlinear waveguides. *Opt. Express 15*, 9, 5382–5387.

Genty G., Lehtonen M., Ludvigsen H. (2004). Effect of cross-phase modulation on supercontinuum generated in microstructured fibers with sub-30 fs pulses. *Opt. Express 12*, 19 (Sept. 20,), 4614–4624.

Gorbach A. V., Skryabin D. V. (2007). Light trapping in gravity-like potentials and expansion of supercontinuum spectra in photonic-crystal fibres. *Nat. Photonics 1*, 11 (NOV), 653–657.

Gordon J. P. (1986). Theory of the soliton self-frequency shift. *Opt. Lett. 11*, 10 (Oct.), 662–664.

Grischkowsky D., Balant A. C. (1982). Optical pulse compression based on enhanced frequency chirping. *Appl. Phys. Lett.* **41**, 1–3.

Hairer E., Nørsett S., Wanner G. (1993). *Solving ordinary differential equations I: nonstiff problems*. Springer, New York.

Harvey J. D., Leonhardt R., Coen S. et al. (2003). Scalar modulation instability in the normal dispersion regime by use of a photonic crystal fiber. *Opt. Lett. 28*, 22 (Nov. 15,), 2225–2227.

Hellwarth R. W. (1977). Third-order optical susceptibilities of liquids and solids. *Prog. Quantum Electron. 5*, 1, 2–68.

Hollenbeck D., Cantrell C. D. (2002). Multiple-vibrational-mode model for fiber-optic Raman gain spectrum and response function. *J. Opt. Soc. Am. B 19*, 12, 2886–2892.

Hult J. (2007). A fourth-order Runge-Kutta in the interaction picture method for, simulating supercontinuum generation in optical fibers. *IEEE J. Lightwave Technol. 25*, 12 (DEC), 3770–3775.

Karasawa N., Nakamura S., Nakagawa N. et al. (2001). Comparison between theory and experiment of nonlinear propagation for a-few-cycle and ultrabroadband optical pulses in a fused-silicafiber. *IEEE J. Quantum Electron. 37*, 3, 398–404.

Kibler B., Dudley J. M., Coen S. (2005). Supercontinuum generation and nonlinear pulse propagation in photonic crystal fiber: influence of the frequency-dependent effective mode area. *Appl. Phys. B* **81**, 337–342.

Kinsler P. (2007). Limits of the unidirectional pulse propagation approximation. *J. Opt. Soc. Am. B* **24**, 2363–2368.

Laegsgaard J. (2007). Mode profile dispersion in the generalised nonlinear Schrödinger equation. *Opt. Express 15*, 24, 16110–16123.

Lin Q., Agrawal G. P. (2006). Raman response function for silica fibers. *Opt. Lett. 31*, 21, 3086–3088.

Mamyshev P., Chernikov S. (1990). Ultrashort-pulse propagation in optical fibers. *Opt. Lett. 15*, 19, 1076–1078.

Mamyshev P. V., Wigley P. G. J., Wilson J. et al. (1993). Adiabatic compression of Schrödinger solitons due to the combined perturbations of higher-order dispersion and delayed nonlinear response. *Phys. Rev. Lett.* **71**, 73–76.

Nishizawa N., Goto T. (2002). Characteristics of pulse trapping by use of ultrashort soliton pulses in optical fibers across the zero-dispersion wavelength. *Opt. Express 10*, 21 (OCT 21), 1151–1159.

Reeves W. H., Skryabin D. V., Biancalana F. et al. (2003). Transformation and control of ultra-short pulses in dispersion-engineered photonic crystal fibres. *Nature*. **424**, 511–515.

Shampine L., Reichelt M. (1997). The MATLAB ODE suite [J]. *SIAM J. Scientific Computing* **18**, 1.

Sinkin O., Holzlohner R., Zweck J., Menyuk C. (2003). Optimization of the split-step Fourier method in modeling optical-fiber communications systems. *IEEE J. Lightwave Technol. 21*, 1, 61–68.

Skryabin D. V., Luan F., Knight J. C., Russell P. St. J. (2003). Soliton self-frequency shift cancellation in photonic crystal fibers. *Science* **301**, 1705–1708.

Skryabin D. V., Yulin A. V. (2005). Theory of generation of new frequencies by mixing of solitons and dispersive waves in optical fibers. *Phys. Rev. E* **72**, 016619/1–10.

Stolen R., Gordon J., Tomlinson W., Haus H. (1989). Raman response function of silica-core fibers. *J. Opt. Soc. Am. B 6*, 6, 1159–1166.

Stolen R. H., Bjorkholm J. E. (1982). Parametric amplification and frequency conversion in optical fibers. *IEEE J. Quantum Electron. QE-18*, 7 (July), 1062–1072.

Stone J. M., Knight J. C. (2008). Visibly "white" light generation in uniform photonic crystal fiber using a microchip laser. *Opt. Express 16*, 4, 2670–2675.

Sulem C., Sulem P. (1999). *The Nonlinear Schrödinger Equation: Self-Focusing and Wave Collapse*. Springer.

Taylor J. R., Ed. (2005). *Optical Solitons: Theory and Experiment*, 1st edn. Cambridge University Press, Cambridge.

Trillo S., Wabnitz S. (1992). Parametric and Raman amplification in birefringent fibers. *J. Opt. Soc. Am. B 9*, 7, 1061–1082.

Tzolov V. P., Fontaine M., Godbout N., Lacroix S. (1995). Nonlinear self-phase-modulation effects: a vectorial first-order perturbation approach. *Opt. Lett. 20*, 5, 456–458.

Voronin A. A., Zheltikov A. M. (2008). Soliton self-frequency shift decelerated by self-steepening. *Opt. Lett. 33*, 15, 1723–1725.

Wai P. K. A., Menyuk C. R., Lee Y. C., Chen H. H. (1986). Nonlinear pulse propagation in the neighborhood of the zero-dispersion wavelength of monomode optical fibers. *Opt. Lett.* **11**, 464–466.

Zhu Z. M., Brown T. G. (2004). Polarization properties of supercontinuum spectra generated in birefringent photonic crystal fibers. *J. Opt. Soc. Am. B 21*, 2 (FEB), 249–257.

4

Fibre supercontinuum generation overview

J. M. Dudley, G. Genty and S. Coen

4.1 Introduction

As discussed in Chapter 1, nonlinear fibre optics is a very mature field of research, and nonlinear propagation effects have been studied under a wide range of conditions. The basic physical processes that cause spectral broadening in fibres were outlined in Chapter 3, and reference was made to particular mechanisms that contribute to supercontinuum generation.

Whilst it is often convenient to discuss nonlinear spectral broadening in terms of effects that occur in either the normal or anomalous dispersion regime of a fibre, supercontinuum generation is more complex as it involves interactions that generate new spectral components on both sides of the zero dispersion wavelength (ZDW). The aim in this chapter is to extend the description of these interactions given in the preceding chapter in order to illustrate in more detail some commonly observed features of fibre supercontinuum generation. We do not intend to reproduce the comprehensive reviews that already exist in the literature (see the references given in Chapter 3); our objective is rather to provide a succinct overview of supercontinuum broadening mechanisms, and to introduce some of the subject matter that will be considered in more detail in subsequent chapters of the book.

The different regimes of supercontinuum generation can be broadly distinguished by considering short (femtosecond) versus long (picosecond, nanosecond and continuous wave) pump pulses. In this chapter, we consider the general features of supercontinuum generation for three particular commonly-observed cases. The first is that of a supercontinuum generated by femtosecond pump pulses, where soliton dynamics play the central role in the spectral broadening process. The second is that of a supercontinuum generated by pump pulses longer than several picoseconds and including continuous wave sources, where it is the spontaneous growth of new frequency components from modulation instability/four-wave mixing (MI/FWM) that dominates the initial spectral broadening. Both these cases have been observed in various types of conventional optical fibre, but have found particularly spectacular application when combined with the dispersion-engineering potential of photonic crystal fibre (PCF) as described in Chapters 5 and 6. We also consider an additional category of picosecond supercontinuum generation which for convenience is called a "telecommunications supercontinuum". This refers to mechanisms underlying the generation of near-uniform supercontinuum spectra around 1550 nm, where work has been motivated by the need to develop broadband sources suitable for spectral slicing in wavelength-division multiplexing. Finally, we briefly consider supercontinuum stability characteristics. More complete treatments of the material touched upon in this chapter can be found in Dudley et al. (2006), Agrawal (2007), Genty et al. (2007) and, of course, in the subsequent chapters of this book.

4.2 Supercontinuum generation with femtosecond pulses

A conceptually clear way to understand the main features of supercontinuum generation in the femtosecond regime is to first consider the case of pumping in the anomalous dispersion regime, but close to the fibre's zero dispersion wavelength. Under typical pumping conditions, the power of the pump pulses is high enough for the input pulses to be considered as solitons of order $N = [L_D/L_{NL}]^{1/2} \gg 1$, where $L_D = T_0^2/|\beta_2|$ and $L_{NL} = (\gamma P_0)^{-1}$. The input pulses then initially undergo the expected higher-order soliton evolution dynamics of spectral broadening and temporal compression (Agrawal 2007) but, because of perturbations such as higher-order dispersion and/or stimulated Raman scattering, the dynamics departs from the recurrent behaviour expected of ideal high-order solitons and the pulses break up (Dianov et al. 1985, Golovchenko et al. 1985). This is the process that has come to be referred to as soliton fission, and in fact it is possible to treat this analytically to show that each higher-order soliton pulse separates into its N distinct fundamental soliton components (Kodama and Hasegawa 1987). When close to the ZDW, however, such idealised soliton fission does not occur in practice, because the same perturbations that induce the breakup in the first place, also modify the subsequent evolution of the ejected solitons. In the initial stage, for example, the energy of the ejected solitons is shed via the generation of dispersive wave spectral components in the normal dispersion regime, whilst continued propagation leads to a continuous shift to longer wavelengths through the Raman soliton self-frequency shift, and cross-phase modulation between the Raman-shifting solitons and the dispersive waves.

This process appears at first sight to be very complex, and the rich dynamical evolution is certainly not apparent by simply measuring the spectrum at the fibre output. However, numerical modelling using techniques as described in Chapter 3

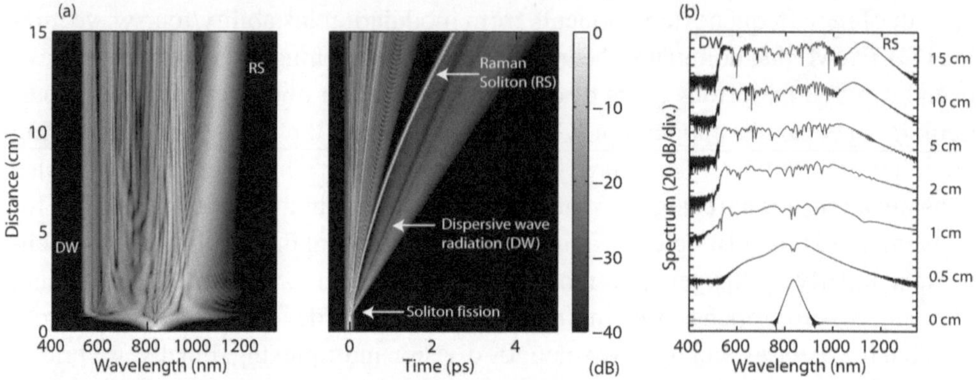

Fig. 4.1. (a) Grayscale and (b) line plot representations of supercontinuum generation in the femtosecond regime. The plots in (a) highlight the point of soliton fission, the characteristic features of dispersive wave radiation, and the frequency and time-domain evolution of an ejected soliton undergoing the Raman soliton self-frequency shift. (b) shows an equivalent line plot of the spectral evolution.

has proven very useful in allowing the spectral broadening to be "deconstructed", and Fig. 4.1 shows typical results of this kind. These simulations correspond to the case of pumping in the anomalous dispersion regime of a PCF with the parameters of Figure 3 of Dudley et al. (2006). Specifically, we consider sech^2 pump pulses of 50 fs duration (FWHM) and 10 kW peak power at 835 nm in a PCF with zero dispersion wavelength around 780 nm. Further simulation details are given in Dudley et al. (2006); see also Section 3.4.

The figure contains both grayscale and line plot representations of the supercontinuum development, and highlights in particular the point of soliton fission, the generation of dispersive wave radiation, and the continual spectral and temporal shifting of an ejected soliton undergoing the Raman self-frequency shift (RS). The grayscale representation is a clearer way in which to visualise the dynamics, but in fact, an even better approach is to show the time and frequency components simultaneously via a projected axis spectrogram as described in Genty et al. (2002), Dudley et al. (2002) and in later chapters of this book.

An important conclusion to draw from Fig. 4.1 is that the initial phase of nonlinear evolution (prior to the point of soliton fission) itself contributes very significantly to the overall spectral bandwidth. Indeed, this phase of what could be termed "clean" higher-order soliton propagation has been extensively studied in the context of generating extra bandwidth for pulse compression, and we can apply insights obtained in this previous work to aid in the practical study of supercontinuum generation. In particular, motivated by studies aiming to optimise the fibre length for compression (Dianov et al. 1986, Chen and Kelley 2002), we can usefully introduce a characteristic length scale to describe the fission point, based on the distance at which the spectral width of the initial soliton reaches its maximum value. This has been termed the fission length and can be written approximately as $L_{\rm fiss} \sim L_{\rm D}/N = (L_{\rm D}\, L_{\rm NL})^{1/2}$. It is approximately after this distance of propagation that the breakup of the initial pulse occurs. The relationship between this initial phase of spectral broadening and the generation of dispersive wave components has also been studied in Cristiani et al. (2004).

Simulations such as those shown in Fig. 4.1 can of course be extended over a wider parameter range, allowing the study of femtosecond propagation where the pump pulse is in the normal GVD regime. In this case, results show that the initial spectral broadening arises primarily from self-phase modulation (SPM). For pump wavelengths approaching the zero dispersion wavelength, the initial spectral broadening due to SPM transfers spectral content into the vicinity of the zero dispersion wavelength and across into the anomalous dispersion regime. Once this occurs, soliton dynamics as described above can also begin to contribute significantly to the spectral and temporal features of the generated supercontinuum (Dudley et al. 2006).

4.3 Supercontinuum generation in the "long pulse" regime

For anomalous dispersion regime pumping and with pulses of longer durations and with high peak power such that the soliton order becomes very large ($N \gg 10$), higher-order soliton evolution and soliton fission as described above become progressively less important. This is because the characteristic length scale over which soliton fission occurs ($L_{\rm fiss} \sim L_{\rm D}/N$) also increases with the pump pulse duration, so that the effect of spontaneous MI/FWM dominates the initial propagation phase.

Under these conditions, the initial MI dynamics leads to the temporal break up of the pump pulse into a large number of sub-pulses. The subsequent evolution of these sub-pulses then leads to additional spectral broadening and supercontinuum formation through a variety of mechanisms, notably soliton dynamical effects such as dispersive wave generation and the Raman soliton self-frequency shift. Naturally, the dispersion characteristics of the fibre used also play an important role with longer pulses, and Chapter 7 discusses this in further detail. In fact, the general features observed in the case of picosecond pulses can also be extended to the continuous wave regime, where the partial coherence of the pump also seeds the input field break up. In this context, the reader can refer to Chapter 8 of this book in which many of the subtleties of noise-driven supercontinuum generation in the continuous wave regime are discussed.

In the case of picosecond pulse excitation, the simulation results shown in Fig. 4.2 illustrate some characteristic features. Specifically, using a fibre with a 780 nm zero dispersion wavelength as above, and for a 20 ps input pulse at 800 nm with 500 W

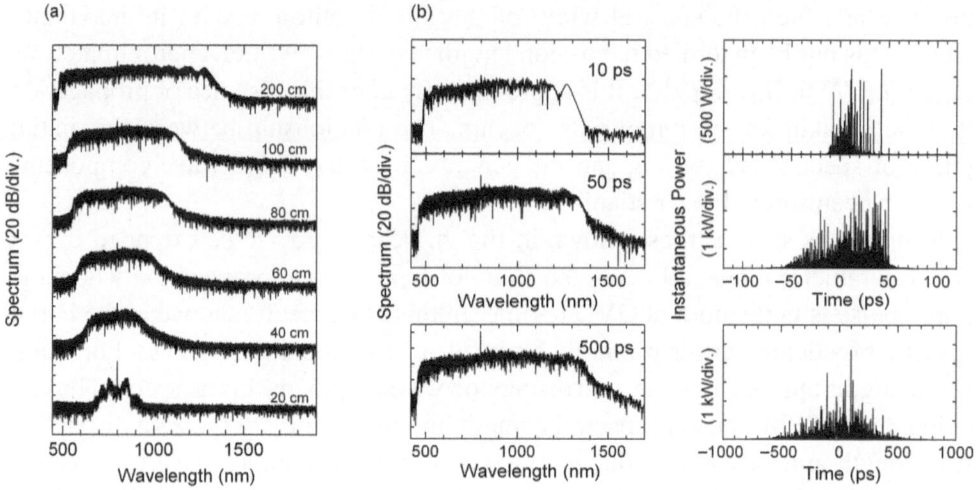

Fig. 4.2. (a) Single-shot simulation results showing spectral evolution of a 500 W peak power 20 ps pulse at 800 nm over 2 m. (b) Similar single-shot results for a range of pulse durations as shown, showing both the output spectra and output temporal profiles.

peak power, the results in Fig. 4.2(a) show the spectral evolution as a function of propagation distance, clearly illustrating the spontaneous growth of MI sidebands from noise, followed by further spectral broadening and the appearance of soliton structure on the long wavelength edge of the spectrum after 2 m. The results in Fig. 4.2(b) present additional simulations for different pulse durations spanning 10–500 ps, showing both output spectra and temporal intensity profiles to illustrate the broad similarity in characteristics observed once the "long pulse" regime is entered. It is very important to note that the simulation results in Fig. 4.2(a) and (b) are single-shot numerical realisations, but in simulations that are performed over an ensemble using different noise seeds, the fine spectral and temporal structure seen in the figure is averaged out and no longer resolved. It is such an averaging effect that also gives rise to apparently smooth spectra in multi-shot experiments using integrating optical spectrum analysers.

With long pulses in the normal dispersion regime, significant initial spectral broadening can develop from four-wave mixing and Raman scattering. Raman effects dominate when pumping far into the normal GVD regime, because the parametric sidebands are too far detuned from the pump. Closer to the ZDW, four-wave mixing becomes progressively more important, since the parametric gain is higher than the Raman gain (Agrawal 2007). In the case where this broadening begins to overlap with the ZDW, soliton effects can again contribute to the overall dynamics as discussed in the preceding paragraph.

4.4 The generation of "telecommunications supercontinuum" spectra

Nonlinear propagation in optical fibres has been studied extensively in the context of optical telecommunications, and there has been much success in applications such as soliton transmission, Raman amplification and wavelength conversion (Agrawal 2001). The possible application of supercontinuum generation to telecommunications has also been a subject of intense research, where work has been particularly motivated by the need to develop broadband sources suitable for spectral slicing in wavelength-division multiplexing. For such an application, a crucial practical requirement is spectral uniformity across as wide a wavelength range as possible around the 1550 nm telecommunications wavelength range.

Initial work in this field was pioneered by Morioka et al. who successfully demonstrated supercontinuum generation by injecting 1.7 W peak power pulses of 3 ps duration into a 3 km length of dispersion-shifted fibre with a small anomalous dispersion at the pump wavelength (Morioka et al. 1994). Spectral broadening over a 200 nm bandwidth of 1440–1640 nm was observed, and stable spectral slicing over four wavelength channels was demonstrated. This initial work was then followed by a comprehensive series of experiments focusing on both increasing the attainable spectral bandwidths and demonstrating practical systems applications. Literature reviews of key results and recent developments in this field can be found in Chapter 1 of this book and in Genty et al. (2007).

The optical fibres used in generating smooth supercontinuum spectra around 1550 nm generally possess dispersion characteristics very different from those considered above. As a result the propagation dynamics and the output spectral characteristics are also qualitatively very different from the results seen in Figs. 4.1 and 4.2. The detailed mechanisms of supercontinuum generation in this regime and their dependence on the fibre dispersion properties have been considered in a number of publications, notably Tamura et al. (2000) and Mori et al. (2001).

A case of particular interest is when the fibre possesses (i) a convex dispersion profile such that two zero dispersion wavelengths bracket the pump wavelength, and (ii) dispersion at the pump wavelength that varies longitudinally from anomalous at the fibre input to normal at the fibre output. With appropriate matching of fibre and source parameters for this case, the output supercontinuum spectra can be simultaneously uniform, symmetric and stable. The propagation dynamics for this regime are considered in considerable detail in Chapter 12 of this book, but for completeness, Fig. 4.3 summarises results from simulations that illustrate the general features observed. Here we show (a) the longitudinal evolution in grayscale as well as (b) the output spectrum after 1 km of propagation in a convex profile dispersion decreasing fibre (see Genty et al. 2007 and Chapter 12 for parameters). In this case, the uniform spectrum at the output develops from a combination of

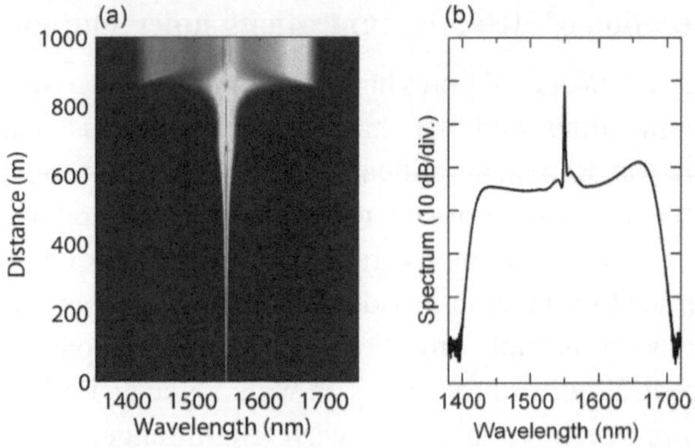

Fig. 4.3. Simulation results illustrating the properties of a "telecommunications supercontinuum". (a) longitudinal evolution and (b) the corresponding output spectrum.

initial soliton compression followed by symmetric dispersive wave generation as the expanding pulse spectrum penetrates into regimes of normal dispersion on both the short wavelength and long wavelength side of the central wavelength.

4.5 Supercontinuum stability

The "stability" of a supercontinuum refers to how the intensity and phase properties of the output spectrum vary with time. Of course, stability or instability in the spectral domain will directly impact on the supercontinuum coherence as well as on the corresponding time-domain characteristics (Dudley and Coen 2002). Stability issues can perhaps be most readily understood when considering supercontinuum generation by pulsed sources. In this case, we consider the variation in the supercontinuum characteristics from shot-to-shot, as successive pulses from a pulse train excite a series of independent spectra. Because there will always be some level of technical and quantum noise on the input pulses, in regimes where the spectral broadening is highly sensitive to noise, this can lead to significant shot-to-shot variation. A consequence here is that the supercontinuum characteristics measured using integrating detectors can appear artificially smooth, not reflecting the fine structure that would be observed if shot-to-shot diagnostics were available. In the case of time-averaged properties of continuous wave supercontinuum generation, the discussion of stability is similar and is considered further in Chapter 8.

Issues of supercontinuum stability and noise were extensively considered in early work studying supercontinuum generation in a telecommunications context, and have since been complemented by work carried out in the framework of optical frequency metrology (Islam et al. 1989a, 1989b, Tamura et al. 2000,

Corwin et al. 2003). As with the underlying spectral broadening mechanisms themselves, the stability characteristics of fibre supercontinuum spectra depend in detail on the combination of particular fibre and source parameters. Although the reader is referred to the original references for a full treatment, we present here a brief summary of the major conclusions for each of the broad propagation regimes presented in the preceding sections of this chapter.

For the case of femtosecond supercontinuum generation in the anomalous dispersion regime (Section 4.2), instability arises because input fluctuations can be amplified along the propagation length through MI gain, introducing a noise source that can compete with and/or drive the otherwise coherent spectral broadening processes involving soliton dynamics. However, when soliton fission dominates the initial dynamics, the input pulses get temporally compressed so quickly that the spectral extent of the emerging supercontinuum can overlap with the frequencies of maximum MI gain before significant amplification of the noise background has taken place. This results in a coherent seeding of the MI gain bandwidth, and the supercontinuum that is generated is thus stable. On the other hand, when the MI amplified noise background becomes a more dominant feature of the initial propagation dynamics, the supercontinuum generated can exhibit spectral and temporal instabilities from pulse-to-pulse. A number of rules of thumb concerning stability have been proposed, generally involving a constraint on the input soliton number. Although it has not been possible to obtain universal conditions applying to any arbitrary combination of initial conditions, it appears that for typical experiments employing photonic crystal fibres of several 10s of cm, input pulses where $N < 20$ possess high coherence, and input pulses where $N > 40$ possess low coherence (Dudley et al. 2006, Genty et al. 2007). More precise matching of fibre length and input soliton number can be determined for specific conditions using simulations, and it is also important to ensure that the desired stability properties are obtained while also generating a bandwidth sufficient for the particular application in question. Of course, as MI does not occur in the normal GVD regime, supercontinuum spectra generated with sub-picosecond pulses with normal dispersion regime pumping are, in principle, always highly coherent. However, the drawback is that the supercontinuum spectral width (at the same peak power) is comparatively much smaller due to the rapid initial temporal spreading of the pump pulses.

The second broad regime to discuss is the case of pumping with longer pulses or with continuous wave radiation. Under typical conditions with fibres possessing only one zero dispersion wavelength over the wavelength range of interest, spectral broadening develops from the spontaneous formation of a fast pump envelope modulation. With normal dispersion regime pumping, this modulation develops from stimulated Raman scattering, while it can arise from both Raman scattering and/or MI in the anomalous dispersion regime. In both cases, however, the fact that the modulation develops from noise at frequencies that do not overlap with the pump bandwidth means that the generated supercontinuum exhibits very large shot-to-shot fluctuations. As a general rule then, it is reasonable to conclude that longer pulses or continuous wave radiation lead to unstable supercontinuum generation. That said, there are some exceptions for particular choices of fibre and source parameters, and recent studies have also begun to examine ways to stimulate supercontinuum generation in the long pulse regime so that the initial envelope modulation develops from a coherent input (Genty et al. 2009). Using two picosecond pulses of slightly different frequencies, this approach appears promising, although whether it can be extended to even longer pulse durations remains an open question.

The third broad regime to consider is that of the telecommunications supercontinuum generated around 1550 nm, and this also provides an example of a further exception to the general conclusion that long pulses necessarily lead to instabilities. With such supercontinuum generation around 1550 nm, the generation of stable spectra arises because the dynamics involve either SPM-broadening in the normal dispersion regime, or a combination of coherent soliton compression and dispersive wave generation (Genty et al. 2007).

4.6 Conclusions

The objective of this chapter was not to present a complete overview of every possible scenario of fibre supercontinuum generation, but rather to discuss a selection of specific results that have been the subject of a great deal of previous study, and which are further extended in subsequent chapters of this book.

A point that it is appropriate to stress again in closing this chapter is that the various interactions that underlie supercontinuum generation depend sensitively on both the fibre and the source parameters, and one should avoid the temptation to propose "one size fits all" explanations that are overly simplistic. Rather, a careful survey of the extensive literature and numerical modelling of the specific system under study should be carried out to ensure a clear understanding of the physics of nonlinear and dispersive propagation for any particular case of interest. Indeed, it is hoped that the content of this book and the numerical code provided in the preceding chapter will assist in helping researchers in this regard.

References

Agrawal, G.P. (2001) *Applications of Nonlinear Fibre Optics*. Academic Press, San Diego.

Agrawal, G.P. (2007) *Nonlinear Fibre Optics*, 4th edition. Academic Press, San Diego.

Chen, C.M. & Kelley, P.L. (2002) Nonlinear pulse compression in optical fibers: Scaling laws and numerical analysis. *J. Opt. Soc. Am. B* **19**, 1961–1967.

Corwin, K.L., Newbury, N.R., Dudley, J.M., Coen, S., Diddams, S.A., Weber, K. & Windeler, R.S. (2003) Fundamental noise limitations to supercontinuum generation in microstructure fiber. *Phys. Rev. Lett.* **90**, 113904:1–4.

Cristiani, I., Tediosi, R., Tartara, L. & Degiorgio, V. (2004) Dispersive wave generation by solitons in microstructured optical fibers. *Opt. Express* **12**, 124–135.

Dianov, E.M., Karasik, A.Ya, Mamyshev, P.V., Prokhorov, A.M., Serkin, V.N., Stelmakh, M.F. & Fomichev, A.A. (1985) Stimulated-Raman conversion of multisoliton pulses in quartz optical fibers. *JETP Lett.* **41**, 294–297.

Dianov, E.M., Nikonova, Z.S., Prokhorov, A.M. & Serkin, V.N. (1986) Optimal compression of multi-soliton pulses in optical fibers. *Pis'ma Zh. Tekh. Fiz.* **12**, 756–760 [*Sov. Tech. Phys. Lett.* **12**, 311–313 (1986)].

Dudley, J.M. & Coen, S. (2002) Coherence properties of supercontinuum spectra generated in photonic crystal and tapered optical fibers. *Opt. Lett.* **27**, 1180–1182.

Dudley, J.M., Gu, X., Xu, L., Kimmel, M., Zeek, E., O'Shea, P., Trebino, R., Coen, S. & Windeler, R. (2002) Cross-correlation frequency-resolved optical gating analysis of broadband continuum generation in photonic crystal fiber: simulations and experiments. *Opt. Express* **10**, 1215–1221.

Dudley, J.M., Genty G. & Coen, S. (2006) Supercontinuum generation in photonic crystal fiber. *Rev. Mod. Phys.* **78**, 1135–1184.

Genty, G., Lehtonen, M., Ludvigsen, H., Broeng, J. & Kaivola, M. (2002) Spectral broadening of femtosecond pulses into continuum radiation in microstructured fibers. *Opt. Express* **10**, 1083–1098.

Genty, G., Coen, S. & Dudley, J.M. (2007) Fiber supercontinuum sources. *J. Opt. Soc. Am. B* **24**, 1771–1785.

Genty, G., Dudley, J.M. & Eggleton, B.J. (2009) Modulation control and spectral shaping of optical fiber supercontinuum generation in the picosecond regime. *Appl. Phys. B* **94**, 187–194.

Golovchenko, E.A., Dianov, E.M., Prokhorov, A.M. & Serkin, V.N. (1985) Decay of optical solitons, *JETP Lett.* **42**, 74–77.

Islam, M.N., Sucha, G., Bar-Joseph, I., Wegener, M., Gordon, J.P. & Chemla, D.S. (1989a) Broad bandwidths from frequency-shifting solitons in fibers. *Opt. Lett.* **14**, 370–372.

Islam, M.N., Sucha, G., Bar-Joseph, I., Wegener, M., Gordon, J.P. & Chemla, D.S. (1989b) Femtosecond distributed soliton spectrum in fibers. *J. Opt. Soc. Am. B* **6**, 1149–1158.

Kodama, Y. & Hasegawa, A. (1987) Nonlinear pulse propagation in a monomode dielectric waveguide. *IEEE J. Quant. Elect.* **QE23**, 510–524.

Mori, K., Takara, H. & Kawanishi, S. (2001) Analysis and design of supercontinuum pulse generation in a single-mode optical fiber. *J. Opt. Soc. Am. B* **18**, 1780–1792.

Morioka, T., Kawanishi, S., Mori, K., & Saruwatari, M. (1994) Nearly penalty-free, < 4 ps supercontinuum Gbit/s pulse generation over 1535–1560 nm. *Electron. Lett.* **30**, 790–791.

Tamura, K.R., Kubota, H. & Nakazawa, M. (2000) Fundamentals of stable continuum generation at high repetition rates. *IEEE J. Quant. Electron.* **36**, 773–779.

5

Silica fibres for supercontinuum generation

J. C. Knight and W. J. Wadsworth

5.1 Introduction

The generic single-mode optical fibre is a fibre drawn from fused silica. Silica fibres are available at very low cost, being produced by sophisticated processes in massive quantity for the global telecommunications network. They are very transparent, and much optical equipment is designed around their transparency windows, making research at these wavelengths particularly convenient. Devices or effects using silica fibres are therefore compatible with a wide range of existing optical fibre technologies, and so easy to incorporate into existing systems should the need or opportunity arise. Silica fibres are available in a wide range of different forms, especially with the development and commercial availability of silica photonic crystal fibres (PCF) and microstructured fibres.

Many of the above arguments rely for their worth on the already widespread deployment of silica optical fibres. However, there are also fundamental reasons to work with silica fibres, many of them the same reasons that they are already so popular. Silica – glassy SiO_2 – can be synthesised with remarkably high levels of purity at very low cost: silicon is an abundant and easily-obtained element in the earth's crust, while the chemical processes used to synthesise silica are efficient and simple. Of course, the fact that these processes have been scaled up to produce large volumes of material further reduces the cost of production. The transparency window of silica is broad, spanning both visible and near-infra-red wavelengths. Practically, silica optical fibres are usually used in the spectral range 300–2500 nm. Although other glasses are transparent at longer wavelengths, these are, and will surely remain, for specialist applications only, because it is harder to generate and detect radiation at these longer wavelengths, there is more interference from thermal noise sources, and many materials are strongly absorbing, making it far harder to develop systems there.

Supercontinuum Generation in Optical Fibers, ed. by J. M. Dudley and J. R. Taylor. Published by Cambridge University Press.

In addition to these optical reasons to work with silica as a material, there is an additional reason to work with silica in the form of an optical fibre – the physical and chemical properties of silica make it very easy to draw to fibres. The chemical processes used in the production of synthetic silica can be effectively adapted to make complex doped structures (e.g. the vapour deposition techniques) or related techniques can be used (as in the various techniques used to incorporate rare earth elements into a silica matrix). Silica has a very high softening temperature, a slow variation of viscosity with temperature, and a large range between the glass transition temperature and the crystallisation temperature. These physical properties make it possible to draw remarkable structures out of silica glass, including the wide range of photonic crystal fibre structures which have been demonstrated using silica over the past decade. It is unlikely that such structures would ever have been realised starting with another glass.

So, what sort of fibres can be created from silica which will form effective supercontinuum fibres? Fibre-based broadband light generation started early in the history of conventional fibres (see Stolen, 2008), making use of the high nonlinearity and long effective length which can be obtained by using a doped-silica core surrounded by a silica cladding. Further development in understanding of the processes involved (Gouveia-Neto *et al.*, 1988, Islam *et al.*, 1989) and the development of applications (e.g. Mori *et al.*, 1993) led to specialised nonlinear fibres based on conventional fibre technology being developed. Such fibres have the refractive index contrast between core and cladding increased, up to around 3%, and well-controlled dispersion characteristics. These enabled improved optical performance: supercontinua spanning hundreds of nanometres based on excitation at 1500 nm wavelength. Further improved performance was reported from fibres which varied along their length (Okuno *et al.*, 2003). Recent rapid progress has resulted from the use of photonic crystal and microstructured fibres (Kaiser and Astle, 1974, Knight *et al.*, 1996, Ranka *et al.*, 2000a) and silica nanowires (Leon-Saval *et al.*, 2004), which use the much larger refractive index difference between silica and air. These not only often provide a higher nonlinear response, but critically cause a strong waveguide contribution to the overall group-velocity dispersion (GVD) of the guided mode (Mogilevtsev *et al.*, 1998, Birks *et al.*, 1999). It was recognised that these structures could provide anomalous group-velocity dispersion at shorter wavelengths than was possible with conventional fibres, and that this would enable solitons to form at shorter wavelengths than was possible using conventional fibre technology (Gander *et al.*, 1999). Silica nanowires (Fig. 5.1(a)) can be considered as a special case (Birks *et al.*, 1999) of a highly nonlinear photonic crystal fibre (see Knight *et al.*, 1996), in which the air-filling fraction of the cladding has reached unity. In this special case, however, the core of the fibre is no longer suspended, and hence nanowires are only useable in a protected environment and in short lengths

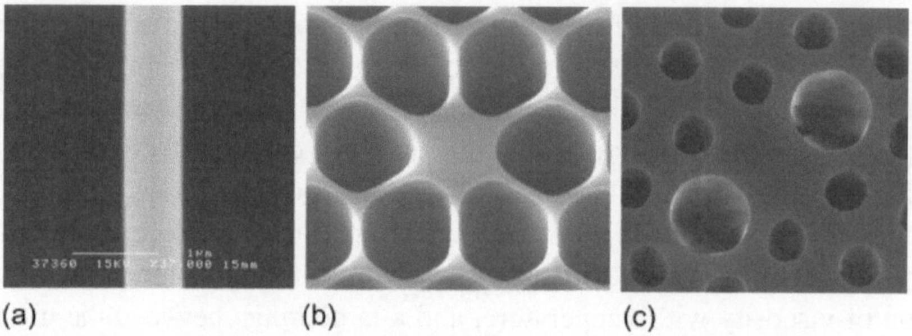

Fig. 5.1. Scanning electron micrographs of novel silica fibres suitable for supercontinuum generation. (a) Silica nanowire, side view. Diameter 630 nm. (See Leon-Saval *et al.* 2004). (b) High-Δ photonic crystal ("suspended strand") fibre, end view. Core diameter 1.2 μm. (c) Highly birefringent ("polarisation maintaining") endlessly single-mode photonic crystal fibre, end view. Core diameter 4.3 μm.

(typically a few to tens of cm). Images of the different silica-based fibre structures useful for supercontinuum generation are shown in Fig. 5.1.

Early researchers identified most of the basic effects which have become so accessible with the development of photonic crystal fibres and fibre-based mode-locked laser systems. Writing in *The Supercontinuum Laser Source*, Johnson and Shank (1989) wrote of the evolution of short pulses in optical fibres that "... the interplay between SPM, SRS and GVD, could easily fill a book chapter." How things have changed. Since then, the field of fibre-based supercontinuum generation has grown rapidly of age, with supercontinuum sources becoming competitive as light sources both in terms of cost and of performance, and with a wide range of applications under investigation. This has been a result mainly of technological rather than purely scientific advances, in both the pump sources and in the fibre design. This chapter explains the most significant features of the current state-of-the-art of silica-based optical fibres for supercontinuum generation.

5.2 Silica for supercontinuum fibres

Several properties of optical fibres are important for supercontinuum generation.

- The dispersion of the fibre, as manifest by the spectral refractive index curve and its derivatives. Worthy of special mention are wavelengths at which the group-velocity dispersion (GVD) passes through zero, and the spectral regions corresponding to normal and anomalous group-velocity dispersion.
- The value of the Kerr intensity-dependent refractive index n_2, where $n = n_0 + n_2 I$.
- The effective area of the fibre mode, which determines the intensity of the light in the fibre core for the available pump power.

- The material-dependent time-delayed Raman response.
- Spectral transparency of the material is obviously important if one hopes to generate a broad continuum spectrum.
- The power-handling ability of the material in fibre form may be important in some cases.

Several key parameters are often defined in order to quantify the nonlinear response of the fibre (Agrawal, 2001), including the effective area, A_{eff}, of the mode and the nonlinear coefficient

$$\gamma = \frac{2\pi n_2}{A_{\text{eff}}\lambda}.$$

So, given this, how does fused silica measure up? It has the lowest demonstrated optical attenuation of any solid material, with the minimum demonstrated fibre loss being below 0.2 dB/km in the spectral range around 1300–1600 nm. The attenuation increases for shorter wavelengths due mainly to Rayleigh scattering from inhomogeneities in the material, with spectral dependence of λ^{-4}. The attenuation at longer wavelengths increases as well, due to intrinsic phonon absorption. In many cases, the spectral transmission of the fibre will be subject to extrinsic effects, and in particular the presence of OH^- ions in the core of the fibre can have a huge effect on its properties. The O–H fundamental mode at 2700 nm gives attenuation of 10,000 dB/(km.ppm) and the overtone at 1380 nm gives an attenuation of 60 dB/(km.ppm) (Humbach *et al.*, 1996). There are also OH^- overtone and combination bands at 1230 nm (2.6 dB/(km.ppm)) and 940 nm (1 dB/(km.ppm)). Of these, the fundamental band is outside the low-loss transmission window of silica, and is not usually encountered. The band around 1380 nm, however, is close to both the minimum attenuation wavelength and also the zero-dispersion wavelength of silica, and is of great importance. Typically, commercially-available low-OH^- silica has OH^- levels in the range of less than 1 ppm.

Associated with this transmission window is a dispersion curve. The refractive index n can be represented by a Sellmeier equation

$$n^2 = 1 + \frac{0.6961663\lambda^2}{\lambda^2 - (0.0684043)^2} + \frac{0.4079426\lambda^2}{\lambda^2 - (0.1162414)^2} + \frac{0.8974794\lambda^2}{\lambda^2 - (9.896161)^2}$$

as shown by Malitson (1965). The resulting curve for n is shown in Fig. 5.2, which also shows the variation of the derived group index, $n_g = \frac{c}{v_g} = n + \omega\frac{\partial n}{\partial\omega}$, with wavelength (where v_g is the group velocity and ω is the angular optical frequency). The figure also shows the group-velocity dispersion (GVD) $D = \frac{-\lambda}{c}\frac{d^2n}{d\lambda^2} = \frac{2\pi c}{v_g^2\lambda^2}\frac{dv_g}{d\omega}$ (on the right-hand scale). The group-index curve changes from being normal (decreasing with increasing wavelength) to anomalous (increasing with wavelength) around

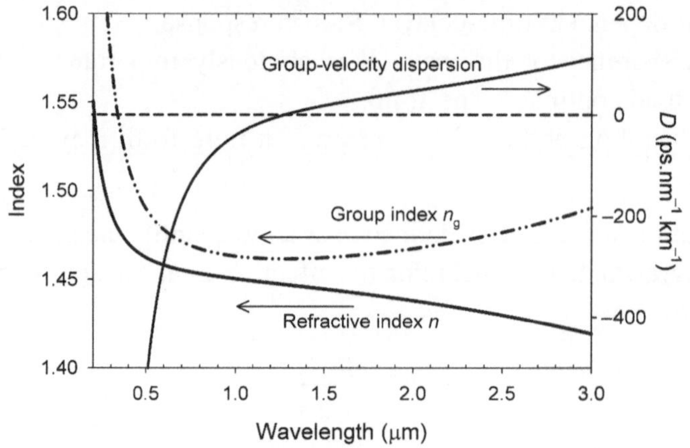

Fig. 5.2. Refractive index, group index and group-velocity dispersion, D, for bulk silica as functions of wavelength in the spectral range 200–3000 nm.

1300 nm, corresponding to the point of inflection in the curve for phase index, n. The group-velocity dispersion D crosses through zero at that wavelength, being positive (anomalous GVD) at longer wavelengths and negative (normal GVD) at shorter wavelengths. Readers should note that the dispersion is also often quoted in terms of derivatives with respect to angular frequency instead of wavelength, with the second derivative, β_2, describing the same property as D. Importantly there is a sign difference between these two quantities, so whilst the terms "normal" and "anomalous" dispersion may be used without ambiguity, "positive" and "negative" dispersion mean different things when referring to D or to β_2. Both terms persist in general usage because D is a natural experimental unit, and β_2 is the natural form in the governing equations of nonlinear fibre optics (see Chapter 3).

Silica has a relatively low positive Kerr response (around $n_2 = 2.4 \times 10^{-20}$ m/W) but this is often more than compensated for by the very long interaction lengths and high confinement attainable in optical fibres. The Raman gain of silica is very broad, and has a peak at 13 THz from the pump and a peak gain of about 10^{-13} m/W (Agrawal, 2001). Optical damage is difficult to quantify in fibre optics, with damage at the fibre input face being most common. Usually, end-face damage is associated with contamination of the face, and direct splicing often enables higher-power and more reliable excitation than using a lens to couple in a free-space beam. The actual peak power or pulse energy values associated with damage depend on the laser characteristics: however, supercontinuum generation in a properly designed fibre is rarely limited by fibre damage. Photodarkening is a more insidious form of damage where prolonged exposure to high power, short wavelength light induces excess optical loss at visible wavelengths. This becomes a serious problem with watt-level supercontinua containing wavelengths below 450 nm.

5.3 Strong waveguiding

Increasing the strength of the waveguiding by using a large difference in refractive index between the core and cladding materials has two effects on the nonlinear response of the waveguide. First, and most obviously, it can increase the nonlinear coefficient (Agrawal, 2001) of the fibre, by decreasing the mode effective area, as a result of the decreased core size needed to maintain single-mode guidance. The resultant increase in intensity (for a given coupled laser power) results in a more rapid build-up of Kerr phase shift and greater Raman gain. Secondly – more subtly, but usually more importantly – it changes the dispersion curve of the fibre. This effect arises because as one changes the wavelength with the core size fixed, the mode has to adapt. In conventional optical fibres, this effect is relatively small: the waveguiding is rather weak because the refractive index contrast between the core and the cladding is low. The effect of waveguide dispersion in a strongly-guiding waveguide is shown in Fig. 5.3(a). The calculations shown were performed by solving the transcendental equations (Senior, 1992) governing the propagation of the fundamental mode of a circular fibre core formed of a glass with refractive index $n_{\text{core}} = 1.5$ (assumed to be dispersion-free), surrounded by an air cladding with $n_{\text{clad}} = 1.0$. When the core diameter is far less than the wavelength, the guided mode effective index $n_{\text{eff}} = \beta/k_0$ must approach the refractive index of the cladding material – which is 1 in the case shown (β is the propagation constant of the guided mode – the magnitude of the guided-mode wavevector along the fibre axis – while k_0 is the free-space wavevector). When the core diameter is much greater than the wavelength, the mode effective index approaches the refractive index of the core material – 1.5 in the case shown. In between these two, there is dispersion which depends on the refractive index difference between the core and cladding. The variation of effective index with the ratio of wavelength over waveguide diameter can be used to calculate a group-velocity dispersion curve (Fig. 5.3(b)). As might be anticipated, the GVD curve passes through zero at a value corresponding to the point of inflection in Fig. 5.3(a). It is anomalous for short wavelengths (or large waveguide diameter), passes through zero when the core diameter is slightly smaller than the wavelength, and becomes strongly normal for longer wavelengths. Comparing Fig. 5.3(b) with Fig. 5.2, one sees that the waveguide dispersion passes from positive to negative with increasing wavelength, whereas the material dispersion passes from negative to positive. Furthermore the magnitude of the waveguide dispersion for the large index step considered in Fig. 5.3 is similar to the magnitude of the material dispersion over a wide range of wavelengths, for core size comparable to the wavelength.

The curves shown are valid for fibres with $n = 1.5$ surrounded with air or vacuum. The overall shapes of the curves will be similar for other values of the refractive

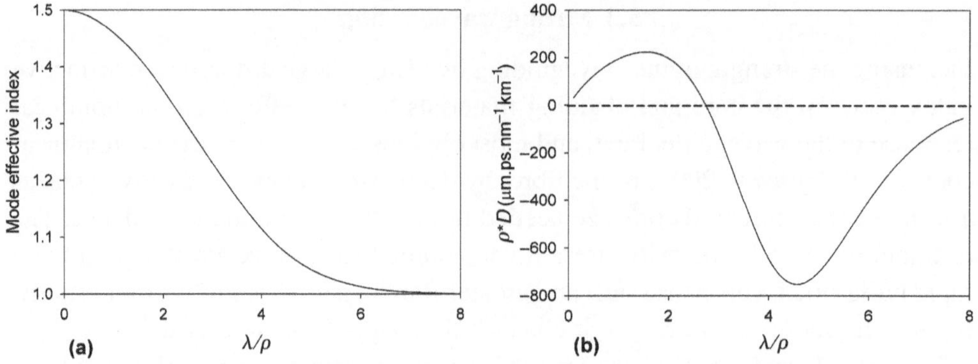

Fig. 5.3. (a) Effective index of a strand of glass (refractive index $n = 1.5$, assumed to be non-dispersive) surrounded by air, as a function of the ratio of wavelength λ over strand radius ρ. (b) Group-velocity dispersion, D, arising from the waveguide dispersion in (a). The values of D are scaled with the core radius ρ, to allow application to any chosen core radius.

index, but rescaled on both axes. In standard fibres where the index contrast between core and cladding is less than 1% the effects of waveguide dispersion are weak (and also occur for larger values of the core size). In this case the waveguide dispersion can have a significant influence on the overall dispersion of the guided mode only in the vicinity of the material zero-GVD wavelength, where the value of D is already low. For example, this is used in the design of some conventional fibres to shift the zero-dispersion wavelength from around 1300 nm to longer wavelengths. For stronger effects, the level of doping in the conventional fibre core can be increased, so that the index contrast between core and cladding is up to around 3%, by doping with around 30 mol % of germanium. This not only increases the strength of the confinement – enabling smaller values of A_{eff} and more control over the dispersion – but simultaneously increases the value of n_2 by up to a factor of about two times. As a result, the nonlinear coefficient γ is increased from around 1.5/W/km in a standard telecommunications fibre (measured at 1550 nm wavelength) up to perhaps 30/W/km. There is generally an increased scattering loss (due partly to the higher refractive index contrast) which results in the minimum attenuation in such fibres approaching perhaps 1 dB/km.

When the index contrast is further increased and the light can be yet more tightly confined (for example, where the core diameter is comparable to the wavelength in Fig. 5.3) the magnitude of the waveguide dispersion can be such as to completely change the nature of the GVD over a much wider spectral range (compare the magnitude of the dispersion shown in Fig. 5.3(b) with that in Fig. 5.2, assuming for example a radius $\rho = 1\,\mu\text{m}$). The only way that these higher values of index contrast can be accessed using silica as a basic material is by incorporating air as one component of the fibre cladding. This can be done using a silica nanowire, or through the use of various designs of photonic crystal fibre (Knight *et al.*, 2000). The direct effect of this increase in index contrast is not fundamentally different to that which can be obtained by using a higher level of doping in a conventional fibre design, but the practical impact is huge. A major reason is that pump laser sources occur in some specific wavelength ranges where suitable gain elements occur – most obviously, Er^{3+} around 1550 nm, Yb^{3+} and Nd^{3+} around 1060 nm, and Ti^{3+}:Al_2O_3 in the 700–1000 nm band. For efficient broadband nonlinear generation in a fibre, it is usually preferable that the dispersion of the fibre guided mode be zero or anomalous at the pump wavelength (see Chapters 3 and 4). However, the dispersion of bulk silica is increasingly strongly normal at wavelengths below 1300 nm, decreasing the effect of pump sources based around 1060 nm and 800 nm. Only by using a fibre with a very high index contrast is the waveguide dispersion strong enough to overcome the normal material dispersion of the silica in the fibre core at these critical wavelength bands. For Er^{3+}-based sources at 1550 nm the nonlinearity and dispersion of conventional step-index silica fibres are appropriate for broadband continuum generation (e.g. Washburn *et al.*, 2004).

5.4 Dispersion of air–silica photonic crystal fibres

Photonic crystal fibres are frequently characterised in terms of the characteristic parameters of the photonic crystal cladding – the inter-hole spacing or pitch Λ and the air hole diameter d. In a general discussion of dispersion in PCF it is useful to consider two special cases. The first is the case where the air-filling fraction is so large that the tiny glass webs between the air holes and the periodicity of the structure become rather unimportant, $d/\Lambda \to 1$ (see Fig. 5.1(b)). The fibre structure and its properties may then be closely approximated by a strand of silica completely surrounded by air (Knight *et al.*, 2000, Birks *et al.*, 1999). Such fibres are typically multimode, although the fundamental mode may be efficiently excited (Ranka *et al.*, 2000b), and they are essentially the same as silica nanowires but supported from the side rather than only at the ends. This type of PCF will be discussed in more detail in Section 5.4.1. A second important case is that of PCFs which are endlessly single-mode (Knight *et al.*, 1996, Birks *et al.*, 1997). This is the case when the relative air-hole diameter $d/\Lambda = 0.4$ or less (see Fig. 5.1(c)). The ability of a PCF to support only a single mode over all wavelengths relies on the periodic array of cladding air holes and cannot be replicated in conventional fibres. In the case of an endlessly single-mode PCF, the fibre can still have reasonably strong waveguide dispersion, and so endlessly single-mode fibres can easily be designed to have zero-dispersion wavelengths around, for example, 1060 nm wavelength. These PCFs will be discussed in more detail in Section 5.4.2. As the relative air-hole size is decreased further, the effects of waveguide dispersion become more subtle, and can no longer compensate the material dispersion of silica at such short wavelengths. For example, with a relative air-hole size of $d/\Lambda = 0.2$, the effect of the waveguide structure is sufficient to manipulate the dispersion profile over a restricted range of longer wavelengths. This gave rise to the "dispersion-flattened" fibres reported by Ferrando *et al.* (2000), which had normalised air-hole diameter $d/\Lambda = 0.2$ and roughly zero GVD over the spectral range 1300–1800 nm (Reeves *et al.*, 2002).

PCFs may also readily be made birefringent by making the cladding air holes not all the same size (Ortigosa-Blanch *et al.*, 2000). Most simply this just gives a polarised supercontinuum output for a polarised laser input, but it is also possible to obtain nonlinear polarisation coupling through the off-diagonal elements of the $\chi^{(3)}$ tensor (see, e.g., Kruhlak *et al.*, 2006 or Luan *et al.*, 2006).

There are many different methods for calculating the dispersion of PCFs including commercial beam propagation or finite element programs, and the freely available software from the CUDOS group in Sydney, Australia (http://www.physics.usyd.edu.au/cudos/mofsoftware/). The PCF dispersion curves in this chapter are all calculated by an implementation of the plane wave method developed at the University of Bath (Pearce, Hedley and Bird, 2005). In many cases the simple empirical formula given by Saitoh and Koshiba (2005) will give an adequate guide to the dispersion without solving for the modes. However, it is important to note that this empirical formula is only valid over the limited range of parameters $0.2 < d/\Lambda < 0.8$, and that the resulting dispersion curves also deviate slightly from the actual dispersion for small values of pitch, $\Lambda < 2$ μm.

5.4.1 The case of high air-filling fraction: a strand of silica in air

The behaviour of the fundamental mode of air–silica PCFs which have a high air-filling fraction in their cladding can be effectively modelled by considering a strand of silica surrounded by air (Birks *et al.*, 1999, Knight *et al.*, 2000). The waveguide dispersion for such a strand of silica was shown in Fig. 5.3. Example dispersion curves (including the material dispersion of silica) for silica strands of a range of diameters are shown in Fig. 5.4. From these curves, we can note the main characteristics:

- The effect of the waveguide dispersion on the GVD as the core diameter is decreased is first noticeable as an increase in D. Where D is anomalous it becomes more anomalous, where it is normal it becomes less normal, and the zero-GVD wavelength is shifted towards shorter wavelengths.
- As the core size is further decreased, the zero-dispersion wavelength continues to decrease. However, on the long-wavelength side of the spectrum, the increase

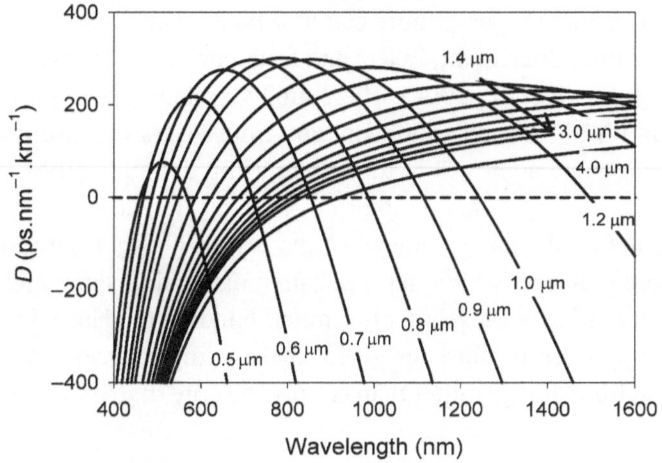

Fig. 5.4. Group-velocity dispersion curves, D, of the fundamental mode of a strand of silica surrounded by air, computed as a function of wavelength λ for different values of the diameter of the strand ranging from 0.5 μm up to 4 μm.

in D slows and eventually stops. At long wavelengths, D then starts to decrease for smaller core sizes (corresponding to moving to the right in Fig. 5.3(b)), due to the waveguide GVD becoming normal.

- For yet smaller core sizes, the strongly normal waveguide GVD causes a second zero-GVD point in the curve at long wavelengths. As the diameter continues to decrease, this second zero-GVD wavelength is swept to shorter wavelengths.
- For sufficiently small values of core size, the maximum effect of the anomalous waveguide dispersion is felt at short wavelengths. Eventually, the maximum waveguide contribution is unable to counter the increasingly strong material contribution at such short wavelengths, and the GVD is normal through the entire spectral range. The shortest wavelength at which anomalous dispersion can be obtained in silica fibres is just below 500 nm.

It is clear that whilst there is a considerable range of dispersion curves available for application to supercontinuum generation, there are also limits to what can be done by varying one parameter – the core diameter. We shall see in later chapters that SC generation generally requires that the pump wavelength be close to the zero-GVD wavelength, and that if the dispersion slope is positive at this point so that the dispersion is anomalous at longer wavelengths it will enable the propagation of self-frequency shifting solitons. The range of zero-GVD wavelengths achievable with high air-filling fraction PCF encompasses the tuning range of the ubiquitous mode-locked Ti:sapphire laser, and propagating these laser pulses through small core diameter PCFs led to the birth of supercontinuum generation in PCF (Ranka *et al.*, 2000a). The same types of fibre can also be used with mode-locked krypton ion lasers at 647 nm (Coen *et al.*, 2001) and frequency doubled Nd^{3+}-based lasers at 532 nm (Leon-Saval *et al.*, 2004). There are two other very important classes of laser: Nd^{3+}- and Yb^{3+}-doped solid state and fibre lasers operating at 1–1.1 μm. Zero dispersion can be achieved in this range using a simple silica strand model, as described above, using either a rather large strand (i.e. core) diameter (6.5 μm) which reduces the nonlinearity somewhat and is also highly multimode, or with a 0.8–0.9 μm core diameter which has a negative dispersion slope and can produce spectral broadening but with relatively limited bandwidth (Harbold *et al.*, 2002). An alternative approach to fibre design for these pump sources is to vary another parameter in the fibre design rather than simply the core diameter: the air-hole size.

5.4.2 Varying the air hole size and endlessly single-mode PCF

The extra degree of freedom in the fibre design offered by changing the diameter of the air holes opens up a vast new array of possible dispersion curves. Because the fibre core diameter relative to the wavelength is an important parameter for dispersion it is sometimes helpful to specify a PCF in terms of the core diameter, $d_{\text{core}} = 2\Lambda - d$, and the relative air-hole size, d/Λ. Comparing fibres of fixed core diameter rather than fixed pitch decouples the effects of fibre core diameter and core to cladding index step. To illustrate the general effect of changing cladding hole diameter, Fig. 5.5 shows the calculated dispersion curves for a fixed core diameter of 2.0 μm. The wavelength and dispersion range is the same as for Fig. 5.4. For

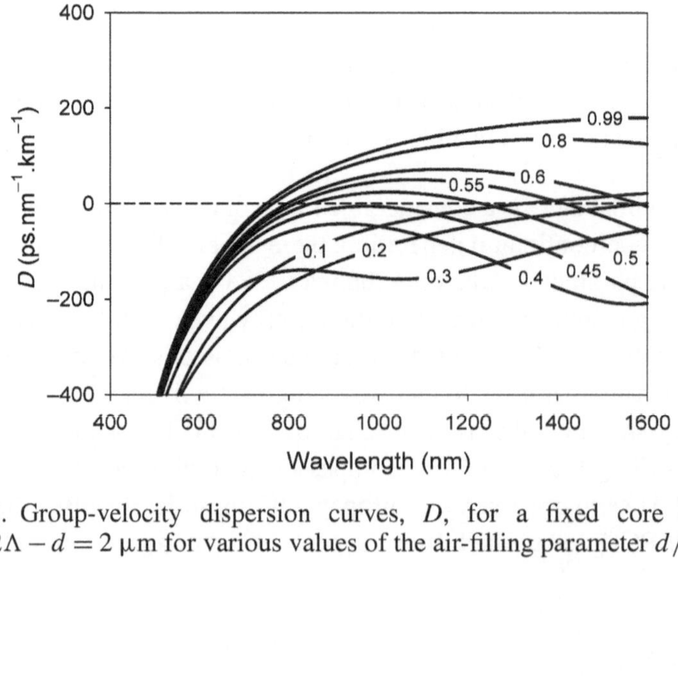

Fig. 5.5. Group-velocity dispersion curves, D, for a fixed core diameter $d_{\text{core}} = 2\Lambda - d = 2$ μm for various values of the air-filling parameter d/Λ.

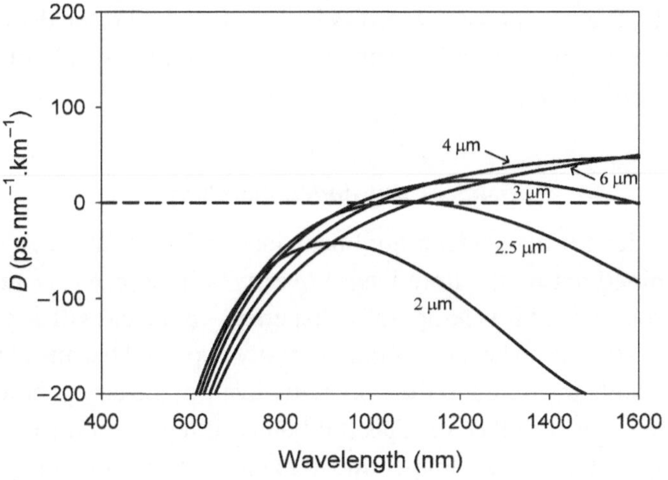

Fig. 5.6. Group-velocity dispersion curves, D, for endlessly single-mode fibre with $d/\Lambda = 0.4$ and different core diameters, $d_{\mathrm{core}} = 2\Lambda - d$.

$d/\Lambda = 0.99$ this is essentially the same as a strand of silica in air of the same diameter (Fig. 5.4). Even as d/Λ reduces to 0.8 (where the cladding is just 60% air) the dispersion still has the general form of the silica strand model, but with a shift to longer wavelengths (or a larger apparent silica strand diameter). For $d/\Lambda = 0.1$ the waveguide is very weak and the dispersion becomes very similar to the bulk dispersion of silica.

Discussion of the whole range of core diameters and hole sizes serves little purpose without a specific application in mind, but there is one particular case which merits attention for supercontinuum generation. The first PCFs made displayed a unique property that for $d/\Lambda < 0.4$ there is only one guided mode, independent of wavelength, so-called endlessly single-mode fibres (Birks *et al.*, 1997). Whilst it is true that for small silica strands which are multimode it is usually possible to excite only the fundamental mode and observe that the continuum output is in the fundamental mode, it is not the same as generating the supercontinuum in a fibre which is strictly single mode at all wavelengths (Wadsworth *et al.*, 2004). Fig. 5.6 shows the dispersion for $d/\Lambda = 0.4$ and several core diameters. The wavelength range is the same as for Figs. 5.4 and 5.5, but the range of dispersion is halved. Just as for the strand of silica in air (Fig. 5.4) the zero dispersion wavelength shifts to shorter wavelengths for smaller core diameters. However, in this case the range of zero-GVD wavelengths spans 980–1100 nm so is suitable for use with Nd^{3+} or Yb^{3+} lasers.

We should also note here that there is no simple analogue for this type of fibre in conventional step-index fibre. One may be able to create a fibre with a zero-GVD wavelength at 1060 nm, but it will be multimode. Only a microstructured fibre such as PCF can achieve a zero-GVD wavelength shorter than the material zero-GVD wavelength (around 1300 nm) and also maintain single-mode guidance (Mogilevtsev *et al.*, 1998).

5.4.3 Group-index matching

Sometimes the length over which nonlinear interactions take place in an optical fibre is determined not by the fibre length over which the beam is attenuated, but by the length over which it is temporally dispersed. In the case of supercontinuum generation, this length can be shorter than initially expected because of the tremendous spectral width generated in the fibre. However, in the case where the GVD is anomalous over some or all of the spectral bandwidth present inside the fibre, the interaction length can be much longer due to the formation and propagation of solitons, which are very common in supercontinuum processes (see Chapters 4 and 9). These solitons can propagate over many dispersion lengths, with only modest pulse broadening. A supercontinuum spectrum can contain none, one, or very many such solitons, which maintain high peak powers even as the spectrum as a whole becomes dispersed. Fig. 5.7 shows that the overall shape of the group-index curve as a 'U' means that the dispersed supercontinuum spectrum will also take on a 'U' shape in a time–wavelength plot (Dudley *et al.*, 2002). Those wavelengths at or near the zero-dispersion point (where the group index is a minimum) will be the first to arrive at any given position along the fibre length (and form the bottom of the 'U'). Both longer and shorter wavelength radiation will be delayed relative to that at the zero-dispersion wavelength. Consequently, any long-wavelength radiation (on one arm of the 'U') will be group-index-matched to another, shorter wavelength (below the zero-dispersion wavelength, and on the other arm of the 'U'). This enables interaction between long and short wavelengths through the instantaneous Kerr response of the medium, which would not be possible without such group-index matching. The continued interaction can lead (Gorbach and Skryabin, 2007) to spectral spreading on the short-wavelength side (see Chapter 9). Consequently, the group-index curve even at the longest-wavelength extent of the generated spectrum can influence the wavelengths generated at visible and ultraviolet wavelengths.

This provides a second design rule if one requires a supercontinuum extending to the shortest possible wavelengths: not only should the zero-dispersion wavelength be close to the pump wavelength (to facilitate the early stages of spectral broadening), but also the structure should be chosen so as to match the group index for the longest wavelengths generated to the shortest possible wavelengths (Stone and Knight, 2008). Group-index curves for two fibres are shown in Fig. 5.7. The two structures shown have almost identical zero-GVD wavelengths, and rather similar core diameters. The curves are virtually indistinguishable at visible and ultraviolet

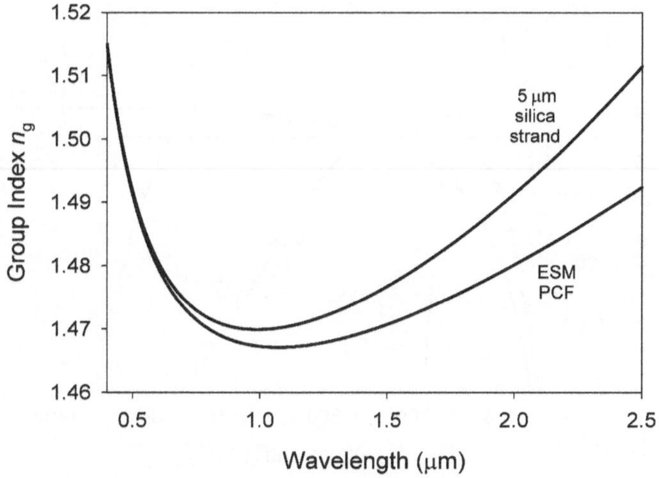

Fig. 5.7. Group index curves for two different fibres with the same zero-GVD wavelength (1060 nm) but different values of the air-filling parameter d/Λ. The curves are virtually indistinguishable at short wavelengths but differ markedly as one moves further into the infra-red.

wavelengths. However, they differ significantly towards the mid-infra-red, with the fibre characterised as a strand of silica in air (corresponding to a PCF with large air holes) having a higher value of group index. This has the effect of generating a supercontinuum which extends to shorter wavelengths in this fibre, as discussed elsewhere in this book. Group-velocity matching can also be important for dual-wavelength pumped supercontinuum generation (Champert *et al.*, 2004).

5.4.4 Higher modes in silica fibres

Although we have spent some time describing fibre structures which actually support more than just the fundamental mode, our analysis so far has focused purely on that mode. Indeed, in the PCFs being discussed, the fundamental mode can be used remarkably effectively, being virtually decoupled from other higher-order confined modes (Ranka *et al.*, 2000b). However, these higher-order modes have some potentially useful or disruptive properties. Usefully, they have dispersion curves which fall outside the range of parameters which can be obtained for the fundamental mode. They can have zero-GVD wavelengths which are shorter than those obtained using the fundamental mode. This can be used to generate otherwise unobtainable supercontinuum spectra (Konorov *et al.*, 2004), or to attain the same spectrum from a larger (and all being well more robust) structure. However, higher-order modes can also be phase-matched to the fundamental mode at some specific wavelengths, which can cause relatively efficient but unwanted conversion

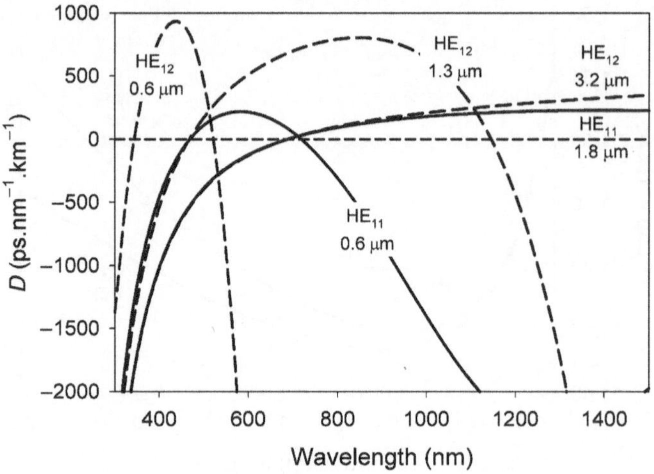

Fig. 5.8. Group-velocity dispersion, D, for the fundamental (HE_{11}, solid lines) and second order (HE_{12}, broken lines) modes of silica nanowires in air for different nanowire diameters.

to higher-order modes (Efimov *et al.*, 2003). Clearly, considering the range of fibre parameters already under investigation, we cannot attempt a comprehensive study of higher-order modes, and we simply include some examples here. Fig. 5.8 shows a comparison of the dispersion of the fundamental HE_{11} mode of a silica strand (as shown in Fig. 5.4) with the dispersion of the HE_{21} mode, one of the family of second order 'doughnut' modes of a circular waveguide. This mode may be cleanly excited in a 'null' fused fibre coupler (Birks *et al.*, 1995). In Fig. 5.4 we noted that the shortest zero-GVD wavelength achievable was just below 500 nm, for a 0.5 or 0.6 μm diameter silica strand. Fig. 5.8 shows that the HE_{21} mode can display zero dispersion as short as 350 nm for a 0.6 μm diameter strand, although the wavelength range over which the dispersion is anomalous is very narrow and the absolute value of the dispersion is high. Conversely, if we wish to compare the two modes for a fixed zero-GVD wavelength, Fig. 5.8 shows that the HE_{21} mode gives a larger diameter strand (1.3 μm from 0.6 μm for zero GVD at 470 nm, 3.2 μm from 1.8 μm for zero GVD at 690 nm). In the case of zero-GVD at 470 nm the HE_{21} mode not only offers a physically bigger waveguide (so easier to fabricate and more robust) but it also offers a wider wavelength range of anomalous dispersion. In the case of zero-GVD at 690 nm the dispersion curves for the two modes are almost identical. Experimental supercontinuum generation in the latter case was presented by Birks *et al.* (2001) but has not appeared in a printed paper. This showed identical supercontinuum spectra for two silica nanowire structures with the *same* dispersion but *different* diameters and different modes. This re-emphasises the importance of dispersion over and above the high nonlinearity which may be achieved in a small diameter waveguide.

5.5 Silica nanowires

Freestanding silica nanowires can readily be fabricated from conventional solid optical fibres. In this context they are also often referred to as tapered fibres. When a fibre is 'tapered' one might expect that its properties (or explicitly, its transverse dimensions) would vary along its length. However, confusingly, the terms 'tapered fibres' and 'fibre tapers' are also used to describe a fibre which has been stretched or re-drawn – it may not be tapered at all, but simply thinner than it originally was. In the context of this book, such fibres are not of interest unless they are so reduced in size that the mode becomes guided by the outer surface of the fibre. The outer diameter should also be less than a few μm to access the useful dispersion shown in Fig. 5.4. In this case, they are frequently referred to as 'nanowires', although of course they are likely to be many hundreds or perhaps even thousands of nanometres in diameter. Silica nanowires (Fig. 5.1(a)) are usually drawn from conventional single-mode fibres (e.g. Leon-Saval *et al.*, 2004). In the standard fibre, of course, light is rather weakly guided in the doped core by total internal reflection from the undoped silica cladding. The core diameter is typically 10 μm while the outer diameter is 125 μm. In the nanowire, the vestiges of the doped core are irrelevant and light is confined by reflection at the air–silica interface. The entire fibre diameter is in the range of 0.5–10 μm, and the mode fills the fibre. The evanescent field extends out of the fibre into the air surrounding the fibre.

If the reduction in transverse dimensions during the transition from the initial fibre down to the nanowire is slow, then the mode of the conventional fibre can be adiabatically transformed into the mode of the nanowire. This is very useful, as trying to efficiently couple powerful laser pulses into a free-standing silica nanowire which may be as small as a micron in diameter is a near-impossible task. Instead, light can be coupled into the conventional fibre, which then acts as a downlead to the nanowire. Of course, the dispersion in the downlead cannot be ignored for short pulses, and there may be significant nonlinear evolution in the conventional fibre or the down-taper as well. Nanowires are very fragile as well as being hard to handle, unless they are packaged in a rigid sealed sleeve in the same way as fused fibre couplers. Recent progress has also been made in embedding silica nanowires in silica aerogel as a rigid low-index cladding material (Xiao *et al.*, 2009).

5.6 Tapered fibres

Of course, fibres and nanowires may also be non-uniform along their length. There is good reason to use tapered fibres as nonlinear elements. Most simply, if the output from a single nonlinear fibre is subsequently propagated through a second, different fibre, the result may be different compared to if the first nonlinear fibre is simply longer (Travers *et al.*, 2005). In general, an adiabatic transformation of the first fibre into the second is then preferable to having two discrete fibres because it reduces coupling losses, reduces the likelihood of end-face damage, and makes the entire system more robust (Xiong *et al.*, 2006, 2009). More profoundly, it enables adiabatic transformation of nonlinearly propagating light, rather than just the adiabatic transformation of linear propagation described above (see Chapter 12). Of course if we consider PCF, the second fibre may not be simply a scaled-down (Liu *et al.*, 2001) or scaled-up version of the first – it may be a fibre with larger or smaller air holes, or a different pattern of holes (Xiong *et al.*, 2006, Lai *et al.*, 2007). Thus, producing tapers for such optical manipulation requires control over various parameters within the fibre, over some fibre length. The length over which the transformation in fibre properties is required will vary hugely depending on the details of the nonlinear optics involved: the pulse length, power, input and output dispersion profiles, fibre attenuation etc. However, the most common scenario is that the length of the transformation will be long on the scale of the optical wavelengths (many hundreds of microns at least) but short on the scale of the fibre attenuation length (so, up to a few hundred metres). Means for fabricating tapered structures over these five orders of magnitude depend on the heat source used to soften the glass of which the fibre is formed. Long taper lengths (see Fig. 5.9) are conveniently formed on a fibre drawing tower while the fibre is being drawn down from a preform. The hot zone length in a fibre-drawing furnace is typically from a few millimetres or so upwards, depending on the preform diameter and other factors, and so variation of fibre parameters on length scales of a metre or so are attainable (Gérôme *et al.*, 2007). Variation in the size of the air holes during the process can be conveniently controlled by pressurising the air holes in the fibre preform – individually if necessary. If variations over shorter length scales are required, these can be obtained by miniaturising the fibre drawing process and post-processing the final fibres using the same type of fibre tapering station used to fabricate fused couplers or silica nanowires. In this case, the hot zone could be a millimetre or less in length, and with only a modest draw-down ratio, variations on a length scale of millimetres to 50 cm are very possible. Indeed, with a fibre

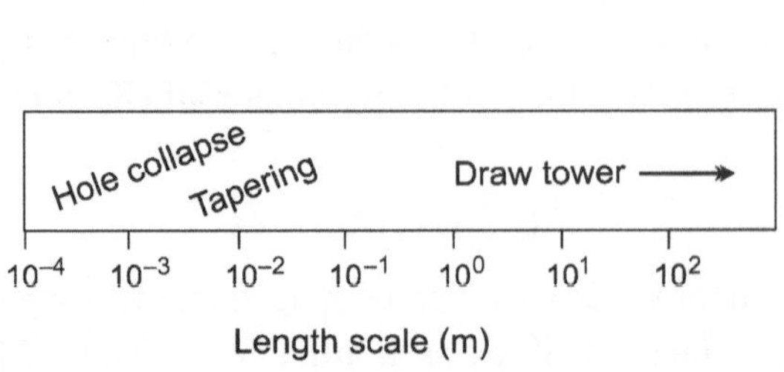

Fig. 5.9. Schematic showing different tapering regimes, and practical ways in which they can be realised.

tapering station some thought is usually required to ensure that variations in the waveguide structure remain adiabatic. There is a range of length scales which is not easily addressed using current techniques – when the transitions required are too long for a standard fibre tapering station but too short for a standard fibre-drawing tower. Transitions on this length scale, of the order of tens of cm up to a few metres, could nonetheless be readily fabricated using a hybrid setup.

5.7 Summary

Whilst the use of silica as the material for supercontinuum fibres may seem at first sight to be a conservative or unimaginative choice, the excellent spectral transparency and the wide range of possible fibre microstructures, combined with the excellent compatibility with conventional fibre and fibre laser systems have made it the material of choice for fibre-based supercontinuum generation. Silica becomes especially attractive when it is recognised that it has just the right combination of refractive index contrast (and hence waveguide dispersion) and material dispersion to enable the zero-dispersion wavelength to be positioned in the spectral band of 700–1100 nm corresponding to the most commonly used short pulse pump lasers. This has resulted in silica fibres being the first choice for both academic researchers and the developers of commercial systems, and the subject of much of the rest of this book.

References

Agrawal, G. (2001). *Nonlinear Fiber Optics*, 3rd edition, Academic Press, San Diego.

Birks, T.A., Culverhouse, D.O., Farwell, S.G. and Russell, P.St.J. (1995). All-fiber polarizer based on a null taper coupler. *Opt. Lett.* **20** 1371–1373.

Birks, T.A., Knight, J.C. and Russell, P.St.J. (1997). Endlessly single-mode photonic crystal fiber. *Opt. Lett.* **22** 961–963.

Birks, T.A., Mogilevtsev, D., Knight, J.C. and Russell, P.St.J. (1999). Dispersion compensation using single-material fibers. *IEEE Photon. Technol. Lett.* **11** 674–676.

Birks, T.A., Wadsworth, W.J. and Russell, P.St.J. (2001) Supercontinuum generation in tapered fibres. *CLEO/QELS 2001* paper CThH1.

Champert, P.A., Couderc, V., Leproux, P., Février, S., Tombelaine, V., Labonté, L., Roy, P., Froehly, C. and Nérin, P. (2004). White-light supercontinuum generation in normally dispersive optical fiber using original multi-wavelength pumping system. *Opt. Express* **12** 4366–4371.

Coen, S., Chan A.H.L., Leonhardt R., Harvey J.D., Knight J.C., Wadsworth W.J. and Russell P.S.J. (2001). White-light supercontinuum generation with 60 ps pump pulses in a photonic crystal fiber. *Opt. Lett.* **26** 1356–1358.

Dudley, J., Gu, X., Xu, L., Kimmel, M., Zeek, E., O'Shea, P., Trebino, R., Coen, S., Windeler, R. (2002). Cross-correlation frequency resolved optical gating analysis of broadband continuum generation in photonic crystal fiber: simulations and experiments. *Opt. Express*, **10** 1215–1221.

Efimov, A., Taylor, A., Omenetto, F., Knight, J.C., Wadsworth, W.J. and Russell, P.St.J. (2003). Nonlinear generation of very high-order UV modes in microstructured fibers. *Opt. Express* **11** 910–918.

Ferrando, A., Silvestre, E., Miret, J.J. and Andres, P. (2000). Nearly zero ultra-flattened dispersion in photonics crystal fibers. *Opt. Lett.* **25** 790–792.

Gander, M.J., McBride, R., Jones, J.D.C., Mogilevtsev, D., Birks, T.A., Knight, J.C. and Russell, P.St.J. (1999). Experimental measurement of group velocity dispersion in photonic crystal fibres. *Electron. Lett.* **35** 63–34.

Gérôme, F., Cook, K., George, A.K., Wadsworth, W.J. and Knight, J.C. (2007). Delivery of sub-100 fs pulses through 8 m of hollow-core fiber using soliton compression. *Opt. Express* **15** 7126–7131.

Gorbach, A. and Skryabin, D.V. (2007). Light trapping in gravity-like potentials and expansion of supercontinuum spectra in photonic-crystal fibres. *Nature Photon.* **1** 653–657.

Gouveia-Neto, A.S., Gomes, A.S.L. and Taylor, J.R. (1988). Femtosecond soliton Raman generation. *IEEE J. Quant. Electron.* **24** 332–340.

Harbold, J.M., Ilday, F.Ö., Wise, F.W., Birks, T.A., Wadsworth, W.J. and Chen, Z. (2002). Long wavelength continuum generation about the second dispersion zero of a tapered fiber. *Opt. Lett.* **27** 1558–1560.

Humbach, O., Fabian, H., Grzesik, U., Haken, U. and Heitman, W. (1996). Analysis of OH absorption bands in synthetic silica. *J. Non-Cryst. Sol.* **203** 19–26.

Islam, M.N., Sucha, G., Bar-Joseph, I., Wegener, M., Gordon, J.P. and Chemla, D.S. (1989). Broad bandwidths from frequency-shifting solitons in fibers. *Opt. Lett.* **14** 370–372.

Johnson, A.M. and Shank, C.V. (1989) in *The Supercontinuum Laser Source*, R.R. Alfano (Ed) Springer-Verlag, New York.

Kaiser, P. and Astle, H.W. (1974). Low-loss single-material fibers made from pure fused silica. *Bell Syst. Tech. J.* **53** 1021–1039.

Knight, J.C., Birks, T.A., Russell, P.St.J. and Atkin, D.M. (1996). All-silica single-mode optical fiber with photonic crystal cladding. *Opt. Lett.* **21** 1547–1549, errata **22** 484 (1997).

Knight, J.C., Arriaga, J., Birks, T.A., Ortigosa-Blanch, A., Wadsworth, W.J. and Russell, P.St.J. (2000). Anomalous dispersion in photonic crystal fiber. *IEEE Photon. Technol. Lett.* **12** 807–809.

Konorov, S., Serebryannikov, E., Zheltikov, A., Zhou, P., Tarasevitch, A. and von der Linde, D. (2004). Mode-controlled colors from microstructure fibers. *Opt. Express* **12** 730–735.

Kruhlak, R.J., Wong, G.K., Chen, J.S., Murdoch, S.G., Leonhardt, R., Harvey, J.D., Joly, N.Y. and Knight, J.C. (2006). Polarization modulation instability in photonic crystal fibers. *Opt. Lett.* **31** 1379–1381.

Lai, K., Leon-Saval, S.G., Witkowska, A., Wadsworth, W.J. and Birks, T.A. (2007). Wavelength-independent all-fiber mode converters. *Opt. Lett.* **32** 328–330.

Leon-Saval, S., Birks, T.A., Wadsworth, W.J., Russell, P.St.J. and Mason, M. (2004). Supercontinuum generation in submicron fibre waveguides. *Opt. Express.* **12** 2864–2869.

Liu, X., Xu, C., Knox, W.H., Chandalia, J.K., Eggleton, B.J., Kosinski, S.G. and Windeler, R.S. (2001). Soliton self-frequency shift in a short tapered air–silica microstructure fiber. *Opt. Lett.* **26** 358–360.

Luan, F., Yulin, A., Knight, J.C. and Skryabin, D.V. (2006). Polarization instability of solitons in photonic crystal fibers. *Opt. Express* **14** 6550–6556.

Malitson, I.H. (1965). Interspecimen comparison of the refractive index of fused silica. *J. Opt. Soc. Am.* **55** 1205–1209.

Mogilevtsev, D., Birks, T.A. and Russell, P.St.J. (1998). Group-velocity dispersion in photonic crystal fibers. *Opt. Lett.* **23** 1662–1664.

Mori, K., Morioka, T. and Saruwatari, M. (1993). Group velocity dispersion measurement using supercontinuum picosecond pulses generated in an optical fibre. *Electron. Lett.* **29** 987–898.

Okuno, T., Hirano, M., Kato, T., Shigematsu, M. and Onishi, M. (2003). Highly nonlinear and perfectly dispersion-flattened fibres for efficient optical signal processing applications. *Electron. Lett.* **39** 972–973.

Ortigosa-Blanch, A., Knight, J.C., Wadsworth, W.J., Arriaga, J., Mangan, B.J., Birks, T.A. and Russell, P.St.J. (2000). Highly birefringent photonic crystal fibers. *Opt. Lett.* **25** 1325–1327.

Pearce, G.J., Hedley, T.D. and Bird, D.M. (2005). Adaptive curvilinear coordinates in a plane-wave solution of Maxwell's equations in photonic crystals. *Phys. Rev. B* **71** 195108.

Ranka, J.K., Windeler, R.S. and Stentz, A.J. (2000a). Visible continuum generation in air-silica microstructure optical fibers with anomalous dispersion at 800 nm. *Opt. Lett.* **25** 25–27.

Ranka, J.K., Windeler, R.S. and Stentz, A.J. (2000b). Optical properties of high-delta air–silica microstructure optical fibers. *Opt. Lett.* **25** 796–798.

Reeves, W.H., Knight, J.C., Russell, P.St.J. and Roberts, P.J. (2002). Demonstration of ultra-flattened dispersion in photonic crystal fibers. *Opt. Expess.* **10** 609–613.

Saitoh, K. and Koshiba, M. (2005). Empirical relations for simple design of photonic crystal fibers. *Opt. Expess.* **13** 267–274.

Senior, J.M. (1992). *Optical Fiber Communications: Principles and Practice*, 2nd edition, Prentice-Hall, Harlow.

Stolen, R.H. (2008). The early years of fiber nonlinear optics. *J. Light. Technol.* **26** 1021–1031.

Stone, J.M. and Knight, J.C. (2008). Visibly "white" light generation in uniform photonic crystal fiber using a microchip laser. *Opt. Express* **16** 2670–2675.

Travers, J.C., Popov, S.V. and Taylor, J.R. (2005). Extended blue supercontinuum generation in cascaded holey fibers. *Opt. Lett.* **30** 3132–3134.

Wadsworth, W.J., Joly, N., Knight, J.C., Birks, T.A., Biancalana, F. and Russell, P.St.J. (2004). Supercontinuum and four-wave mixing with Q-switched pulses in endlessly single-mode photonic crystal fibres. *Opt. Express* **12** 299–309.

Washburn, B.R., Diddams, S.A., Newbury, N.R., Nicholson, J.W., Yan, M.F. and Jørgensen, C.G. (2004). Phase-locked, erbium-fiber-laser-based frequency comb in the near infrared. *Opt. Lett.* **29** 250–252.

Xiao, L.M., Grogan, M.D.W., Leon-Saval, S.G., Williams, R., England, R., Wadsworth, W.J., and Birks, T.A. (2009). Tapered fibres embedded in silica aerogel. *Opt. Lett.* **34** 2724–2726.

Xiong, C., Witkowska, A., Leon Saval, S.G., Birks, T.A. and Wadsworth, W.J. (2006). Enhanced visible continuum generation from a microchip 1064 nm laser. *Opt. Express* **14** 6188–6193.

Xiong, C., Chen, Z., and Wadsworth, W.J. (2009). Dual-wavelength-pumped supercontinuum generation in an all-fiber device. *J. Light. Technol.* **27** 1638–1643.

6

Supercontinuum generation and nonlinearity in soft glass fibres

Jonathan H. V. Price and David J. Richardson

6.1 Introduction

There are many applications for broad bandwidth infrared laser sources, including optical frequency metrology (Udem *et al.*, 2002), precision spectroscopy (Holzwarth *et al.*, 2000) and optical tomography (Hartl *et al.*, 2001), and moving into the mid-infrared (mid-IR), uses for wavelengths beyond 2 μm include LIDAR, molecular spectroscopy and active hyperspectral imaging. Fibre-based supercontinuum sources are attractive for these applications due to their combination of high brightness and broad bandwidth in comparison to alternative thermal or laser sources. Current high brightness mid-IR sources are typically based on optical parametric oscillators (OPO) or quantum cascade lasers (QCL). While OPOs achieve excellent performance they require large pump lasers and can be rather complex and costly to maintain, and QCLs are hard to scale up in power and cannot at present be used to access the important 2–3 μm regime. New fibre-based technology could create an important additional source of robust and lower cost broad bandwidth mid-IR light for the future.

Beyond a wavelength of 2 μm, due to the onset of losses in silica, it is necessary to consider the use of non-silica glasses. The fundamental material properties of these glasses can enhance supercontinuum generation across the mid-IR since these glasses can have intrinsic nonlinearities $\sim$10 × to 100 × that of silica. However, the zero-dispersion wavelengths of these materials are generally longer than for silica, implying the need for longer wavelength pump lasers.

Soft glass fibre research has been ongoing for many years for sensing and imaging applications and for CO_2 laser beam delivery, where their low mid-IR loss is critical. In addition, tellurite and fluoride glass fibres have been developed for both

Supercontinuum Generation in Optical Fibers, ed. by J. M. Dudley and J. R. Taylor. Published by Cambridge University Press. © Cambridge University Press 2010.

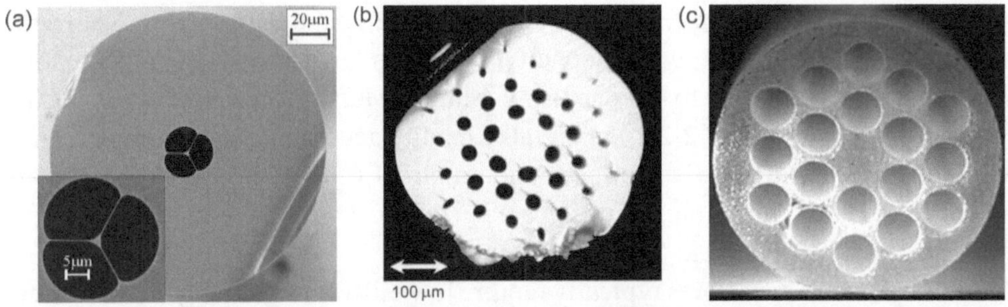

Fig. 6.1. Soft glass MOFs and preforms fabricated at the Optoelectronics Research Centre, University of Southampton, UK. (a) Small core extruded fibre (from Ebendorff-Heidepriem *et al.*, 2004; © 2004 OSA). (b) Single mode tellurite fibre with extremely large mode area (from Feng *et al.*, 2008a; © 2008 OSA). (c) Preform structure created by drilling (from Feng *et al.*, 2005a; © 2005 IEEE).

Raman and rare-earth doped telecommunication amplifiers to provide gain across greater transmission windows (see Mori, 2008, Mori *et al.*, 1997), as well as to obtain laser operation at visible wavelengths using upconversion schemes based on rare earth ions which is enabled by the low phonon energies of such glasses (see e.g. Smart *et al.*, 1991). The field of soft glass fibre research gained new impetus after the development of microstructured optical fibre (MOF) technology, and tapering led to the promise of enhanced nonlinearity and dispersion-tailored structures both for telecommunications switching (Fu *et al.*, 2005, Lamont *et al.*, 2006, Leong *et al.*, 2005b, Pelusi *et al.*, 2008, Petropoulos *et al.*, 2001) and for supercontinuum generation (SCG). Examples of soft glass MOF and preform structures are shown in Fig. 6.1.

MOF technology also restarted research into visible supercontinuum generation, and since silica MOF produced visible continuum using seed pulses directly from a Ti:sapphire oscillator (Ranka *et al.*, 2000), there has been tremendous progress towards explaining this process (Dudley *et al.*, 2006). When working with soft glasses the material dispersion is such that the zero dispersion wavelength (ZDW) is at a longer wavelength than for silica, and the ZDW is strongly dependent on the choice of glass. An attractive option is to use the unique waveguide dispersion characteristics of small core MOF to shift the fibre ZDW to below 2 μm. Such dispersion tailoring would enable the fibres to be combined with diode pumped solid-state lasers operating at wavelengths close to 2 μm for supercontinuum generation, and the tight mode-confinement provides a high effective nonlinearity thereby reducing the optical power requirements. Pumping at 1.55 μm from an Er-fibre pump system is a very attractive option enabled by using a dispersion-tailored MOF, since a wide selection of lasers and amplifiers has been developed for telecommunications. A nanosecond pulsed Er-fibre source and a ZBLAN fibre have been used to generate supercontinuum extending from 0.8 to 4.5 μm using 1.55 μm pumping (Xia *et al.*, 2006a, 2009). A 1.55 μm femtosecond OPO system has also seeded supercontinuum extending from 0.79 to 4.9 μm using tellurite MOF (Domachuk *et al.*, 2008). Seed wavelengths of 2.0–2.25 μm are also readily accessible from commercially available optical parametric oscillator and amplifier (OPO/OPA) systems as well as Tm^{3+} and Ho^{3+} fibre sources which are becoming ever more efficient, powerful and commercially available (Moulton *et al.*, 2009).

Although small core fibres typically require low pulse energies to generate supercontinuum, which relaxes the source power requirements, for applications requiring high average powers, a small core limits the maximum pulse energy that can be used, and producing high average power necessitates vastly increasing the repetition rate of the source laser. In order to increase the pulse energy requirement to satisfy average power needs for moderate pulse repetition rates it becomes necessary to work with fibres with a larger core area. In this instance the dispersion of the fibre is dominated by the material dispersion and it is not possible to shift the fibre ZDW to match the pump wavelength. Consequently one can envisage a need for two sorts of fibre and two sorts of pump laser suited for either high or low power applications.

The SCG research with soft glass MOFs has benefitted from the understanding of the mechanisms leading to the formation of the continuum obtained from prior work with silica fibres. The modelling tools used to characterise the loss and modal dispersion properties of soft glass MOFs are essentially those developed previously for silica structures, as briefly reviewed in Section 6.4. From an applications viewpoint there is a broad understanding of the appropriate choice of pump source and fibre type required to produce a supercontinuum with particular desirable characteristics. However, additional nonlinear effects can become important in soft glass fibres such as two-photon absorption (Lamont *et al.*, 2006). The key difference compared to silica fibres is the relationship between the photon energy at the pump wavelength compared to the band-gap of the material because in soft glasses the seed photon energy is typically close to, or in excess of, half the band-gap.

Due to the range of potential uses of mid-IR SCG sources, they can be considered as a key enabling technology. The aims of the present chapter are to provide the reader new to this exciting field with an overview of the ongoing experimental supercontinuum generation work and to provide information about the material characteristics of a range of soft glasses that are promising candidates for future applications. This chapter includes numerical modelling and experimental results for mid-IR SCG both from small core and from large core fibres. As an introduction to the breadth of this field, we note that the more traditional core-clad fibres from mid-IR glasses such as fluoride glasses have been used for mid-IR SCG (Xia *et al.*, 2006a). In MOF format, lead silicate and Bi glasses were used for several initial demonstrations of soft-glass SCG. Recently there has been significant progress with tellurite fibres (Domachuk *et al.*, 2008, Feng *et al.*, 2008a), and this glass appears to have a combination of robust physical characteristics such that it should offer further investigation in the future. Chalcogenide MOF has also been used to demonstrate mid-IR SCG (Shaw *et al.*, 2005).

Given the wide variety of emerging technologies the scope of the chapter is inevitably limited to a defined subset of materials and waveguide structures. In this chapter, the focus is on soft glass MOF, but the alternative structures are briefly considered and it is noted that crystalline materials in either fibre or planar waveguide format may be useful alternatives (Dadap *et al.*, 2008, Kim *et al.*, 2008, Sazio *et al.*, 2006). Tight mode confinement for enabling both dispersion management and high effective nonlinearity over extended interaction lengths can be achieved in a number of waveguide structures including rib-waveguides, tapered fibres and MOFs. For fibres, and to a lesser extent tapered fibres, the combination of loss and nonlinearity leads to scaling of device length in order to achieve an optimum nonlinear figure of merit (Agrawal, 1995). Planar waveguides are usually longer than typical nonlinear crystal devices but shorter than fibre-based nonlinear devices. Therefore, for planar waveguides, high nonlinearity and dispersion tailoring are perhaps more critical than loss. Fibre tapering is an attractive option since it enables post-processing to fine tune the dispersion characteristics without requiring new fibres to be drawn and recent advances have enabled very long tapers to be fabricated (Vukovic *et al.*, 2008). Mid-IR continuum generation has been demonstrated using a bismuth fibre taper (Brambilla *et al.*, 2005). Using silica MOF, tapering has also been used for dispersion micro-management starting with a small core fibre and then tapering to shift the ZDW to shorter wavelengths (Lu *et al.*, 2005). This technique enabled enhancement of spectral power generation in a variety of wavelength ranges, and should also be applicable to non-silica MOFs in the future. Note that there are new research areas emerging that are based on tapering to produce nanowires (Xu *et al.*, 2008, Yeom *et al.*, 2008).

In the next section the material properties of soft glasses are presented. Section 6.3 focuses on the fabrication technologies that have been developed for soft glass microstructured fibres. Section 6.4 provides information about numerical design techniques used to create MOF structures that have desirable dispersion and nonlinear characteristics. Section 6.5 is devoted to modelling predictions for the supercontinuum spectra with 200 fs pump pulses in small core and in large mode area soft glass MOFs. Section 6.6 focuses on experimental results in small core fibres with femtosecond seed pulses. Section 6.7 considers experimental results with large mode area fibres with femtosecond seed pulses. Section 6.8 focuses on experimental results from nanosecond seed pulses. Section 6.9 provides conclusions and an outlook on possible future research areas.

6.2 Material properties

This section concentrates on the optical properties of the materials, because the high nonlinearity of some of the soft glasses enables <10 cm device lengths at practical power levels and consequently the mechanical requirements are less demanding than those for telecommunications cables. Clearly from a practical perspective, the mechanical and chemical durability of the fibres is important and for guidance about which glass types are most robust for particular applications the reader is referred to reviews elsewhere such as Harrington's chapter in the book by Méndez (Méndez and Morse, 2007). Here, the range of refractive indices and dispersion properties of several non-silica glasses are reviewed. Then, the loss characteristics in the mid-IR wavelength region are described, and compared to silica this is the main benefit of non-silica glasses for mid-IR supercontinuum generation. Finally, the Raman responses of different types of glass are considered.

6.2.1 Refractive index, dispersion and loss characteristics

Soft glasses have a relatively high bound electronic nonlinearity. This property when combined with their reasonably low two-photon-absorption (TPA) makes these glasses highly suited to both SCG and ultra-fast nonlinear switching applications. The linear and nonlinear indices, and the predicted zero-dispersion wavelengths ($D_{MAT} = 0$) of different glasses are listed in Table 6.1. Fluoride glasses have linear and nonlinear refractive indices similar to those of silica, and ZBLAN has a lower loss than silica in the mid-IR, which enabled its use in some recent high power SCG experiments (Xia *et al.*, 2007). Heavy-metal oxide glasses (lead silicate, bismuth oxide, tellurite) have linear indices in the range 1.8–2.0, nonlinear indices ~10 times higher than silica, material ZDWs of 2–3 μm. Chalcogenide glasses (GLS, As_2S_3) have linear indices of 2.2–2.4, nonlinear indices significantly greater than those of the oxide glasses, and ZDWs larger than 4 μm.

Although experimental characterisation is necessary for accurate values of n_2 to be obtained, a high intrinsic nonlinearity is predicted for glasses with a high linear refractive index according to the empirical Miller's rule. Figure 6.2(a) shows a summary of the relation between the linear refractive index n and the nonlinear refractive index n_2 in various glasses (Feng *et al.*, 2005a). Introducing heavy-metal compounds, or introducing the chalcogen elements S, Se and Te to replace oxygen, acts to increase the polarisibility of the components in the glass matrix and also increases the nonlinear index n_2.

Supercontinuum generation depends on the fibre dispersion not only at the pump wavelength, but also over a broad wavelength range. A key consideration is therefore how the refractive index and its variation with wavelength depend on glass composition. Although MOF structures which achieve a high degree of mode

Table 6.1. *Properties of different glasses*

Glass type	Code	Main components	n_0	$n_2 \times 10^{20}$ (m^2/W)	ZDW (μm)
Silica	Si	SiO_2	1.45	2.7 (Vogel *et al.*, 1991)	1.26 (Fujino and Morinaga, 1997)
Lead silicate	SF57	$PbO\text{-}SiO_2$	1.81	41 (Friberg and Smith, 1987)	2.00
Bismuth oxide	Bi	Bi_2O_3	2.02	32 (Kikuchi *et al.*, 2002)	2.29
Germanate	PbGe	$PbO\text{-}GeO_2$	1.80	22 (Hall *et al.*, 1989)	1.78^b (Fujino and Morinaga, 1997)
Tellurite	ZnTe	$ZnO\text{-}TeO_2$	2.03	51^a (Vogel *et al.*, 1991)	2.24^a (Fujino and Morinaga, 1997)
Fluoride	ZBLAN	$ZrF_4\text{-}BaF_2$	1.50	3.3 (Vogel *et al.*, 1991)	1.62 (France *et al.*, 1987)
Chalcogenide	AsS	As_2S_3	2.44	594 (Harbold *et al.*, 2002)	4.81
	GLS	$Ga_2S_3\text{-}La_2S_3$	2.41	216 (Requejo-Isidro *et al.*, 2003)	
	GLSO	$Ga_2S_3\text{-}La_2O_3$	2.25	177 (Requejo-Isidro *et al.*, 2003)	4.64

Reproduced from Price *et al.*, (2007 © 2007 IEEE).
Linear and nonlinear indices n_0 and n_2 at 1.06 μm (Si, SF57, PbGe, ZnTe, ZBLAN) and 1.5 μm (Bi, AsS, GLS, GLSO), and zero dispersion wavelength (ZDW).
^{a} value for $Na_2O\text{-}TeO_2$ glass , ^{b}value for $Na_2O\text{-}GeO_2$ glass.

confinement could tailor the dispersion by moving the ZDW to shorter wavelengths or allow flattening of the dispersion profile over a limited wavelength range (Knight, 2003), an inherently favourable material dispersion is still required to enable a broad and flat dispersion profile to be obtained. As refractive index data are generally available only in the visible and near-IR, the refractive indices of the glasses at wavelengths above 2 μm have been calculated by fitting that data to a generalised Sellmeier equation (data from www.amorphousmaterials.com, Asahi Glass Co., www.schott.com and www.irphotonics.com). The material dispersion, D_{MAT}, was then derived for several glasses as shown in Fig. 6.2(b). In general the zero dispersion wavelength (ZDW) of a glass shifts to longer wavelengths with increasing linear refractive index.

The long wavelength transmission limit in glasses is determined by the multiphonon absorption edge and it can be moved to longer wavelengths if the lattice vibration frequencies are reduced by using compositions containing heavier atoms or having weaker chemical bonds in the glass network. The multiphonon absorption losses of several glasses are shown in Fig. 6.2(c) as calculated using the method

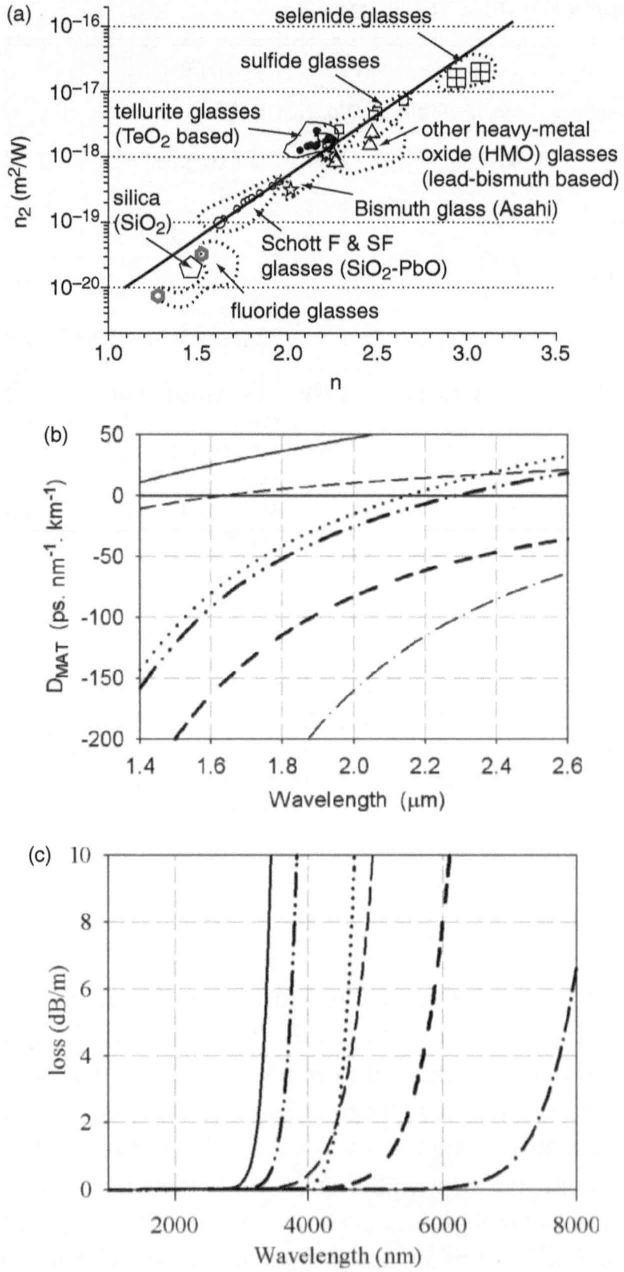

Fig. 6.2. (a) Relation between the linear refractive index n and nonlinear refractive index n_2 in various glasses. (b) Material dispersion curves of silica, bismuth oxide glass (Bi), tellurite glass (NaTeZn), fluoride glass (ZBLAN) and chalcogenide glasses (GLSO, AsS). Line-styles: silica = solid, ZBLAN = short dash, bismuth = dash – dot – dot, tellurite = dotted, GLSO = short double-dash, AsS = dash – dot. (c) Multiphonon edge of the glasses (theoretical fit). (Based on Price *et al.*, 2007; © 2007 IEEE.)

described in Brady *et al.* (1998). Due to the strong Si-O bonding, the oxide glasses containing SiO_2 cannot be transparent at wavelengths longer than 5 μm, whilst glasses containing any oxide have a transparency range limited to wavelengths below ~8 μm.

Extrinsic losses due to scattering and impurity absorption can affect the glass transmission below the multiphonon absorption edge. The main impurity that is difficult to remove from oxide glasses is the OH-group, which can result in losses of more than 1000 dB/m at certain wavelengths. Together with the weaker mechanical properties compared to silica, these extrinsic losses have been one of the limiting factors that have prevented wider use of soft glass fibres for nonlinear devices, and has left their uses confined to areas where the mid-IR loss of silica has made the use of non-silica glass essential. However, dehydration by dry/reactive gas treatment or the addition of fluoride can decrease the OH loss substantially. Fluoride glasses are generally less susceptible to OH impurities, and dehydration can then reduce the OH losses to as low as 0.001 dB/m (France *et al.*, 1987). We note that the OH bands are very broad in oxide glasses, and span over the whole mid-IR range, whereas the OH bands of fluoride and chalcogenide glasses are within the 2.8–3.4 μm range. The impurity levels in chalcogenide glasses depend on the melting conditions during fabrication. GLSO microstructured fibre has been fabricated with OH absorption as low as 20 dB/m at 3 μm (Fig. 6.6). Commercial As_2S_3 fibres exhibit OH absorption of <2 dB/m at 3 μm and SH absorption of 2–10 dB/m at 4 μm (data from www.amorphousmaterials.com, www.coractive.com).

Although non-silica fibre fabrication technology is less mature than that of silica fibre fabrication, the near-IR waveguide losses of soft glass MOFs are typically ~2 dB/m (in addition to the material loss), due to scattering from core surface roughness and contaminants accumulated during fabrication (Ebendorff-Heidepriem *et al.*, 2007). In addition, all modes of a MOF have an associated confinement loss (due to the finite extent of the cladding region) (White *et al.*, 2001), but these losses can be reduced to low values by ensuring a sufficiently large microstructured cladding region. Therefore, the total fibre losses of the short lengths of fibre used for SCG are often dominated by the rather high material losses of impurity related absorption bands and at the multiphonon absorption edge. However, we note that in the case of ZBLAN fibre, both the fibre losses due to bending and those due to the material losses were considered to be a limitation to the extent of the mid-IR SCG (Xia *et al.*, 2006b).

6.2.2 Raman response

The almost instantaneous electronic nonlinearity of the glasses considered here is accompanied by a delayed Raman contribution to the nonlinearity and it is necessary to understand this contribution in order to predict the performance of soft glass fibre

Raman amplifiers. The Raman response is also important for SCG because when the dispersion is anomalous, solitons can form and these are continuously frequency down-shifted due to the soliton self frequency shift (Gordon, 1986, Mitschke and Mollenauer, 1986). With nanosecond or CW pumping, new frequency components are also generated by the formation of Stokes/anti-Stokes sidebands.

When modelling the process of SCG in silica fibres, the nonlinear response to the applied field in the time domain has usually been written as $R(t) = (1 - f_R)\delta(t) + f_R h(t)$, where the δ-function represents the instantaneous electronic response and $h(t)$ represents the delayed Raman response of the ions (Agrawal, 1995). For silica fibre, the Raman temporal response $h(t)$ has been previously determined from the shape of the Raman gain in the frequency domain, and $f_R = 0.18$ was determined by combining that data with measurements of the absolute value of the Raman gain (Stolen *et al.*, 1989).

For soft-glass characterisation, it is more efficient to perform measurements on bulk-glass samples in order to avoid the additional steps necessary to fabricate fibres. Spontaneous Raman scattering measurements are easily performed on bulk samples and are useful for efficiently screening these materials. When Raman scattering measurements are made with a tightly focused pump beam, as is often the case in Raman microscopes, the detected scattered signal comes from a small volume just below the sample surface and is not sample-size dependent. Therefore if polished samples are tested and a silica sample is included in the data collection as a calibration such micro-Raman measurements can be used to calculate the absolute Raman gain of the comparison materials (for details see Rivero *et al.*, 2005). To complete the characterisation, the nonlinear refractive index and TPA coefficient of glasses are usually measured using the Z-scan technique (Said *et al.*, 1992, Sheik-bahae *et al.*, 1990). It is then possible to calculate the delayed temporal response by using the procedure developed for silica by Stolen *et al.* (1989), and as long as n_2 is also known then it is possible to calculate f_R.

Stolen's calculations for silica fibres obtained the Raman gain by dividing the spontaneous Raman spectrum by $(1 + n_B)$, where n_B is the thermal phonon number or Bose–Einstein occupation number: $[n_B = (\exp(h\nu/k_B T) - 1)^{-1}]$. A temperature of $T = 300$ K corresponds to $(k_B T/h) \sim 6$ THz, and therefore dividing by $(1 + n_B)$ results in a large difference between the phonon spectrum and Raman gain in approximately this frequency range. The temporal response is calculated from the Raman gain using the equation:

$$h(\tau - t) \quad \propto \int_0^\infty d\Omega \, g_{\text{Raman}}(\Omega) \sin[\Omega(\tau - t)]. \tag{6.1}$$

Calculations of the delayed Raman response of Bi glasses have been reported using Raman scattering data measured at our institute (Price *et al.*, 2006). In that instance, a calibrating reference sample of fused silica was not used, and in the related SCG modelling studies it was assumed that $f_R = 0.2$, based on the known fraction of 0.18 for silica. The Raman temporal response has since been calculated for SF57 glass (Leong *et al.*, 2006b), for tellurite glasses (Domachuk *et al.*, 2008, Serebryannikov *et al.*, 2007), and for SF6 (Kalashnikov *et al.*, 2007, Moeser *et al.*, 2007). For SF6 a silica reference enabled f_R to be calculated, and it was found to be variously 0.13 (Kalashnikov *et al.*, 2007) or 0.10 (Moeser *et al.*, 2007). A homogeneous Raman gain can be assumed if there is a single dominant peak in the Raman spectrum that can be attributed to a mode of the glass lattice. In that case the Raman response has sometimes been modelled using a damped oscillator approximation rather than including the continuum of resonances contained in the Raman spectra.

It is difficult to confirm the accuracy of the Raman temporal response derived from spontaneous scattering data using direct time-domain experimental measurements, because measurements of the Raman gain of soft glasses are generally made using pico second or longer pump pulses, which do not enable the sub-picosecond response to be resolved. A few studies have used shorter pulses to measure the nonlinearity either with a time-resolved output signal (Aber *et al.*, 2000, Kang *et al.*, 1996, Montant *et al.*, 1998) or without a time-resolved output (Kang *et al.*, 1995), and the former measurements were consistent with the delayed nonlinear response calculated from the Raman spectra.

Finally, one further difference between the nonlinear response of soft glasses and silica which should be appreciated is that the relatively small band-gap of these materials leads to a significant dispersion of the nonlinearity (DeSalvo *et al.*, 1996, Kang *et al.*, 1995). Indeed, a study using spectrally resolved two-beam coupling to investigate the nonlinearity of several multi-component glasses found that for photon energies above half the linear absorption edge energy-gap, the nonlinearity was no longer purely third order, and that $\chi^{(5)}$ and $\chi^{(7)}$ components were observable in the response (Chen *et al.*, 2006). Furthermore, Raman-scattering experiments on tellurite glasses with Raman resonances of comparable strength from several subcomponents have shown that the shape of the Raman gain depends on the pump wavelength such that the strength of each Raman response increased or decreased depending on the proximity of the pump photon energy to the band-gap of each component (Rivero *et al.*, 2005). To date, the change in Raman gain shape with wavelength has not been included in simulations of soft glass based SCG and hence leads to additional uncertainties in those results. This area perhaps requires further data before appropriate modelling tools can be developed.

6.3 Fabrication technologies

The fabrication strategies used for soft glass fibres are dependent on the thermomechanical properties of each glass. The emergence of MOF fabrication technology is well suited to multi-component glasses as it eliminates the need for two thermally, chemically and optically compatible glasses to form the fibre core and cladding as required for conventional fibres. MOF technology thus provides a simple and convenient route to realising fibres in glasses that otherwise might not be able to be drawn into fibre form. Several techniques have been used to fabricate compound glass MOFs (Feng *et al.*, 2005a). One approach is to manually stack capillary tubes to produce the structured preform. This is the approach that is routinely used to produce silica microstructured fibres, and the resulting fibres typically consist of a hexagonal lattice of air holes surrounding the fibre core. Soft glass MOFs fabricated using this method have been reported by several groups (Mairaj *et al.*, 2001, Shaw *et al.*, 2005, Troles *et al.*, 2007). In addition tellurite glass MOFs prepared by stacking have been reported with rare earth dopants (Cordeiro *et al.*, 2006). It is anticipated that using the flexibility of this approach fibre geometries could be designed to tailor the dispersion characteristics (for example to achieve lower dispersion slopes).

Typical compound glass softening temperatures are ~500°C, as opposed to ~2000°C for silica. The reduction in processing temperature enables the use of extrusion for soft glass fibre preform manufacture (Feng *et al.*, 2005a, Kiang *et al.*, 2002, Monro, 2006). However, a challenge is that the non-silica glasses generally have steep viscosity curves compared to silica and the temperature range for drawing soft glass fibre is much less than that for pure silica glass fibre (Feng *et al.*, 2005a). A schematic of the system used for preform extrusions at the University of Southampton is shown in Fig. 6.3. This technique has been used to produce a broad range of fibre geometries in a number of compound glass materials including lead silicate (Leong *et al.*, 2005a, 2006a, Petropoulos *et al.*, 2003), bismuth oxide (Ebendorff-Heidepriem *et al.*, 2004) and tellurite glass (Feng *et al.*, 2005b, 2008a, Kumar *et al.*, 2003).

The extrusion geometry used for many of the early fibres is the so-called "wagon-wheel" (WW) fibre, which is a micron-scale solid core suspended by three or more long fine glass struts that optically isolate the core within a robust solid jacket. A scanning electronic microscope (SEM) image of a WW fibre is shown in Fig. 6.1 (a), and the die used for fabricating the structure is shown inset to Fig. 6.3. The size of the core can be adjusted during the drawing process by modifying the outer fibre diameter. These fibres can readily be made with small cores and high numerical aperture (NA), leading to tight mode confinement, and thus high fibre nonlinearity.

In order to achieve greater control over the dispersion profile of the fibres, it is desirable to have several rings of holes. To combine this requirement with the advantages of the preform extrusion technique, a hybrid approach has been demonstrated whereby several structured elements are prepared by extrusion, then stacked to provide a complex hole arrangement that can be drawn to create the final fibre. A schematic of the stacked extruded elements is shown in Fig. 6.4 (from Leong,

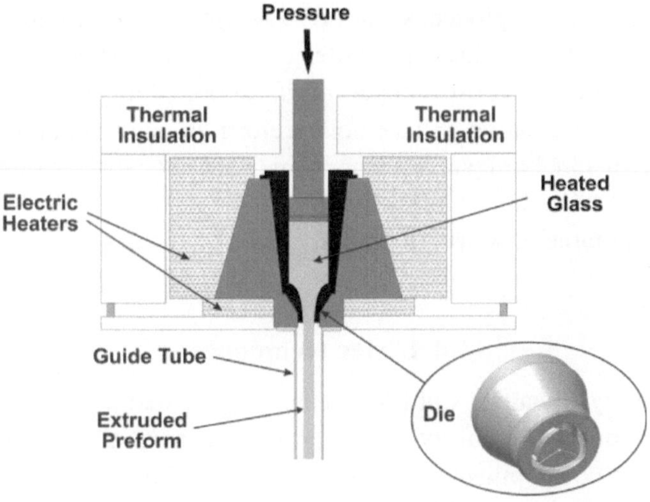

Fig. 6.3. Schematic of soft glass preform extrusion apparatus. Inset: a typical die structure. (A SEM image of the fibre produced by such a die is shown in Figure 6.1(a).)

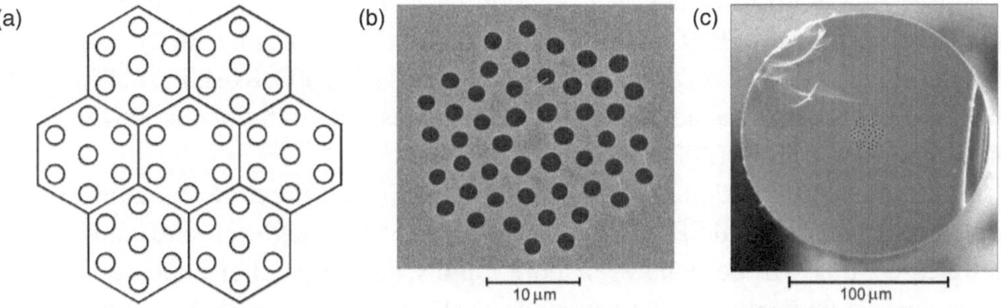

Fig. 6.4. Design and SEM images of fibres made by stacking complex sub-structures, each of which was fabricated using the extrusion technique. (a) Schematic of stacked elements. (b) Fibre core. (c) Final fibre. (From Leong, 2007.)

2007). The figure also shows an example of the resulting fibre. It is to be appreciated that improvements in die design have also more recently allowed the direct extrusion of preforms with many more features than hitherto possible (Ebendorff-Heidepriem and Monro, 2007). As noted by other reviews (Seddon *et al.*, 2008), a method reported by Feng *et al.* used thin discs of high and low index glasses that were stacked in multiple layers and then extruded to form concentric rings. Low and high refractive index glass rings provided guidance in an all-solid structure rather than using air holes to provide index contrast between core and cladding (Feng *et al.*, 2005c).

Other demonstrated approaches for MOF fabrication with compound glasses include drilling the desired pattern of holes in the preform (Feng *et al.*, 2003) and casting the structured preform (Guiyao *et al.*, 2006, Mori *et al.*, 2004). Drilling has the advantage of creating a multi-hole preform in one step, and achieves high dimensional precision, but it suffers from somewhat poor surface quality, and usually has a low yield. Casting enables high dimensional precision, and high air-filling fraction, but is suitable only with a limited choice of glasses.

6.4 Design techniques

One of the primary advantages of MOF technology is that it provides for an enormous amount of design flexibility. However, in order to be able to properly exploit this flexibility the development of suitable numerical fibre design tools is essential. Over the years a number of different approaches have been developed that use a given transverse fibre structure to directly predict the key modal optical properties of microstructured fibres (e.g. effective area, dispersion, confinement loss, birefringence etc.). These include the modal expansion methods (plane wave (Silvestre *et al.*, 1998) or orthogonal function basis approaches (Mogilevtsev *et al.*, 1998, Monro *et al.*, 1999)), multipole methods (White *et al.*, 2002), and FEM based techniques (Brechet *et al.*, 2000, Cucinotta *et al.*, 2002, Koshiba, 2002). Each have their own particular advantages, disadvantages and limitations as conveniently summarised in Monro (2006).

It is fair to say that calculating the properties of a given structure (either idealised, or from an experimental SEM image) is now relatively straightforward, and as time progresses, this will be done ever more rapidly due to parallel improvements in technique implementation and computer hardware. The main challenge now is to find the best ways of exploiting these direct calculation approaches to investigate the very large structural parameter space (and indeed possible material parameter space) in order to better identify and design optimised fibres for a given application.

We identify two basic classes of method used to design fibres that we feel particularly worthy of mention which may be broadly defined as:

1 *inverse design methods* (Kerrinckx *et al.*, 2004, Poletti *et al.*, 2005) whereby a particular fibre property, or combination of properties is specified, and an algorithm is implemented that automatically explores the design space (subject obviously to some form of structural parameterisation), to establish an optimum design or range of suitable designs; and
2 *improved visualisation methods* (Poletti *et al.*, 2007, Tse *et al.*, 2008) that allow one to explore the design space more readily in order to identify the most interesting regions worthy of more detailed consideration.

As an example of the use of inverse design tools, we refer to the work reported in Poletti *et al.* (2005) in which a genetic algorithm identified appropriate designs for a dispersion-shifted and flattened highly nonlinear silica microstructured fibre operating at 1550 nm. In this work fibre designs with up to six free parameters were considered and the parameters optimised to maximise objective functions that parameterised the desired dispersive properties. This early work highlighted the fact that similar fibre properties can be obtained using seemingly quite different structural designs – underlining the need for automated design tools to ensure that optimum overall performance is obtained. Subsequently, similar approaches have been applied to a number of fibre design problems (Tse *et al.*, 2008, Varshney *et al.*, 2007), including the design of complex, highly nonlinear compound glass microstructured fibres (Leong *et al.*, 2006a, Riishede and Sigmund, 2008). Note that the use of genetic algorithm based optimisation approaches is not mandatory and that other standard optimisation methods including simulated annealing and simplex methods are also applicable and can often be more computationally efficient.

With regard to improved visualisation tools we would like to highlight the very simple yet very helpful multi-contour plot method first reported by Poletti *et al.* (2007). According to this approach the basic structural parameters of the fibre, e.g. hole-spacing Λ and diameter d, are used to define the grid axes and relevant contour plots of the key properties of the optical mode under consideration, e.g. dispersion, effective area and dispersion slope, can be simultaneously plotted (evaluated on a grid using the direct design tools previously discussed). Fig. 6.5 shows a plot

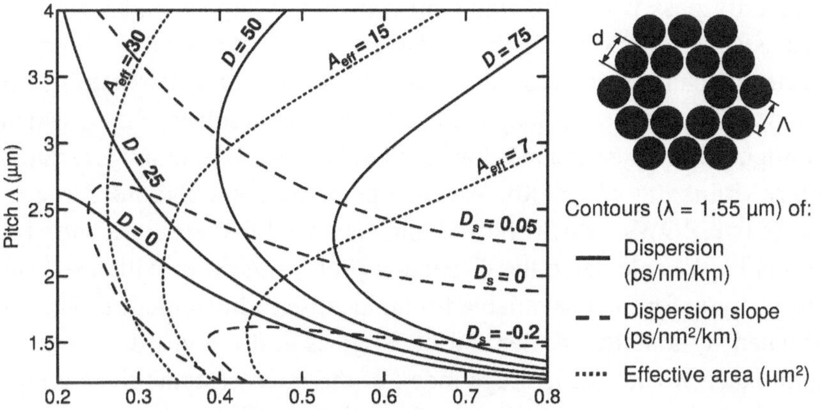

Fig. 6.5. Contour map showing dispersion, dispersion slope and effective area as a function of structural parameters for a triangular lattice silica MOF at 1550 nm. (From Tse *et al.*, 2008; © 2008 IEEE.)

from Tse *et al.* (2008) to illustrate the value of this approach and it shows the variation of the key nonlinear/linear parameters of a lead-silicate MOF with a triangular lattice structure. It is clear from the plot that there is a region of design space corresponding to zero dispersion and zero dispersion slope at 1550 nm as required for nonlinear optical processing applications within telecommunications. The multi-contour mapping approach has also been used to design fibres with a longitudinally varying structure providing uniform dispersion properties but with an enhanced Brillouin threshold (Poletti *et al.*, 2007).

The examples described above provide just a glimpse of the design tools available to MOF fibre fabricators and whilst it is true that to date these have mostly been applied to silica based fibres it is to be appreciated that they are fully applicable to compound glass MOFs and will provide a valuable aid to the design of fibres for mid-IR SC generation.

6.5 Modelling

To assess how non-silica MOFs could be of benefit for mid-IR SCG, a numerical modelling survey was performed considering a range of fibre parameters to provide an indication of what an optimised source could produce. In previous publications we have presented the results of numerical simulations to indicate the optimum glass-type and fibre-design combinations for mid-IR supercontinuum generation when pumped at near-IR wavelengths with 200 fs seed pulses (Price *et al.*, 2006, 2007). We summarise the key findings in this section. (For an introduction to CW SCG see Chapter 8.)

For the small core fibre structures, we chose a pump wavelength of 1.55 μm, and a small core Bi glass WW fibre with core diameter of 2.6 μm and ZDW at 1.5 μm was used as a starting-point for the investigation. We studied how the continuum was affected by changes in the seed wavelength and the fibre dispersion profile. We considered first a pump wavelength close to the fibre's lowest ZDW, which is the usual configuration for demonstrations of visible supercontinuum generation using Ti:sapphire (Ranka *et al.*, 2000). We then performed simulations for small core MOFs with two ZDWs, which showed that enhanced mid-IR continuum broadening is possible. We calculated the dispersion profiles of fibres with a second ZDW using the analytic formulae available for an idealised fibre structure. The core was approximated by a solid circular rod of Bi-glass and the cladding was taken to be a solid material with the average index of the microstructured cladding region (see Snyder and Love, 1983). Similar dispersion profiles would be possible with real MOF structures fabricated using the stacking technique. The lower ZDW was maintained at 1.55 μm in order to maintain one parameter constant. Similar studies of the influence of dispersion variation in silica MOF for visible and near-IR continuum generation have been reported in the literature (Frosz *et al.*, 2005, Tse *et al.*, 2006a).

By varying each fibre parameter with the other parameters held constant, we found that nonlinear compression, soliton fission, and rapid spectral broadening occurred over a typical "threshold length" of less than 40 mm (Herrmann *et al.*, 2002). The pulses then broaden temporally, causing a reduction of the peak power and the supercontinuum width saturates (Chang *et al.*, 2003). Once the seed-pulse energy was above a threshold level necessary for effective continuum generation, further increasing the energy or fibre nonlinearity led to only modest increases in the continuum bandwidth. We used a fibre length above this threshold length, and therefore differences between spectra are attributed to different dispersion profiles, not length scales. Note that longer fibre lengths could be required with seed pulse durations above a few pico seconds, e.g. >50 ps, because in a wavelength range where the fibre has anomalous dispersion initial processes such as modulation instability can lead to pulse compression and to soliton generation before a continuum develops. Other mechanisms for spectral broadening include self-phase-modulation (SPM) and four-wave-mixing (FWM) (Agrawal, 1995, Coen *et al.*, 2002). A separate modelling study would be required to predict the appropriate fibre length and choice of soft glass if long seed pulses were to be used.

We also performed simulations for large model area (LMA) GLSO fibres that would be suitable for higher power operation. The predicted spectra in all graphs that follow are shown using a dBm/nm scale, assuming a pulse repetition rate of 1 Hz. This scale would show an average power spectral density of -30 dBm/nm for 1 μJ pulses with 1 nm spectral widths. Scaling the power spectra to different pulse repetition rates can be done as follows: e.g. starting with -90 dBm/nm at 1 Hz, and converting to 1 MHz (10^6 Hz) gives $(-90+60) = -30$ dBm/nm.

6.5.1 Numerical model

As with silica, we have assumed that the amorphous multi-component glasses are centro-symmetric, and therefore have no intrinsic second order nonlinear susceptibility ($\chi^{(2)}$), and have a homogeneous third order nonlinear susceptibility ($\chi^{(3)}$). The third order nonlinear susceptibility is assumed to be small compared to the linear susceptibility, and is also assumed to be wavelength independent over the wavelength range considered. The total refractive index therefore includes a small intensity dependent nonlinear contribution ($n = n_0 + n_2 I$). Effects such as two-photon absorption have not been included. Our experience is that 2–3 μm core fibres can be operated in an effectively single mode fashion and the low NA large mode area fibres that we consider here could be rigorously single mode structures throughout the mid-IR (i.e. above the pump wavelength). Hence, only processes occurring within the fundamental fibre mode are modelled, and mode-mixing is not considered. These simplifications enable modelling of the pulse-propagation using the modified nonlinear Schrödinger equation (NLSE) with loss (Agrawal, 1995), as shown below:

$$\frac{\partial A}{\partial z} - i\sum_{k\geq 2}\frac{i^k \beta_k}{k!}\frac{\partial^k A}{\partial t^k} + \frac{\alpha(\omega)}{2}A$$
$$= i\gamma\left(1+\frac{i}{\omega_0}\frac{\partial}{\partial t}\right)\left(A(z,t)\int_{-\infty}^{+t} R(t')|A(z,t-t')|^2 dt'\right), \qquad (6.2)$$

where $A(z,t)$ is the electric field envelope, ω_0 is the centre frequency, $\gamma = (n_2\omega_0/cA_{\text{eff}})$, β_k are the dispersion coefficients at the centre frequency, and $\alpha(\omega)$ is the frequency dependent fibre loss. For the small core fibre simulations, we used the wavelength dependent dispersion data shown in the figures below, and for the large mode fibres we used the bulk-glass dispersion from Fig. 6.2(b), i.e. the smaller waveguide dispersion was not included. We used the A_{eff} data from mode calculations at a seed wavelength of 1.55 μm. To solve the propagation equation, we used a standard split-step Fourier algorithm treating dispersion in the frequency domain and the nonlinearity in the time domain, apart from the temporal derivative for the self-steepening effect, which was evaluated using Fourier transforms.

The model includes both the instantaneous electronic response (responsible for the Kerr effect), and the delayed ionic response (responsible for Raman scattering) in the nonlinear component of the refractive index, n_2. We used $h(t)$ calculated from spontaneous Raman spectra measured at the University of Southampton, and following the procedure described in Section 6.2.2. We have assumed that $f_R = 0.2$, based on the known fraction of 0.18 for silica. Due to our measurement setup, no spontaneous Raman data was available for the frequency range 0–6 THz (0–200 cm^{-1}) and we have assumed a linear increase in the spontaneous Raman spectrum in this range. We believe this assumption is reasonable, particularly since the phonon data shows a decreasing power at the lowest measured frequency. However, we note that 0–6 THz is an important frequency range because the frequency width (FWHM) for a 200 fs transform limited Gaussian pulse is 2.2 THz, and it is therefore this part of the Raman spectrum that provides the long wavelength gain for soliton self-frequency shifting. In our previous study of multi-component glass MOFs, we showed the spectra from simulations performed for Bi glass WW fibre for the same seed pulse parameters and with: (i) SPM term only, (ii) SPM and self-steepening terms, and (iii) including the Raman response (Price *et al.*, 2006). The overall shape of the continuum was similar in all cases and we concluded that the initial continuum generation is not highly dependent on the detailed Raman response in the fibre. Therefore whilst the exact form of the Raman response is uncertain for these glasses, small differences should lead to only small changes in our predicted spectra.

We note that there are further refinements that can be made to the above NLSE to include effects such as polarisation coupling, but the simulations as they stand should provide a reasonable estimate of the expected spectra. Modelling of the wavelength variation of the effective area ($A_{\text{eff}}(\lambda)$) has also been reported recently (Canat *et al.*, 2008, Hult, 2007, Laegsgaard, 2007, Travers *et al.*, 2008) and nonlinear loss has been included in simulations by others (Lamont *et al.*, 2006). Our simulation results show the expected fine structuring (Dudley and Coen, 2002) and we have applied a rolling average to smooth the spectra. These smoothed spectra should be approximately comparable to the time average over several pulses that would be measured in experiments in which the seed pulses have small energy fluctuations. The simulated seed pulse was taken to have a transform limited Gaussian profile at the relevant centre wavelength.

We expect that our mid-IR simulations should provide a reasonable guide to the spectra that might be possible from future experimental work, although they should not be considered exact predictions. The modified NLSE has been used by other authors for the numerical study of supercontinuum generation (Dudley and Coen, 2002, Genty *et al.*, 2002). In addition, we have previously published a study of visible/UV supercontinuum generation from a silica MOF (Price *et al.*, 2003) and a study of visible/NIR supercontinuum from an extruded SF57 MOF as shown in Leong *et al.* (2006b). Both of these studies combined experimental and numerical results, and the simulations and experiments were in qualitative agreement.

6.5.2 Simulation results in small core fibres

For the small core fibre simulations we fixed the pulse energy at 1 nJ, $T_0 = 0.2$ ps (FWHM), and fibre length to be 40 mm. In an experiment, the maximum pulse energy will be limited either by the available pump laser or by the material damage threshold. The experimentally measured absorption profiles of the bismuth oxide glass and GLSO glass used for the numerical modelling is shown in Fig. 6.6. We show simulations that use a pump wavelength equal to the short wavelength ZDW at 1.55 μm with two different fibre dispersion profiles. We also compare results from dispersion profiles with ZDWs spaced close together and wider apart. Simulations are then presented comparing seed wavelengths of 1.55 μm and 2.0 μm, and dispersion profiles with ZDWs spaced close together and wider apart. The 2.0 μm seed wavelength may in future be realised by a Ho/Tm fibre laser, although the pulse duration used in practice may be somewhat longer than the 200 fs pulses modelled.

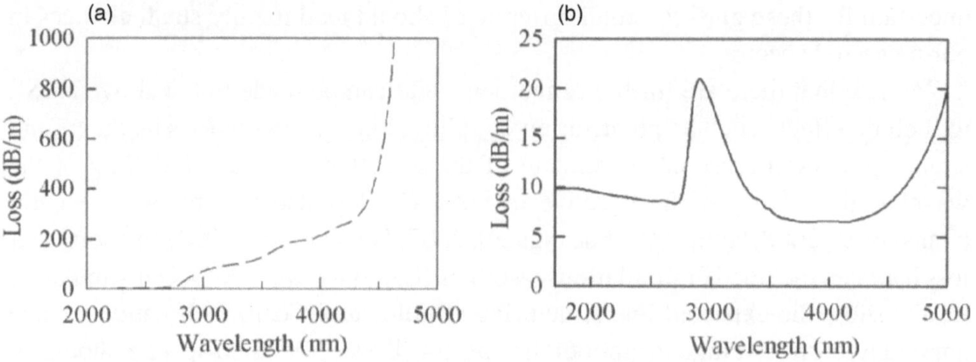

Fig. 6.6. Loss spectra of (a) bismuth (sample had been dehydrated to remove OH impurities), and (b) GLSO glasses (used in simulations). Losses were measured using a >1 meter length of unclad Bi fibre and the cutback technique for wavelengths below 1.5 microns or 1–30 mm thickness bulk glass samples with a commercial spectrophotometer for bismuth at wavelengths above 1.5 μm, and for GLSO at all wavelengths shown. (Based on Price *et al.*, 2007; © 2007 IEEE.)

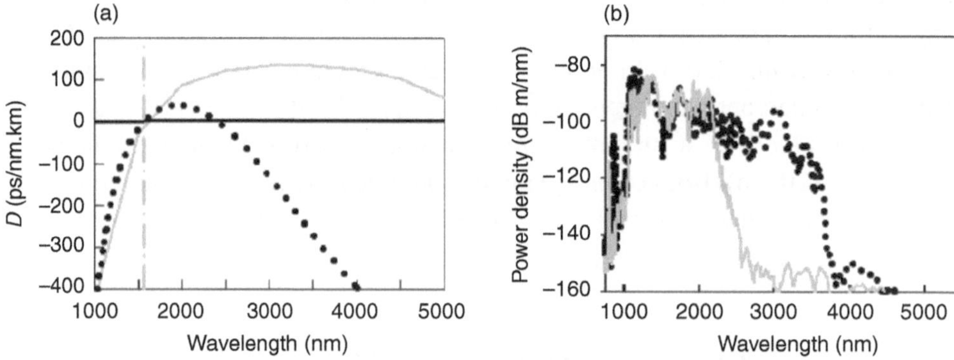

Fig. 6.7. (a) Dispersion profile of Bi glass WW fibre and of Bi glass circular rod (fibre core) with radius of 0.9628 μm and surrounding average refractive index $n_{clad} = 1.6649$. The pump wavelength is at the lower ZDW ($\lambda_{pump} = 1.55$ μm). (b) Spectra for Bi glass MOF using the dispersion profiles shown in (a). Dotted (circular rod core) and solid line (WW) styles show corresponding spectra and dispersion profiles. ($E_0 = 1$ nJ, $T_0 = 0.2$ ps, fibre length = 40 mm, loss = Bi glass shown in Figure 6.6(a).) (Based on Price *et al.*, 2007; © 2007 IEEE.)

The dispersion profiles and spectra predicted for the Bi WW fibre and a Bi rod with two closely spaced ZDWs with a 1.55 μm seed wavelength are shown in Fig. 6.7. For the Bi WW fibre, the supercontinuum extended only to $\lambda = 2.2$ μm, due to the steeply rising dispersion and very high peak dispersion of >100 ps/(nm.km), whereas for the dispersion profile with closely spaced ZDWs the spectra broadened to ~3.5 μm by phase-matching to wavelengths above the second ZDW where the

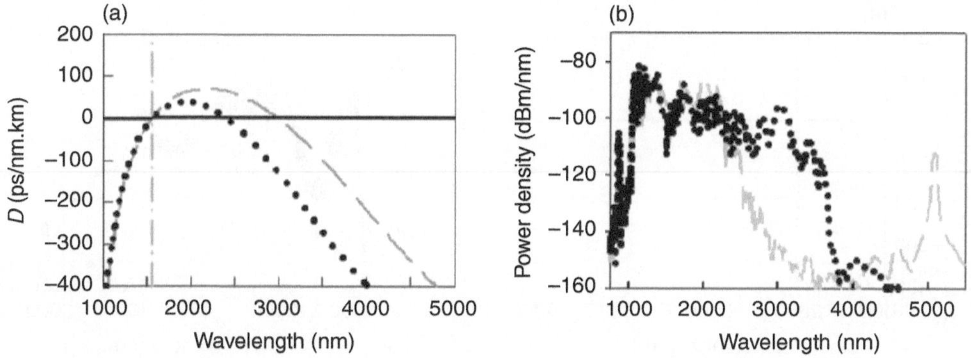

Fig. 6.8. (a) Dispersion profile of Bi glass circular rods (fibre cores) with radii of 0.9628 μm or 1.1529 μm and surrounding average refractive indices of $n_{\text{clad}} =$ 1.6649 or 1.6206. The pump wavelength is at the lower ZDW ($\lambda_{\text{pump}} = 1.55$ μm). (b) Spectra for Bi glass MOF using the dispersion profiles shown in (a). Dotted and dashed line-styles show corresponding spectra and dispersion profiles. ($E_0 = 1$ nJ, $T_0 = 0.2$ ps, fibre length = 40 mm, loss = Bi glass shown in Figure 6.6(a).) (Based on Price *et al.*, 2007; © 2007 IEEE.)

dispersion is normal. At the shorter wavelength edge of the continuum there is minimal variation between the fibres which is probably due to the steep dispersion profiles that naturally result from the tight mode confinement required to create the appropriate waveguide dispersion needed to compensate the bulk material ZDW of >2 μm and provide an overall fibre ZDW at the Er-pump wavelength of ~1.55 μm.

The dispersion profiles and spectra predicted for the Bi rod with two closely spaced and wider spaced ZDWs with a 1.55 μm seed wavelength are shown in Fig. 6.8. For the fibre with closely spaced ZDWs the supercontinuum extended to $\lambda \sim 3.5$ μm by phase-matching to wavelengths above the second ZDW where the dispersion is normal. The dispersion profile with wider spaced ZDWs has a more limited continuum bandwidth, despite offering the possibility of phase-matching to wavelengths above 3.5 μm. The explanation is that due to the increase in the fibre dispersion above the pump wavelength the initial solitons broaden temporally and they do not have sufficient bandwidth to transfer power across the ZDW to the mid-IR dispersive wave.

Figure 6.9 shows a comparison of the results obtained by varying the seed wavelength from the lower ZDW to midway between the two ZDWs. The dispersion profile with wider spaced ZDWs has been used ($r_{\text{core}} = 1.1529$ μm and upper ZDW at ~3 μm). Whereas a seed wavelength of 1.55 μm produced a spectrum close to the lower ZDW but not extending past the more distant ZDW, a seed wavelength of 2.0 μm produced a much broader spectrum which extended across both ZDWs to ~4 μm. The soliton self-frequency shift (SSFS) occurs in conjunction with

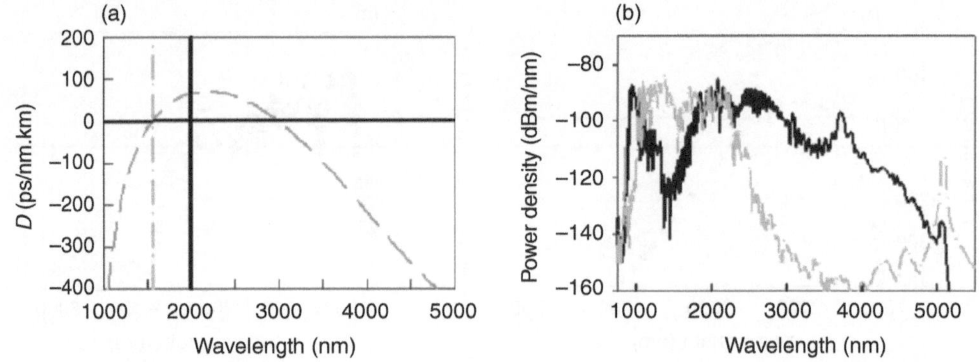

Fig. 6.9. (a) Dispersion profile of Bi glass circular rod (fibre core) with radius of 1.1529 μm and surrounding average refractive index of $n_{clad} = 1.6206$. The pump wavelength is either at the lower ZDW (= 1.55 μm) or between the two ZDWs ($\lambda_{pump} = 2.0$ μm). (b) Spectra for Bi glass MOF using the dispersion profile and pump wavelengths shown in (a). Dash-dot and solid line-styles show corresponding spectra and pump wavelengths. ($E_0 = 1$ nJ, $T_0 = 0.2$ ps, fibre length = 40 mm, loss = Bi glass shown in Figure 6.6(a).) (Based on Price *et al.*, 2007; © 2007 IEEE.)

decreasing fibre dispersion such that the short duration and broad bandwidth of the initial solitons is maintained and hence more power is transferred to dispersive waves at long wavelengths.

Having established that a 2.0 μm seed wavelength is advantageous, a comparison of the spectra predicted for the Bi rod with two closely spaced and wider spaced ZDWs is shown in Fig. 6.10. In contrast with the 1.55 μm seeded results shown in Fig. 6.8, the position of the upper ZDW has clearly influenced the extent of the mid-IR continuum. In the fibre with upper ZDW at ~3 μm, there is phase-matched power transfer across that ZDW to the 4–5 μm range, whereas in the fibre with upper ZDW below 2.5 μm, the continuum stops abruptly at a wavelength of ~3.5 μm.

The evolution of pulses propagating in the anomalous dispersion region commences with soliton formation and is followed by the generation of dispersive waves in the normal dispersion region. The wavelength of the dispersive waves is determined by a phase matching condition (Husakou and Herrmann, 2001) and we calculated the phase-matched wavelengths and found that they were consistent with the predicted extent of the SCG shown above. We assumed that the seed pulse initially generates a high-order soliton, which subsequently decays into fundamental solitons because of higher order dispersion. The most powerful of the fundamental solitons has a peak power of $P_0 \times (2N_{sol} - 1)^2/N_{sol}^2$, where P_0 is the peak power of the input pulse and N_{sol} is the soliton order corresponding to the input pulse. The peak power of the most powerful fundamental soliton was used for the

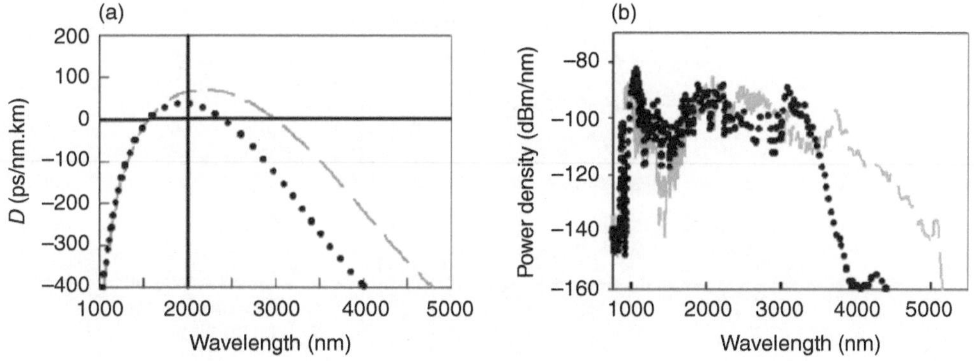

Fig. 6.10. (a) Dispersion profile of Bi glass circular rods (fibre cores) with radii of 0.9628 μm or 1.1529 μm and surrounding average refractive indices of n_{clad} = 1.6649 or 1.6206. The pump wavelength is between the lower ZDW and the upper ZDW (λ_{pump} = 2.0 μm). (b) Spectra for Bi glass MOFs using the dispersion profiles shown in (a). Dotted and dashed line-styles show corresponding spectra and dispersion profiles. (E_0 = 1 nJ, T_0 = 0.2 ps, fibre length = 40 mm, loss = Bi glass shown in Figure 6.6(a).) (Based on Price *et al.*, 2007; © 2007 IEEE.)

nonlinear phase matching term since it was this soliton which is mainly responsible for the generation of dispersive waves. The soliton experiences a recoil effect, which ensures energy conservation, and after sufficient propagation length, soliton recoil would be expected to cancel the SSFS (Skryabin *et al.*, 2003) so the dip in power close to the ZDW at 3 μm would remain. However, extreme slowing of the SSFS due to soliton recoil is associated with a threshold length, and was not observed for the short lengths of fibre simulated.

The negative dispersion slope coupled with anomalous group velocity dispersion (GVD) experienced near the long wavelength ZDW in MOFs with two ZDWs can lead to red-shifted dispersive waves (Dudley *et al.*, 2006). As demonstrated by the simulations above, if the seed pulse experiences a negative dispersion slope, the mid-IR SCG bandwidth is enhanced. Research in silica fibre based SCG has also considered fibres with two ZDWs for CW pumping and to maximise the power in particular wavelength ranges (e.g. Boucon *et al.*, 2008, Hilligsoe *et al.*, 2004, Mussot *et al.*, 2007, Travers *et al.*, 2008, Tse *et al.*, 2006a, 2006b). Fig. 6.11 shows a schematic of how the spectra produced by soliton propagation along a fibre change depending on whether the initial soliton wavelength experiences a positive or negative dispersion slope. With a positive dispersion slope, solitons tend to broaden temporally and this rapidly reduces the pulse bandwidth. If the solitons do not have any overlap with the phase-matched wavelengths above the second ZDW, then the limited benefit of the second ZDW is the slight reduction in the maximum dispersion value, and the continuum will not be broadened by phase-matching to

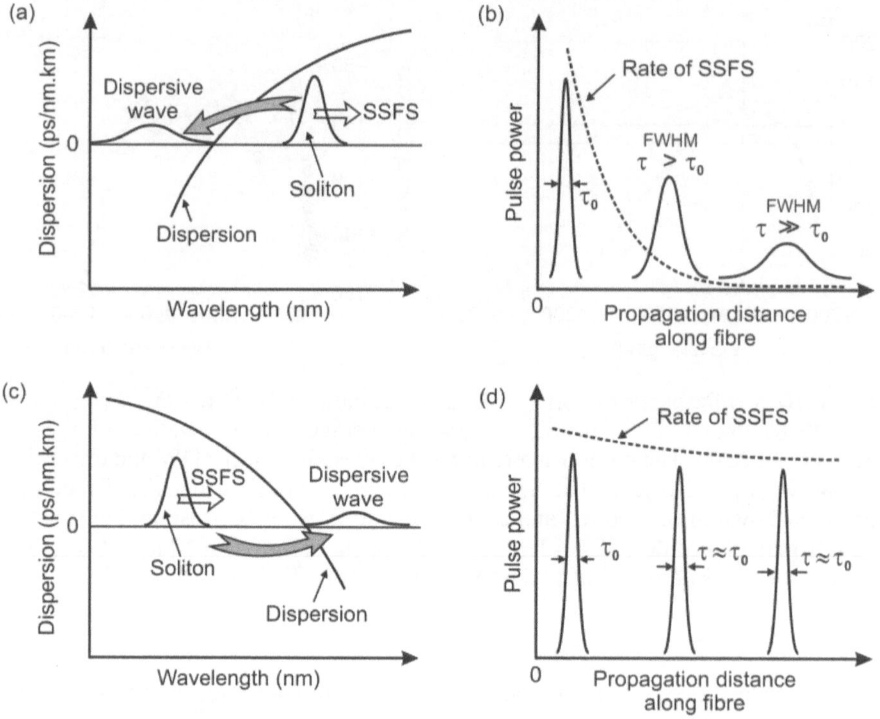

Fig. 6.11. (a) and (c) Illustration of soliton dynamics near the shorter and longer wavelength ZDW for the small core fibres discussed above. (b) and (d) Illustration of the pulse duration and rate of SSFS for the dispersion profiles that have positive or negative slopes of (a) and (b) near to the seed wavelength.

longer wavelengths. In contrast, with a negative dispersion slope, a short pulse duration can be maintained as losses that tend to broaden initially created solitons are compensated by the reduced dispersion. The large bandwidth solitons and SSFS generate dispersive radiation and rapidly transfer energy to mid-IR wavelengths.

In conclusion, modelling for small core Bi fibre illustrated that use of fibre with two appropriately spaced ZDWs can enable efficient mid-IR continuum generation from convenient pump wavelengths by phase-matching across the upper ZDW. More careful optimisation of the seed pulse characteristics and fibre dispersion should enable improved results and future work may consider engineered dispersion profiles along the fibre length obtained either by splicing short sections of fibres with different core dimensions or by using tapered fibres.

6.5.3 Large mode area fibre simulations

In order to increase the pulse energy it becomes necessary to work with fibres with a larger core area in order to avoid material damage. Since the dispersion in an LMA fibre is almost equal to the material dispersion, changing the dispersion profile for a given pump wavelength equates to the use of a different glass, which would also have different intrinsic nonlinearity as discussed in Section 6.2. Simulations for Bi glass and GLSO glass have been performed since these glasses have different nonlinear indices and different ZDWs. The simulated supercontinuum spectra are for fibres with $A_{\mathrm{eff}} = 100\ \mu\mathrm{m}^2$, which is $\sim$30 times greater than the small core Bi glass fibres considered above, and the LMA fibres have correspondingly reduced effective nonlinearity. We used the same seed pulse duration of 0.2 ps as for the small core simulations, and we maintained approximately the same initial peak intensity by scaling the pulse energy by the ratio of the mode areas: $(100\ \mu\mathrm{m}^2/3.05\ \mu\mathrm{m}^2) \times 1\,\mathrm{nJ} \approx 30$ nJ. Although the $A_{\mathrm{eff}} = 100\ \mu\mathrm{m}^2$ fibre does not correspond to a particular design, it is representative of what might be suitable for a low NA fibre providing approximately single mode operation (Feng *et al.*, 2004), and the results should be applicable to larger fibres provided that the pulse energy is scaled appropriately.

The supercontinuum spectra shown in Fig. 6.12 include the losses from Fig. 6.6. The chosen seed wavelength was 2.25 μm which is close to the ZDW of Bi glass (see Fig. 6.2(b)), and would be available from commercial OPOs. The spectrum for Bi glass extends only to $\sim$3.5 μm because the dispersion profile has a positive slope which prevents more extended broadening. Multiphonon losses rise rapidly to very high levels at wavelengths above $\sim$4.5 μm so high losses will dominate the spectral evolution if longer seed wavelengths enabled the supercontinuum to extend to these wavelengths. The supercontinuum spectrum of the GLSO fibre extends to $\sim$3.8 μm. The seed wavelength is significantly below the ZDW of GLSO so the dispersion of GLSO fibre in the continuum wavelength range is normal and

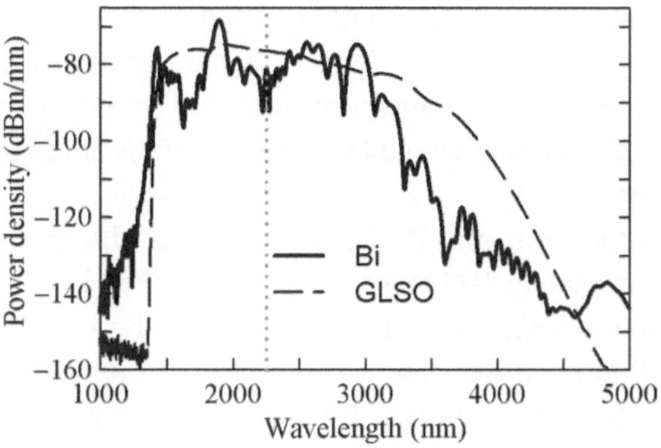

Fig. 6.12. Supercontinuum spectra predicted for large mode area fibres made from Bi glass and GLSO. Simulations include losses shown in Fig. 6.6. (Fibre length = 40 mm, $T_0 = 0.2$ ps, $E_0 = 30$ nJ, $\lambda_{\mathrm{seed}} = 2.25$ μm as shown by dotted vertical line.) (Based on Price *et al.*, 2007; © 2007 IEEE.)

SPM leads to a continuum that would have good temporal coherence. The spectral broadening in this fibre would be expected to increase steadily with the initial pulse peak power. This is different from our findings for anomalously dispersive fibres in which continuum broadening saturates above a given seed pulse energy. Pumping large mode GLSO fibre at longer wavelengths that are closer to the ZDW should produce a broader spectrum.

These results suggest that soft glass LMA fibres are promising candidates for high power mid-IR sources. When compared to small core MOFs, higher pulse energies are required due to the larger mode area and it is to be stressed that the dispersion profiles of LMA fibres are very similar to the dispersion profiles of the bulk materials. It is interesting that even with near-IR seed pulses, where the dispersion profiles of oxide glasses appear more favourable than those of chalcogenide glasses, the high nonlinearity of chalcogenide glasses is the dominant factor and ensures that the broadest continuum is obtained using these materials within our simulations.

6.6 Experimental demonstration of mid-IR supercontinuum generation with femtosecond pumping in high nonlinearity fibres

As an example of current state-of-the-art results with small core soft glass fibres, we show the broad bandwidth, mid-IR supercontinuum that has been demonstrated using a sub-cm (8 mm) length of highly nonlinear tellurite MOF (Domachuk *et al.*, 2008). The fibre was fabricated by extrusion and had a "wagon wheel" design with six very fine struts 16 μm long and 120 nm wide that supported a 2.5 μm diameter core. The 100 fs pump pulses from an OPO had an energy of 1.9 nJ at a wavelength of 1550 nm. Fig. 6.13 shows that the supercontinuum bandwidth of 4080 nm extended from 789 to 4870 nm measured at 20 dBm below the peak spectral power. The authors suggested that the short fibre length enabled smoother SC spectra, lower dispersion, and reduced material absorption at longer wavelengths. In addition, the pump wavelength is compatible with telecommunications hardware such as erbium doped fibre amplifiers.

The first demonstration of mid-IR SCG in chalcogenide glass MOF also used seed pulses of approximately 100 fs which were launched into a 1 metre length of an As-Se MOF (Shaw *et al.*, 2005). Fig. 6.14 shows the bandwidth of the initial pump beam at 2500 nm and the supercontinuum that extends from ~2100 nm to ~3200 nm. The As-Se MOF was fabricated by stacking hollow As-Se tubes and rods around a central As-Se rod and then that preform was drawn down to produce the As-Se MOF. The fibre diameter was ~130 μm and the core size was ~10 μm. The loss of the fibre at 1.5 μm was ~4.8 dB/m and the NA was ~0.4 to 0.5. The zero dispersion point for the fibre was estimated to be ~6 μm.

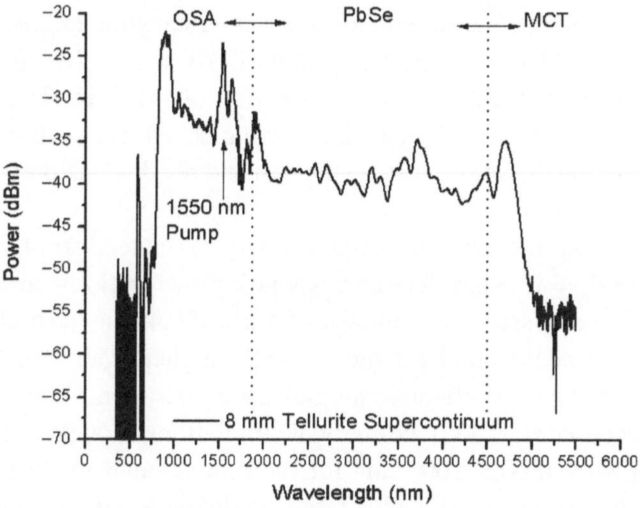

Fig. 6.13. Spectrum generated by 8 mm segment of tellurite PCF. The wavelength regions measured by an OSA and with a monochromator with lead selenide (PbSe) or mercury cadmium telluride (MCT) detectors are indicated. Also indicated is the pump wavelength of 1550 nm. (From Domachuk *et al.*, 2008; © 2008 OSA.)

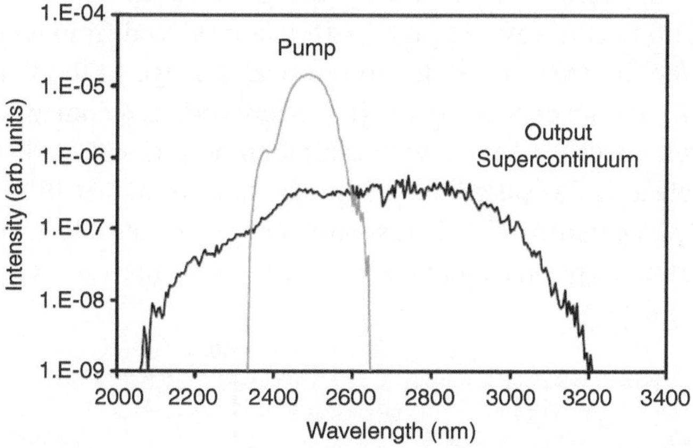

Fig. 6.14. Spectra of IR supercontinuum generated from 1 m of As-Se photonic crystal fibre. (From Shaw *et al.*, 2005; © 2005 OSA.)

6.7 Experimental demonstration of mid-IR supercontinuum generation with femtosecond pumping in large core fibres

Recent work at the University of Southampton reported the fabrication of a single-mode tellurite glass MOF with extremely large mode area (Feng *et al.*, 2008a, 2008b). The large mode tellurite MOF was fabricated from an extruded preform, with a core diameter of ~80 μm. Robust single-mode guidance at 1.55 μm was confirmed by both optical measurements and numerical simulations. The fibre had an attenuation of 2.9 dB/m at 1.55 μm and its ZDW was at 2.15 μm. Fig. 6.1(b) shows an optical photograph of the cross-section of the fibre. The uniform hole spacing of the fibre, Λ, was 53 μm, but owing to effects such as temperature gradients in the radial direction, surface tension and residual air-pressure inside the holes during the fibre-drawing process, the average hole diameter d_i (i is the ring number counted from the core outwards, $i = 1$ to 3) reduced from $d_1 = 28.1$ μm, $d_2 = 23.3$ μm, to $d_3 = 13.8$ μm. The corresponding d_i/Λ ratios were 0.53, 0.44 and 0.26 respectively, compared to the initial d/Λ ratio of 0.454 in the preform. The loss ratio between the fundamental LP_{01} mode and the higher order modes was $\sim 10^{-4}$, so differential loss enabled effective single-mode operation to be realised (Dong *et al.*, 2007, Grassi *et al.*, 2007). Numerical simulations showed that the effective mode area A_{eff} was 3000 ± 200 μm^2 for the fundamental mode LP_{01}, which is believed to be the first report of such a large single mode area in a non-silica glass fibre and was comparable to the record LMA of 3160 μm^2 reported by Dong *et al.* in a silica MOF (Dong *et al.*, 2007). The effective nonlinearity, γ, of the tellurite LMA MOF was calculated to be 0.23 W^{-1} km^{-1} at 1.55 μm.

Figure 6.15(b) shows the broadband SC between 1.0–2.5 μm observed using a 9 cm length of this fibre with 6 mW output (6 μJ pulse energy). The pump pulse duration was 120 fs at a wavelength of 2.15 μm produced from an OPA. During the course of the SC measurements, no physical damage to the fibre facets was observed. It was not understood why self-focusing-induced damage did not occur in the fibre even under the MW-level peak pump power, and while the reason(s) for this are unclear, it was suggested that it may have been that IR glasses such as tellurite and chalcogenides, with their smaller band-gap than silica, could generate a low-density electron plasma more readily under strong optical excitation, thereby

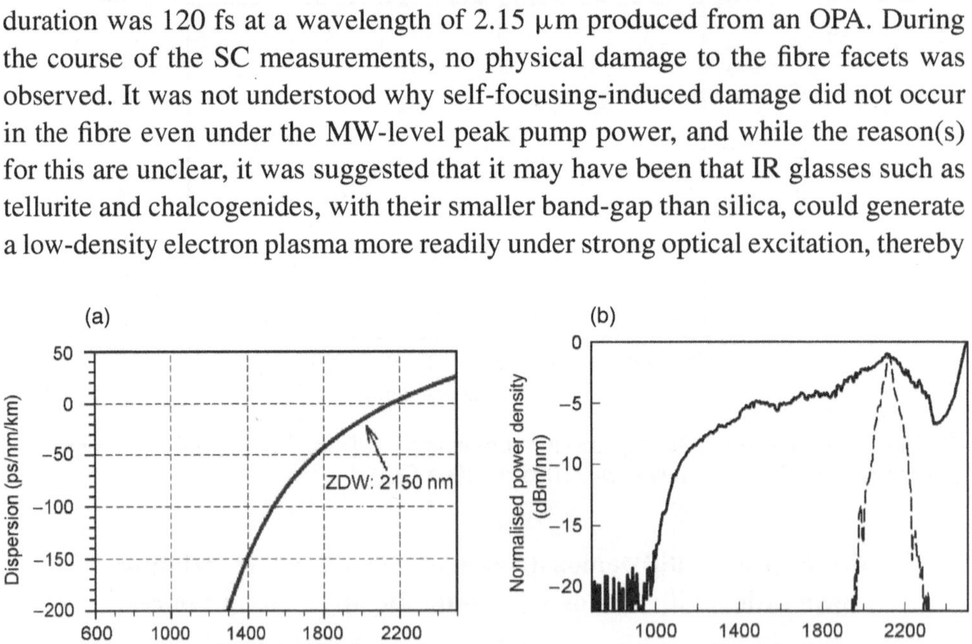

Fig. 6.15. (a) Dispersion profile of the tellurite LMA MOF. (b) SC spectra from a 9 cm long tellurite LMA MOF at both low incident power and for an incident pump power of 15.2 mW. (From Feng *et al.*, 2008a; © 2008 OSA.)

counteracting the self-focusing effect and preventing catastrophic optical damage (Ashcom *et al.*, 2006).

6.8 Experimental demonstration of mid-IR supercontinuum generation with nanosecond pumping

Mid-IR SCG results have been demonstrated using nanosecond pumping with a relatively low nonlinearity ZBLAN fibre. The fibre had a conventional core/clad structure and it is possible that further enhancements could be enabled if a dispersion tailored MOF structure were used. Fig. 6.16 shows the mid-infrared supercontinuum ranging from 0.8 μm to ~4.0 μm (Xia *et al.*, 2007). The 2 ns duration pump pulses were generated from an electrically modulated seed diode laser and a high power fibre amplifier cascade with a large mode erbium/ytterbium co-doped cladding-pumped fibre used in the final amplifier. By increasing the repetition rate of the system to 300 kHz, the average supercontinuum output power was as high as 1.3 W. The continuum was generated in a two-stage process. In the first stage, a length of single mode silica fibre enabled break up of the nanosecond pulses into femtosecond pulses by modulation instability. The spectrum was then broadened through the interplay of self-phase modulation, parametric four-wave mixing, and stimulated Raman scattering in the ZBLAN (ZrF_4-BaF_2-LaF_3-AlF_3-NaF...) fluoride fibre. The figure also shows the spectrum from similar duration pulses but with a 25% higher

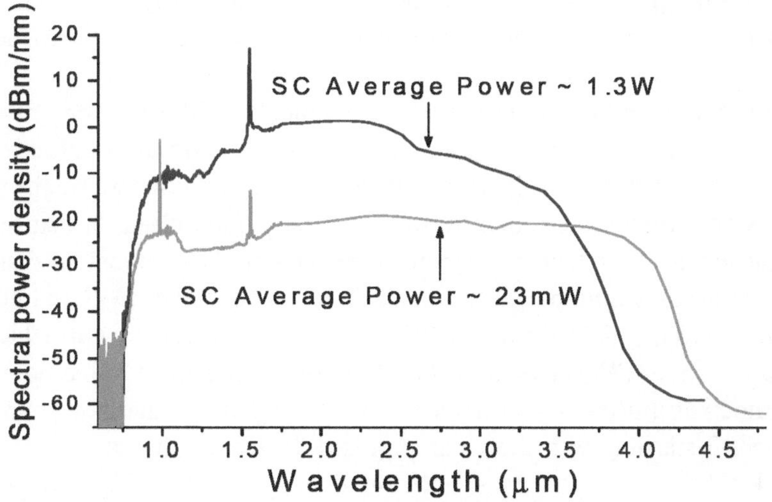

Fig. 6.16. SC spectrum from 3 m SMF fibre followed by 13 m ZBLAN fibre in high power setup, and from 1 m SMF fibre followed by 8 m ZBLAN fibre in low power setup. (From Xia *et al.*, 2007; © 2007 OSA.)

peak power, at lower repetition rate produced with a different choice of final fibre amplifier. Xia *et al.* suggest that the long wavelength bend-loss is an important consideration for the ZBLAN fibre design in this application. The variation of effective mode area at long wavelengths is also important since an enlarged mode area reduces the effective nonlinearity (Xia *et al.*, 2009).

6.9 Conclusions

This chapter has explained how soft glass fibre based sources with their highly directional output and moderate to high spectral power density across a broad bandwidth could become practical sources for a wide range of sensing and imaging applications. However, there is much research to be done in this emerging field, and reflecting that, the chapter contained sections which introduced the optical properties of the soft glasses and the fibre fabrication technology. A section was also devoted to the design techniques for these microstructured fibres.

The discussion of SCG began with several numerical simulations that showed how dispersion tailored fibres with two ZDWs enabled ultra-short seed pulses to evolve rapidly into solitons that could transfer energy across the upper ZDW to generate broad continuum spectra in short fibre lengths. Large mode fibres have a higher surface damage threshold pulse energy compared to small core fibres, and are potentially useful for high power applications. Simulations for larger mode area fibres suggested that they were also suitable for producing continuum provided that higher pulse energies are used due to the reduced nonlinearity compared to small core fibre designs and that the pump wavelength is above 2.25 μm due to the longer wavelength of the zero-dispersion point in large mode fibres.

Selected experimental results have been included to illustrate progress in the field. Whilst the majority of the results used femtosecond pump pulses, we note that the nanosecond pumped experimental results are in a regime where the material response is on a time-scale much shorter than the pump pulses. In that regard, if the seed pulse wavelength is in a region where the fibre has anomalous dispersion, the pulse evolution would be subject to modulation-instability (MI) and nonlinear pulse compression, which is similar to the build-up of pulses from noise that occurs in continuous wave (CW) pumped SCG. CW pumped mid-IR SCG has not yet been demonstrated yet it offers the attraction of much higher average power and lower cost than SCG starting with short pump pulses. However, the nanosecond pulses still have kW level peak powers, so moving to study CW pumping with powers in the range 1–10 W will require further research.

Moving beyond this study, the silica fibre based supercontinuum literature suggests that for a phase-stable mid-IR continuum, a pump source producing pulses shorter than perhaps 100 fs and using a sub-centimetre length fibre would be desirable. Then with a seed wavelength close to the ZDW in a fibre with a single ZDW, a broad bandwidth with high coherence could be produced. As with silica fibres, spectral fine-structuring would be expected, and any requirement to achieve strong conversion to particular wavelengths could be assisted by using sections of fibres with different dispersion properties, or by using tapered fibres. In contrast, when weak coherence is desirable, then a broad supercontinuum could be produced by soliton fission starting from longer duration seed pulses, and using longer lengths of MOF.

Future challenges relate to the uncertainties in the optical properties of the non-silica glasses at mid-IR wavelengths. The refractive index data is not widely available and this may be important at a detailed level when incorporating the variation of effective mode area in future numerical simulations. An important difference compared to silica fibre based SCG is the relationship between the photon energy at the pump wavelength compared to the band-gap of the material because for soft glasses the seed photon energy is typically close to, or in excess of, half the band-gap. Resonant enhancement of the electronic nonlinearity, and perhaps free carrier generation could therefore become important at high pump intensities, and IR photons can be subject to two-photon absorption (TPA). More glass characterisation data measured with pump wavelengths across the 2–5 μm range are required.

Looking ahead, the nonlinear applications of soft glasses is an interesting area of research that is gaining substantial attention from scientists in physics, chemistry, healthcare and defence. In addition, for communications applications, experiments have shown that tellurite-based glass is a promising candidate for slow light generation via stimulated Raman scattering (SRS) (Qin *et al.*, 2007). These glasses thus offer the prospect of providing sources that would enable applications in a wide range of scientific and industrial settings.

Acknowledgements

We thank Dr R. Paschotta of RP Photonics Consulting (www.rp-photonics.com) for providing his ProPulse software for pulse propagation modelling. We thank our collaborators in this work including Professor Tanya M. Monro, Dr Heike Ebendorff-Heidepriem, Dr Francesco Poletti, Dr Peter Horak, Dr Vittoria Finazzi, Dr Julie Y. Y. Leong, Dr Periklis Petropoulos, Dr Joanne C. Flanagan, Dr Gilberto Brambilla, Dr Wei H. Loh, Dr Angela Camerlingo, Dr Sonali Dasgupta, Ken E. Frampton, Nicholas M. White, Professor Harvey N. Rutt and Dr Xian Feng. We also acknowledge helpful discussions with Professor D. Hewak and Dr A. Favre. This work was supported in part by BAE SYSTEMS Virtual University (UK). J. H. V. Price is supported by a Royal Academy of Engineering/EPSRC (UK) research fellowship.

References

Aber, J. E., Newstein, M. C. & Garetz, B. A. (2000) Femtosecond optical Kerr effect measurements in silicate glasses. *Journal of the Optical Society of America B-Optical Physics*, **17**, 120–127.

Agrawal, G. P. (1995) *Nonlinear Fiber Optics*, San Diego, Academic Press.

Ashcom, J. B., Gattass, R. R., Schaffer, C. B. & Mazur, E. (2006) Numerical aperture dependence of damage and supercontinuum generation from femtosecond laser pulses in bulk fused silica. *Journal of the Optical Society of America B-Optical Physics*, **23**, 2317–2322.

Boucon, A., Alasia, D., Beugnot, J. C., Melin, G., Lempereur, S., Fleureau, A., Maillotte, H., Dudley, J. M. & Sylvestre, T. (2008) Supercontinuum generation from 1.35 to 1.7 microns by nanosecond pumping near the second zero-dispersion wavelength of a microstructured fiber. *IEEE Photonics Technology Letters*, **20**, 842–844.

Brady, D. J., Schweizer, T., Wang, J. & Hewak, D. W. (1998) Minimum loss predictions and measurements in gallium lanthanum sulphide based glasses and fibre. *Journal of Non-Crystalline Solids*, **242**, 92–98.

Brambilla, G., Koizumi, F., Finazzi, V. & Richardson, D. J. (2005) Supercontinuum generation in tapered bismuth silicate fibres. *Electronics Letters*, **41**, 795–797.

Brechet, F., Marcou, J., Pagnoux, D. & Roy, P. (2000) Complete analysis of the characteristics of propagation into photonic crystal fibers, by the finite element method. *Optical Fiber Technology*, **6**, 181–191.

Canat, G., Laverre, T., Lombard, L., Jolivet, V. & Bourdon, P. (2008) Influence of the wavelength dependence of the effective area on infrared supercontinuum generation. *2008 Conference on Lasers and Electro-Optics (CLEO).*

Chang, G. Q., Norris, T. B. & Winful, H. G. (2003) Optimization of supercontinuum generation in photonic crystal fibers for pulse compression. *Optics Letters*, **28**, 546–548.

Chen, Y. F., Beckwitt, K., Wise, F. W., Aitken, B. G., Sanghera, J. S. & Aggarwal, I. D. (2006) Measurement of fifth- and seventh-order nonlinearities of glasses. *Journal of the Optical Society of America B-Optical Physics*, **23**, 347–352.

Coen, S., Chau, A. H. L., Leonhardt, R., Harvey, J. D., Knight, J. C., Wadsworth, W. J. & St. J. Russell, P. (2002) Supercontinuum generation by stimulated Raman scattering and parametric four-wave mixing in photonic crystal fibers. *Journal of the Optical Society of America B-Optical Physics*, **19**, 753–764.

Cordeiro, C. M. B., Chillcce, E. F., Barbosa, L. C. & Cruz, C. H. B. (2006) Tellurite photonic crystal fiber made by a stack-and-draw technique. *Journal of Non-Crystalline Solids*, **352**, 3423–3428.

Cucinotta, A., Selleri, S., Vincetti, L. & Zoboli, M. (2002) Holey fiber analysis through the finite-element method. *IEEE Photonics Technology Letters*, **14**, 1530–1532.

Dadap, J. I., Panoiu, N. C., Chen, X. G., Hsieh, I. W., Liu, X. P., Chou, C. Y., Dulkeith, E., Mcnab, S. J., Xia, F. N., Green, W. M. J., Sekaric, L., Vlasov, Y. A. & Osgood, R. M. (2008) Nonlinear-optical phase modification in dispersion-engineered Si photonic wires. *Optics Express*, **16**, 1280–1299.

DeSalvo, R., Said, A. A., Hagan, D. J., Vanstryland, E. W. & Sheikbahae, M. (1996) Infrared to ultraviolet measurements of two-photon absorption and n(2) in wide bandgap solids. *IEEE Journal of Quantum Electronics*, **32**, 1324–1333.

Domachuk, P., Wolchover, N. A., Cronin-Golomb, M., Wang, A., George, A. K., Cordeiro, C. M. B., Knight, J. C. & Omenetto, F. G. (2008) Over 4000 nm bandwidth of mid-IR supercontinuum generation in sub-centimeter segments of highly nonlinear tellurite PCFs. *Opt Express*, **16**, 7161–7168.

Dong, L., Peng, X. & Li, J. (2007) Leakage channel optical fibers with large effective area. *Journal of the Optical Society of America B-Optical Physics*, **24**, 1689–1697.

Dudley, J. M. & Coen, S. (2002) Coherence properties of supercontinuum spectra generated in photonic crystal and tapered optical fibers. *Optics Letters*, **27**, 1180–1182.

Dudley, J. M., Genty, G. & Coen, S. (2006) Supercontinuum generation in photonic crystal fiber. *Reviews of Modern Physics*, **78**, 1135.

Ebendorff-Heidepriem, H. & Monro, T. M. (2007) Extrusion of complex preforms for microstructured optical fibers. *Optics Express*, **15**, 15086–15092.

Ebendorff-Heidepriem, H., Li, Y. & Monro, T. M. (2007) Reduced loss in extruded soft glass microstructured fibre. *Electronics Letters*, **43**, 1343–1345.

Ebendorff-Heidepriem, H., Petropoulos, P., Asimakis, S., Finazzi, V., Moore, R. C., Frampton, K., Koizumi, F., Richardson, D. J. & Monro, T. M. (2004) Bismuth glass holey fibers with high nonlinearity. *Optics Express*, **12**, 5082–5087.

Feng, X., Loh, W., Flanagan, J. C., Camerlingo, A., Dasgupta, S., Petropoulos, P., Horak, P., Frampton, K., White, N. M., Price, J. H., Rutt, H. N. & Richardson, D. J. (2008a) Single-mode tellurite glass holey fiber with extremely large mode area for infrared nonlinear applications. *Optics Express*, **16**, 13651–13656.

Feng, X., Loh, W. H., Camerlingo, A., Dasgupta, S., Flanagan, J. C., Petropoulos, P., Frampton, K. E., White, N. M., Rutt, H. N. & Richardson, D. J. (2008b) Broadband supercontinuum using single-mode/dual-mode tellurite glass holey fibers with large mode area. *2008 Conference on Lasers and Electro-Optics*. QELS.

Feng, X., Mairaj, A. K., Hewak, D. W. & Monro, T. M. (2004) Towards high-index glass based monomode holey fibre with large mode area. *Electronics Letters*, **40**, 167–169.

Feng, X., Mairaj, A. K., Hewak, D. W. & Monro, T. M. (2005a) Nonsilica glasses for holey fibers. *Journal of Lightwave Technology*, **23**, 2046–2054.

Feng, X., Monro, T. M., Finazzi, V., Moore, R. C., Frampton, K., Petropoulos, P. & Richardson, D. J. (2005b) Extruded singlemode, high-nonlinearity, tellurite glass holey fiber. *Electronics Letters*, **41**, 835–837.

Feng, X., Monro, T. M., Petropoulos, P., Finazzi, V. & Hewak, D. (2003) Solid microstructured optical fiber. *Optics Express*, **11**, 2225–2230.

Feng, X. A., Monro, T. M., Petropoulos, P., Finazzi, V. & Richardson, D. J. (2005c) Extruded single-mode high-index-core one-dimensional microstructured optical fiber with high index-contrast for highly nonlinear optical devices. *Applied Physics Letters*, **87**, 81110.

France, P. W., Carter, S. F., Moore, M. W. & Day, C. R. (1987) Progress in fluoride fibres for optical communications. *British Telecommunications Technology Journal*, **5**, 28–44.

Friberg, S. R. & Smith, P. W. (1987) Nonlinear optical-glasses for ultrafast optical switches. *IEEE Journal of Quantum Electronics*, **23**, 2089–2094.

Frosz, M. H., Falk, P. & Bang, O. (2005) The role of the second zero-dispersion wavelength in generation of supercontinua and bright-bright soliton-pairs across the zero-dispersion wavelength. *Optics Express*, **13**, 6181–6192.

Fu, L. B., Rochette, M., Ta'eed, V. G., Moss, D. J. & Eggleton, B. J. (2005) Investigation of self-phase modulation based optical regeneration in single mode As_2Se_3 chalcogenide glass fiber. *Optics Express*, **13**, 7637–7644.

Fujino, S. & Morinaga, K. (1997) Material dispersion and its compositional parameter of oxide glasses. *Journal of Non-Crystalline Solids*, **222**, 316–320.

Genty, G., Lehtonen, M., Ludvigsen, H., Broeng, J. & Kaivola, M. (2002) Spectral broadening of femtosecond pulses into continuum radiation in microstructured fibers. *Optics Express*, **10**, 1083–1098.

Gordon, J. P. (1986) Theory of the soliton self-frequency shift. *Optics Letters*, **11**, 662–664.

Grassi, A. M., Casagrande, F., D'Alessandro, M. & Marinoni, S. (2007) Single-modeness of short large mode area fibers: An experimental study. *Optics Communications*, **273**, 127–132.

Guiyao, Z., Zhiyun, H., Shuguang, L. & Lantian, H. (2006) Fabrication of glass photonic crystal fibres with die-cast process. *Applied Optics*, **45**, 4433–4436.

Hall, D. W., Newhouse, M. A., Borrelli, N. F., Dumbaugh, W. H. & Weidman, D. L. (1989) Nonlinear Optical Susceptibilities of High-Index Glasses. *Applied Physics Letters*, **54**, 1293–1295.

Harbold, J. M., Ilday, F. O., Wise, F. W., Sanghera, J. S., Nguyen, V. Q., Shaw, L. B. & Aggarwal, I. D. (2002) Highly nonlinear As-S-Se glasses for all-optical switching. *Optics Letters*, **27**, 119–121.

Hartl, I., Li, X. D., Chudoba, C., Ghanta, R. K., Ko, T. H., Fujimoto, J. G., Ranka, J. K. & Windeler, R. S. (2001) Ultrahigh-resolution optical coherence tomography using continuum generation in an air–silica microstructure optical fiber. *Optics Letters*, **26**, 608–610.

Herrmann, J., Griebner, U., Zhavoronkov, N., Husakou, A., Nickel, D., Knight, J. C., Wadsworth, W. J., St. J. Russell, P. & Korn, G. (2002) Experimental evidence for supercontinuum generation by fission of higher-order solitons in photonic fibers. *Physical Review Letters*, **88**, 173901.

Hilligse, K. M., Andersen, T. V., Paulsen, H. N., Nielsen, C. K., Molmer, K., Keiding, S., Kristiansen, R., Hansen, K. P. & Larsen, J. J. (2004) Supercontinuum generation in a photonic crystal fiber with two zero dispersion wavelengths. *Optics Express*, **12**, 1045–1054.

Holzwarth, R., Udem, T., Hansch, T. W., Knight, J. C., Wadsworth, W. J. & St.J.Russell, P. (2000) Optical frequency synthesizer for precision spectroscopy. *Physical Review Letters*, **85**, 2264–2267.

Hult, J. (2007) A fourth-order Runge–Kutta in the interaction picture method for, simulating supercontinuum generation in optical fibers. *Journal of Lightwave Technology*, **25**, 3770–3775.

Husakou, A. V. & Herrmann, J. (2001) Supercontinuum generation of higher-order solitons by fission in photonic crystal fibers. *Physical Review Letters*, **87**, 203901.

Kalashnikov, V. L., Sorokin, E. & Sorokina, I. T. (2007) Raman effects in the infrared supercontinuum generation in soft-glass PCFs. *Applied Physics B-Lasers and Optics*, **87**, 37–44.

Kang, I., Krauss, T. D., Wise, F. W., Aitken, B. G. & Borrelli, N. F. (1995) Femtosecond measurement of enhanced optical nonlinearities of sulfide glasses and heavy-metal-doped oxide glasses. *Journal of the Optical Society of America B-Optical Physics*, **12**, 2053–2059.

Kang, I., Smolorz, S., Krauss, T., Wise, F., Aitken, B. G. & Borrelli, N. F. (1996) Time-domain observation of nuclear contributions to the optical nonlinearities of glasses. *Physical Review B*, **54**, 12641–12644.

Kerrinckx, E., Bigot, L., Douay, M. & Quiquempois, Y. (2004) Photonic crystal fiber design by means of a genetic algorithm. *Optics Express*, **12**, 1990–1995.

Kiang, K. M., Frampton, K., Monro, T. M., Moore, R., Tucknott, J., Hewak, D. W., Richardson, D. J. & Rutt, H. N. (2002) Extruded singlemode non-silica glass holey optical fibres. *Electronics Letters*, **38**, 546–547.

Kikuchi, K., Taira, K. & Sugimoto, N. (2002) Highly nonlinear bismuth oxide-based glass fibres for all-optical signal processing. *Electronics Letters*, **38**, 166–167.

Kim, J. H., Chen, M. K., Yang, C. E., Lee, J., Yin, S., Ruffin, P., Edwards, E., Brantley, C. & Luo, C. (2008) Broadband IR supercontinuum generation using single crystal sapphire fibers. *Optics Express*, **16**, 4085–4093.

Knight, J. C. (2003) Photonic crystal fibres. *Nature*, **424**, 847–851.

Koshiba, M. (2002) Full-vector analysis of photonic crystal fibers using the finite element method. *IEICE Transactions on Electronics*, **E85C**, 881–888.

Kumar, V., George, A. K., Knight, J. C. & Russell, P. St. (2003) Tellurite photonic crystal fiber. *Optics Express*, **11**, 2641–2645.

Laegsgaard, J. (2007) Mode profile dispersion in the generalized nonlinear Schrodinger equation. *Optics Express*, **15**, 16110–16123.

Lamont, M. R. E., Rochette, M., Moss, D. J. & Eggleton, B. J. (2006) Two-photon absorption effects on self-phase-modulation-based 2R optical regeneration. *IEEE Photonics Technology Letters*, **18**, 1185–1187.

Leong, J. Y. Y. (2007) Fabrication and applications of lead-silicate glass holey fibre for 1–1.5 microns: Nonlinearity and dispersion trade offs, PhD Thesis, University of Southampton.

Leong, J. Y. Y., Asimakis, S., Poletti, F., Petropoulos, P., Feng, X., Moore, R. C., Frampton, K. E., Monro, T. M., Ebendorff-Heidepriem, H., Loh, W. H. & Richardson, D. J. (2005a) Towards zero dispersion highly nonlinear lead silicate glass holey fibres at 1550 nm by structured-element-stacking. *ECOC, 31st European Conference on Optical Communications, Post Deadline paper TH.4.4.5.* Glasgow, UK, IEE.

Leong, J. Y. Y., Asimakis, S., Poletti, F., Petropoulos, P., Feng, X., Moore, R. C., Frampton, K. E., Monro, T. M., Ebendorff-Heidepriem, H., Loh, W. H. & Richardson, D. J. (2006a) Nonlinearity and dispersion control in small core lead silicate holey fibers by structured element stacking. *2006 Optical Fiber Communication Conference/National Fiber Optic Engineers Conference, Vols 1–6*, 199–201.

Leong, J. Y. Y., Petropoulos, P., Asimakis, S., Ebendorff-Heidepriem, H., Moore, R. C., Frampton, K., Finazzi, V., Feng, X., Price, J. H. V., Monro, T. M. & Richardson, D. J. (2005b) A lead silicate holey fiber with $\gamma = 1860\ \mathrm{W}^{-1}\mathrm{km}^{-1}$ at 1550 nm. *Optical Fiber Communications Conference (OFC).* Los Angeles, CA, paper PDP22.

Leong, J. Y. Y., Petropoulos, P., Price, J. H. V., Ebendorff-Heidepriem, H., Asimakis, S., Moore, R. C., Frampton, K. E., Finazzi, V., Feng, X., Monro, T. M. & Richardson, D. J. (2006b) High-nonlinearity, dispersion-shifted lead-silicate holey fibers for efficient 1-μm pumped supercontinuum generation. *Journal of Lightwave Technology*, **24**, 183–190.

Lu, F., Deng, Y. J. & Knox, W. H. (2005) Generation of broadband femtosecond visible pulses in dispersion-micromanaged holey fibers. *Optics Letters*, **30**, 1566–1568.

Mairaj, A. K., Petrovich, M. N., West, Y. W., Fu, A., Harwood, D. W. J., Ng, L. N., Monro, T. M., Broderick, N. G. & Hewak, D. W. (2001) Advances in gallium lanthanum sulphide glass for optical fibre and devices. *Fiber Optic Sensor Technology II*, **4204**, 278–286.

Méndez, A. & Morse, T. F. (2007) *Specialty Optical Fibers Handbook*, Academic Press.

Mitschke, F. M. & Mollenauer, L. F. (1986) Discovery of the soliton self-frequency shift. *Optics Letters*, **11**, 659–661.

Moeser, J. T., Wolchover, N. A., Knight, J. C. & Omenetto, F. G. (2007) Initial dynamics of supercontinuum generation in highly nonlinear photonic crystal fiber. *Optics Letters*, **32**, 952–954.

Mogilevtsev, D., Birks, T. A. & Russell, P. St. (1998) Group-velocity dispersion in photonic crystal fibers. *Optics Letters*, **23**, 1662–1664.

Monro, T. M. (2006) Progress in non-silica microstructured fibres. *Photonics West, paper 6116–01.* SPIE.

Monro, T. M., Richardson, D. J., Broderick, N. G. R. & Bennett, P. J. (1999) Holey optical fibers: An efficient modal model. *Journal of Lightwave Technology*, **17**, 1093–1102.

Montant, S., Le Calvez, A., Freysz, E., Ducasse, A. & Couzi, M. (1998) Time-domain separation of nuclear and electronic contributions to the third-order nonlinearity in glasses. *Journal of the Optical Society of America B-Optical Physics*, **15**, 2802–2807.

Mori, A. (2008) Tellurite-based fibers and their applications to optical communication networks. *Journal of the Ceramic Society of Japan*, **116**, 1040–1051.

Mori, A., Ohishi, Y. & Sudo, S. (1997) Erbium-doped tellurite glass fibre laser and amplifier. *Electronics Letters*, **33**, 863–864.

Mori, A., Shikano, K., Enbutsu, W., Oikawa, K., Naganuma, K., Kato, M. & Aozasa, S. (2004) 1.5 μm band zero-dispersion shifted tellurite photonic crystal fibre with a nonlinear coefficient γ of 675 $W^{-1}km^{-1}$. *European Conference on Optical Communications (ECOC), paper Th3.3.6*. Sweden, IEE.

Moulton, P. F., Rines, G. A., Slobodtchikov, E. V., Wall, K. F., Frith, G., Samson, B. & Carter, A. L. G. (2009) Tm-doped fiber lasers: Fundamentals and power scaling. *IEEE Journal of Selected Topics in Quantum Electronics*, **15**, 85–92.

Mussot, A., Beaugeois, M., Bouazaoui, M. & Sylvestre, T. (2007) Tailoring CW supercontinuum generation in microstructured fibers with two-zero dispersion wavelengths. *Optics Express*, **15**, 11553–11563.

Pelusi, M. D., Ta'eed, V. G., Fu, L. B., Magi, E., Lamont, M. R. E., Madden, S., Choi, D. Y., Bulla, D. A. P., Luther-Davies, B. & Eggleton, B. J. (2008) Applications of highly-nonlinear chalcogenide glass devices tailored for high-speed all-optical signal processing. *IEEE Journal of Selected Topics in Quantum Electronics*, **14**, 529–539.

Petropoulos, P., Ebendorff-Heidepriem, H., Finazzi, V., Moore, R. C., Frampton, K., Richardson, D. J. & Monro, T. M. (2003) Highly nonlinear and anomalously dispersive lead silicate glass holey fibers. *Optics Express*, **11**, 3568–3573.

Petropoulos, P., Monro, T. M., Belardi, W., Furusawa, K., Lee, J. H. & Richardson, D. J. (2001) 2R-regenerative all-optical switch based on a highly nonlinear holey fiber. *Optics Letters*, **26**, 1233–1235.

Poletti, F., Finazzi, V., Monro, T. M., Broderick, N. G. R., Tse, V. & Richardson, D. J. (2005) Inverse design and fabrication tolerances of ultra-flattened dispersion holey fibers. *Optics Express*, **13**, 3728–3736.

Poletti, F., Furusawa, K., Yusoff, Z., Broderick, N. G. R. & Richardson, D. J. (2007) Nonlinear tapered holey fibers with high stimulated Brillouin scattering threshold and controlled dispersion. *Journal of the Optical Society of America B-Optical Physics*, **24**, 2185–2194.

Price, J. H., Monro, T. M., Ebendorff-Heidepriem, H., Poletti, F., Horak, P., Finazzi, V., Leong, J. Y. Y., Petropoulos, P., Flanagan, J. C., Brambilla, G., Feng, X. & Richardson, D. J. (2007) Mid-IR supercontinuum generation from non-silica microstructured optical fibers. *IEEE Journal of Selected Topics in Quantum Electronics*, **3**, 738–749.

Price, J. H., Monro, T. M., Furusawa, K., Belardi, W., Baggett, J. C., Coyle, S. J., Netti, C., Baumberg, J. J., Paschotta, R. & Richardson, D. J. (2003) UV generation in a pure silica holey fiber. *Applied Physics B-Lasers and Optics*, **77**, 291–298.

Price, J. H. V., Monro, T. M., Ebendorff-Heidepriem, H., Poletti, F., Finazzi, V., Leong, J. Y. Y., Petropoulos, P., Flanagan, J. C., Brambilla, G., Feng, X. & Richardson, D. J. (2006) Non-silica microstructured optical fibers for mid-IR supercontinuum generation from 2μm–5μm (Invited Paper). In Brown, A. J. W., Nilsson, J., Harter, D. J. & Tünnermann, A. (Eds.) *Fiber Lasers III, Photonics West, paper 6102-A1020*. USA, SPIE-Int. Society Optical Engineering.

Qin, G. S., Jose, R. & Ohishi, Y. (2007) Stimulated Raman scattering in tellurite glasses as a potential system for slow light generation. *Journal of Applied Physics*, **101**, 093109.

Ranka, J. K., Windeler, R. S. & Stentz, A. J. (2000) Visible continuum generation in air–silica microstructure optical fibers with anomalous dispersion at 800 nm. *Optics Letters*, **25**, 25–27.

Requejo-Isidro, J., Mairaj, A. K., Pruneri, V., Hewak, D. W., Netti, M. C. & Baumberg, J. J. (2003) Self refractive non-linearities in chalcogenide based glasses. *Journal of Non-Crystalline Solids*, **317**, 241–246.

Riishede, J. & Sigmund, O. (2008) Inverse design of dispersion compensating optical fiber using topology optimization. *Journal of the Optical Society of America B-Optical Physics*, **25**, 88–97.

Rivero, C., Stegeman, R., Couzi, M., Talaga, D., Cardinal, T., Richardson, K. & Stegeman, G. (2005) Resolved discrepancies between visible spontaneous Raman cross-section and direct near-infrared Raman gain measurements in TeO_2-based glasses. *Optics Express*, **13**, 4759–4769.

Said, A. A., Sheikbahae, M., Hagan, D. J., Wei, T. H., Wang, J., Young, J. & Vanstryland, E. W. (1992) Determination of bound-electronic and free-carrier nonlinearities in ZnSe, GaAs, CdTe, and ZnTe. *Journal of the Optical Society of America B-Optical Physics*, **9**, 405–414.

Sazio, P. J. A., Amezcua-Correa, A., Finlayson, C. E., Hayes, J. R., Scheidemantel, T. J., Baril, N. F., Jackson, B. R., Won, D. J., Zhang, F., Margine, E. R., Gopalan, V., Crespi, V. H. & Badding, J. V. (2006) Microstructured optical fibers as high-pressure microfluidic reactors. *Science*, **311**, 1583–1586.

Seddon, A., Furniss, D., Zheng, L., Bekker, E. & Benson, T. (2008) Review of novel glass photonic crystal fibres. *2008 IEEE/LEOS Winter Topical Meeting Series*, 208–209.

Serebryannikov, E. E., Rivero, C., Stegeman, R. & Zheltikov, A. A. (2007) Soliton transients and supercontinuum generation in high-Raman-gain materials. *Journal of the Optical Society of America B-Optical Physics*, **24**, 137–141.

Shaw, L. B., Thielen, P. A., Kung, F. H., Nguyen, V. Q., Sanghera, J. S. & Aggarwal, I. D. (2005) IR Supercontinuum Generation in As-Se Photonic Crystal Fiber. *Conference on Advanced Solid State Lasers (ASSL), Paper TuC5.*

Sheikbahae, M., Said, A. A., Wei, T. H., Hagan, D. J. & Vanstryland, E. W. (1990) Sensitive measurement of optical nonlinearities using a single beam. *IEEE Journal of Quantum Electronics*, **26**, 760–769.

Silvestre, E., Andres, M. V. & Andres, P. (1998) Biorthonormal-basis method for the vector description of optical-fiber modes. *Journal of Lightwave Technology*, **16**, 923–928.

Skryabin, D. V., Luan, F., Knight, J. C. & Russell, P. St. (2003) Soliton self-frequency shift cancellation in photonic crystal fibers. *Science*, **301**, 1705–1708.

Smart, R. G., Hanna, D. C., Tropper, A. C., Davey, S. T., Carter, S. F. & Szebesta, D. (1991) CW room-temperature upconversion lasing at blue, green and red wavelengths in infrared-pumped Pr^{3+}-doped fluoride fiber. *Electronics Letters*, **27**, 1307–1309.

Snyder, A. W. & Love, J. D. (1983) *Optical Waveguide Theory*. Chapman & Hall, London.

Stolen, R. H., Gordon, J. P., Tomlinson, W. J. & Haus, H. A. (1989) Raman response function of silica-core fibers. *Journal of the Optical Society of America B-Optical Physics*, **6**, 1159–1166.

Travers, J. C., Rulkov, A. B., Cumberland, B. A., Popov, S. V. & Taylor, J. R. (2008) Visible supercontinuum generation in photonic crystal fibers with a 400 W continuous wave fiber laser. *Optics Express*, **16**, 14435–14447.

Troles, J., Brilland, L., Smektala, F., Traynor, N., Houizot, P. & Desevedavy, F. (2007) Chalcogenide photonic crystal fibers for near and middle infrared applications. *2007 International Conference on Transparent Optical Networks*, 297–300.

Tse, M. L. V., Horak, P., Poletti, F., Broderick, N. G., Price, J. H. V., Hayes, J. R. & Richardson, D. J. (2006a) Supercontinuum generation at 1.06 microns in holey fibers with dispersion flattened profiles. *Optics Express*, **14**, 4445–4451.

Tse, M. L. V., Horak, P., Poletti, F. & Richardson, D. J. (2008) Designing tapered holey fibers for soliton compression. *IEEE Journal of Quantum Electronics*, **44**, 192–198.

Tse, M. L. V., Horak, P., Price, J. H. V., Poletti, F., He, F. & Richardson, D. J. (2006b) Pulse compression at 1.06 μm in dispersion-decreasing holey fibers. *Optics Letters*, **31**, 3504–3506.

Udem, T., Holzwarth, R. & Hansch, T. W. (2002) Optical frequency metrology. *Nature*, **416**, 233–237.

Varshney, S. K., Florous, N. J., Saltolt, K., Koshiba, M. & Fujisawa, T. (2007) Numerical investigation and optimization of a photonic crystal fiber for simultaneous dispersion compensation over S+C+L wavelength bands. *Optics Communications*, **274**, 74–79.

Vogel, E. M., Weber, M. J. & Krol, D. M. (1991) Nonlinear optical phenomena in glass. *Physics and Chemistry of Glasses*, **32**, 231–254.

Vukovic, N., Broderick, N. G. R., Petrovich, M. N. & Brambilla, G. (2008) Novel method for the fabrication of long optical fiber tapers. *IEEE Photonics Technology Letters*, **20**, 1264–1266.

White, T. P., Kuhlmey, B. T., Mcphedran, R. C., Maystre, D., Renversez, G., De Sterke, C. M. & Botten, L. C. (2002) Multipole method for microstructured optical fibers. I. Formulation. *Journal of the Optical Society of America B-Optical Physics*, **19**, 2322–2330.

White, T. P., Mcphedran, R. C., De Sterke, C. M., Botten, L. C. & Steel, M. J. (2001) Confinement losses in microstructured optical fibers. *Optics Letters*, **26**, 1660–1662.

Xia, C., Kumar, M., Cheng, M.-Y., Kulkarni, O. P., Alexander, V., Islam, M., Galvanauskas, A., Terry, F. L., Freeman, M. J., Polulain, M. & Mazé, G. (2006a) 0.8–4.5 microns supercontinuum generation in ZBLAN fluoride fibers scaled up to 1.25 W power. *Conference on Lasers and Electro-optics.* Longbeach, CA, Postdeadline paper CPDA10, OSA.

Xia, C., Kumar, M., Cheng, M. Y., Hegde, R. S., Islam, M. N., Galvanauskas, A., Winful, H. G., Terry, F. L., Freeman, M. J., Poulain, M. & Maze, G. (2007) Power scalable mid-infrared supercontinuum generation in ZBLAN fluoride fibers with up to 1.3 watts time-averaged power. *Optics Express*, **15**, 865–871.

Xia, C. A., Kumar, M., Kulkarni, O. R., Islam, M. N., Terry, F. L. & Freeman, M. J. (2006b) Mid-infrared supercontinuum generation to 4.5 μm in ZBLAN fluoride fibers by nanosecond diode pumping. *Optics Letters*, **31**, 2553–2555.

Xia, C. A., Xu, Z., Islam, M. N., Terry (JR.), F. L., Freeman, M. J., Zakel, A. & Mauricio, J. (2009) 10.5 W time-averaged power mid-IR supercontinuum generation extending beyond 4 μm with direct pulse pattern modulation. *IEEE Journal of Selected Topics in Quantum Electronics*, **15**, 422–434.

Xu, F., Pruneri, V., Finazzi, V. & Brambilla, G. (2008) An embedded optical nanowire loop resonator refractometric sensor. *Optics Express*, **16**, 1062–1067.

Yeom, D. I., Maegi, E. C., Lamont, M. R. E., Roelens, M. A. F., Fu, L. B. & Eggleton, B. J. (2008) Low-threshold supercontinuum generation in highly nonlinear chalcogenide nanowires. *Optics Letters*, **33**, 660–662.

7

Increasing the blue-shift of a picosecond pumped supercontinuum

M. H. Frosz, P. M. Moselund, P. D. Rasmussen,
C. L. Thomsen and O. Bang

7.1 Introduction

The first experiments with supercontinuum generation in a photonic crystal fibre (PCF) demonstrated impressive spectra spanning from 400 nm to 1500 nm using 100 fs pulses (Ranka et al., 2000). Often, one does not require the use of the entire supercontinuum bandwidth, or the spectrum needs to be concentrated in a specific spectral region where other lasers are not readily available. One method is to use the soliton self-frequency shift to simply red-shift a laser pulse over a desired wavelength range, which can be done over 900 nm (Chan et al., 2008). This provides a basis for tunable lasers with applications including broadband spectroscopy (Walewski et al., 2004), and coherent anti-Stokes Raman scattering (CARS) microspectroscopy (Andresen et al., 2007). ZBLAN fluoride (a mixture of zirconium, barium, lanthanum, aluminium, and sodium fluorides) fibres have been used to extend a supercontinuum spectrum beyond 4.5 μm with potential applications in spectroscopy (Xia et al., 2006). Besides these examples of generating light in the near- or mid-infrared, one also finds examples of generating light in the ultraviolet–blue region of the spectrum. This wavelength region is highly interesting for several reasons. Primarily, many fluorescent molecules are excited in a wavelength range from ~600 nm down to ~350 nm (Prasad, 2003). Supercontinuum light sources covering this wavelength range are highly useful for fluorescence microscopy. In particular, a high spectral density over a broad wavelength range removes the need for using several lasers, each corresponding to the excitation wavelength of a specific fluorescent molecule. Commercial systems based on supercontinuum generation have become available, allowing continuous tuning of the excitation wavelength (Jalink et al., 2008). Another application is absorption spectroscopy in the ultraviolet–visible region.

Supercontinuum Generation in Optical Fibers, ed. by J. M. Dudley and J. R. Taylor. Published by Cambridge University Press.

In this chapter we consider different approaches to achieve a supercontinuum with high spectral density in the blue region of the spectrum. Section 7.2 briefly reviews some of the recent work on the subject and the physical mechanisms described in the literature. The propagation equation used for numerical simulations in Sections 7.4 and 7.5 is also described. Building on the understanding of the physical mechanisms involved, Section 7.3 investigates how modifying the PCF structure can enhance the blue part of the supercontinuum spectrum. In Section 7.4 we show examples of how modifying the glass composition of the fibre material can also be used to achieve this. Section 7.5 considers the use of two coupled fibres with significantly different dispersion profiles. Finally, in Section 7.6 we conclude this chapter and also take a look at the most recent innovations that have been used to shift the supercontinuum spectrum towards shorter wavelengths.

7.2 Physical mechanisms

Considering the complexity of supercontinuum generation in general, it is not surprising that there is more than one way of achieving particular spectral features, such as significant spectral power in the visible region of the spectrum. Exploiting the high index contrast of pure silica PCFs with air-holes, one method is to use multimode phase-matched harmonic generation (Omenetto et al., 2001; Price et al., 2003; Efimov et al., 2003).

We will here focus on a method which is inherently linked to the group-velocity profile of the PCF. The physical mechanisms underlying this method were first tentatively explained by Tartara et al. (2003), and recently this explanation has been firmly supported experimentally in the work by Stone and Knight (2008). We assume the use of a picosecond pump, since these sources typically have a lower cost than femtosecond sources, and also provide a high average power. The first step then involves breaking the pump pulse up into solitons no longer than ~100 fs, so that their red-shift is reasonably fast along the fibre (Agrawal, 2007). This can be achieved by pumping at a wavelength with anomalous dispersion but close to the zero-dispersion wavelength (ZDW) of the PCF. As described in Chapter 3, modulation instability leads to the build-up of fast temporal oscillations on the picosecond pulse, with an oscillation period given by (Agrawal, 2007)

$$T_{\mathrm{MI}} = \sqrt{\frac{2\pi^2 |\beta_2|}{\gamma P_0}}, \tag{7.1}$$

where β_2 is the group-velocity dispersion at the pump wavelength, γ is the nonlinear parameter of the fibre (defined in Chapter 3), and P_0 is the pump peak power. The modulated picosecond pulse can then break up into a train of solitons with durations approximately given by T_{MI}. If the soliton duration is sufficiently short the solitons can immediately start to red-shift due to intrapulse Raman scattering, also known as soliton self-frequency shift (Mitschke and Mollenauer, 1986; Gordon, 1986). If the solitons do not initially have sufficiently short duration for significant red-shift, they can undergo soliton collisions which transfer energy (due to Raman scattering) between solitons so that some of them obtain higher peak power, become shorter, and then undergo a larger red-shift (Islam et al., 1989; Frosz et al., 2006a). The rate of frequency shift per propagation length for solitons with duration T_0 is given by (Gordon, 1986; Agrawal, 2007)

$$\frac{\mathrm{d}\nu_0}{\mathrm{d}z} = -\frac{8\,|\beta_2|\,T_{\mathrm{R}}}{2\pi\,15T_0^4}, \tag{7.2}$$

where $T_{\mathrm{R}} \approx 3$ fs is related to the slope of the Raman gain spectrum (Gordon, 1986; Agrawal, 2007).

While the solitons are formed from the pump pulse in the anomalous dispersion region, the solitons transfer energy into the normal dispersion region in the form of dispersive waves (Akhmediev and Karlsson, 1995). Assuming that the pump wavelength is located close to a ZDW where there is normal dispersion on the short-wavelength side of the ZDW, we now have red-shifting solitons on the long-wavelength side of the pump and dispersive waves on the short-wavelength side. It is found that for the cases investigated here, the dispersive waves initially have a lower group-velocity than the solitons and will therefore lag behind the solitons. However, as the solitons gradually red-shift they will also experience a reduction in group-velocity and eventually meet with the dispersive waves. The temporal overlap between a soliton and a dispersive wave allows them to interact nonlinearly. Two mechanisms have been suggested for this interaction. First, the temporal overlap with the high power soliton leads to cross-phase modulation (XPM) (Agrawal, 2007) of the dispersive wave. Since the dispersive wave is located on the trailing edge of the soliton, the XPM leads to a blue-shift of the dispersive wave (Genty et al., 2004, 2005). Second, a type of four-wave mixing (FWM) can occur between the soliton and the dispersive wave which leads to the generation of energy at wavelengths shorter than the dispersive wave (Gorbach et al., 2006). For both of the two methods, the blue-shifted radiation experiences a decrease in group-velocity and will therefore lag behind the soliton. But as long as the soliton can continue its red-shift, it will also decrease its group-velocity until it again meets with the blue-shifted radiation. The blue-shift mechanism can therefore take place continuously, as long as the soliton is able to red-shift and decrease its group-velocity (note that solitons exist in the anomalous dispersion region, where a red-shift implies a decrease in group-velocity). Since the blue-shifted radiation cannot escape from

Table 7.1. *Structural parameters for five different PCFs investigated.*

Pull	Pitch Λ [μm]	Relative hole size d/Λ
1	2.42	0.46
2	2.18	0.45
3	2.01	0.45
4	1.81	0.45
5	2.63	0.48

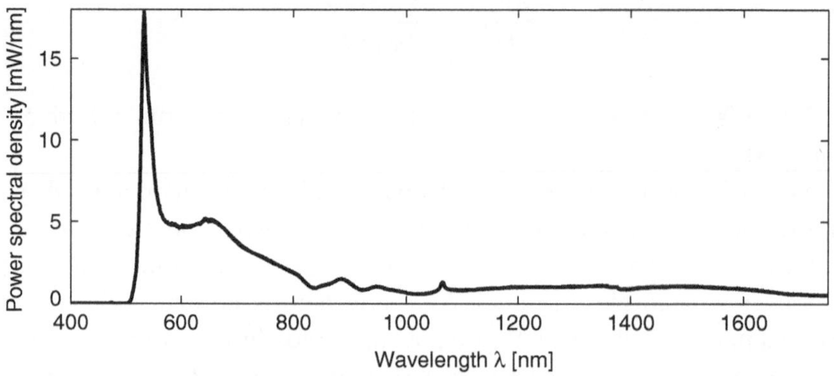

Fig. 7.1. Experimentally measured spectrum using 5 m of Pull 1. The input pulses have a centre wavelength of 1064 nm, 8 ps FWHM duration, and 9.7 kW peak power. A strong peak is clearly seen at 533 nm.

the soliton as long as this cycle continues, this process has also been termed a "trapping effect" (Nishizawa and Goto, 2002; Gorbach and Skryabin, 2007). With this mechanism the maximum blue-shift is limited by the ability of the solitons to red-shift. The red-shift can be limited by the increase in both material loss and confinement loss for wavelengths in the near-infrared. Depending on the structural design of the PCF, the existence of a second ZDW will also limit the soliton red-shift, since the red-shift is halted gradually by spectral recoil as the solitons red-shift to the vicinity of the second ZDW (Skryabin et al., 2003).

The principle and limits for generation of blue-shifted light as described above can be demonstrated experimentally. Since the blue-shift continues as long as the solitons can red-shift while maintaining a group-velocity match with the short-wavelength radiation, it should be possible to predict the blue-edge of the supercontinuum from the group-velocity profile. An example of the output supercontinuum spectrum from a particular PCF (Pull 1 in Table 7.1) is shown in Fig. 7.1.

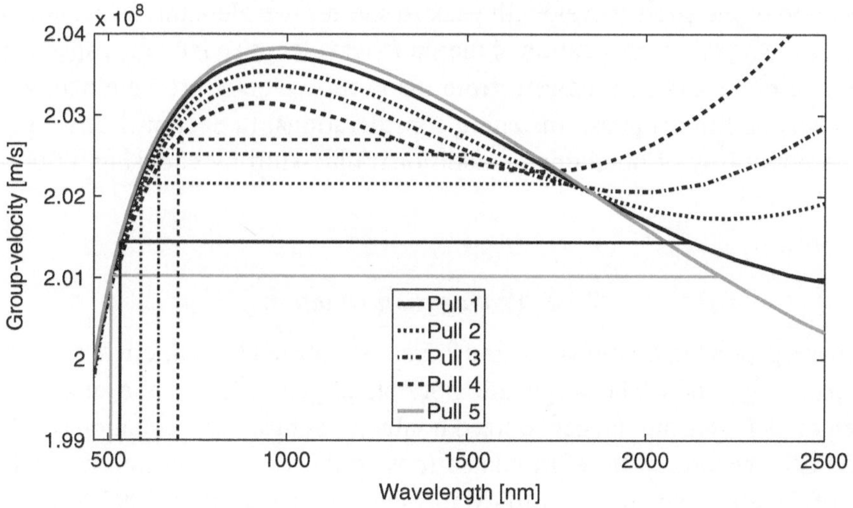

Fig. 7.2. Calculated group-velocity profiles for the five fibres listed in Table 7.1. For each fibre the vertical lines indicate the experimentally measured location of a strong peak on the short-wavelength side of the output supercontinuum spectrum. The horizontal lines show the group-velocity match between this peak and the group-velocity on the long-wavelength side of the pump.

The spectrum shows a clear peak at 533 nm and a broad flat spectrum extending into the near-infrared (limited to 1750 nm by the optical spectrum analyser). Similar spectra were measured for the four other fibres listed in Table 7.1. For each fibre the group-velocity dispersion profile was calculated using a finite element method mode solver (COMSOL Multiphysics 3.4, 2007). The calculated group-velocity profiles are shown in Fig. 7.2.

The figure also shows the experimentally measured spectral location of the strong blue-shifted peak for each fibre (vertical lines) and how these wavelengths are group-velocity matched to a wavelength in the near-infrared. For each fibre the location of the strong blue-shifted peak is seen to be group-velocity matched to the wavelength at which the group-velocity profile has approximately zero curvature. This indicates that when the solitons have red-shifted to this particular wavelength, their group-velocity can no longer decrease sufficiently to ensure continued interaction with the blue-shifted radiation. Note that the decrease in group-velocity experienced by blue-shifting radiation on the short-wavelength side of the spectrum is quite steep due to the inherent material dispersion of silica. This figure thus demonstrates how the short-wavelength edge of the spectrum is limited by the group-velocity profile for these particular five fibres. It is seen from the figure that shifting the location of the second ZDW (the higher wavelength at which the group-velocity profile has zero slope) to longer wavelengths simultaneously shifts the location of the short-wavelength peak to shorter wavelengths. Especially noteworthy is the fact that the location of the short-wavelength edge of the spectrum for all the five fibres can be predicted from the group-velocity profile alone, without making any additional phase-matching considerations. In Section 7.3, we investigate the possibility of obtaining an additional blue-shift by choosing a fibre with practically only one ZDW.

7.2.1 Propagation equation

The investigations in Sections 7.4 and 7.5 are supported by numerical simulations of propagating pulses. The propagation of optical pulses in a nonlinear waveguide is often modelled using the generalized nonlinear Schrödinger equation (GNLSE). The GNLSE was recently reformulated to account for a strong frequency dependence of the effective area $A_{\text{eff}}(\omega)$ (Mamyshev and Chernikov, 1990; Lægsgaard, 2007). This modified GNLSE is written here in a form very similar to the standard GNLSE (see Chapter 3):

$$\begin{aligned}\frac{\partial \tilde{C}}{\partial z} &= i \sum_{m=2}^{\infty} \beta_m \frac{[\omega - \omega_0]^m}{m!} \tilde{C}(z,\omega) \\ &+ i\gamma(\omega) \left[1 + \frac{\omega - \omega_0}{\omega_0}\right] \mathcal{F}\left\{ C(z,t) \int_{-\infty}^{\infty} R(t-t_1)\, |C(z,t)|^2 \, \mathrm{d}t_1 \right\}, \end{aligned} \tag{7.3}$$

where the nonlinear coefficient $\gamma(\omega)$ is given by

$$\gamma(\omega) = \frac{n_2 n_0 \omega_0}{c n_{\text{eff}}(\omega) \sqrt{A_{\text{eff}}(\omega) A_{\text{eff}}(\omega_0)}}. \tag{7.4}$$

n_2 is the nonlinear-index coefficient of the waveguide material set to the value corresponding to fused silica: $n_2 = 2.6 \cdot 10^{-20}$ m^2/W (Agrawal, 2007). $n_{\text{eff}}(\omega)$ is the frequency dependent effective index of the guided mode and $n_0 = n_{\text{eff}}(\omega_0)$. The variation of $n_{\text{eff}}(\omega)$ is usually much smaller than the variation of $A_{\text{eff}}(\omega)$ and therefore neglected here. $\tilde{C}(z,\omega)$ is the Fourier transform of $C(z,t)$, and is related to the Fourier transform of the pulse envelope $A(z,t)$ by

$$\tilde{C}(z,\omega) = \left[\frac{A_{\text{eff}}(\omega)}{A_{\text{eff}}(\omega_0)}\right]^{-1/4} \tilde{A}(z,\omega). \tag{7.5}$$

The symbol $\mathcal{F}$ denotes Fourier transform, and $R(t)$ is the Raman response of the nonlinear waveguide. For this work the standard approximation for silica glass was used (Blow and Wood, 1989; Agrawal, 2007):

$$R(t) = (1 - f_R)\delta(t) + f_R \frac{\tau_1^2 + \tau_2^2}{\tau_1 \tau_2^2} \exp(-t/\tau_2) \sin(t/\tau_1) \Theta(t), \tag{7.6}$$

where $\delta(t)$ is the Dirac delta function, $f_R = 0.18$ is the fractional Raman response, $\tau_1 = 12.2$ fs, $\tau_2 = 32$ fs, and $\Theta(t)$ is the Heaviside step function.

Note that Eqn. (7.3) is reduced to the standard GNLSE (see Chapter 3) if (1) C and $\tilde{C}$ are replaced by A and $\tilde{A}$, respectively, and if (2) $\gamma(\omega)$ is approximated by

$$\gamma(\omega) \approx \gamma(\omega_0) = \frac{n_2 \omega_0}{c A_{\text{eff}}(\omega_0)}. \tag{7.7}$$

This means that current implementations of numerical solutions to the standard GNLSE only need small modifications to instead solve the modified GNLSE, Eqn. (7.3). The modified GNLSE was recently found to more accurately take into account the wavelength dependence of A_{eff} (Lægsgaard, 2007).

7.3 Modifying the photonic crystal fibre structure

As shown in the previous section, one can shift the short-wavelength edge of the supercontinuum to shorter wavelengths by using a fibre which allows the solitons to red-shift further into the infrared. As a first step the fibre should therefore have a region of anomalous dispersion extending far into the near-infrared. Secondly, the shape of the group-velocity profile may also provide the possibility for further optimisation, as demonstrated by Stone and Knight (2008). Stone and Knight compared PCFs with pitches varying from $\Lambda = 3.71$ μm to 3.38 μm with the same $d/\Lambda = 0.77$. Decreasing the pitch shifted the shape of the group-velocity profile, so that shorter wavelengths were achieved in the resulting supercontinuum. However, decreasing the pitch also shifted the ZDW away from the pump, so that less energy was transferred from the pump (at 1064 nm) to shorter wavelengths at the initial stage of spectral broadening. The optimisation therefore successfully shifted the spectrum to shorter wavelengths, but also reduced the power generated in the visible region.

In this section we investigate two fibre designs for generation of blue light. Both meet the requirement of having an anomalous dispersion region extending far into the near-infrared. The calculated group-velocity profiles are shown in Fig. 7.3 (left). The supercontinuum output spectra were experimentally measured under similar pumping conditions (input pulse at 1064 nm of ~10 ps FWHM duration, ~7.6 kW peak power, 80 MHz repetition rate) and are shown in Fig. 7.3 (right). Both fibres are multimoded, but it was verified that the generated supercontinuum remained in the fundamental mode.

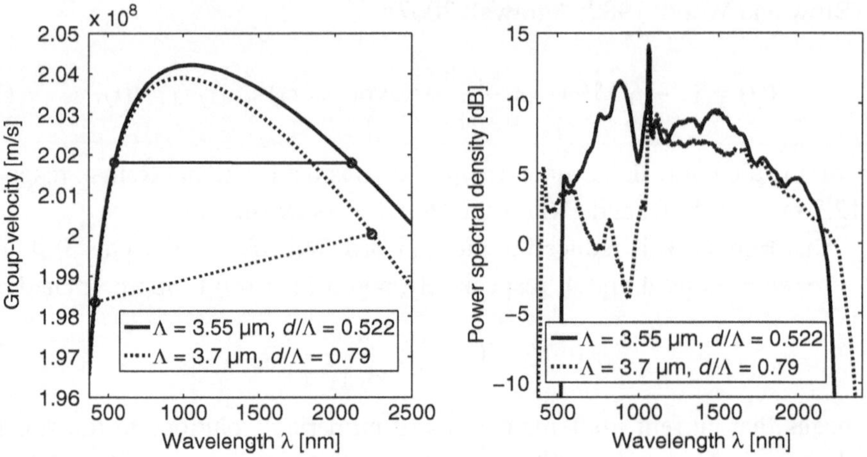

Fig. 7.3. *Left:* Calculated group-velocity profiles for the two fibre designs investigated in this section. Circles indicate position of edges of experimentally measured spectra seen on the right. *Right:* Experimentally measured spectra for two different fibres (structural parameters indicated) with a length of 17.3 m each. The spectra are measured from 350 nm to 1600 nm using an Ando OSA, combined with measurements from 1600 nm to 2500 nm using an Acton spectrograph.

The circles in Fig. 7.3 (left) indicate the visible and near-infrared edges of the supercontinuum spectrum. For the fibre with 3.55 μm pitch the circles are connected by a practically horizontal line, supporting that the blue edge of the spectrum arises from dispersive waves blue-shifted by red-shifting solitons under continuous group-velocity matching, as described in Section 7.2. The red-shifting solitons apparently stop at approximately 2100 nm. The soliton red-shift, and the accompanying blue-shift of the spectrum, could ideally be increased by using a longer fibre or higher input power. Experiments with doubling the input peak power did lead to a shift in the blue peak from 545 nm to 534 nm. However, even if one continues to increase the fibre length or input power, the intrinsic fibre losses [both material absorption (Kobayashi et al., 1978) and confinement loss (White et al., 2001)] rise steeply with increasing wavelength beyond ~1600 nm. For the 3.55 μm fibre we calculated that confinement loss is negligible compared to the material absorption, which is ~0.1 dB/m at 2100 nm and ~1 dB/m at 2365 nm (Hansen, 2007). We can therefore not expect solitons to red-shift much further than ~2300 nm.

To achieve a larger blue-shift the fibre with 3.7 μm pitch was designed and drawn. As seen in Fig. 7.3 (left), the fibre with 3.7 μm pitch has essentially the same group-velocity profile in the visible as the fibre with 3.55 μm pitch. This is because at short wavelengths the contribution from material dispersion is much larger than the contribution from waveguide dispersion. At longer wavelengths (~1000 nm and upwards) the fibre with 3.7 μm pitch provides significantly lower group-velocity than the fibre with 3.55 μm pitch. This means that group-velocity match can be achieved at shorter wavelengths than in the 3.55 μm pitch fibre by solitons red-shifted to the same wavelength in the near-infrared. In other words, this fibre should have a short-wavelength edge further down in the blue, even if the long-wavelength edge is the same as for the fibre with $\Lambda = 3.55$ μm. The fibre design also has one more advantage: the core is smaller (the fibre with 3.7 μm pitch has an effective area at $\lambda = 2000$ nm approximately 1.4 times smaller than the fibre with 3.55 μm pitch) resulting in more efficient nonlinear effects. Indeed, it is seen from Fig. 7.3 (right) that the solitons in the 3.7 μm pitch fibre achieve larger red-shift than in the 3.55 μm pitch fibre. The increased red-shift and the more optimal group-velocity profile of the 3.7 μm pitch fibre together result in a peak at shorter wavelengths (~413 nm) than in the 3.55 μm pitch fibre (~545 nm).

The line connecting the circles for the 3.7 μm fibre in Fig. 7.3 (left) is not horizontal. This could in principle be due to mechanisms providing additional blue-shift without group-velocity match to the near-infrared solitons. However, we expect that it is caused by the fact that the calculated group-velocity profile deviates from the actual group-velocity profile at long wavelengths, because the waveguide dispersion becomes more important for longer wavelengths, and the structural PCF parameters are not known exactly. Further, the calculated group-velocity profile assumes an ideal structure, but in reality the size of the air-holes may vary both across the fibre cross-section and along the fibre length.

The examples of this section demonstrate how the blue-edge of the supercontinuum spectrum can be shifted by modifying the structure of the PCF. In the next section, we consider whether the same can be achieved by modifying the fibre material.

7.4 Modifying the fibre material

7.4.1 Modifying the group-velocity profile by modifying the glass composition

As seen in the previous section, modifying the PCF structure mostly affects the group-velocity profile $v_g(\lambda)$ for long wavelengths ($\lambda \gtrsim 1000$ nm), while $v_g(\lambda)$ at shorter wavelengths is only slightly affected by a change in the PCF structure. This is because material dispersion provides the dominant contribution to the total dispersion at short wavelengths. In this section we consider whether it would be possible to shift the supercontinuum to shorter wavelengths by directly modifying the material dispersion (Frosz et al., 2008). This could be achieved in practice by changing the glass material from which the fibre is drawn.

In the previous section it was seen that shifting the group-velocity profile downwards for the long-wavelength part of the spectrum led to an increased blue-shift. This was because the red-shifting solitons were then able to match the

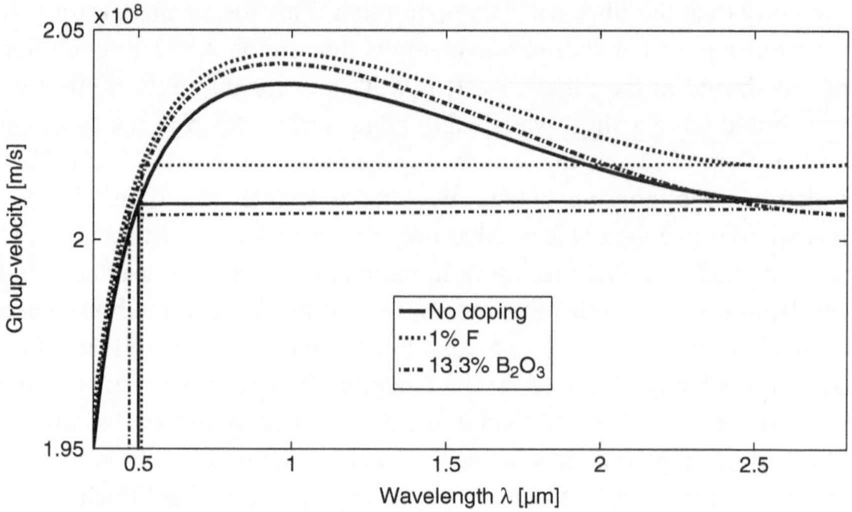

Fig. 7.4. Calculated group-velocity profiles for Pull 1 in Table 7.1 ($\Lambda = 2.42$ μm, $d/\Lambda = 0.46$) for different glass compositions: 100% fused silica (solid line), 1% F (dotted), and 13.3% B_2O_3 (dash-dot). The vertical lines indicate the calculated location of the short-wavelength peak of the corresponding spectra (Fig. 7.5). From the intersections between these lines and the corresponding group-velocities, there are also drawn almost horizontal lines intersecting the group-velocities at the long-wavelength edges of the spectra.

group-velocity of dispersive waves at shorter wavelengths. Likewise, upshifting the group-velocity profile for the short-wavelength part of the spectrum should also lead to an increased blue-shift of the supercontinuum. To investigate whether a suitable manipulation of the group-velocity profile is possible by using an alternative glass material, we calculated $v_g(\lambda)$ for a range of different glass compositions. The Sellmeier coefficients for the alternative glass compositions were obtained from Fleming (1978, 1979); for pure fused silica we used the data from Malitson (1965), which is used widely (Agrawal, 2007). In Fig. 7.4 we show the resulting group-velocity profiles for a particular PCF structure (Pull 1 in Table 7.1) composed of either pure fused silica, silica with 1% F, or silica with 13.3% B_2O_3.

It is seen in Fig. 7.4 that using silica glass containing 1% F upshifts the entire group-velocity profile with a diminishing shift towards shorter wavelengths. Using silica glass with 13.3% B_2O_3 results in a smaller upshift of $v_g(\lambda)$, and from a wavelength of approximately 2.3 μm and upwards the group-velocity is smaller than for a pure silica fibre. Also, using glass with 13.3% B_2O_3 shifts the ZDW further into the near-infrared. Based on the understanding of the physical mechanism behind the short-wavelength part of the spectrum, presented in Section 7.2, we can now use the group-velocity profiles to make the following predictions about the short-wavelength edge of supercontinua generated in the three fibres investigated here.

One could initially think that the upshift of $v_g(\lambda)$ at short wavelengths for the fibre with 1% F would lead to a larger blue-shift. However, since $v_g(\lambda)$ is also shifted upwards for longer wavelengths, the red-shifting solitons are not expected to be able to slow down sufficiently to match their group-velocity to radiation at shorter wavelengths than in the pure silica fibre. On the other hand, red-shifting solitons in the 13.3% B_2O_3 fibre eventually experience a lower group-velocity than in the pure silica fibre, and are thus able to match their group-velocity to radiation at shorter wavelengths than in the pure silica fibre. From these considerations we expect that even though using glass with 1% F upshifts the group-velocity at the short-wavelength part of the spectrum, this will not result in an increased blue-shift of the spectrum. Instead, using glass with 13.3% B_2O_3 is expected to result in a larger blue-shift.

In the following, we use numerical simulations (described in Section 7.2.1) of supercontinuum generation to test these expectations.

7.4.2 Simulation results

All the simulations in this section were made with 2^{17} sampling points. The temporal resolution was set to 1.54 fs, resulting in a temporal window of 201.85 ps and a frequency resolution of 4.95 GHz. The adaptive step size method was used (Sinkin et al., 2003) with a local goal error of $\delta_G = 10^{-6}$ resulting in a maximum change in photon number (a measure of the numerical error which is ideally zero) (Blow and Wood, 1989; Lægsgaard, 2007) of $-5.9 \cdot 10^{-6}\%$ after 2.5 m of propagation. For validation, simulations were also performed up to a length of 0.25 m with $\delta_G = 10^{-8}$ resulting in a maximum change in photon number of $-5.0 \cdot 10^{-8}\%$; the resulting spectra showed insignificant differences compared to the spectra using $\delta_G = 10^{-6}$, which at this length had a maximum change in photon number of $-4.5 \cdot 10^{-6}\%$. Since the change in photon number is larger from $z = 0$ m to $z = 0.25$ m than from $z = 0.25$ m to $z = 2.5$ m and there was no significant difference in the spectra calculated with either $\delta_G = 10^{-6}$ or $\delta_G = 10^{-8}$ at 0.25 m, the numerical accuracy is assumed to be sufficient. The input pulse has a sech shape, peak power of 9 kW, intensity FWHM (full-width at half-maximum) duration of 8 ps, and centre wavelength of 1064 nm. To realistically model the finite linewidth of the pump laser, we used the same phase noise model as Frosz et al. (2006a), since the underlying phase-diffusion model is physically well founded (Cavalcanti et al., 1995; Mussot et al., 2004). The linewidth in the simulations was set to 530 GHz ($\sim$2 nm) FWHM. The centre wavelength of the simulation frequency window was set to 750 nm. The power spectra $S(\lambda)$ shown are scaled so that $\int_{\lambda_{min}}^{\lambda_{max}} S(\lambda) d\lambda$ is equal to the average pulse power, assuming a pulse repetition frequency of 80 MHz. To focus on the effect of changing the group-velocity profile by modifying the fibre material,

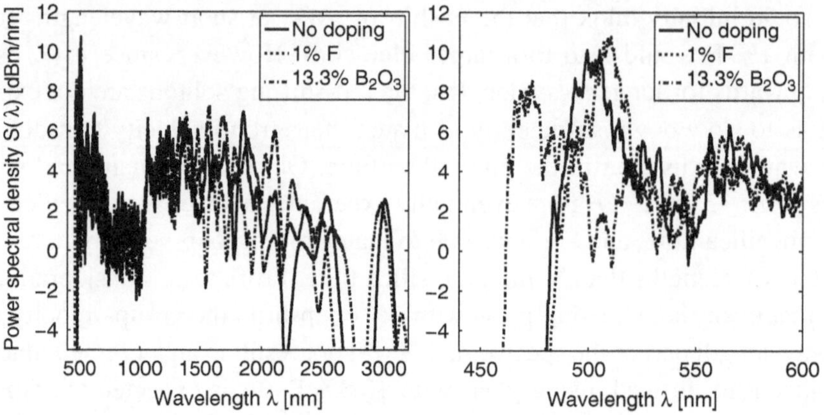

Fig. 7.5. Calculated spectra after 2.5 m of propagation in pure silica and alternative glass composition PCFs with structural parameters $\Lambda = 2.42$ μm and $d/\Lambda = 0.46$. The spectra have been smoothed using Savitzky–Golay filtering (Press et al., 2002).

we have not included losses in the simulations. Nor have we considered how the nonlinearity of the material could be increased due to the alternative glass material.

The resulting spectra are shown in Fig. 7.5. The pure silica fibre and the fibre with 1% F both have a ZDW at ∼2.6–2.7 μm. It is seen in Fig. 7.5 that in these fibres the solitons have red-shifted to ∼2500 nm and generate dispersive waves at ∼3000 nm. Note again that losses are not included in the simulations; in a real fibre the dispersive waves at ∼3000 nm would be highly attenuated due to material absorption and confinement loss. The fibre with 13.3% B_2O_3 has a ZDW at ∼2.9 μm, and the solitons have here red-shifted to ∼2.6 μm and generate dispersive waves at ∼3.1 μm.

At the short-wavelength edge of the spectra, we see that using a fibre glass material with 1% F can actually result in a slightly smaller blue-shift than when using a pure silica fibre. As mentioned in Section 7.4 this is expected because $v_g(\lambda)$ is upshifted more in the wavelength region of red-shifting solitons (between the pump wavelength and the higher ZDW) than in the wavelength region of the dispersive waves. The fibre with 13.3% B_2O_3 has a smaller upshift of $v_g(\lambda)$ in the wavelength region of dispersive waves, but this is more than compensated by the downshift of $v_g(\lambda)$ in the wavelength region of red-shifting solitons (see Fig. 7.4). This explains why the short-wavelength edge of the spectrum calculated for the 13.3% B_2O_3 fibre is more blue-shifted than in the pure silica fibre.

In Fig. 7.4 we have also drawn lines between $v_g(\lambda)$ at the short-wavelength edge of the spectrum and the wavelength to which the solitons have red-shifted. These lines are almost horizontal indicating that there is almost perfect group-velocity match between the red-shifting solitons and the blue-shifted dispersive waves. This again supports the understanding of the physical mechanism behind the blue-shift described in Section 7.2.

In conclusion, this section has demonstrated that modifying the group-velocity profile through using an alternative glass composition of the nonlinear fibre can be used to increase the blue-shift of the supercontinuum. It was found that care must be taken to ensure that the alternative glass composition does not just lead to an upshift of the entire group-velocity profile, but that the resulting group-velocity at the long-wavelength part of the spectrum must be lowered relative to the short-wavelength part of the spectrum. One could also combine choosing an alternative glass material to raise $v_g(\lambda)$ at short wavelengths, with a modification of the PCF structure to lower $v_g(\lambda)$ at long wavelengths. This is feasible because the material dispersion is dominant for short wavelengths and the waveguide dispersion is dominant for long wavelengths.

Finally, we should also mention that experiments of visible supercontinuum generation in Ge-doped fibres were recently presented (Leproux et al., 2008; Tombelaine et al., 2008a, 2008b) but the objective of the doping was to increase the second-harmonic generation in the fibre, and not to modify the group-velocity profile.

7.5 Combining different fibres: Cascading

7.5.1 Combining two fibres for increased blue-shift

Another method of manipulating the supercontinuum spectrum involves using non-linear mechanisms in two different fibres coupled together. This approach has been used to achieve increased blue-shift (Travers et al., 2005), but also to obtain a supercontinuum extending into the mid-IR region (Xia et al., 2006). The basic idea in this approach for increased blue-shift is to pump close to the zero-dispersion wavelength in the first fibre ($\lambda_{ZDW,1}$), so that energy is shifted to a region centred at $\lambda_N < \lambda_{ZDW,1}$ in the normal dispersion region. The second fibre has a zero-dispersion wavelength $\lambda_{ZDW,2} < \lambda_{ZDW,1}$, and the objective of the cascaded fibre approach is to transfer energy to below $\lambda_{ZDW,2}$. The energy transfer in the first fibre can be FWM or amplification of dispersive waves, followed by blue-shift through XPM, as described in Section 7.2. In previous work the energy transfer in the second fibre was attributed to FWM (Travers et al., 2005). The investigation in this section shows an example where the spectral broadening in the second fibre is caused by amplification of dispersive waves.

7.5.2 Simulation results

Fig. 7.6 shows the calculated dispersion profiles for three fibres used here (note that we now only consider fibres made of pure silica). We will here use the fibre with

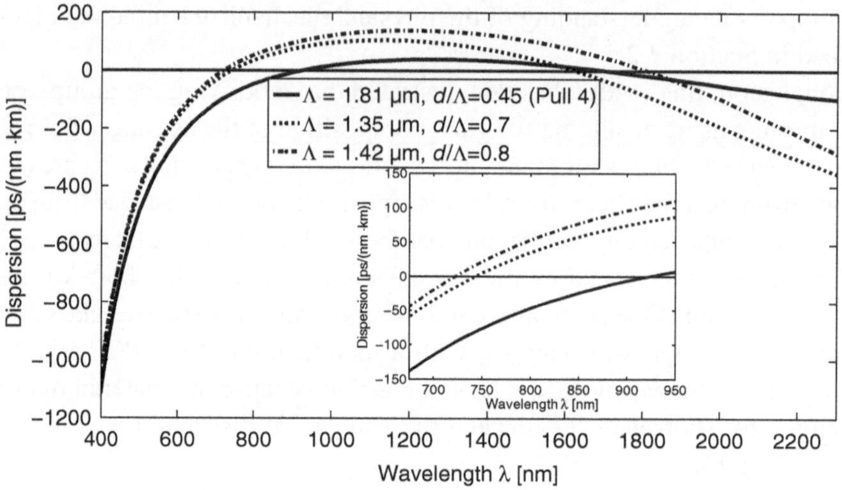

Fig. 7.6. Calculated dispersion profiles for the three fibres used for investigation of a cascaded fibre setup.

$\Lambda = 1.81$ μm and $d/\Lambda = 0.45$ as the first fibre in a cascaded setup. We investigate the use of either $\Lambda = 1.35$ μm with $d/\Lambda = 0.7$, or $\Lambda = 1.42$ μm with $d/\Lambda = 0.8$, as Fibre 2.

First, the propagation of a sech-pulse with 8 ps FWHM duration, centre wavelength 1064 nm, 9 kW peak power and 530 GHz FWHM spectral linewidth (~2 nm), is simulated over a propagation length of 20 cm in Fibre 1. 2^{17} points were used, local goal error $\delta_G = 10^{-6}$, temporal resolution 0.95 fs, and the centre wavelength of the spectral window was set to 500 nm. The output from this simulation was then used as the input for simulating propagation in 2 cm of Fibre 2, neglecting any losses in the coupling between Fibres 1 and 2. The resulting spectra are shown in Fig. 7.7.

It is seen from the spectrum for Fibre 1 in Fig. 7.7 that a peak just below 600 nm is formed on the short-wavelength side of the spectrum. It is also seen that after 2 cm of continued propagation in Fibre 2, a significant amount of spectral energy is transferred to shorter wavelengths. When using the $\Lambda = 1.35$ μm, $d/\Lambda = 0.7$ fibre as Fibre 2, the resulting spectrum extends down to ~415 nm; on the long-wavelength side of the spectrum there is also a growth in spectral energy at ~2100 nm. Using the $\Lambda = 1.42$ μm, $d/\Lambda = 0.8$ as Fibre 2 causes the spectrum to extend down to ~380 nm, and there is also growth in spectral energy at ~3200 nm (not shown in Fig. 7.7).

7.5.3 Physical mechanisms

To understand the spectral broadening mechanisms occurring in Fibre 2 of the cascaded fibres, we first consider the calculated group-velocity profiles $v_g(\lambda)$ shown

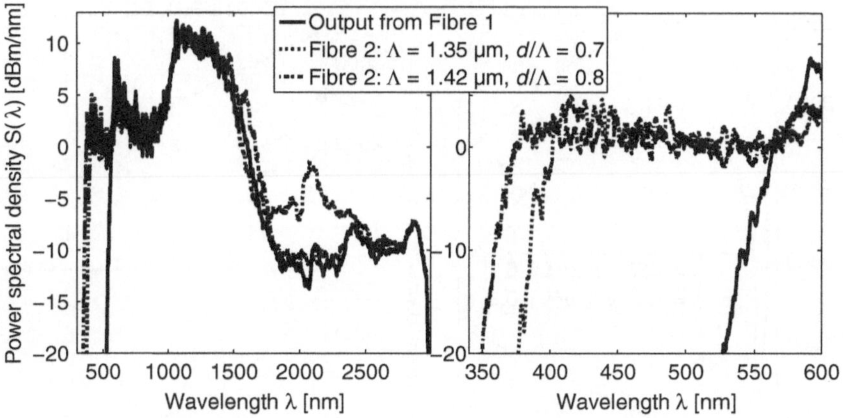

Fig. 7.7. Calculated spectra after 20 cm of propagation in Pull 4 ($\Lambda = 1.81$ μm, $d/\Lambda = 0.45$, solid line), followed by 2 cm of propagation in either $\Lambda = 1.35$ μm, $d/\Lambda = 0.7$ (dotted line) or $\Lambda = 1.42$ μm, $d/\Lambda = 0.8$ (dash-dotted line).

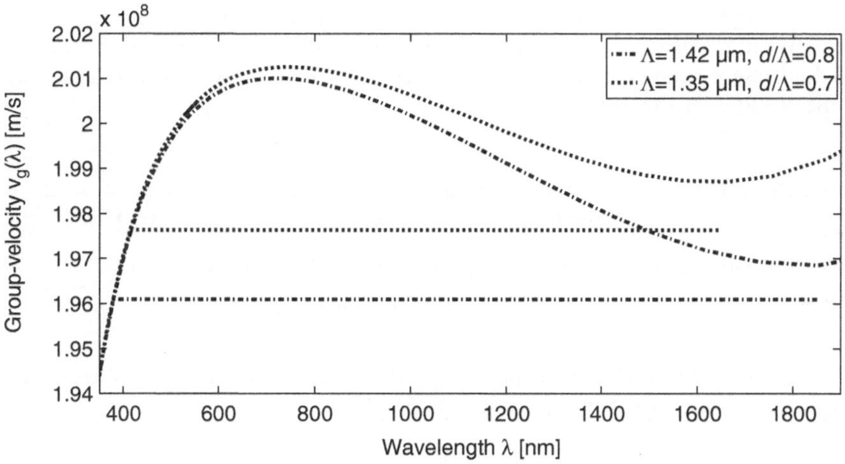

Fig. 7.8. Calculated group-velocity profiles $v_g(\lambda)$ for the two fibres used as the second fibre in a cascaded setup: $\Lambda = 1.42$ μm, $d/\Lambda = 0.8$ (dash-dotted) and $\Lambda = 1.35$ μm, $d/\Lambda = 0.7$ (dotted). The horizontal lines intersect $v_g(\lambda)$ at the short-wavelength edge of the corresponding calculated spectra (Fig. 7.7).

in Fig. 7.8. Horizontal lines intersecting $v_g(\lambda)$ at the short-wavelength edge of the corresponding calculated spectra (Fig. 7.7) have been added. Since for each fibre the horizontal lines cannot intersect $v_g(\lambda)$ at longer wavelengths, it is clear that the blue-shift in Fibre 2 cannot be caused by a group-velocity matched interaction (e.g. XPM) between red-shifting solitons and blue-shifting dispersive waves, as was the case in previous sections.

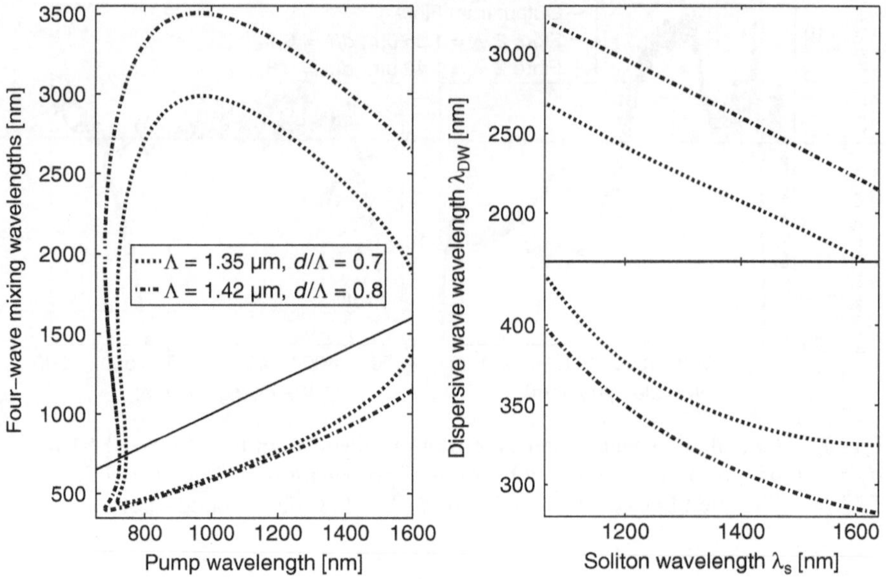

Fig. 7.9. *Left:* Four-wave mixing wavelengths as a function of pump wavelength in the two fibres used as the second fibre in a cascaded setup. *Right:* The wavelength to which energy can be transferred as dispersive waves from solitons, as a function of the soliton wavelength λ_S.

As mentioned previously, the blue-shift in Fibre 2 of the cascaded setup investigated by Travers et al. (2005) was attributed to degenerate FWM in which two pump photons at angular frequency ω_p are converted to a photon at the anti-Stokes frequency $\omega_{aS} > \omega_p$ and a photon at the Stokes frequency $\omega_S < \omega_p$. To investigate whether FWM plays a role in the configuration investigated here, we have in Fig. 7.9 (left) plotted the solutions to the degenerate FWM phase-matching condition (Agrawal, 2007; Frosz et al., 2006b)

$$\kappa = 2\gamma P_0 + 2\sum_{m=1}^{\infty} \frac{\bar{\beta}_{2m}}{(2m)!}\Omega^{2m} = 0, \tag{7.8}$$

where the $\bar{\beta}_m$ dispersion coefficients for a pump frequency ω_p are estimated from a polynomial fit to the calculated dispersion profile $\beta_2(\omega)$:

$$\beta_2(\omega) = \bar{\beta}_2 + \bar{\beta}_3[\omega - \omega_p] + \frac{1}{2}\bar{\beta}_4[\omega - \omega_p]^2 + \frac{1}{6}\bar{\beta}_5[\omega - \omega_p]^3 + \dots \tag{7.9}$$

After solving Eqn. (7.8) for $\Omega = \omega_p - \omega_S = \omega_{aS} - \omega_p$ over a range of ω_p, the corresponding values of the anti-Stokes and Stokes wavelengths were plotted in Fig. 7.9 (left). We found that including the $2\gamma P_0$ term in Eqn. (7.8) only slightly decreased the largest obtainable shift between pump and FWM-wavelengths, and therefore we show the plot for $P_0 = 0$.

From Fig. 7.9 (left) it is found that the shortest wavelengths generated by FWM occur when $\lambda_p \sim 680$ nm, $\lambda_{aS1} \sim 400$ nm, $\lambda_{aS2} \sim 420$ nm, $\lambda_{S1} \sim 1600$ nm, and $\lambda_{S2} \sim 2400$ nm, for the fibre with $\Lambda = 1.42$ μm, $d/\Lambda = 0.8$. For the $\Lambda = 1.35$ μm, $d/\Lambda = 0.7$ fibre, the shortest wavelengths generated by FWM are obtained when $\lambda_p \sim 725$ nm, $\lambda_{aS1} \sim 440$ nm, $\lambda_{aS2} \sim 515$ nm, $\lambda_{S1} \sim 1230$ nm and $\lambda_{S2} \sim 2100$ nm. FWM can therefore explain that there is a larger blue-shift in the $\Lambda = 1.42$ μm, $d/\Lambda = 0.8$ fibre, but it does not accurately predict the short-wavelength edge of the spectra (~380 nm and ~415 nm in the $\Lambda = 1.42$ μm, $d/\Lambda = 0.8$ and $\Lambda = 1.35$ μm, $d/\Lambda = 0.7$ fibre, respectively), nor does it correctly predict the growth of spectral energy in the long-wavelength part of the spectrum (~3200 nm and ~2100 nm in the $\Lambda = 1.42$ μm, $d/\Lambda = 0.8$ and $\Lambda = 1.35$ μm, $d/\Lambda = 0.7$ fibre, respectively).

We have therefore examined whether the spectral changes in Fibre 2 can instead be explained by soliton amplification of dispersive waves. Since this amplification can occur when the soliton wavenumber matches the wavenumber of a dispersive wave (Akhmediev and Karlsson, 1995), the condition for amplification of dispersive waves can be written as (Frosz et al., 2005)

$$\frac{|\beta_2(\omega_{sol})|}{2T_{sol}^2} = \frac{\gamma P_{sol}}{2} = \sum_{m \geq 2} \frac{\beta_m(\omega_{sol})}{m!} [\omega_{DW} - \omega_{sol}]^m. \tag{7.10}$$

The first equality is simply shown because either of the two expressions can be used for the wavenumber of a fundamental soliton with peak power P_{sol} and temporal width T_{sol} at the angular frequency ω_{sol}. The right-hand side of the second equality expresses the wavenumber of a dispersive wave at angular frequency ω_{DW}. The solutions to Eqn. (7.10) are plotted in Fig. 7.9 (right). The term $\gamma P_{sol}/2$ was set to zero when solving Eqn. (7.10), because it is negligible for large values of $\omega_{DW} - \omega_{sol}$. We found that setting P_{sol} as high as 100 kW did not shift the phase-matching curves significantly.

From Fig. 7.9 (right) it is seen that in the $\Lambda = 1.35$ μm, $d/\Lambda = 0.7$ fibre, the growth of spectral power in the range ~1800–2400 nm (Fig. 7.7) can be caused by solitons in the range ~1210–1560 nm. On the short-wavelength side, solitons between the pump (1064 nm) and ~1095 nm can amplify dispersive waves in the range ~ 415–430 nm. For the $\Lambda = 1.42$ μm, $d/\Lambda = 0.8$ fibre, there is growth of spectral power in the range ~3000–3500 nm (not shown in Fig. 7.7); solitons in the range ~1064–1180 nm can transfer energy to ~3000–3200 nm. On the short-wavelength side, solitons between 1064 nm and ~1110 nm can amplify dispersive waves in the range ~ 380–400 nm. Amplification of dispersive waves can therefore explain both the growth of spectral power around 2100 nm and 3200 nm, respectively (the fibre with $\Lambda = 1.35$ μm and $\Lambda = 1.42$ μm, respectively), and the extension of the spectrum down to ~415 nm and ~380 nm, respectively. One can, however, based on Fig. 7.9 (right) ask why the blue-shift of the dispersive waves does not continue below 415 nm and 380 nm, respectively, since it is known from the spectrum that the solitons have red-shifted further than 1100 nm. This can be caused by the fact that it is necessary for a soliton to have temporal overlap with the dispersive wave seed to which it transfers energy (Akhmediev and Karlsson, 1995). Since the group-velocity of the dispersive waves at ~400 nm is much lower than the group-velocity of the solitons at ~1100 nm (Fig. 7.8), the energy transfer process is quickly arrested as the solitons lose temporal overlap with energy at ~400 nm, which could otherwise have acted as a seed for energy transfer to lower wavelengths. Nevertheless, the energy transfer is sufficient to generate more than 1 mW/nm at ~400 nm.

To further investigate the role of amplification of dispersive waves, we have also simulated direct propagation of the input pulses in the $\Lambda = 1.42$ μm, $d/\Lambda = 0.8$ fibre, without a preceding fibre. The result is seen in Fig. 7.10. After 5 cm of propagation one sees distinctly generated peaks at ~380 nm and ~3100 nm, practically the same locations where spectral energy is generated when using the cascaded setup, indicating that the same mechanism is at play. Upon further propagation up to 17.5 cm the red-shifting solitons transfer energy to wavelengths increasingly shorter than the initial ~3100 nm and slightly shorter than the initial ~380 nm. This corresponds well with what one would expect from the phase-match plot in Fig. 7.9 (right), thus supporting that amplification of dispersive waves is playing a major role. Since the peak spectral density around 380 nm slightly decreases from 2 cm to 17.5 cm, the energy transfer from the solitons seems to have stopped quickly and some of the additional blue-shift is probably due to nonlinear interaction with

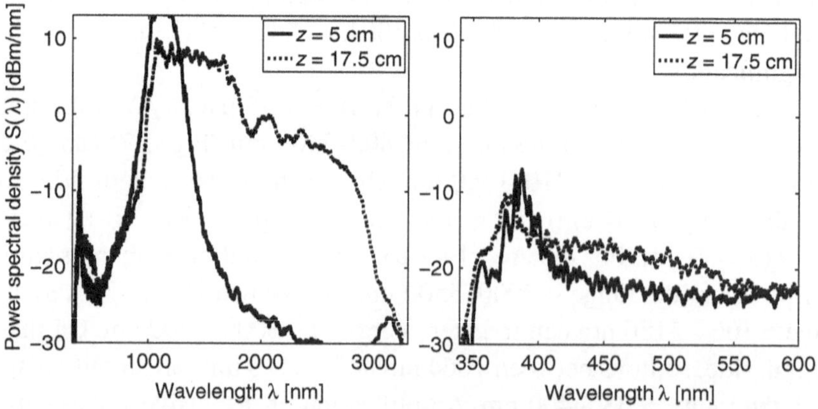

Fig. 7.10. Calculated spectra after 5 cm (solid line) and 17.5 cm (dashed) of propagation in the $\Lambda = 1.42$ μm, $d/\Lambda = 0.8$ fibre, without a preceding fibre.

the solitons (e.g. XPM). The interesting thing to note is that when pumping the $\Lambda = 1.42$ μm, $d/\Lambda = 0.8$ fibre directly without a preceding fibre, the maximum spectral density around 380 nm is only about −7 dBm/nm; in the cascaded setup it was about 3 dBm/nm (Fig. 7.7) even though the 20 cm of propagation in Fibre 1 had shifted spectral energy away from the pump to longer wavelengths, before the light was input to Fibre 2. It is therefore clear that the success of the cascaded setup in generating high spectral power in the blue region of the spectrum is caused by the intermediate energy transfer in Fibre 1 to ~600 nm. The energy in this region acts as a seed for dispersive waves in Fibre 2, thereby greatly increasing the power (by 10 dB in this case) that can be generated at 380 nm.

These results show that designing the dispersion profile of Fibre 2 can be used to shift the phase-matching wavelength to a desired spectral region in the blue–UV region of the output spectrum. The dispersion profile design should ensure that dispersive waves can be generated at the desired wavelength [Fig. 7.9 (right)] from solitons at a wavelength where there is initially temporal overlap with a seed for dispersive waves.

7.5.4 Advantages and disadvantages of the cascaded setup

As shown in the last section, the blue-shift in Fibre 2 of the cascaded setup does not require a group-velocity match between solitons and the blue-shifting dispersive waves (although a large group-velocity mismatch is what ultimately seems to limit the blue-shift). This means that the blue-shift mechanism is not limited by losses in the near-infrared, as is expected to be the case for the approaches investigated in Sections 7.3–7.4. On the other hand, one disadvantage of using cascaded fibres is that it can be difficult to splice together PCFs with significantly different core size without considerable loss.

7.6 Conclusion and outlook

The investigations of this chapter have demonstrated three distinct approaches to shift the spectrum of a supercontinuum to shorter wavelengths: modifying the PCF structure, modifying the glass material, and cascading of fibres. For the first two approaches it was confirmed that the spectral broadening towards the blue can be explained by a nonlinear interaction (e.g. XPM) between red-shifting solitons and blue-shifting dispersive waves. This approach required a group-velocity match between the long-wavelength edge and the short-wavelength edge of the spectrum. For the cascaded fibres it was found that FWM could to some degree explain the spectral features, but phase-matched amplification of dispersive waves provided a better agreement with the properties of the spectra for both a cascaded setup, and for a single fibre of the same structure as Fibre 2 in the cascaded setup. It should be noted that one could also combine two or all of the three approaches to achieve even larger blue-shift.

The problem with splicing together fibres for a cascaded setup can be avoided by instead using tapered fibres. This still assures that light is generated in one spectral region in the beginning of the fibre, followed by an energy transfer to lower wavelengths as the dispersion profile gradually changes along the fibre. This has been demonstrated with both 100 fs pulses (Lu et al., 2005) and pulses as long as 600 ps (Kudlinski et al., 2006). The blue-shift mechanism requiring group-velocity match between the solitons and the dispersive waves has also been observed in a 1.1 m long tapered fibre (Travers et al., 2008). Due to the simplicity of tapering a fibre and the impressive blue–UV light generation already demonstrated, we expect that this approach will be preferred over both the more complicated approach of using an alternative glass material (Section 7.4), and the splicing problems involved when using a cascaded setup (Section 7.5).

As briefly mentioned in Section 7.4, another novel approach to increase light generation in the visible is to dope the glass material in order to increase the second-harmonic generation in the fibre (Leproux et al., 2008; Tombelaine et al., 2008a, 2008b).

Finally, it should also be mentioned that it has been shown theoretically that the group-velocity dispersion of a tapered fibre can be tuned by immersing the taper in different liquids (Zhang et al., 2004). This could potentially be used in the same way as modifying the PCF structure or the glass material, to obtain a dispersion profile suitable for generating light at a desired wavelength.

Acknowledgements

P. M. Moselund is funded by the Photonics Academy Denmark. The authors wish to thank Kent Mattsson and Kim P. Hansen, Crystal Fibre A/S, for enlightening discussions and for kindly providing the fibres for the experimental measurements.

References

Agrawal G. P. (2007). *Nonlinear Fiber Optics*, 4th edn. Burlington, MA: Academic Press.

Akhmediev N., Karlsson M. (1995). Cherenkov radiation emitted by solitons in optical fibers. *Phys. Rev. A 51,* 3 (March), 2602–2607.

Andresen E. R., Nielsen C. K., Thøgersen J., Keiding S. R. (2007). Fiber laser-based light source for coherent anti-Stokes Raman scattering microspectroscopy. *Opt. Express 15,* 8, 4848–4856.

Blow K. J., Wood D. (1989). Theoretical description of transient stimulated Raman scattering in optical fibers. *IEEE J. Quantum Electron.* **25**, 2665–2673.

Cavalcanti S. B., Agrawal G. P., Yu M. (1995). Noise amplification in dispersive nonlinear media. *Phys. Rev. A 51,* 5 (May), 4086–4092. http://dx.doi.org/10.1103/PhysRevA.51.4086.

Chan M.-C., Chia S.-H., Liu T.-M. et al. (2008). 1.2- to 2.2-μm tunable Raman soliton source based on a Cr:forsterite laser and a photonic-crystal fiber. *IEEE Photonics Technology Letters 20,* 11, 900–902.

COMSOL Multiphysics 3.4. (2007). www.comsol.com.

Efimov A., Taylor A. J., Omenetto F. G., Knight J. C., Wadsworth W. J., St. J. Russell P. (2003). Phase-matched third harmonic generation in microstructured fibers. *Opt. Express 11,* 20, 2567–2576.

Fleming J. W. (1978). Material dispersion in lightguide glasses. *Electronics Letters 14,* 11, 326–328.

Fleming J. W. (1979). Material dispersion in lightguide glasses [erratum]. *Electronics Letters 15,* 16, 507.

Frosz M. H., Bang O., Bjarklev A. (2006a). Soliton collision and Raman gain regimes in continuous-wave pumped supercontinuum generation. *Opt. Express 14,* 20 (Oct.), 9391–9407. www.opticsinfobase.org/abstract.cfm?URI=oe-14-20-9391.

Frosz M. H., Falk P., Bang O. (2005). The role of the second zero-dispersion wavelength in generation of supercontinua and bright-bright soliton-pairs across the zero-dispersion wavelength. *Opt. Express 13,* 16 (Aug.), 6181–6192. http://www.opticsexpress.org/abstract.cfm?URI=OPEX-13-16-6181. Erratum: ibid, **15**(8), 5262–5263 (2007). www.opticsinfobase.org/abstract.cfm?URI=oe-15-8-5262.

Frosz M. H., Moselund P. M., Rasmussen P. D., Thomsen C. L., Bang O. (2008). Increasing the blue-shift of a supercontinuum by modifying the glass composition. *Opt. Express 16,* 25 (Dec.), 21076–21086. http://www.opticsexpress.org/abstract.cfm?URI=oe-16-25-21076.

Frosz M. H., Sørensen T., Bang O. (2006b). Nanoengineering of photonic crystal fibers for supercontinuum spectral shaping. *J. Opt. Soc. Am. B 23,* 8 (Aug.), 1692–1699. www.opticsinfobase.org/abstract.cfm?URI=josab-23-8-1692.

Genty G., Lehtonen M., Ludvigsen H. (2004). Effect of cross-phase modulation on supercontinuum generated in microstructured fibers with sub-30 fs pulses. *Opt. Express 12,* 19, 4614–4624. www.opticsexpress.org/abstract.cfm?URI=OPEX-12-19-4614.

Genty G., Lehtonen M., Ludvigsen H. (2005). Route to broadband blue-light generation in microstructured fibers. *Opt. Lett. 30,* 7, 756–758.

Gorbach A. V., Skryabin D. V. (2007). Light trapping in gravity-like potentials and expansion of supercontinuum spectra in photonic-crystal fibres. *Nature Photonics 1,* 11, 653–657.

Gorbach A. V., Skryabin D. V., Stone J. M., Knight J. C. (2006). Four-wave mixing of solitons with radiation and quasi-nondispersive wave packets at the short-wavelength edge of a supercontinuum. *Opt. Express 14,* 21, 9854–9863.

Gordon J. P. (1986). Theory of the soliton self-frequency shift. *Opt. Lett. 11,* 10 (Oct.), 662–664.

Hansen K. P. (2007). Crystal Fibre A/S, Personal communication.

Islam M. N., Sucha G., Bar-Joseph I., Wegener M., Gordon J. P., Chemla D. S. (1989). Femtosecond distributed soliton spectrum in fibers. *J. Opt. Soc. Am. B 6,* 6 (June), 1149–1158. www.opticsinfobase.org/abstract.cfm?URI=josab-6-6-1149.

Jalink K., Diaspro A., Caorsi V., Bianchini P. (2008). Leica TCS SP5 X – White Light Laser. Application Letter 29, Leica Microsystems. Aug. www.leica-microsystems.com.

Kobayashi S., Shibata N., Shibata S., Izawa T. (1978). Characteristics of optical fibers in infrared wavelength region. *Review of the Electrical Communication Laboratories 26,* 3-4, 453–467.

Kudlinski A., George A. K., Knight J. C. et al. (2006). Zero-dispersion wavelength decreasing photonic crystal fibers for ultraviolet-extended supercontinuum generation. *Opt. Express 14,* 12, 5715–5722.

Lægsgaard J. (2007). Mode profile dispersion in the generalised nonlinear Schrödinger equation. *Opt. Express 15,* 24 (Nov.), 16110–16123.

Leproux P., Buy-Lesvigne C., Tombelaine V. et al. (2008). Methods for visible supercontinuum generation in doped/undoped holey fibres. *Proceedings of the SPIE – The International Society for Optical Engineering* **6990**, 699007–1–4.

Lu F., Deng Y., Knox W. H. (2005). Generation of broadband femtosecond visible pulses in dispersion-micromanaged holey fibers. *Opt. Lett. 30,* 12, 1566–1568.

Malitson I. H. (1965). Interspecimen comparison of the refractive index of fused silica. *Journal of the Optical Society of America 55,* 10, 1205–1209.

Mamyshev P. V., Chernikov S. V. (1990). Ultrashort-pulse propagation in optical fibers. *Opt. Lett. 15,* 19, 1076–1078.

Mitschke F. M., Mollenauer L. F. (1986). Discovery of the soliton self-frequency shift. *Opt. Lett. 11,* 10 (Oct.), 659–661.

Mussot A., Lantz E., Maillotte H., Sylvestre T., Finot C., Pitois S. (2004). Spectral broadening of a partially coherent CW laser beam in single-mode optical fibers. *Opt. Express 12,* 13 (Jun.), 2838–2843. www.opticsexpress.org/abstract.cfm?URI=OPEX-12-13-2838.

Nishizawa N., Goto T. (2002). Characteristics of pulse trapping by ultrashort soliton pulses in optical fibers across the zero-dispersion wavelength. *Opt. Express 10,* 21, 1151–1160.

Omenetto F. G., Taylor A. J., Moores M. D. et al. (2001). Simultaneous generation of spectrally distinct third harmonics in a photonic crystal fiber. *Opt. Lett. 26,* 15, 1158–1160.

Prasad P. N. (2003). *Introduction to Biophotonics*. John Wiley & Sons.

Press W. H., Teukolsky S. A., Vetterling W. T., Flannery B. P. (2002). *Numerical Recipes in C++: The Art of Scientific Computing*, 2nd edn. Cambridge: Cambridge University Press. www.nr.com.

Price J., Monro T., Furusawa K. et al. (2003). UV generation in a pure-silica holey fiber. *Applied Physics B 77,* 2-3, 291–298.

Ranka J. K., Windeler R. S., Stentz A. J. (2000). Visible continuum generation in air–silica microstructure optical fibers with anomalous dispersion at 800 nm. *Opt. Lett. 25,* 1 (Jan.), 25–27.

Sinkin O. V., Holzlöhner R., Zweck J., Menyuk C. R. (2003). Optimization of the split-step Fourier method in modeling optical-fiber communications systems. *J. Lightwave Technol. 21,* 1 (Jan.), 61–68. http://dx.doi.org/10.1109/JLT.2003.808628.

Skryabin D. V., Luan F., Knight J. C., Russell P. S. J. (2003). Soliton self-frequency shift cancellation in photonic crystal fibers. *Science* **301**, 1705–1708.

Stone J. M., Knight J. C. (2008). Visibly "white" light generation in uniform photonic crystal fiber using a microchip laser. *Opt. Express 16,* 4, 2670–2675.

Tartara L., Cristiani I., Degiorgio V. (2003). Blue light and infrared continuum generation by soliton fission in a microstructured fiber. *Applied Physics – Section B – Lasers and Optics 77,* 2-3, 307.

Tombelaine V., Buy-Lesvigne C., Couderc V. et al. (2008a). Second harmonic generation in Ge-doped silica holey fibres and supercontinuum generation. *Proceedings of the SPIE – The International Society for Optical Engineering* **6990**, 69900N–1–7.

Tombelaine V., Buy-Lesvigne C., Leproux P., Couderc V., Mélin G. (2008b). Optical poling in germanium-doped microstructured optical fiber for visible supercontinuum generation. *Opt. Lett. 33,* 17, 2011–2013.

Travers J. C., Popov S. V., Taylor J. R. (2005). Extended blue supercontinuum generation in cascaded holey fibers. *Opt. Lett. 30,* 23, 3132–3134.

Travers J. C., Popov S. V., Taylor J. R. (2008). Trapping of dispersive waves by solitons in long lengths of tapered PCF. *2008 Conference on Lasers and Electro-Optics and 2008 Conference on Quantum Electronics and Laser Science*, 1–2.

Walewski J., Borden M., Sanders S. (2004). Wavelength-agile laser system based on soliton self-shift and its application for broadband spectroscopy. *Applied Physics B 79,* 8, 937–940.

White T. P., McPhedran R. C., de Sterke C. M., Botten L. C., Steel M. J. (2001). Confinement losses in microstructured optical fibers. *Opt. Lett. 26,* 21, 1660–1662.

Xia C., Kumar M., Kulkarni O. P. et al. (2006). Mid-infrared supercontinuum generation to 4.5 μm in ZBLAN fluoride fibers by nanosecond diode pumping. *Opt. Lett. 31,* 17, 2553–2555.

Zhang R., Teipel J., Zhang X., Nau D., Giessen H. (2004). Group velocity dispersion of tapered fibers immersed in different liquids. *Optics Express 12,* 8, 1700–1707.

8

Continuous wave supercontinuum generation

J. C. Travers

8.1 Introduction

It is perhaps not surprising that using extremely high power and short pulse duration pump sources leads to dramatic nonlinear processes in optical fibres; in contrast, the generation of a supercontinuum from a continuous pump wave of relatively meagre power is at first sight, astounding. Yet supercontinua spanning over 1000 nm have been generated with pump powers of a few tens of watts – orders of magnitude lower than pulse pumped systems.

The key to continuous wave (CW) supercontinuum generation is the utilisation of modulation instability (MI). This is inherent to any anomalously dispersive, nonlinear medium, and has been observed in a wide range of systems. This instability can enable the creation of the extremes of peak power and pulse duration necessary for dramatic nonlinear processes to occur, even from very low power CW pump lasers. But although MI from CW pump lasers was observed in the 1980s by Itoh et al. (1989), other factors required for efficient continuum generation were missing, causing another decade to pass before such results were obtained.

A full review of experimental results will be presented later, but for reference, some examples of continuous wave supercontinua are shown in Fig. 8.1, which illustrate the high spectral power and the spectral smoothness and flatness which are characteristic of CW continuum generation.

Although significantly different from the physical mechanisms involved in ultra-short – femtosecond based – supercontinuum generation, the basic physical processes underpinning CW continuum generation are the same as for longer pump pulses (greater than a few picoseconds) and the observations and conclusions developed in the 1980s surrounding these type of sources apply to the CW regime. In particular, the formation of fundamental solitons from MI, the effect of Raman

Supercontinuum Generation in Optical Fibers, ed. by J. M. Dudley and J. R. Taylor. Published by Cambridge University Press.

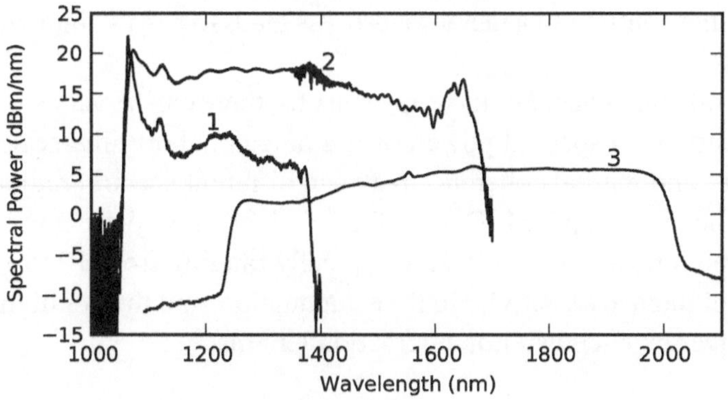

Fig. 8.1. Examples of the extent of spectral power achieved with continuous wave supercontinuum generation: (1) using 100 m of PCF with a 10 W Yb fibre laser, by Avdokhin et al. (2003); (2) using 25 m of PCF with a 50 W Yb fibre laser, by Cumberland et al. (2008a); and (3) using 600 m of highly nonlinear conventional fibre with a 10 W Er fibre laser, by Popov et al. (2002).

self-scattering of the solitons, the role of inter-soliton collisions, and the formation of dispersive waves around points of zero group velocity dispersion, are collectively referred to in the early literature as the Raman-soliton continuum (Gouveia-Neto et al., 1988; Taylor, 1992; Dianov et al., 1992; Islam et al., 1989a).

Despite these common fundamentals, however, there are significant differences between the high peak power, long pulse pumped systems, and continuous wave pumped systems. Some notable features of CW supercontinua are:

- Simple experimental realisation. Usually the setup consists of a continuous wave fibre laser spliced directly to a suitable optical fibre, leaving no mechanically tunable components and resulting in extremely stable operation.
- High average spectral power. As high power levels are required to initiate the nonlinear processes, and we have a duty factor of 1, higher average powers than those used with pulse lasers, are required. This leads to very high spectral powers in the resulting continuum, typically greater than 10 mW/nm.
- Dependence on pump noise. Continuous wave pump sources have very different noise properties to pulsed systems, and can exhibit significant intensity fluctuations, which have a significant influence on the initiation of the MI process.
- Low peak power. Even with pump intensity fluctuations, the peak pump powers involved are relatively low, which means that for efficient MI and Raman continuum development there is a far greater sensitivity to the dispersion landscape around the pump wavelength and the fibre nonlinearity.
- Smooth continuum spectra. As we shall discuss below, the stochastic nature of the modulation instability, and the very large number of interacting solitons

involved in the continuum generation process leads to a smoothing of the spectral components.

- Low temporal coherence. For the same reasons, there can be very significant temporal fluctuations of spectral power, over a range of short time-scales (generally picosecond), and mutual coherence between different spectral regions is low.
- Long fibre lengths required. Due to the relatively low peak powers involved, larger interaction length is called for, typically ranging from 10 to 1000 m. This leads to a heightened sensitivity to fibre attenuation, and the possibility of slower dynamics over extended soliton interaction lengths.

8.1.1 Structure of this chapter

These and other features will be considered in the following sections of this chapter. We begin in Section 8.1.2 with a brief outline of the types of experimental arrangement used for CW continuum generation. In Section 8.2 we provide an overview of the continuum formation mechanism, followed by in-depth studies of the most important processes, how they interact and some illustrative numerical examples. This is followed by a historical review of the literature and some recent experimental results in Section 8.3.

8.1.2 Experimental arrangements

The typical experimental arrangement for continuous wave supercontinuum generation consists simply of a continuous wave pump source and a suitable fibre.

The pump source is usually a fibre laser. Around 1400–1600 nm either an Er-doped fibre laser or a Raman fibre laser are commonly used, whereas Yb-doped laser sources are generally used for pumping around 1060 nm. Alternatively Er or Yb-doped fibre amplifiers can be cascaded with inter-stage filtering to create an amplified spontaneous emission source. All of these systems can be scaled to the region of a few tens of watts with commercial components. Typically 5–20 W is used for CW continuum generation, although some groups regularly use 50 W and over 200 W of fibre coupled power has been demonstrated by Travers et al. (2008b).

The laser output is either collimated and then focused into the fibre end, or spliced directly to the fibre. The latter arrangement brings advantages in terms of robustness and stability, and can also allow more readily for power scaling if sufficient care is taken over thermally managing the splice to prevent the creation of a fibre fuse. The main disadvantage is that it requires a low loss splice to the fibre, which in the case of some photonic-crystal fibres can be challenging, although manageable with sufficient effort.

The choice of fibre is complex and properly requires a careful consideration of the mechanism as we will outline below. However, broadly speaking one requires relatively small values of anomalous dispersion at the pump wavelength (say up to 50 ps $nm^{-1}km^{-1}$) and reasonable nonlinear coefficients, commonly over 10 $W^{-1}km^{-1}$, although lower values are possible when compensating with longer fibre lengths.

For pumping around 1400–1600 nm these parameters can be met with conventional dispersion shifted fibres or speciality highly nonlinear fibre. Around 1060 nm they can only be achieved with the use of PCF. Details of experimental results are given in Section 8.3.

8.2 Mechanism

8.2.1 Overview of the mechanism

The process of CW continuum formation begins with the modulation instability of the continuous pump wave in the anomalous dispersion region (Hasegawa and Brinkman, 1980; Tai et al., 1986; Itoh et al., 1989; Mussot et al., 2004). Through this process the continuous pump wave breaks up into fundamental solitons. The rate of breakup, and the duration of the solitons formed depend on the pump conditions and fibre parameters such as dispersion and nonlinearity. Given the correct conditions, the solitons can be short enough in duration to have the spectral width necessary for Raman self-scattering, leading to the soliton self-frequency shift (Dianov et al., 1985; Gordon, 1986). As modulation instability is a noise seeded process, the solitons formed will have variable bandwidths and durations and thus a range of soliton frequency shifts occurs, creating a smooth red-shifted continuum. Another, very important contributor to the CW continuum development is inelastic inter-soliton collisions, also mediated through Raman, which leads to the enhancement of the higher energy solitons, driving the red-shift of the continuum (Islam et al., 1989a; Korneev et al., 2008; Frosz et al., 2006). The long wavelength extent of the continuum is eventually limited by either the dispersion magnitude increasing or nonlinearity decreasing, at lower frequencies, such that the soliton bandwidth is compressed and Raman self-scattering reduced; or a second zero dispersion wavelength leading to dispersive wave generation and spectral recoil of the solitons (Biancalana et al., 2004; Cumberland et al., 2008a). Alternatively, the continuum will simply be limited by the length of fibre available or the level of attenuation. Short wavelength expansion of the supercontinuum is mediated through the excitation of phase-matched dispersive waves (Wai et al., 1986; Akhmediev and Karlsson, 1995) and soliton trapping mechanisms (Beaud et al., 1987; Nishizawa and Goto, 2002; Genty et al., 2004; Gorbach and Skryabin, 2007a). In the following sections we study these processes in detail.

8.2.2 What is a CW pump source?

Before we begin it is worth stepping back and considering what exactly our CW pump sources consist of. As we noted above, the majority of pump sources used experimentally are fibre lasers of a few tens of watts. Although single longitudinal mode lasers with these powers are possible they are almost never used and, as we shall see below, would be sub-optimal. Instead the pump sources usually have a spectral width of around one nanometre. This means that they have thousands of longitudinal modes. A single mode would be purely CW, but thousands of modes without any mode-locking element will consist of significant intensity fluctuations, because each mode will not have a precisely fixed phase with respect to another. Even if at some point the fluctuations are zero, they will immediately re-emerge due to dispersion. An amplified spontaneous emission (ASE) source will be similar. In fact, the major contributor to spectral broadening in a CW laser or ASE source is self-phase modulation which is known to decrease the coherence of a partially coherent source (Agrawal, 2004). The only information we have about these CW sources consists of their average power and their average spectral content. It is also possible to measure the average fluctuation duration (coherence time) through autocorrelation or interferometry. But a full reconstruction of the field is, at present, impossible. Modelling such a field poses some problems. Some recent theoretical work has quantified the average statistical behaviour of such systems with concepts of wave turbulence (Babin et al., 2007). But for numerical simulations, we require a representative single shot model of the noise, towards which a number of attempts have been made (Frosz et al., 2006; Barviau et al., 2006; Vanholsbeeck et al., 2005; Travers et al., 2008a; Mussot et al., 2004).

In what follows, for analytic treatments we consider a pure CW source with no fluctuations. This means that real experimental results can be expected to differ slightly depending on the strength of the pump noise fluctuations.

For numerical work, based on the generalised nonlinear Schrödinger equation,[1] our initial conditions are necessarily going to be approximate. This means we will not fully understand the initial noise conditions of the continuum evolution which are so important to the dynamics. The technique we use in what follows (unless otherwise stated) is an ASE model based on iterating an initial white noise field through an amplifier stage and a spectral filter stage until the average power

[1] Although the physics of CW supercontinuum generation are well modelled by the generalised nonlinear Schrödinger equations described by Travers et al. in an earlier chapter, difficulties, apart from the initial conditions, arise due to the time scales involved. To make the simulations tractable we can only simulate a snapshot of the field as it propagates. We, therefore, have to carefully choose a time window which contains sufficient information to accurately reproduce experimental observations. In what follows we use a 256 ps time window with 2^{17} grid points and a central wavelength of 1070 nm, to simulate the supercontinua. For each set of parameters we average the simulations over 10 sets of initial conditions to more accurately reproduce experimental observations which typically average on the scale of milliseconds.

and spectral width match our chosen pump conditions; this scheme is designed to replicate the important components in a real fibre laser system. This leads to the development of a triangular spectrum (on a logarithmic scale),[2] similar to those experimentally measured for fibre lasers and ASE sources, while creating a reasonable representation of the temporal and spectral noise fluctuations.

8.2.3 Modulation instability

In many physical systems nonlinear waves suffer an instability which leads to the spread of energy from the central wave frequency to neighbouring frequencies. In the time domain this is characterised by the splitting of the wave into shorter temporal features. This process is termed modulation instability or in some other systems, Benjamin–Feir instability, after its discoverers (Lighthill, 1965; Benjamin and Feir, 1967).

The physical parameters required for such instabilities to occur are the same as those for soliton propagation or, more generally, self-trapping of a wave. This is when the dispersive or diffractive effect of the medium acts opposed to the nonlinear wave spreading. The process can be understood as follows: fluctuations of the optical intensity lead to small local modifications of the refractive index through the Kerr effect, which then form localised refractive index wells which further trap the intensity fluctuations, leading to an intensity enhancement. This process naturally leads to the formation of solitons (Karpman, 1967; Hasegawa and Brinkman, 1980).

The equivalent frequency domain picture is that of four-wave mixing, with the dominant phase-matching contribution provided by the nonlinear refractive index. This process amplifies the background noise at side-bands either side of the pump, corresponding to the frequency of modulation in the time domain.

To illustrate this process, Fig. 8.2 shows the evolution of an initially continuous wave into a train of solitons in an optical fibre. It is immediately apparent that the resulting peak powers are much higher than the input pulse, and that the resulting soliton durations can be short. This then provides the seeds for a CW continuum.

The frequency separation between the pump and maximum side-band gain is given by Eqn. 8.1 (Hasegawa and Brinkman, 1980),

$$\Delta\omega_{\mathrm{mi}} = \sqrt{\frac{2\gamma P_{\mathrm{cw}}}{|\beta_2|}}, \tag{8.1}$$

and the maximum gain is simply $2\gamma P_{\mathrm{cw}}$. In these equations, P_{cw} is the average continuous pump power, β_2 is the group velocity dispersion coefficient and γ is

[2] A triangular spectrum on a logarithmic scale corresponds to a sech^2 shaped linear scale spectrum.

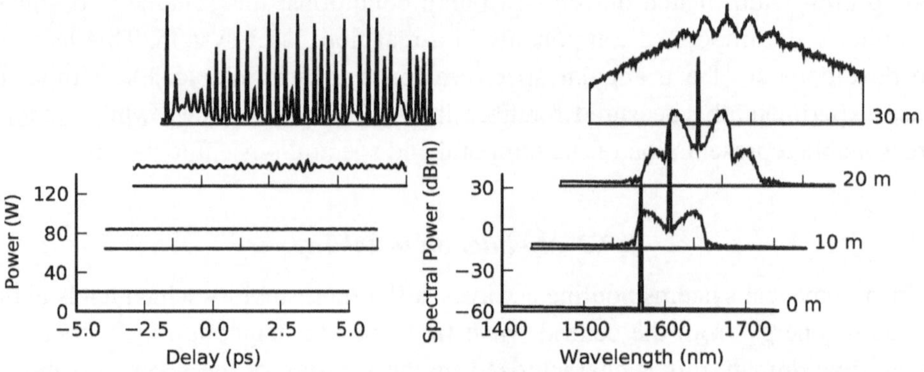

Fig. 8.2. The development of MI from a 20 W pure CW pump laser with additional shot noise (not intensity fluctuations). The left panel shows the time domain evolution into intensity spikes, the right panel show the amplification of frequency side-bands from background noise. The parameters were $\gamma = 20\ \mathrm{W^{-1}km^{-1}}$, $D \sim 2\ \mathrm{ps\ nm^{-1}km^{-1}}$.

the nonlinear coefficient. These side-bands correspond to a modulation in the time domain over the MI period, obtained from Eqn. 8.2,

$$T_{\mathrm{mi}} = \frac{2\pi}{\Delta\omega_{\mathrm{mi}}} = \sqrt{\frac{2\pi^2|\beta_2|}{\gamma P_{\mathrm{cw}}}}. \tag{8.2}$$

This period is important for several reasons. Firstly, as we will see in Section 8.2.6, it determines a lower limit of the coherence time of a pump source which will experience MI. Secondly it determines the number of solitons generated per unit time. Finally it allows one to calculate the energy that each soliton will contain (E_{sol}). This is done by noting that under conditions of energy conservation the energy contained within one modulation period (E_{mi}) must entirely end up in the soliton that emerges from that period (Dianov et al., 1992; Kutz et al., 2005):

$$E_{\mathrm{sol}} = 2P_0\tau_0 = E_{\mathrm{mi}} = P_{\mathrm{cw}}T_{\mathrm{mi}} \tag{8.3}$$

where P_0 and τ_0 are the soliton peak power and duration respectively.

Eqn. 8.3 suggests that the soliton duration and peak power are related to the fibre properties; in particular if we minimise the MI period (i.e. maximise its bandwidth) we can expect to generate the shortest solitons. This is illustrated in Fig. 8.3 which shows the pulse structures obtained from a 20 W CW pump source propagating through fibres with a range of nonlinear and dispersive parameters. In this figure we have indicated the MI period with the horizontal bars and we can see that the pulse structures appear to follow these periods closely.

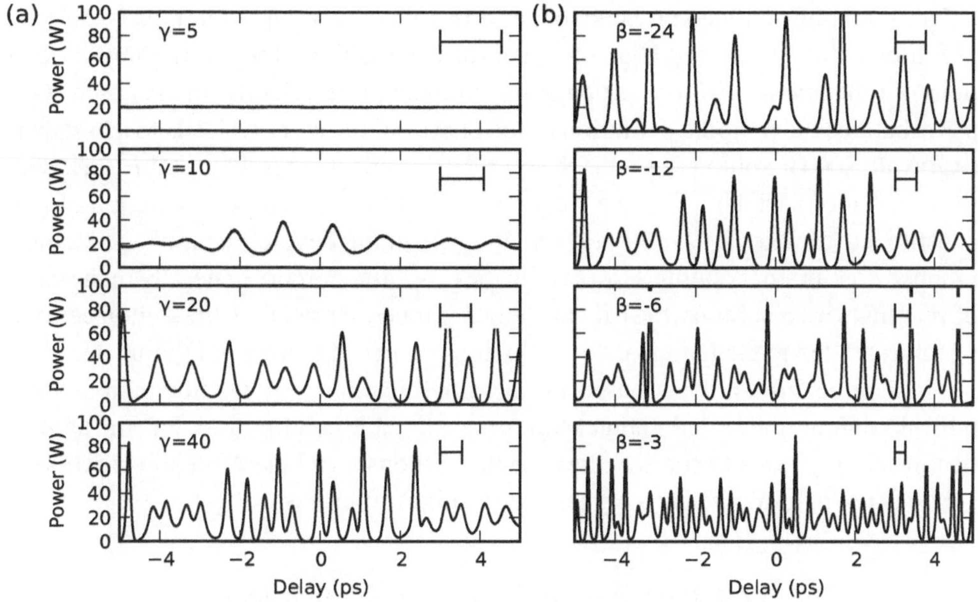

Fig. 8.3. Examples of the formation of pulse structures due to MI in 50 m of fibre with 20 W pump power. (a) For $\beta_2 = -12$ ps^{2}km^{-1} and the indicated value of γ in units of W^{-1}km^{-1}. (b) For $\gamma = 40$ W^{-1}km^{-1} and the indicated value of β_2 in units of ps^{2}km^{-1}. The initial conditions were a pure CW pump field (no intensity fluctuations). The bars show the calculated MI period.

We can estimate the duration of the solitons emitted from MI as follows. First we substitute for P_0 in Eqn. 8.3 using the relation between τ_0 and P_0 from the soliton equation (Agrawal, 2004) to obtain:

$$\frac{2|\beta_2|}{\gamma \tau_0} = P_{\mathrm{cw}} T_{\mathrm{mi}} \tag{8.4}$$

which after some manipulation gives

$$\tau_0 = \frac{1}{\pi^2} T_{\mathrm{mi}}, \tag{8.5}$$

i.e. the MI full width half maximum duration is approximately $T_{\mathrm{mi}}/5$. A similar result was shown by Dianov et al. (1992).

The soliton period (T_{mi}) is not a fixed quantity when evolving from background noise as can be seen from the bandwidth of the MI gain (see e.g. Agrawal, 2004). Instead there will be some jitter on the period and hence also on the soliton duration (and hence peak power). (Note that if seeded with a narrow Stokes line MI can give very precise pulse trains.) So a range of soliton durations will be produced, which is the key requirement for a smooth continuum evolution.

One common confusion related to the formation of CW supercontinua is the role of soliton fission processes. High order solitons, or multi-solitons, are simply collections of fundamental solitons propagating with the same velocity. In the absence of perturbations the periodic evolution of the phases of the solitons leads to a breather (Agrawal, 2004). Soliton fission is the break-up of this inherently unstable state due to perturbations which affect the frequency, amplitude or temporal location of the constituent solitons. It usually plays a vital role in the development of short pulse pumped continuum systems. However in the CW case, or more generally in the case of very high order solitons, the MI gain length is much shorter than the soliton period, which is related to the length at which fission occurs. Therefore MI dominates.

Modulation instability is based on the amplification of noise fluctuations into pulse-like structures. Fundamental solitons are a stable condition – adiabatic amplification simply leads to shorter fundamental solitons – and therefore the process of MI does not naturally lead to higher order soliton solutions.

8.2.4 Raman scattering and soliton collisions

Although modulation instability can provide the majority of the spectral expansion for systems with high peak power and long pulse duration, under available CW scale pump powers it can only provide the seed solitons for supercontinuum formation. The majority of the spectral expansion is actually provided by Raman scattering.

8.2.4.1 Raman self-scattering

Raman self-scattering of solitons, also known as the soliton self-frequency shift, can lead to very significant red-shift for short duration solitons. The equation for soliton frequency shift through a fibre is given approximately by Eqn. 8.6 (Gordon, 1986),

$$\frac{\partial \omega}{\partial z} \propto -\frac{|\beta_2|}{\tau_0^4}. \tag{8.6}$$

The strong dependence on the soliton duration reflects the requirement for the soliton bandwidth to be a significant fraction of the Raman gain spectrum, for self-scattering to occur. Note that in this relation we have neglected high order dispersion and the dispersion of γ, which we will account for in later sections.

From this relation and Eqns. 8.2 and 8.5 above we have:

$$\frac{\partial \omega}{\partial z} \propto -\frac{|\beta_2|}{T_{\mathrm{mi}}^4} \propto -\frac{\gamma^2 P_{\mathrm{cw}}^2}{|\beta_2|}. \tag{8.7}$$

Thus we can expect the greatest spectral expansion when pumping highly nonlinear, low dispersion fibres with high pump power. But despite these observations, we will see below that one should not simply minimise the dispersion, as this will have an unavoidable effect on higher order dispersion, which can completely prevent continuum development.

8.2.4.2 Soliton collisions

Any temporal overlap between a soliton or by the undepleted pump and another further red-shifted soliton, within the spectral range of Raman scattering, will enhance the energy of the red-shifted soliton – this is known as a soliton collision. Red-shifted solitons travel more slowly than those at higher frequencies. Therefore any two frequency separated solitons may eventually collide while propagating through a fibre. As they do so, some energy will be transferred to the most red-shifted soliton. This results in a further red-shift for that soliton and an increase in its energy. The energy increase leads to a corresponding compression and so the red-shifted soliton is now also shorter in duration, leading to further Raman self-scattering. The increased red-shift also slows the soliton down, leading it to pass by more solitons and hence experience more collision events. In combination, these effects lead to a significantly enhanced red-shift for the solitons, and also for a number of solitons to become especially dominant and particularly red-shifted.

The energy transferred in a collision will in general not be precisely the amount required for a soliton to maintain its shape on propagation, and therefore some energy is shed in the form of dispersive waves, after a collision event. This provides a clear signature of collision events in the spectrograms, which we shall observe in later sections.

Soliton collisions can considerably increase the extent of a supercontinuum compared to Raman self-scattering alone. Firstly, the spectral separation between the solitons can coincide with the peak Raman gain, whereas for Raman self-scattering the soliton bandwidth is often not broad enough for this to occur (around 50 nm at 1070 nm). Secondly, in CW pumped systems the number of solitons can be very large, leading to a large number of scattering events. They can also initiate CW supercontinuum generation for parameter regimes where the soliton bandwidths generated by MI are themselves insufficient for Raman self-scattering to significantly occur.

The role of soliton collision in supercontinuum generation as a whole was identified by Islam et al. (1989b), and for CW continua in particular by Frosz et al. (2006) and Korneev et al. (2008). In these works it has been shown that soliton collisions are in fact the dominant mechanism for spectral expansion in CW pumped supercontinua.

8.2.5 Influence of group velocity dispersion and nonlinearity

The variation of supercontinua generated by a CW pump in fibres with differing values of dispersion and nonlinearity is shown in Fig. 8.4. From Fig. 8.4(a) we

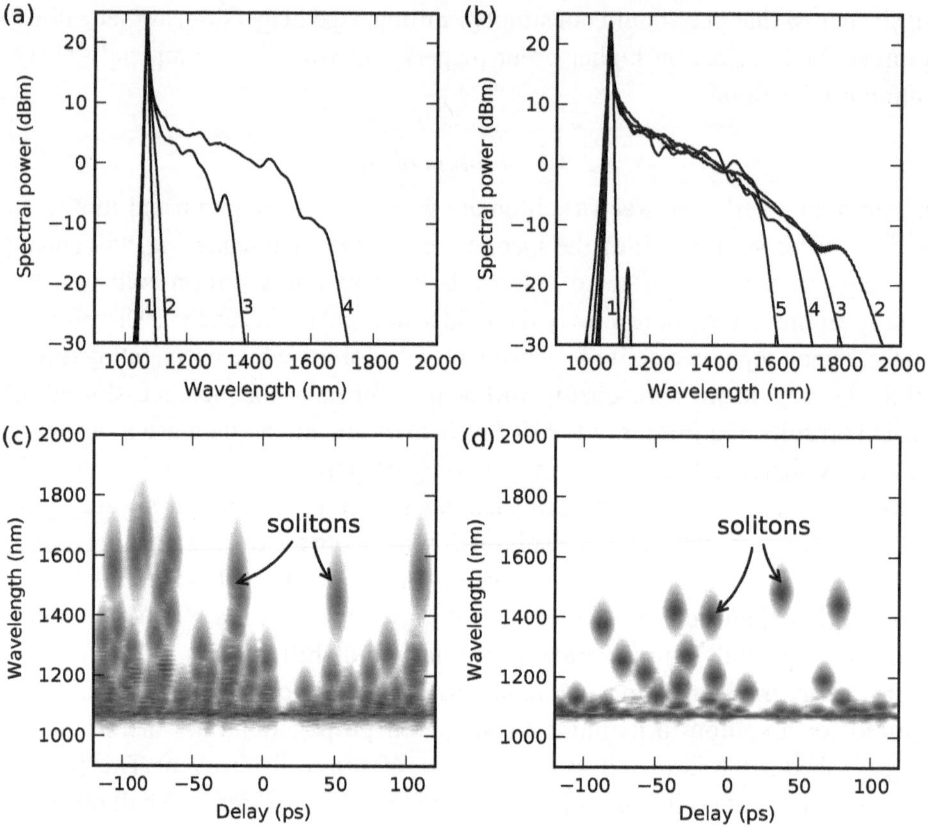

Fig. 8.4. (a and b) The simulated spectra after propagation of a 1 nm bandwidth 10 W CW field through 20 m of fibre with: (a) a dispersion of $\beta_2 = -12.0\,\mathrm{ps^2km^{-1}}$ and nonlinearity (1) 5, (2) 10, (3) 20 (4) 40 $\mathrm{W^{-1}km^{-1}}$; (b) a nonlinearity of 40 $\mathrm{W^{-1}km^{-1}}$ and dispersion of (1) +3, (2) −3, (3) −6, (4) −12, (5) −24 $\mathrm{ps^2km^{-1}}$. (c and d) Spectrograms of the field out of the 20 m fibre for $\gamma = 40\,\mathrm{W^{-1}km^{-1}}$ and (c) $\beta_2 = -3\,\mathrm{ps^2km^{-1}}$ and (d) $\beta_2 = -24\,\mathrm{ps^2km^{-1}}$. The full scale of the density plots, from white to black, is 50 dB.

see that increasing nonlinearity clearly broadens the spectrum. This is due to both the greater MI gain and Raman gain. Similarly the continuum width significantly extends for smaller values of anomalous dispersion as shown in Fig. 8.4(b). These observations are explained well by Eqn. 8.6. One further point is the number and duration of the solitons formed through MI. The spectrograms (also referred to as XFROG traces) shown in Fig. 8.4(c and d) show that many of the solitons formed through MI have red-shifted (the solitons are the regularly shaped structures in the spectrograms). Fig. 8.4(c) is for a smaller value of dispersion and shows many more solitons than Fig. 8.4(d) and they are of shorter duration. Consequently they have red-shifted further. Note that the curve marked (1) in Fig. 8.4(b) is for a small value of normal dispersion and shows little spectral broadening, but rather a Raman Stokes line has emerged because in this case MI cannot occur.

8.2.6 Effect of pump bandwidth

Fig. 8.5 shows the dependence of supercontinuum spectral width on the pump bandwidth for one set of simulation parameters. It is clear from Fig. 8.5(a) that there is an optimum value for the pump bandwidth to obtain the broadest continuum

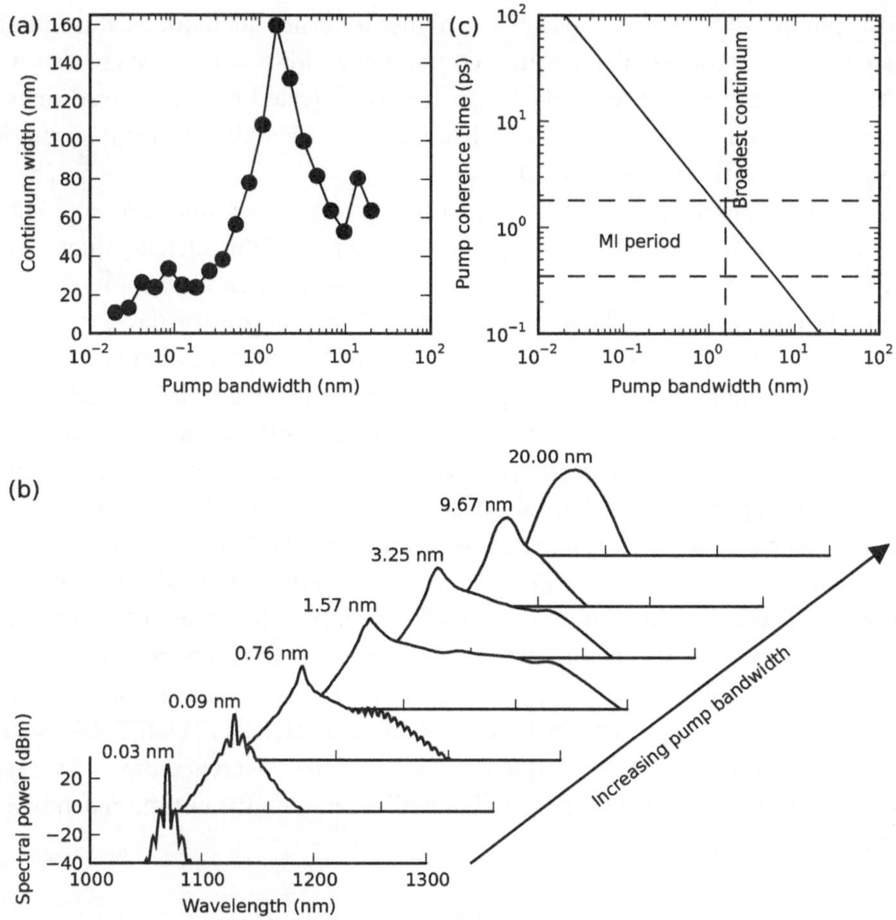

Fig. 8.5. Numerical results showing the dependence of continuum width on pump source spectral bandwidth. The results are for 10 W average power in 10 m of fibre with $\beta_2 = -12\,\mathrm{ps^2km^{-1}}$ and $\gamma = 40\,\mathrm{W^{-1}km^{-1}}$. Each point or plot is the ensemble average of ten simulations with a Gaussian spectrum with random spectral phase as initial conditions. (a) A summary plot, each point is the 30 dB spectral width of the continuum produced for the given pump FWHM bandwidth; (b) some selected spectra showing the dependence on pump bandwidth. Panel (c) shows the temporal coherence time of the pump as a function of bandwidth (solid line), along with the bandwidth producing the broadest continuum width, and the duration of the MI periods for the fluctuation power levels seen in the pump noise (horizontal dashed lines).

spectral width. From Fig. 8.5(b) it is clear that for a narrow pump bandwidth we see clean MI sidelobes, relatively close to the pump frequency, indicating a relatively slow modulation, whereas for broader bandwidths a broad continuum is formed. But it is interesting to note that further increasing the pump bandwidth actually leads to narrower continuum spectra. This has been confirmed experimentally by Martin-Lopez et al. (2006). This can be understood by considering the coherence time of the pump light. As illustrated in Fig. 8.5(c), the coherence time decreases with the pump bandwidth. As this leads to higher intensity fluctuations, this initially enhances the MI process. The optimum pump bandwidth corresponds to that with a coherence time very close to the MI period as indicated by the crossed lines in Fig. 8.5(c). When the coherence time becomes significantly shorter than the MI period, then the MI efficiency is reduced.

An intuitive understanding of the effect of the coherence time can be obtained from Fig. 8.6(a). This shows some single-shot examples of the intensity fluctuations for the 1.57 nm and 20 nm pump bandwidths, along with an indication of the MI period. The longer duration fluctuations of the 1.57 nm pump are on the same scale as the MI period and hence are sufficiently coherent for MI to occur, whereas the 20 nm pump exhibits fluctuations that change many times during one MI period and are hence too incoherent for MI. Fig. 8.6(b) clearly illustrates this as the 20 nm pump has generated little soliton amplification, not reaching over 100 W peak power, whereas the 1.57 nm pump has generated a 1.4 kW peak power soliton with sub 30 fs duration.

As long as the pump coherence time is longer than the MI period, the intensity noise fluctuations enhance the soliton energies formed through MI and hence the continuum. Therefore, the optimum pump bandwidth is that which has the maximum bandwidth while still being sufficiently coherent for MI. The relation between bandwidth and coherence time depends on the specific pulse shape; for a Gaussian spectrum it is approximately given by $\Delta\nu \sim 0.66/\tau_c$ (Saleh and Teich, 2007), where $\Delta\nu$ is the pump bandwidth (in frequency) and τ_c is the coherence time. If we take a coherence time of twice the MI period as sufficient, then the optimum bandwidth is approximately given by:

$$\Delta\lambda_{\mathrm{opt}} \sim \Delta\nu_{\mathrm{opt}}\frac{\lambda^2}{c} \sim \frac{0.66}{2T_{\mathrm{mi}}}\frac{\lambda^2}{c} \sim 0.07\sqrt{\frac{\gamma P_{\mathrm{cw}}}{|\beta_2|}}\frac{\lambda^2}{c}. \tag{8.8}$$

For the parameters of Fig. 8.5 this gives $\Delta\lambda_{\mathrm{opt}} = 1.5$ nm.

8.2.7 Influence of third order dispersion

All of the analysis until now has neglected higher order dispersion terms. However, the continuum bandwidths we have considered propagate far outside the region

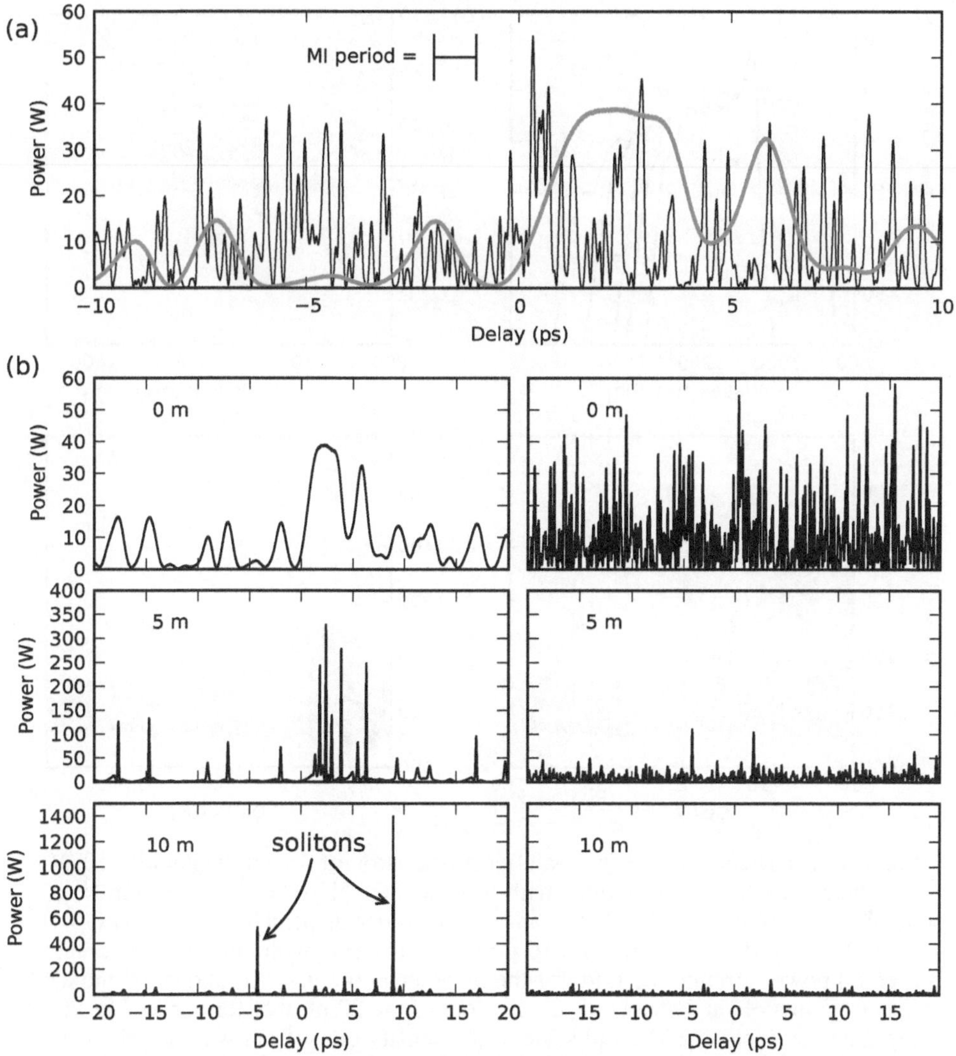

Fig. 8.6. (a) Single-shot examples of the intensity corresponding to the 1.57 nm pump bandwidth (thick grey) and 20 nm bandwidth pumps (thin black); also marked is the MI period for the conditions of Fig. 8.5. (b) The evolution of a single shot propagating through the same fibre as Fig. 8.5, for the 1.57 nm pump bandwidth (left-hand side) and the 20 nm pump bandwidth (right-hand side).

of flat dispersion. In this section we look at the role of the dispersion slope, also called the third order dispersion as it is parameterised through the β_3 term of the propagation constant expansion.

Figure 8.7(a) shows how the supercontinuum is curtailed by positive third order dispersion which means that the dispersion magnitude (β_2 or D) is increasing with

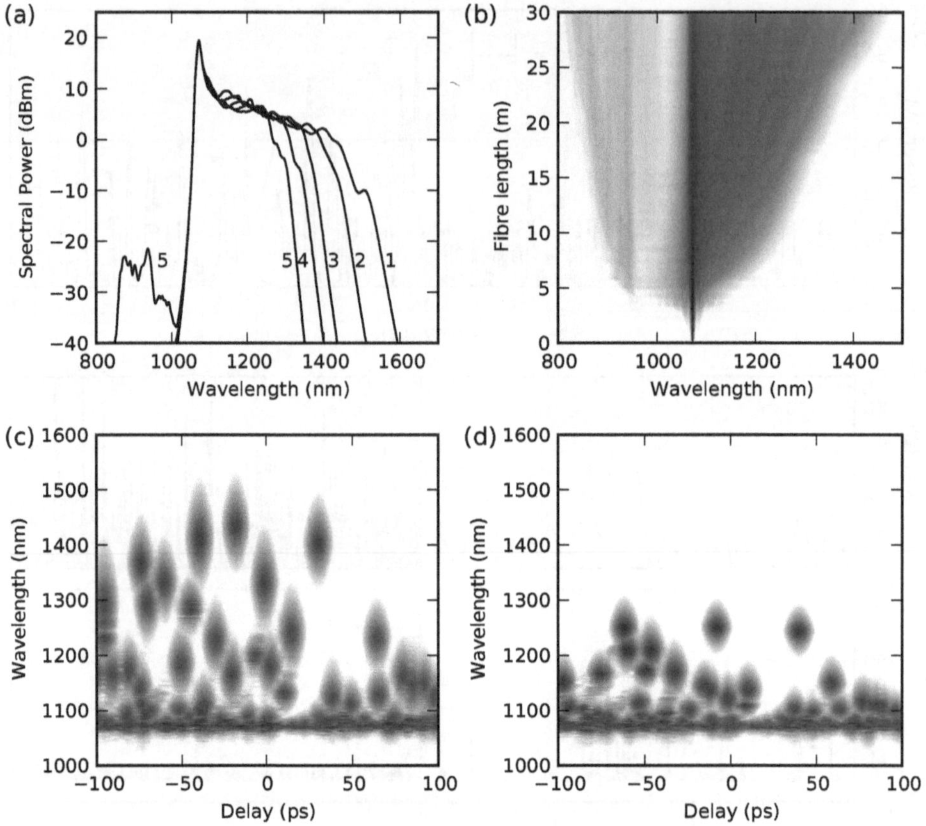

Fig. 8.7. (a) The simulated spectra after propagation of a 1 nm bandwidth 10 W CW field through 20 m of fibre with a dispersion of $\beta_2 = -12.0$ ps^{2}km^{-1}, a nonlinearity of $\gamma = 40$ W^{-1}km^{-1} and a third order dispersion (β_3) of (1) 0.01, (2) 0.02, (3) 0.04, (4) 0.08, (5) 0.16 ps^{3}km^{-1}. (b) The evolution of the average spectral power through the fibre for the parameters of curve (5) above. (c and d) Spectrograms of single shots of the field out of the 20 m fibre for curves (1) and (5) above respectively. The full scale of the density plots, from white to black, is 80 dB for the spectral evolution (b) and 50 dB for the spectrograms.

increasing wavelength (negative third order dispersion is rarely experienced at common pump wavelengths). As the magnitude of the dispersion slope increases, the continuum is restricted further. The reason for this is due to the soliton condition. All else being equal (i.e. no loss, constant γ etc.) the soliton duration has an inverse relationship with dispersion. That means that as the solitons red-shift, they become longer as they climb the dispersion 'hill' formed by the positive dispersion slope. This means that they are reduced in spectral bandwidth and their Raman self-scattering rate is reduced. The increased duration and reduced bandwidth of the red-shifting solitons can be observed by comparing Figs. 8.7(c) and (d). We can include this effect in Eqn. 8.6 by substituting for $|\beta_2|$ the frequency dependent value including the effect of β_3: $\beta_2(\omega) = \beta_2 + (\omega - \omega_0)\beta_3$, with all of the β expansion terms given at ω_0, the pump frequency. This gives the following expression for the Raman self-scattering driven spectral shift:

$$\frac{\partial \omega}{\partial z} \propto -\frac{(\frac{\omega}{\omega_0}\gamma)^2 P_{\mathrm{cw}}^2}{|\beta_2 + (\omega - \omega_0)\beta_3|}, \tag{8.9}$$

from which it is clear that for positive third order dispersion the Raman self scattering of solitons to lower frequencies is restricted. In Eqn. 8.9 we have also included the effect of Stokes losses on the nonlinearity (included via the shock or self-steepening term in the GNLSE), which also decrease the rate of soliton shift. Typically the effective area of a fibre also increases towards lower frequencies further reducing the shift, and this can also be included in Eqn. 8.9.

8.2.8 Dispersive wave excitation and four-wave mixing

Another interesting effect of the third order dispersion is that it allows for the excitation of dispersive waves at spectral regions in the normal dispersion regime. This can be seen, at a very low level, for curve (5) in Fig. 8.7(a) and the spectral evolution in Fig. 8.7(b), where dispersive waves have been excited.

If the solitons are spectrally broad enough such that they overlap with the normal dispersion region, then they can excite a dispersive wave at certain phase-matched wavelengths (Wai et al., 1986; Akhmediev and Karlsson, 1995). This phase-matching is unique to solitons as it depends on the extra nonlinear phase a soliton possesses. It can be understood as a tunnelling effect from the self-trapping soliton to free propagating radiation (Karpman, 1993) and is analogous to Cherenkov radiation (Akhmediev and Karlsson, 1995). Full details of this process are given in other chapters of this book. The phase-matching condition is simply $\beta_{\mathrm{sol}} = \beta_r$, with $\beta_{\mathrm{sol}} = \gamma P_0/2$ the propagation constant of the soliton and β_r that of the dispersive wave. When just considering the expansion of β_2 and β_3 (about the soliton frequency, ω_{sol}) the phase-matching condition is given by:

$$\frac{\gamma P_0}{2} = \frac{(\omega_r - \omega_{\mathrm{sol}})^2}{2}\beta_2 + \frac{(\omega_r - \omega_{\mathrm{sol}})^3}{6}\beta_3. \tag{8.10}$$

As β_2 must be negative for soliton propagation, we see that without any higher order dispersion terms the phase-matching condition cannot be met. However, if β_3 is non-zero phase-matching can be achieved. When β_3 is positive, i.e. the cases discussed above, then phase-matching to higher frequencies is possible. This is the conventional situation where wavelengths in the normal dispersion region at shorter wavelengths than the zero dispersion are excited, as seen in Fig. 8.7(a and b) and in the next section. Another possibility is for β_3 to be negative and the dispersive wave to be at lower frequencies than our soliton. This case is found when we meet a second zero dispersion wavelength at the long edge of the supercontinuum and will be discussed in Section 8.2.10.

The rate of energy transfer from the soliton to the dispersive wave is proportional to the magnitude of the soliton at the dispersive wave frequency and this decreases exponentially in amplitude as they are phase-matched further and further from the soliton (Biancalana et al., 2004). Therefore there is a requirement for significant spectral overlap between the soliton and the phase-matched wavelength. For CW pumped systems where the solitons generated are often not as short as in pulse pumped systems, this requirement makes the generation of dispersive waves relatively rare, and only when propagating very close to the zero dispersion wavelength.

Other four-wave mixing processes can occur between solitons and dispersive waves, or purely between dispersive waves (Skryabin and Yulin, 2005).

Usually the MI side-bands are not broad enough to generate dispersive waves directly in the normal dispersion region, but the phase-matched process discussed above is required. Noting the importance of spectral overlap, we see that in the fibres used in Fig. 8.7, only the one with the largest β_3 was able to achieve this. For that fibre the zero-dispersion wavelength is at 1026 nm. We show more optimal examples shortly.

With regard to generating blue-shifted radiation from our CW pump sources, especially with an eye towards visible generation, there is one restriction of all four-wave mixing processes, which is that they excite only certain fixed blue-shifted wavelengths based on a strict phase-matching condition. Further blue shift depends on soliton trapping effects.

8.2.9 Soliton trapping

If dispersive waves have been successfully excited from the initial CW pump, then further blue shift can be caused by soliton trapping which has been described in a range of papers (Beaud et al., 1987; Nishizawa and Goto, 2002; Genty et al., 2004; Gorbach and Skryabin, 2007a; 2007b) and by Travers (2009) specifically for CW pump conditions. The most important requirement for soliton trapping of dispersive waves to occur is for group velocity matching between the red-shifting soliton and dispersive wave to be maintained, so that as the soliton shifts to longer wavelengths, the dispersive wave shifts to shorter wavelengths with the same group velocity. This process does occur with remarkable robustness, but will eventually

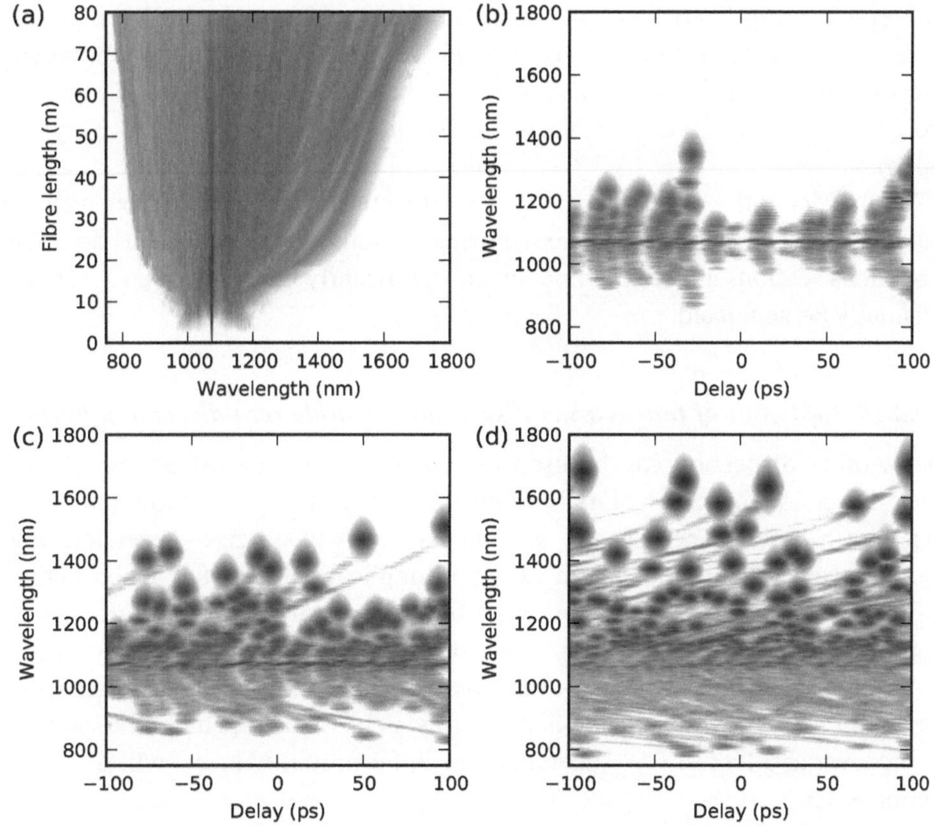

Fig. 8.8. Simulations of CW continuum generation in a fibre with $\gamma = 10\ \mathrm{W^{-1}km^{-1}}$, $\beta_2 = -0.1\ \mathrm{ps^2km^{-1}}$ and $\beta_3 = 0.06\ \mathrm{ps^3km^{-1}}$ with a 50 W pump laser. (a) The evolution of spectral power through the fibre; (b–d) the spectrograms of the field at (b) 20 m, (c) 40 m, (d) 80 m. The full scale of the density plots, from white to black, is 80 dB for the spectrum evolution (a) and 50 dB for the spectrograms.

be limited, usually by a second zero-dispersion wavelength, or simply extremely high losses on the long wavelength edge preventing further soliton red-shift. Other chapters of this book also describe this process in more detail, here we simply describe some results based on numerical experiments.

Fig. 8.8(a) shows the spectral evolution of a 50 W pump laser source in fibre designed for efficient visible generation through a strong third order dispersion and low dispersion parameter which position the zero dispersion wavelength at 1069 nm. The spectrum extends significantly short of the 1070 nm pump wavelength, as the long wavelength edge expands. Fig. 8.8(b) shows the spectrogram of the field at 20 m of propagation. From here it is clear that solitons are splitting from the pump wave and starting to red shift. Clear excitation of dispersive waves is observed in Fig. 8.8(b to d) as compared to Fig. 8.7(c or d). In Fig. 8.8(c and d) we see that some portions of these dispersive waves are moving synchronously with strongly red-shifting solitons, which is characteristic of the soliton trapping effect (Beaud et al., 1987; Nishizawa and Goto, 2002; Gorbach and Skryabin, 2007b).

This process will eventually be limited by either the breaking of the group velocity matching or by the halting of the red-shifting solitons due to the reasons described in previous sections above. But note that a significantly blue-shifted spectrum can potentially be generated.

8.2.10 Influence of fourth order dispersion – double zero-dispersion fibres

Fourth order dispersion can be used to somewhat mitigate the effects of third order dispersion in terms of it limiting the red-shift of the continuum. However, fourth order dispersion can also introduce a second zero-dispersion point at longer wavelengths which has a variety of implications. Fig. 8.9(a) shows three dispersion curves (in engineering units) for three positive values of β_4. As the magnitude of β_4 increases, the second zero-dispersion point moves to shorter wavelengths. The curves between the two zero-dispersions are much flatter than when the fourth order dispersion is negligibly small. This fact has led to some of the broadest and flattest CW supercontinua so far produced as we will discuss in Section 8.3.5.

As solitons can only propagate in anomalous dispersion, the second zero-dispersion prevents extension of the continuum beyond that point. This is evident in the simulated spectra shown in Figs. 8.9(b) and (c). As the solitons shift towards the second zero they excite dispersive waves in the normal dispersion region, through the process discussed in Section 8.2.8 and are themselves spectrally recoiled due to conservation of energy and their red-shift is cancelled. This has been studied in detail (Akhmediev and Karlsson, 1995; Skryabin et al., 2003; Biancalana et al., 2004). The highest energy solitons shift most rapidly, generating a dispersive wave and becoming frequency-locked before the second zero-dispersion wavelength. These solitons continue to propagate along the fibre and thus affect the rate of Raman shift of subsequent solitons, which lose energy to the further frequency shifted solitons, through collisions, and thus their own red-shift slows down. As the number of solitons at the second zero-dispersion increases, so the number of solitons across the continuum increases and the overall spectral power between the two zero-dispersion wavelengths is enhanced. Despite the collisions reducing the velocity of subsequent solitons, eventually a significant number of solitons come to rest before the second zero-dispersion wavelength. This increase in spectral power density is clearly visible in the simulated spectra of Fig. 8.9(b and c). The bunched

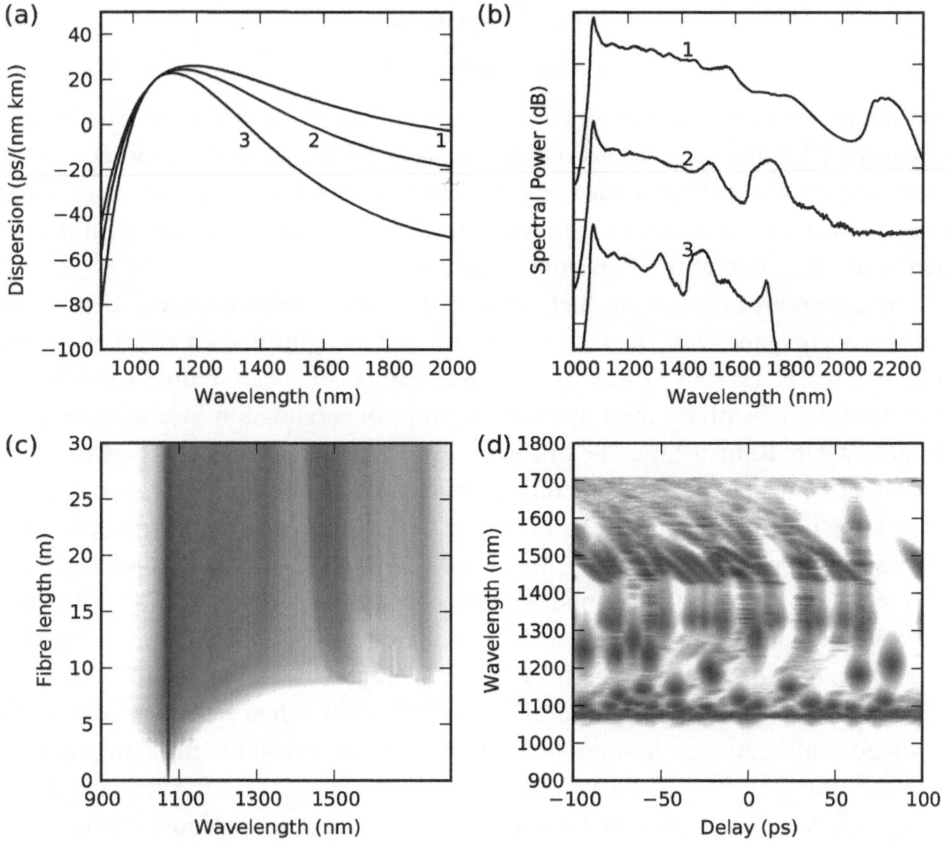

Fig. 8.9. (a) The dispersion curves including fourth order dispersion. (b) The simulated spectra (each an average of 10 simulations) after propagation of a 1 nm bandwidth 10 W field through 20 m of fibre with a dispersion of $\beta_2 = -12.0\ \mathrm{ps^2km^{-1}}$, a nonlinearity of $\gamma = 40\ \mathrm{W^{-1}km^{-1}}$, a third order dispersion of $\beta_3 = 0.06\ \mathrm{ps^3km^{-1}}$ and a fourth order dispersion (β_4) of (1) 0.0002, (2) 0.0003 and (3) 0.0004 $\mathrm{ps^4km^{-1}}$. These curves are offset by 40 dB and the vertical scale is 20 dB/division. (c) The evolution of the average spectral power through the fibre for the parameters of curve (3) above. (d) The spectrograms of a single shot of the field out of 30 m of the fibre for curve (3) above. The full scale of the density plots, from white to black, is 80 dB for the spectrum evolution (c) and 50 dB for the spectrogram (d).

solitons are also visible in the spectrogram of Fig. 8.9(d) along with the chirping of the dispersive waves beyond the second zero-dispersion wavelength. Beyond the dispersive wave, there is no mechanism for further spectral shift except for the generation of Raman Stokes waves which can also be seen in Fig. 8.9(b, c and d). Signatures of collisions between the solitons can be observed in the low power dispersive traces between the solitons in Fig. 8.9(d).

8.3 Experiment

8.3.1 Early work

Continuous wave continuum sources in the form of lamps (such as tungsten filament or xenon arc lamps) are perhaps one of the most widely used light sources in the laboratory, for applications such as spectroscopy (Johnson et al., 1974). But scaling of the average power levels and the very low spatial coherence severely limit such sources as compared to modern supercontinua.

Supercontinuum generation had been studied for several decades before CW continua were generated. In fact a detailed and insightful theoretical work by Golovchenko et al. (1991) was published a number of years before experimental realisation, describing the important regimes of modulation instability and its interaction with Raman gain. The first experimental report of spectral broadening of a CW laser, to which the term supercontinuum could be reasonably applied, was by Persephonis et al. (1996). In this work, an Yb:Er fibre laser was used to pump 2.3 km of Ge-doped fibre with 1 W at 1565 nm. A 100 mW supercontinuum spanning from 1500 to 1950 nm was obtained with good spectral power between 1500–1700 nm. Several years later a 1 W supercontinuum with a bandwidth of 100 nm, centred at 1480 nm, pumped with a Raman fibre laser was reported (Prabhu et al., 2000).

The rate of progress changed dramatically in 2002 when Popov et al. (2002) presented conference results demonstrating a 2.2 W supercontinuum spanning over 1000 nm, pumped with a 10 W erbium laser, the spectrum of which is shown in Fig. 8.1. Very soon after that the first demonstration in photonic crystal fibre (PCF) (Avdokhin et al., 2003) showed that multi-watt average powers are achievable in a simple experimental configuration. To achieve this a 15 W master oscillator, power fibre amplifier (MOPFA), Ytterbium based source was used to pump a 100 m PCF with a zero dispersion wavelength around 860 nm. Approximately 9 W was coupled into the fibre resulting in a 3.8 W continuum with 324 nm bandwidth and a very flat spectrum of ±3.5 dB between 1090 and 1375 nm. The resulting spectral power was up to 12 mW/nm. The continuum bandwidth was limited by a dramatic reduction of power after the spectrum reached the water loss region.

Subsequently a large body of work has been produced, which we discuss in the following sections.

8.3.2 Highly nonlinear conventional fibre

Most CW pumped continuum results can be split between those pumped with Raman or erbium lasers around 1400–1600 nm, usually utilising highly nonlinear conventional fibre (HNLF) and those in PCF, usually pumped with Ytterbium systems around 1060 nm. The basic physics and processes are the same between the two. Here we consider the former, and in the following sections we consider results in PCF.

Following from Popov et al. (2002), a significant number of results were reported in HNLF. Prabhu et al. (2003) used higher level pumping of essentially the same setup as in Prabhu et al. (2000) to obtain a 2 W supercontinuum spanning 1430 to 1530 nm. Nicholson et al. (2003) used a 4.5 km fibre, pumped with 0.9 W at a wavelength of 1594 nm leading to a supercontinuum of over 247 nm bandwidth with 210 mW of average power. González-Herráez et al. (2003) used a 2.1 W fibre Raman laser at 1455 nm to pump a 7 km dispersion shifted fibre, with a zero dispersion at 1453 nm, resulting in a 207 nm continuum (20 dB level). Another result with a Raman laser was reported by Abeeluck et al. (2004), where a 4 W pump at 1486 nm led to a broad supercontinuum in a highly nonlinear fibre. Popov et al. (2004) used either a high power Raman laser at 1480 nm or a Yb:Er laser at 1550 nm, depending on the zero dispersion wavelength of the fibre, to produce supercontinua spanning from 1480 to 2050 nm with up to 10 W of output power. Subsequently, Champert et al. (2004) used a tunable fibre laser amplified through a 5 W erbium amplifier to pump 1 km of HNLF at 1550 nm just above the zero dispersion wavelength. A continuum width of 450 nm was achieved with 2.5 W of output power.

A comparison between normal and anomalous dispersion pump wavelengths was made by Abeeluck and Headley (2005). The effect of pump bandwidth on supercontinuum extent was then studied by Martin-Lopez et al. (2006). They found that too broad a pump quenched the MI and reduced the continuum and that a too-narrow bandwidth also reduced the continuum. More recently, Abrardi et al. (2007) used a 1315 nm pump to obtain a CW supercontinuum centred around 1300 nm which is a useful wavelength for OCT. They used 8 W of pump power in 17 km of standard fibres split into four steps with decreasing dispersion to obtain a 260 nm (20 dB level) continuum. Also Rulkov et al. (2008) developed a CW pumped continuum source for OCT systems.

8.3.3 PCF

Work in PCF was going on in parallel with the HNLF results. de Matos et al. (2004) made the first study of the noise characteristics of both laser and ASE pumped supercontinua. A supercontinuum of 5.5 W average power and over 10 mW/nm spectral power was generated from both ASE and laser based Yb pump sources at 1060 nm with up to 10 W pump power and similar $\sim$1 nm pump linewidths. A very low noise was measured for the ASE pumped continuum and much higher noise for the laser pumped continuum, although the continuum spectra were very similar. Again, the supercontinuum was strongly limited by water loss absorption. This source was then used for OCT with $< 5\ \mu$m axial resolution (Hsiung et al.,

2004). Results in low water-loss PCF were reported by Travers et al. (2005b) and later power scaling was demonstrated by Cumberland et al. (2008a) and Travers et al. (2008b).

A number of papers have explicitly, experimentally, confirmed the role of solitons in the continuous wave supercontinuum process. In particular de Matos et al. (2004) measured the autocorrelation of spectrally filtered parts of the supercontinuum. Pulse durations ranging from 0.95 to 0.55 ps were found between 1150 and 1320 nm respectively. At 1060 nm a coherence spike was observed. Similar results were confirmed by Cumberland et al. (2008a), and reasonable agreement between numerically calculated autocorrelations from numerical simulations of the soliton continuum and experiment was found.

Further results in PCF are discussed in more detail below (Cumberland et al., 2008b; Kudlinski et al., 2008a, 2008b; Traynor et al., 2009; Travers et al., 2008b; Mussot and Kudlinski, 2009; Mussot et al., 2007, 2008).

8.3.4 Water loss

The air-holes along the length of photonic crystal fibres provide a large air–silica interface to which water molecules readily become bound. The effect of this water is to cause attenuation at the overtones of the Si-OH bonds. Loss values of several hundred dB/km are common at 1380 nm and are particularly significant for high air-filling fraction PCFs. Fig. 8.10(a) shows the effect of this loss on the expansion of a supercontinuum with two different pump sources. It is clear that although a relatively flat continuum can expand up to 1380 nm, further spectral expansion is curtailed.

A significant advance was made by Travers et al. (2005a) when a PCF which was particularly carefully developed to be free of water during manufacturing, resulting in a water loss peak of 70 dB/km, was used for extended CW pumped supercontinuum. Further improvement was achieved by Travers et al. (2006) and Traynor et al. (2009), the results of which are shown in Fig. 8.10(b). To reduce the water loss a careful manufacturing procedure is required, while still based on the conventional stack-and-draw method, extra chemical cleaning and polishing techniques to avoid the formation of Si-OH liaisons at defect points around the air–silica interfaces are used. The fibre draw is then performed under controlled dry atmosphere conditions in order to avoid extrinsic contamination during this step. The resulting absorption losses achieved by Traynor et al. (2009) using this technique were independent of the fibre mode field diameter, indicating that no excess Si-OH liaisons are formed at the air–silica interfaces. The resulting fibres, as used for the results in Fig. 8.10(b) can have a water loss as low as 30 dB/km.

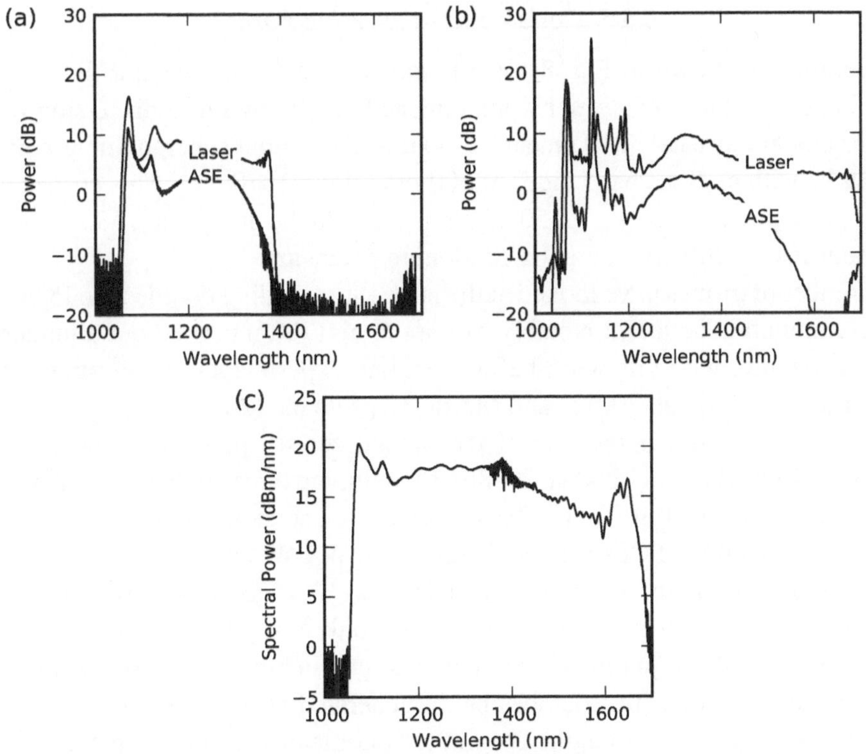

Fig. 8.10. (a) The measured continuum spectra obtained from 100 m of non-specialised photonic crystal fibre for 13 W pump power with the indicated sources. (b) The measured continuum spectra obtained from 200 m of special low-water-loss photonic crystal fibre under the same pump conditions. (c) The spectral power obtained from pumping just 20 m of the same fibre as (a) but with 44 W of launched power from the laser. After Cumberland et al. (2008a) and Traynor et al. (2009).

One point of interest in Fig. 8.10(a and b) is that the laser pump generated a broader continuum than the ASE pump source, at exactly equal pump powers. There are two reasons for this. One is that the ASE source had a broader bandwidth, of 5 nm, as compared to 1.2 nm for the laser, and as discussed in Section 8.2.6, this can significantly affect the continuum bandwidth. Secondly the laser used showed intensity fluctuations at multiples of the laser cavity round-trip time, which may have enhanced the peak pump power.

Apart from reducing the water loss of the fibre, a different approach to reducing the effect of water loss is to pump shorter fibre lengths with higher powers. This approach was taken by Cumberland et al. (2008a) and the results are shown in Fig. 8.10(c). Note that this result was taken with exactly the same fibre as Fig. 8.10(a), but with 20 m rather than 200 m and a pump power of 44 W instead of 13 W.

8.3.5 Double zero-dispersion fibres

The continuum shown in Fig. 8.10(c) is remarkable for its spectral flatness and spectral power. It was achieved by pumping a fibre with two zero-dispersion wavelengths at approximately 810 nm and 1730 nm. The output supercontinuum power was 29 W with an 8 dB bandwidth of 600 nm (1060–1670 nm). This corresponds to a spectral power density over 50 mW/nm between the pump and 1380 nm where the water loss slightly perturbs the continuum extension.

A number of groups have looked at the issues surrounding double zero PCFs and CW continuum generation. Notably Mussot et al. (2007) carried out a numerical study in parallel with this work before separate experimental demonstrations by Kudlinski et al. (2008a, 2008b) and Martin-Lopez et al. (2008).

To investigate further the role of the second zero-dispersion wavelength, the results by Kudlinski et al. (2008a, 2008b), for pumping three double zero-dispersion fibres are shown in Fig. 8.11. The calculated dispersion curves are shown in Fig. 8.11(a). All three PCFs were spliced to a 50 W CW Yb fibre laser.

The results for fibre A are shown in Fig. 8.11(b). For this fibre, the second zero-dispersion is very close to the first, with only a small region of anomalous dispersion around the pump wavelength. The continuum generation in this fibre is dominated by MI, with sidelobes being generated at wavelengths short of the first zero-dispersion wavelength. The limited anomalous region limits the Raman-soliton continuum.

Fibre B has a slightly expanded range of anomalous dispersion, and Fig. 8.11(c) clearly shows that a Raman-soliton continuum extends towards the second zero-dispersion wavelength just short of 1375 nm. There is a significant build-up of spectral power before the second zero, indicating the bunching of the frequency locked solitons, and a strong dispersive wave is generated in the normal dispersion region. Fibre C extends these observations, although water loss around 1380 nm perturbs the spectrum considerably. Interestingly there is actually a build-up of solitons just before the water loss region, indicating that the loss also acts to cancel the soliton Raman self-scattering by reducing the soliton energy such that their bandwidths are insufficient for this process.

The results in Fig. 8.10 and Fig. 8.11 confirm the results in Section 8.2.10, and were obtained only after considerable effort to design optimal fibre conditions for continuum extension. The advantage of a double zero fibre for achieving spectral extent and flatness will be clear when we consider other fibres in the next section, with strong third order dispersion.

Clearly the broadest infrared continuum would be obtained with very high nonlinearity and very low dispersion, with no high order dispersion terms, i.e. a flat dispersion curve. However, such a fibre is very hard to realise in practice. To increase nonlinearity one normally shrinks the core size and this can dramatically alter the

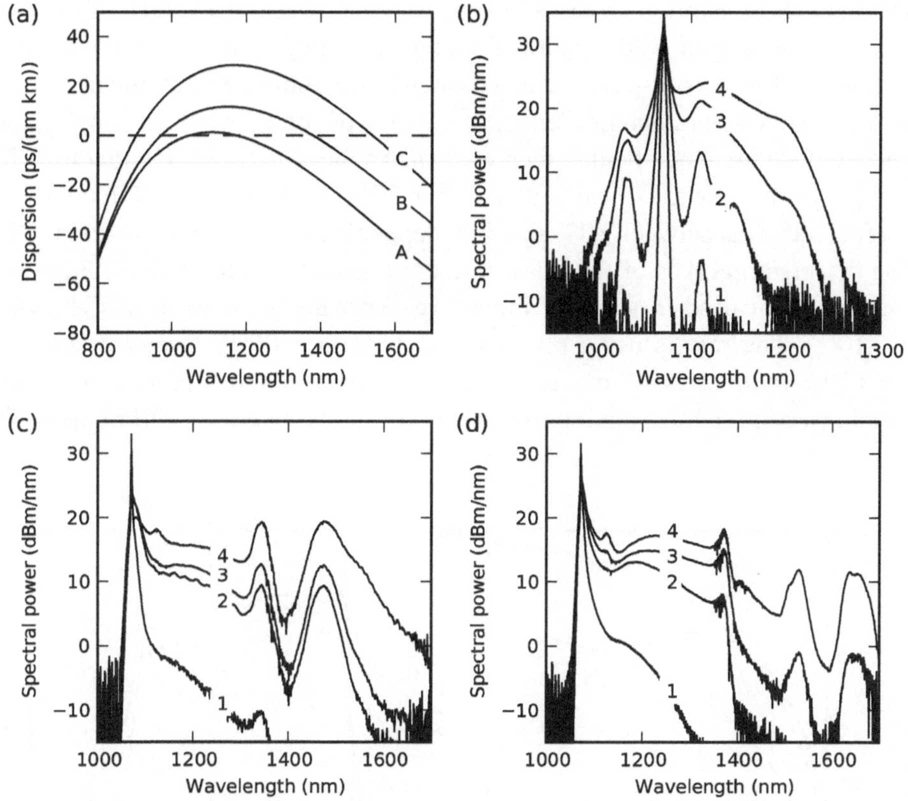

Fig. 8.11. (a) The dispersion curves for the double zero-dispersion fibres used. (b–d) The measured spectra for pump powers of (1) 9 W, (2) 25 W, (3) 30 W, (4) 56 W in : (b) 60 m of fibre A, (c) 55 m of fibre B and (d) 40 m of fibre C. After Kudlinski et al. (2008a, 2008b).

dispersion, usually strongly increasing the third order dispersion. It turns out that perhaps the optimal design is to use a double zero fibre (fourth order dispersion) to cancel the third order dispersion leading to a flatter continuum, although this will inevitably lead to the generation of a dispersive wave and the halting of the continuum extension at the second zero-dispersion wavelength. The requirements for short wavelength extension are different, as discussed in Section 8.2.8.

8.3.6 Power scaling

Continuous wave pumped supercontinua naturally lend themselves to high average spectral power applications. The results in Fig. 8.10(c) have already shown very high spectral powers of over 50 mW/nm. Further power scaling was recently achieved by Travers et al. (2008b) by utilising an industrial class, 400 W, CW laser.

The pump laser emitted up to 432 W of average power at 1070 nm with random polarisation and a spectral linewidth of 3.6 nm. The single mode output of the laser was interfaced to a collimating unit producing a 7 mm ($1/e^2$) collimated beam. This was coupled via a singlet lens into a series of mode matching single mode fibres to reduce the mode field, which were then spliced to the PCF under test. The free-space coupling was typically greater than 70% efficient, the total free-space to PCF efficiency was between 30% and 50% depending on the splice loss to the PCF.

The dispersion curves of the fibres tested are shown in Fig. 8.12(a). Fibre A is the same as that used in some of the above experiments, with results shown in Fig. 8.10(c). The results using this fibre but with the 400 W laser are shown in Fig. 8.12(b). In this case 17 m were used with a maximum of 170 W coupled into it, forming a continuum spanning from 1060 to 1900 nm with ~10 dB flatness. The

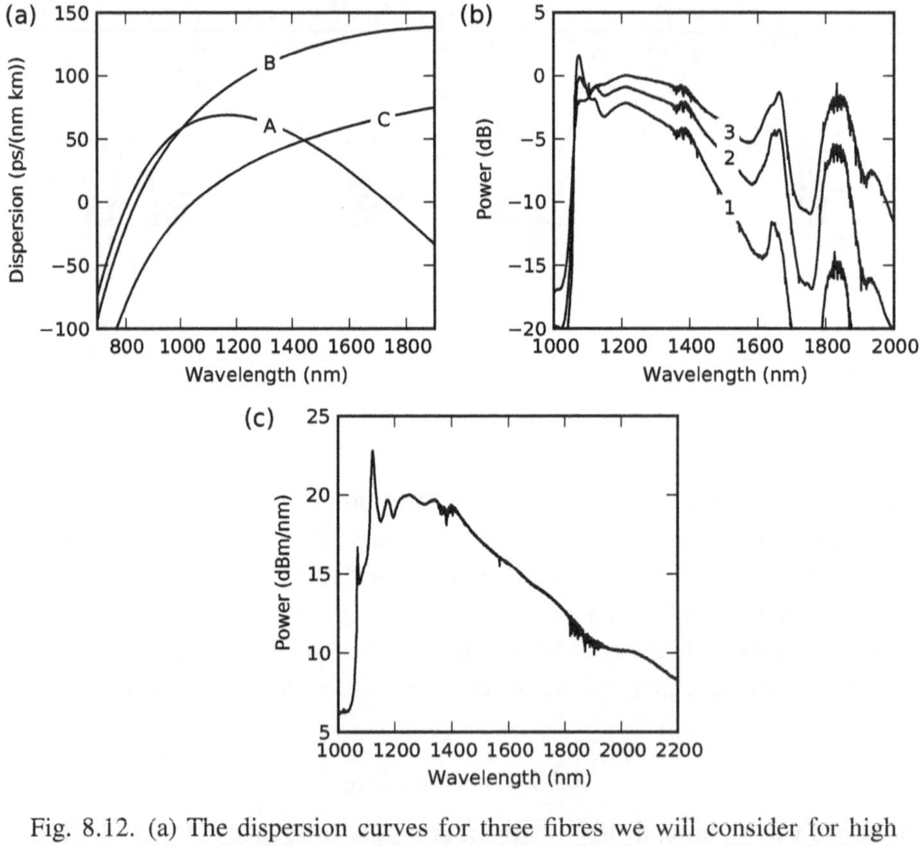

Fig. 8.12. (a) The dispersion curves for three fibres we will consider for high power pumping. (b) The supercontinuum spectra for pumping 17 m of fibre A with (1) 50 W, (2) 85 W, (3) 170 W. (c) The spectral power of the supercontinuum generated in 20 m of fibre B for 170 W equivalent pump power. The total output power was 50 W. After Travers et al. (2008b).

role of the second zero-dispersion wavelength is again clearly evidenced by the dip in spectral power around 1700 nm. The qualitative form of the spectra is very similar to those shown above for other double zero-dispersion fibres. The spectral power densities between 1060 nm and 1400 nm are over 50 mW/nm as the output powers were around 27 W.

Fibre B has a very different dispersion curve. From Fig. 8.12(a) we see it has a similar first zero-dispersion wavelength to fibre A, but has a very strong third order dispersion, and little fourth order dispersion to flatten the curve. The supercontinuum spectrum obtained at the output of 20 m of this fibre is shown in Fig. 8.12(c) for a pump power of 170 W. The total average output power was over 50 W and the continuum extended from 1060 to 2220 nm. The 10 dB width of the spectrum was over 900 nm, between 1140 and 2050 nm. Across this range the spectral power exceeds 100 mW/nm, with half of the range between 50 and 100 mW/nm. Water loss causes a significant fall off in spectral power, although the power density at the long edge of the continuum around 2100 nm is still over 5 mW/nm.

The continuous roll-off of spectral power shown in Fig. 8.12(c) confirms the restrictive role of strong third order dispersion discussed above. A fibre with a dispersion curve between A and B would clearly be optimal to further extend the continuum while maintaining spectral power.

8.3.7 Noise

Because CW continuum generation is a noise driven process, one may wonder what the intensity noise of the spectrum is. First, it should be noted that the noise driving the MI process is at optical frequencies, beyond the range noticeable for many applications. In addition it is clear that, being MI driven, the coherence of the spectrum is very low, as has been shown for long pulse pumped systems dominated by MI (Genty et al., 2007). In this section, we look at a slightly different area, which is the relative intensity noise at radio frequencies, important for a number of applications. The major one driving this work was for CW continuum sources intended for optical coherence tomography systems, where intensity noise in the kHz to MHz band is important.

The relative intensity noise of a supercontinuum produced by pumping fibre A in Fig. 8.12(a) was measured under both CW and ASE pump conditions. The supercontinuum at the output of the PCF was passed through a fused fibre coupler selected to filter the pump wavelengths from the supercontinuum. This allows us to monitor the noise of the actual supercontinuum in the absence of the strong pump power. The resulting spectrum was approximately Gaussian with a centre wavelength of 1230 nm and bandwidth of 150 nm. We also measured the relative intensity noise of the two pump sources alone (without the coupler).

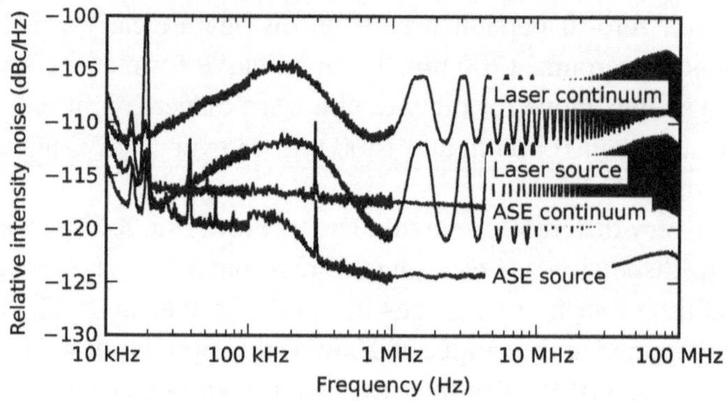

Fig. 8.13. The measured relative intensity noise of the laser and ASE pump sources and the resulting supercontinua.

The results are shown in Fig. 8.13. The ASE pump source has a much lower RIN than the laser pump source. The noise of the pump sources arises from beating between longitudinal modes (in a laser) and between different frequencies outside the laser cavity or for an ASE source. It can be shown (Derickson, 1998) that the noise of an ASE source is inversely related to its bandwidth, because the DC power increases faster with respect to bandwidth than the beat noise between the extra spectral components. For the ASE pump source, the bandwidth was 6 nm or 1.6 THz; for this source we measured a RIN of −125 dB/Hz up to 200 kHz and −137 dB/Hz at higher frequencies, which corresponds reasonably well with calculations. The laser source has a much higher RIN due to the longitudinal mode beating and significant oscillations in the RIN spectrum are observed at the cavity frequency. Launching either source through the PCF to generate a supercontinuum results in a broad amplification of the RIN. The noise amplification has been shown to be due to the nonlinear amplification of two quantum noise sources: the input shot noise and the spontaneous Raman scattering down the fibre. Of these two effects, shot noise amplification has been shown to be dominant (Corwin et al., 2003). The laser source oscillations are carried through and amplified into the supercontinuum. Due to the higher initial absolute RIN value of the laser, the resulting continuum has a relative intensity noise more than 10 dB higher than the ASE pumped supercontinuum, despite the fact that the relative gain to the continuum noise is lower for the laser pump source. This can be significant for some applications, in particular OCT, where these sources have been used by collaborators (Povazay et al., 2005; Bizheva et al., 2006).

8.3.8 Visible generation

The excitation of wavelengths short of the pump wavelength and in the normal dispersion regime has a different set of requirements compared to infrared extension, which we looked at in Section 8.2.8. Such results have been readily observed when pumping HNLF with erbium laser systems, as demonstrated clearly by Popov et al. (2002) and shown in Fig. 8.1, and also by other authors (Abeeluck and Headley, 2005; Vanholsbeeck et al., 2005). Until 2008 such results had not been reproduced in PCF pumped at 1060 nm, although such a result is highly desirable as it may pave the way for CW pumped visible supercontinuum generation. A number of groups have begun to achieve this aim (Travers et al., 2008b; Cumberland et al., 2008b; Kudlinski and Mussot, 2008; Mussot and Kudlinski, 2009; Mussot et al., 2008).

There have been three main techniques to achieve this. Following the HNLF results and considering the analysis in Section 8.2.8, pumping a PCF very close to its zero dispersion wavelength is an obvious route to follow. Cumberland et al. (2008b) demonstrated the feasibility of this, pumping a fibre with a zero dispersion wavelength of 1068 nm with a pump laser at 1071 nm – clearly in the low dispersion, high third order dispersion regime. The results obtained clearly demonstrated the excitation of dispersive waves, down to 900 nm with pure CW pumping and to 600 nm with enhanced pump powers obtained by modulating the pump laser.

A second technique, demonstrated by Travers et al. (2008b) is to simply pump a suitable fibre with significantly enhanced powers, as discussed in Section 8.3.6. The results of pumping 50 m of fibre C in Fig. 8.12(a) with the 400 W laser described in Section 8.3.6 are shown in Fig. 8.14. The pump power was 230 W. The total average output power was 28 W; the power reduction was mainly attributed to very high losses of the PCF in the visible and infrared spectral region, including

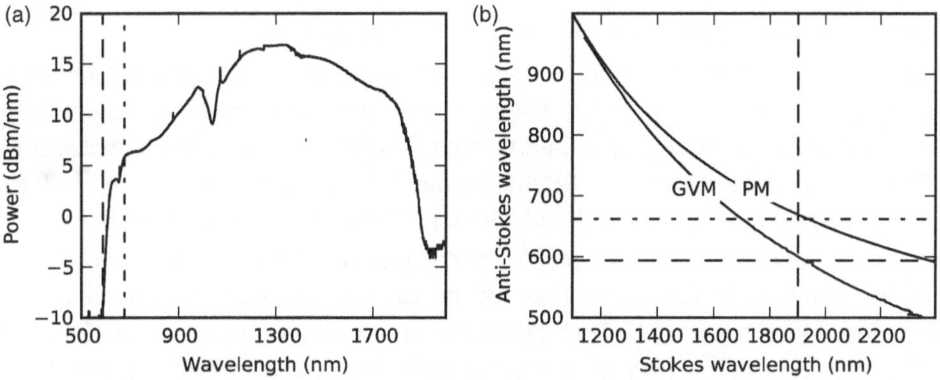

Fig. 8.14. (a) The spectral power of the supercontinuum generated in 50 m of fibre C in Fig. 8.12(a) for approximately 220 W of CW pump power. (b) The calculated phase matching (PM) and group velocity matching (GVM) curves for this fibre. After Travers et al. (2008b).

the water loss. It is clear that the supercontinuum extends to the visible spectral region, down to 600 nm. The continuum appeared bright red to the eye. The long wavelength edge of the continuum was 1900 nm. The spectral powers were over 2 mW/nm in the short wavelength side, which is competitive with the highest power pulse pumped visible supercontinua, and over 100 mW/nm in the infrared region, with substantial spectral regions between 20 and 30 mW/nm. This result shows that visible supercontinua similar to those obtained with pulse pumped systems can be achieved with extremely high power CW pump conditions, but with additional benefits due to scalable spectral power and flatness.

The mechanism responsible for the short wavelength extension can be determined by examining the phase-matching and group velocity matched curves for fibre C, as shown in Fig. 8.14(b). For short Stokes wavelengths it is hard to distinguish whether phase matching, and hence a four-wave mixing process, is the main active dynamic, or whether soliton trapping, based on group velocity matching is key. At longer Stokes wavelengths the matching curves diverge and some discrimination can be achieved. The long wavelength edge of the continuum is at approximately 1900 nm and is marked by the vertical dashed line in Fig. 8.14(b). The two horizontal lines interacting with it mark the corresponding anti-Stokes wavelengths for the two processes. From the spectrum in Fig. 8.14(a) we see that the shortest wavelength in the visible, around 590 nm, must be generated by the soliton trapping mechanism as that region matches in group velocity to the long edge of the continuum. But there is also a feature in the spectrum around 670 nm which approximately coincides with the phase-matching curve indicating that four-wave mixing does also play a role in the short wavelength generation.

It should be possible to extend the continuum to the blue using this relatively brute-force technique, by carefully designing the fibre dispersion and pump conditions to optimise the soliton trapping mechanism.

The third technique to generate visible supercontinuum, and perhaps the most promising, is the use of tapered PCF. Long PCF tapers have been very successfully used to enhance visible supercontinuum generation for picosecond and nanosecond pump systems (Kudlinski et al., 2006) and have also now been shown to work for CW pumped systems (Kudlinski and Mussot, 2008). In this result the fibre had an initial constant cross-section to enable the early development of MI and a Raman-soliton continuum to form. The first 100 m had a zero dispersion wavelength of 1053 nm, while the second 100 m is tapered with a final zero-dispersion at 950 nm. A 20 W CW laser at 1.064 μm was used as the pump source and a continuum extending from 670 to 1350 nm was produced. The effect of the taper is to simultaneously increase the nonlinearity and also change the dispersion parameters. Critically it changes the group velocity matching curve as described by Travers and Taylor (2009) to enhance the soliton trapping effect significantly.

8.4 Conclusion

Continuous wave supercontinuum generation provides a rich set of dynamics and some unique properties compared to other supercontinuum schemes. In this chapter we have discussed the most important mechanisms for the spectral development and how the fibre dispersion and nonlinearity can be used to control and understand the dynamics.

Experimentally a wide variety of results have been achieved, and some key examples have been shown here, including broad and flat, high spectral power sources across the near infrared; low noise sources, for example OCT applications; the role of a second zero dispersion wavelength in the continuum development; the range of power scaling opportunities; and the extension of CW continuum sources into the visible spectral region.

Much work remains to be done including improvements to the models of the initial CW pump conditions; further extension into the visible using careful optimisation of the fibre dispersion and tapering techniques; further spectral power scaling and more detailed investigation into the dynamics – including the role of polarisation; and further work on the effect of pump bandwidth, and possibly even single-shot diagnostics of the MI dynamics. In the longer term, expansion further into the mid-infrared may be considered if suitable pump sources and high quality novel glass fibres can be manufactured.

8.5 Acknowledgements

Much of the work which constitutes this chapter was only possible because of my current and past colleagues, in particular: J. R. Taylor, S. V. Popov, B. A. Cumberland, A. B. Rulkov, R. E. Kennedy and E. H. Kelleher. Others who have either collaborated on experimental projects or provided useful discussions include A. Kudlinksi, A. Mussot, J. M. Dudley and N. Traynor. I also want to thank A. V. Jausovec for her endless support.

References

Abeeluck A., Headley C. (2005). Continuous-wave pumping in the anomalous and normal-dispersion regimes of nonlinear fibers for supercontinuum generation. *Opt. Lett.* **30**, 61–63.

Abeeluck A. K., Headley C., Jørgensen C. G. (2004). High-power supercontinuum generation in highly nonlinear, dispersion-shifted fibers by use of a continuous-wave Raman fiber laser. *Opt. Lett.* **29**, 2163–2165.

Abrardi L., Martin-Lopez S., Carrasco-Sanz A., Corredera P., Hernanz M. L., González-Herráez M. (2007). Optimized all-fiber supercontinuum source at 1.3 μm generated in a stepwise dispersion-decreasing-fiber arrangement. *J. Lightwave Technol.* **25**, 2098–2102.

Agrawal G. P. (2004). *Nonlinear Fiber Optics*. Academic Press.
Akhmediev N., Karlsson M. (1995). Cherenkov radiation emitted by solitons in optical fibers. *Phys. Rev. A* **51**, 2602–2607.
Avdokhin A. V., Popov S. V., Taylor J. R. (2003). Continuous-wave, high-power, Raman continuum generation in holey fibers. *Opt. Lett.* **28**, 1353–1355.
Babin S. A., Churkin D. V., Ismagulov A. E., Kablukov S. I., Podivilov E. V. (2007). Four-wave-mixing-induced turbulent spectral broadening in a long Raman fiber laser. *J. Opt. Soc. Am. B* **24**, 1729–1738.
Barviau B., Randoux S., Suret P. (2006). Spectral broadening of a multimode continuous-wave optical field propagating in the normal dispersion regime of a fiber. *Opt. Lett.* **31**, 1696–1698.
Beaud P., Hodel W., Zysset B., Weber H. (1987). Ultrashort pulse propagation, pulse breakup, and fundamental soliton formation in a single-mode optical fiber. *IEEE J. Quantum Electron.* **23**, 1938–1946.
Benjamin T., Feir J. (1967). The disintegration of wave trains on deep water. I: Theory. *J. Fluid Mech.* **27**, 30.
Biancalana F., Skryabin D. V., Yulin A. V. (2004). Theory of the soliton self-frequency shift compensation by the resonant radiation in photonic crystal fibers. *Phys. Rev. E* **70**, 016615.
Bizheva K., Pflug R., Hermann B. et al. (2006). Optophysiology: Depth-resolved probing of retinal physiology with functional ultrahigh-resolution optical coherence tomography. *Proc. Natl. Acad. Sci. USA* **103**, 5066–5071.
Champert P. A., Couderc V., Leproux P., Février S., Tombelaine V., Labonté L., Roy P., Froehly C., Nérin P. (2004). White-light supercontinuum generation in normally dispersive optical fiber using original multi-wavelength pumping system. *Opt. Express* **12**, 4366–4371.
Corwin K., Newbury N., Dudley J., Coen S., Diddams S. A., Weber K., Windeler R. (2003). Fundamental noise limitations to supercontinuum generation in microstructure fiber. *Phys. Rev. Lett.* **90**, 113904.
Cumberland B. A., Travers J. C., Popov S. V., Taylor J. R. (2008a). 29 W High power CW supercontinuum source. *Opt. Express* **16**, 5954–5962.
Cumberland B. A., Travers J. C., Popov S. V., Taylor J. R. (2008b). Towards visible CW pumped supercontinua. *Opt. Lett.* **33**, 2122–2124.
de Matos C. J. S., Popov S. V., Taylor J. R. (2004). Temporal and noise characteristics of continuous-wave-pumped continuum generation in holey fibers around 1300 nm. *Appl. Phys. Lett.* **85**, 2706.
Derickson D. (1998). *Fiber optic test and measurement*. Prentice Hall.
Dianov E., Grudinin A., Prokhorov A., Serkin V. (1992). Non-linear transformations of laser radiation and generation of Raman solitons in optical fibers. *Cambridge Studies in Modern Optics; Optical Solitons – Theory and Experiment*. Ed. J.R. Taylor, Cambridge University Press, Chapter 7, 197.
Dianov E. M., Karasik A. Y., Mamyshev P. V., Prokhorov A. M., Serkin V. N., Stelmakh M. F., Fomichev A. A. (1985). Stimulated-Raman conversion of multisoliton pulses in quartz optical fibers. *JETP Lett.* **41**, 294.
Frosz M. H., Bang O., Bjarklev A. (2006). Soliton collision and Raman gain regimes in continuous-wave pumped supercontinuum generation. *Opt. Express* **14**, 9391–9407.
Genty G., Coen S., Dudley J. M. (2007). Fiber supercontinuum sources (Invited). *J. Opt. Soc. Am. B* **24**, 1771–1785.
Genty G., Lehtonen M., Ludvigsen H. (2004). Effect of cross-phase modulation on supercontinuum generated in microstructured fibers with sub-30 fs pulses. *Opt. Express* **12**, 4614–4624.

Golovchenko E. A., Mamyshev P. V., Pilipetskii A. N., Dianov E. M. (1991). Numerical-analysis of the Raman-spectrum evolution and soliton pulse generation in single-mode fibers. *J. Opt. Soc. Am. B* **8**, 1626–1632.

González-Herráez M., Martín-López S., Corredera P., Hernanz M. L., Horche P. R. (2003). Supercontinuum generation using a continuous-wave Raman fiber laser. *Opt. Commun.* **226**, 323–328.

Gorbach A. V., Skryabin D. V. (2007a). Light trapping in gravity-like potentials and expansion of supercontinuum spectra in photonic-crystal fibres. *Nature* **1**, 653–657.

Gorbach A. V., Skryabin D. V. (2007b). Theory of radiation trapping by the accelerating solitons in optical fibers. *Phys. Rev. A* **76**, 053803.

Gordon J. P. (1986). Theory of the soliton self-frequency shift. *Opt. Lett* **11**, 662–664.

Gouveia-Neto A. S., Gomes A. S. L., Taylor J. R. (1988). Femto soliton Raman generation. *IEEE J. Quantum Electron.* **24**, 332–340.

Hasegawa A., Brinkman W. F. (1980). Tunable coherent IR and FIR sources utilizing modulational instability. *IEEE J. Quantum Electron.* **16**, 694–697.

Hsiung P. L., Chen Y., Ko T. H., Fujimoto J. G., de Matos C. J. S., Popov S. V., Taylor J. R., Gapontsev V. P. (2004). Optical coherence tomography using a continuous-wave, high-power, Raman continuum light source. *Opt. Express* **12**, 5287–5295.

Islam M. N., Sucha G., Bar-Joseph I., Wegener M., Gordon J. P., Chemla D. S. (1989a). Broad bandwidths from frequency-shifting solitons in fibers. *Opt. Lett* **14**, 370–372.

Islam M. N., Sucha G., Bar-Joseph I., Wegener M., Gordon J. P., Chemla D. S. (1989b). Femtosecond distributed soliton spectrum in fibers. *J. Opt. Soc. Am. B* **6**, 1149–1158.

Itoh H., Davis G. M., Sudo S. (1989). Continuous-wave-pumped modulational instability in an optical fiber. *Opt. Lett.* **14**, 1368–1370.

Johnson D. J., Plankey F. W., Wineford J. D. (1974). Comparison of continuous wave and pulsed continuum sources for atomic fluorescence flame spectrometry. *Anal. Chem.* **46**, 1892–1902.

Karpman V. (1967). Self-modulation of nonlinear plane waves in dispersive media. *JETP Lett.* **6**, 277–279.

Karpman V. I. (1993). Radiation by solitons due to higher-order dispersion. *Phys. Rev. E* **47**, 2073–2082.

Korneev N., Kuzin E. A., Ibarra-Escamilla B., Flores-Rosas M. B. J. A. (2008). Initial development of supercontinuum in fibers with anomalous dispersion pumped by nanosecond–long pulses. *Opt. Express* **16**, 2636–2645.

Kudlinski A., Mussot A. (2008). Visible cw-pumped supercontinuum. *Opt. Lett.* **33**, 2407–2409.

Kudlinski A., Bouwmans G., Quiquempois Y., Mussot A. (2008a). Experimental demonstration of multiwatt continuous-wave supercontinuum tailoring in photonic crystal fibers. *App. Phys. Lett.* **92**, 141103.

Kudlinski A., Cumberland B. A., Travers J. C., Bouwmans G., Quiquempois Y., Mussot A. (2008b). CW supercontinuum generation in photonic crystal fibres with two zero-dispersion wavelengths. *1st Workshop on Specialty Optical Fibers and Their Applications* **1055**, 15–18.

Kudlinski A., George A. K., Knight J. C., Travers J. C., Rulkov A. B., Popov S. V., Taylor J. R. (2006). Zero-dispersion wavelength decreasing photonic crystal fibers for ultraviolet-extended supercontinuum generation. *Opt. Express* **14**, 5715–5722.

Kutz J. N., Lyngå C., Eggleton B. J. (2005). Enhanced supercontinuum generation through dispersion-management. *Opt. Express* **13**, 3989–3998.

Lighthill M. (1965). Contributions to the theory of waves in non-linear dispersive systems. *IMA J. Appl. Math.* **1**, 269–306.

Martin-Lopez S., Abrardi L., Corredera P., González-Herráez M., Mussot A. (2008). Spectrally-bounded continuous-wave supercontinuum generation in a fiber with two zero-dispersion wavelengths. *Opt. Express* **16**, 6745–6755.

Martin-Lopez S., Carrasco-Sanz A., Corredera P., Abrardi L., Hernānz M. L., González-Herráez M. (2006). Experimental investigation of the effect of pump incoherence on nonlinear pump spectral broadening and continuous-wave supercontinuum generation. *Opt. Lett.* **31**, 3477–3479.

Mussot A., Kudlinski A. (2009). 19.5 W CW-pumped supercontinuum source from 0.65 to 1.38 μm. *Electron. Lett.* **45**, 29–30.

Mussot A., Beaugeois M., Bouazaoui M., Sylvestre T. (2007). Tailoring CW supercontinuum generation in microstructured fibers with two-zero dispersion wavelengths. *Opt. Express* **15**, 11553–11563.

Mussot A., Kudlinski A., Beugnot J. C., Sylvestre T., González-Herráez M., Bouwmans G. (2008). Extended blue side of flat supercontinuum generation in PCFs with a CW Yb fiber laser. *2008 IEEE/LEOS Winter Topical Meeting Series*, 178–179.

Mussot A., Lantz E., Maillotte H., Sylvestre T., Finot C., Pitois S. (2004). Spectral broadening of a partially coherent CW laser beam in single-mode optical fibers. *Opt. Express* **12**, 2838–2843.

Nicholson J. W., Abeeluck A. K., Headley C., Yan M. F., Jorgensen C. G. (2003). Pulsed and continuous-wave supercontinuum generation in highly nonlinear, dispersion-shifted fibers. *Appl. Phys. B* **77**, 211–218.

Nishizawa N., Goto T. (2002). Characteristics of pulse trapping by ultrashort soliton pulse in optical fibers across zerodispersion wavelength. *Opt. Express* **10**, 1151–1160.

Persephonis P., Chernikov S. V., Taylor J. R. (1996). Cascaded CW fibre Raman laser source 1.6–1.9 μm. *Electron. Lett.* **32**, 1486–1487.

Popov S. V., Champert P. A., Solodyankin M. A., Taylor J. R. (2002). Seeded fibre amplifiers and multi-watt average power continuum generation in holey fibers. *Proceedings OSA Annual Meeting*, 117.

Popov S. V., Taylor J. R., Rulkov A. B., Gapontsev V. P. (2004). Multi-watt, 1.48–2.05 μm range CW Raman-soliton continuum generation in highly-nonlinear fibres. *Conference on Lasers and Electro-Optics* **2**, TuA6.

Povazay B., Bizheva K., Gasparoni S. et al. (2005). Ultrahigh resolution, functional OCT with all-fiber broadband Raman-continuum light-source. *Quantum Electronics and Laser Science Conference* **3**, JThE34.

Prabhu M., Kim N. S., Ueda K. (2000). Ultra-broadband CW supercontinuum generation centered at 1483.4 nm from Brillouin/Raman fiber laser. *Jpn. J. Appl. Phys* **39**, 291–293.

Prabhu M., Taniguchi A., Hirose S., Lu J., Musha M., Shirakawa A., Ueda K. (2003). Supercontinuum generation using Raman fiber laser. *Appl. Phys. B* **77**, 205–210.

Rulkov A. B., Ferin A. A., Travers J. C., Popov S. V., Taylor J. R. (2008). Broadband, low intensity noise CW source for OCT at 1800 nm. *Opt. Commun.* **281**, 154–156.

Saleh B. E. A., Teich M. C. (2007). *Fundamentals of Photonics*. Wiley-Interscience.

Skryabin D. V., Yulin A. V. (2005). Theory of generation of new frequencies by mixing of solitons and dispersive waves in optical fibers. *Phys. Rev. E* **72**, 016619.

Skryabin D. V., Luan F., Knight J. C., Russell P. S. J. (2003). Soliton self-frequency shift cancellation in photonic crystal fibers. *Science* **301**, 1705–1708.

Tai K., Hasegawa A., Tomita A. (1986). Observation of modulational instability in optical fibers. *Phys. Rev. Lett.* **56**, 135–138.

Taylor J. R. (1992). Soliton-Raman effects. *Cambridge Studies in Modern Optics; Optical Solitons – Theory and Experiment.* Ed. J.R. Taylor, Cambridge University Press, Chapter 11, 409.

Travers J. C. (2009). Blue solitary waves from infrared continuous wave pumping of optical fibers. *Opt. Express* **17**, 1502–1507.

Travers J., Taylor J. (2009). Soliton trapping of dispersive waves in tapered optical fibers. *Opt. Lett.* **34**, 115–117.

Travers J. C., Cumberland B. A., Popov S. V., Taylor J. R., Traynor N. J., Monteville A. (2006). Pump format influence on noise characteristics and spectral extent of CW continuum generation in new low-water-loss holey fibers. *Conference on Lasers and Electro-Optics* CMGG5.

Travers J. C., Popov S. V., Taylor J. R. (2005a). Extended blue supercontinuum generation in cascaded holey fibers. *Opt. Lett.* **30**, 3132–3134.

Travers J. C., Popov S. V., Taylor J. R. (2005b). Extended CW supercontinuum generation in a low water-loss Holey fiber. *Opt. Lett* **30**, 3132.

Travers J. C., Popov S. V., Taylor J. R. (2008a). A new model for CW supercontinuum generation. *Conference on Lasers and Electro-Optics* CMT3.

Travers J. C., Rulkov A. B., Cumberland B. A., Popov S. V., Taylor J. R. (2008b). Visible supercontinuum generation in photonic crystal fibers with a 400 W continuous wave fiber laser. *Opt. Express* **16**, 14435–14447.

Traynor N. J., Monteville A., Provino L., Landais D., Le Goffic O., Nguyen T. N., Chartier T., Tregoat D., Travers J. C. (2009). Fabrication and applications of low loss nonlinear holey fibers. *Fiber Integrated Opt.* **28**, 51–59.

Vanholsbeeck F., Martin-Lopez S., González-Herráez M., Coen S. (2005). The role of pump incoherence in continuous-wave supercontinuum generation. *Opt. Express* **13**, 6615–6625.

Wai P. K. A., Menyuk C. R., Lee Y. C., Chen H. H. (1986). Nonlinear pulse propagation in the neighborhood of the zero-dispersion wavelength of monomode optical fibers. *Opt. Lett.* **11**, 464–466.

9

Theory of supercontinuum and interaction of solitons with dispersive waves

D. V. Skryabin and A. V. Gorbach

9.1 Motivation

Chapter 1 of this book and a number of previous publications (see, e.g., Russell, 2003; Dudley et al., 2006; Smirnov et al., 2006; Knight and Skryabin, 2007; Skryabin and Wadsworth, 2009) give excellent bibliographic and historical accounts of the supercontinuum and other nonlinear effects observed in photonic crystal fibres since, and to some extent before, the seminal results by Ranka et al. (Ranka et al., 2000) appeared in 1999–2000. Here we present a focused account of our understanding of the fibre supercontinuum based on the theory developed by the Bath group over recent years (Skryabin et al., 2003; Yulin et al., 2004; Skryabin and Yulin, 2005; Gorbach et al., 2006; Gorbach and Skryabin, 2007a,b,c) and on the experiments carried out in Los Alamos and Bath (Skryabin et al., 2003; Efimov et al., 2004, 2005, 2006; Gorbach et al., 2006; Stone and Knight, 2008), which have closely followed our theoretical work. Concepts and results systematically described below are centred around the problem of the frequency conversion due to interaction between solitons and dispersive waves (Skryabin and Yulin, 2005). In the end our approach leads to a qualitative understanding and quantitative description of the expansion of the femtosecond supercontinua (Gorbach and Skryabin, 2007b).

We start this chapter by introducing the deterministic model of supercontinuum generation, discuss its limitations and move on to the soliton self-frequency shift problem, presenting it in the way trimmed for our goals. Then we describe emission and interaction of dispersive waves with solitons and elaborate on the role of this process in supercontinuum generation. Finally we explain formation of soliton–radiation bound states, crucial for supercontinuum expansion, and introduce the related concept of the gravity-like force acting on photons. The format of a book chapter has given us an opportunity to emphasize the links between the different sub-topics of our research and to mention a few of the rarely discussed problems.

Supercontinuum Generation in Optical Fibers, ed. by J. M. Dudley and J. R. Taylor. Published by Cambridge University Press.

9.2 Numerical model for fibre supercontinuum

The physical system which we consider is an optical fibre with a sufficiently high nonlinear parameter $\gamma \sim 0.01$ W/m and the zero of the group velocity dispersion (GVD) located close to the pump wavelength. These conditions are found in fibres with silica cores of a few ($\sim$1–5) microns in diameter, which can be, e.g., either photonic crystal or tapered fibres pumped with a variety of sources. The latter include femtosecond pulses of a mode-locked Ti:sapphire laser with wavelengths around 800 nm (Ranka et al., 2000) and nanosecond pulses from a microchip laser close to 1 μm (Stone and Knight, 2008). An important aspect of the dispersion profile allowing us to achieve the widest supercontinua is the normal GVD extending towards the shorter wavelengths, see Fig. 9.1(a). Use of the opposite GVD slope for spectral broadening has also been considered (see, e.g. Skryabin et al., 2003; Efimov et al., 2004; Harbold et al., 2002).

The widely accepted model, reproducing well experimental measurements of supercontinuum, is

$$\begin{aligned}\partial_z A &= ik(i\partial_t)A \\ &\quad + i\gamma(1-\theta)|A|^2 A + i\gamma\theta A \int_0^\infty dt' R(t')|A(t-t')|^2. \end{aligned} \tag{9.1}$$

Here z is the propagation length and t is time. $\theta = 0.18$ measures the strength of the Raman nonlinearity with respect to the Kerr one and $R(t)$ is the Raman response function (Gorbach et al., 2006; Agrawal, 2001). A is the modal amplitude, which

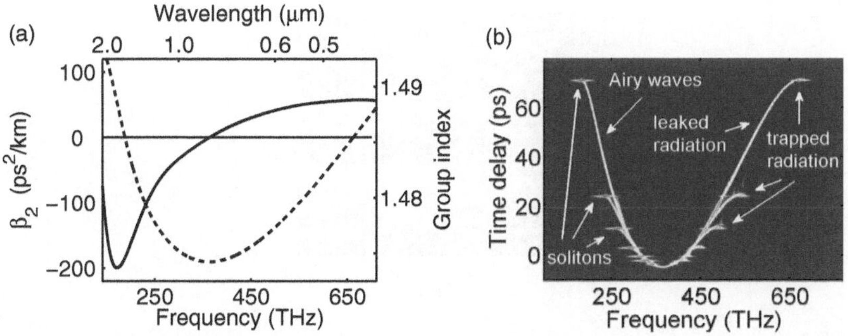

Fig. 9.1. (a) Group index, that is, speed of light divided by the group velocity $n_g = c\partial_\delta k$ (dashed line), and GVD $\partial_\delta^2 k$ (full line) typical for photonic crystal fibres used in supercontinuum experiments. The zero GVD point is at $\simeq$790 nm (380 THz). (b) XFROG spectrogram showing simultaneous frequency and time domain pictures of the supercontinuum from Fig. 9.2 at $z = 1.5$ m. Pump wavelength is 850 nm (350 THz). For more details on use of XFROG for characterization of a supercontinuum see Gu et al. (2003); Hori et al. (2004); Efimov et al. (2004).

carries all the spectrum we are interested in. The dispersion profile of a linear fibre is found by taking $A = e^{ik(\delta)z - i\delta t}$. $k(\delta)$ is obtained by replacing $i\partial_t$ with δ. $k(\delta)$ is routinely taken as $k = \sum_{n=2}^{N} \frac{1}{n!}\beta_n\delta^n$, which is a polynomial fit (not Taylor expansion) of the fibre dispersion. δ^0 and δ^1 terms in $k(\delta)$ can always be removed by the appropriate phase shift and choice of the moving reference frame. Positive δ corresponds to larger absolute frequencies.

The model (9.1) includes only three simple ingredients:

- *dispersion* (1st term in the right-hand side of Eq. (9.1))
- *nonlinear phase modulation* due to the instantaneous Kerr nonlinearity (2nd term)
- *Raman scattering* (3rd term).

Taken separately, the action of these terms is well known. Notably, the last two lead to generation of new spectral content acting on their own. However, conversion into discrete Raman side-bands has not manifested itself in typical supercontinuum experiments. Also self- (SPM) and cross- (XPM) phase modulations (as particular cases of the generic nonlinear phase modulation) generate spectra, which are usually relatively narrow (Agrawal, 2001), and cannot explain fully developed supercontinua. It is only when dispersion comes into play (i.e., after a sufficient propagation length) and all three of the above effects act together in a symphony, that the qualitatively new phenomenon of supercontinuum generation occurs. A typical supercontinuum experiment with a femtosecond pump leads to spectra covering the range from 400 nm to 2000 nm after propagation in a ~1 m long photonic crystal fibre (Ranka et al., 2000; Gu et al., 2003; Wadsworth et al., 2002). Figs. 9.1(b) and

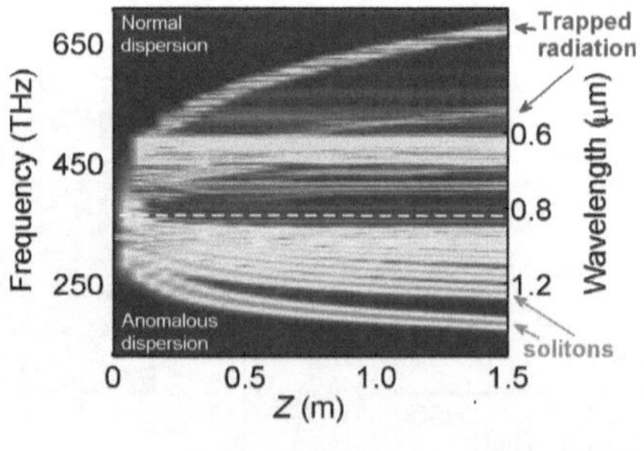

Fig. 9.2. Numerical simulation of supercontinuum generation in a photonic crystal fibre pumped with 200-fs pulses at 850 nm and having 6-kW peak power. The fibre dispersion as in Fig. 9.1(a). One can see three stages in the supercontinuum expansion. Symmetric spectral broadening due to SPM happens over the first few cm. Antisymmetric spectral broadening due to soliton fission, accompanied by the emission of the resonant Cherenkov radiation and soliton–radiation FWM, develop from ~5 cm to ~15 cm. After that the dynamics of the short-wavelength edge is determined by the intrapulse FWM and radiation trapping.

9.2 illustrate results of the modelling of this process using Eq. (9.1). Following our previous work (Gorbach and Skryabin, 2007b, 2007c) we will argue here that the formation of coupled soliton–radiation states with continuously blue shifting radiation components and red shifting solitons plays a paramount role in a supercontinuum expansion.

Results of modelling of Eq. (9.1) have been directly compared with many experimental measurements of femtosecond pulse propagation in photonic crystal fibres in both time and frequency domains, and we have established excellent agreement (Skryabin et al., 2003; Efimov et al., 2004, 2005, 2006; Gorbach et al., 2006). However, as with any model, Eq. (9.1) has limits to its applicability. In particular, it is not applicable for description of sharp field variations happening over a few femtoseconds and less. So far the role of such ultrashort features in typical fibre supercontinuum experiments has not been revealed, and the octave spectral broadening happens through nonlinear interactions of dispersive waves and soliton pulses which are well described by Eq. (9.1).

The relative smoothness of the temporal field profile corresponding to supercontinuum is one of the reasons why the self-steepening does not make a notable quantitative and qualitative impact if included or excluded from Eq. (9.1). Therefore we have opted for its exclusion. In addition, going beyond the amplitude model is likely to be needed to correctly describe the self-steepening effect. In our experience, the most noticeable disagreements of Eq. (9.1) with experimental measurements are due to omission of the frequency dependent loss term. Other neglected effects are possible excitation of higher order modes of the fibre, third harmonic generation and noise. Spontaneous Raman noise and input noise are unavoidably present in experiments. Spectra averaged over the statistical ensemble may, in some cases, be better suited for comparison with measurements, where fine spectral features of a single pulse excitation can be obscured due to a resolution and a long (over many pulses) integration time of a spectrum analyzer (Dudley and Coen, 2002; Corwin et al., 2003). It is important to realize, however, that small to moderate pulse to pulse fluctuations do nothing to the existence of deterministic nonlinear processes behind the supercontinuum generation in optical fibres, which can and shall be understood without the stochastic terms included.

9.3 Soliton self-frequency shift and soliton dispersion

During the first stage of a supercontinuum development SPM induces a spectral broadening and associated chirping of a femtosecond pulse, which are compensated by the anomalous GVD ($\partial_\delta^2 k < 0$). These processes initiate formation of multiple solitons from an initial high power pump pulse (soliton fission), see Fig. 9.2. Amongst the host of soliton related effects, there are two which are particularly important for us. These are the soliton self-frequency shift induced by the Raman scattering (Mitschke and Mollenauer, 1986) and emission of dispersive radiation. The latter comes from the overlap of a soliton spectrum with the range of frequencies where the fibre GVD is normal (Wai et al., 1990; Karpman, 1993; Akhmediev and Karlsson, 1995).

Spectra of femto- or picosecond solitons are sufficiently narrow, so that the Raman gain profile of silica peaking at 13 THz can be approximated by a straight line rising across the soliton spectrum from negative values (damping) at the short-wavelength end of the soliton spectrum to positive values (gain) at the long-wavelength end (Agrawal, 2001; Luan et al., 2006; Gorbach and Skryabin, 2008). The net result on the soliton is that its spectral centre of mass is shifted towards redder frequencies (Mitschke and Mollenauer, 1986). The group index (ratio of the vacuum speed of light c to the group velocity, $n_g = c\partial_\delta k$) increases with wavelength provided the GVD is anomalous. This leads to continuous negative acceleration (deceleration) of solitons in the presence of the Raman scattering.

An approximate soliton solution of Eq. (9.1) moving with a constant acceleration can be derived under the assumption that the fibre group index varies linearly in frequency (i.e., $k \sim \delta^2$ and $\partial_\delta k \sim \delta$). Assuming zero initial frequency, the soliton is given by (Gagnon and Belanger, 1990; Gorbach and Skryabin, 2007b, 2007c):

$$A_s = \psi(t - t_s)\exp[-i\delta_s t + i\varphi(z)], \tag{9.2}$$

$$\psi(t) = \sqrt{2q/\gamma}\ \mathrm{sech}\left(\sqrt{\frac{2q}{|k''|}}\,t\right),\ \varphi(z) = qz + \frac{1}{3}k''\delta_s^2 z, \tag{9.3}$$

$$T = \int_0^\infty tR(t)dt,\ \delta_s = \frac{g_0 z}{k''},\ k'' = \partial_\delta^2 k < 0,\ g_0 = \frac{32Tq^2}{15}. \tag{9.4}$$

The soliton delay t_s caused by the Raman effect is expressed as

$$t_s = \int \partial_\delta k|_{\delta=\delta_s} dz = \frac{1}{2}g_0 z^2. \tag{9.5}$$

$q > 0$ is the soliton wavenumber shift proportional to its intensity. The soliton trajectory in the (t,z)-plane is a parabola given by $t = t_s$ (see Fig. 9.3(a)). δ_s is the soliton frequency, which decreases linearly with propagation distance z (see Fig. 9.3(b)).

The plane wave expansion of the soliton is given by

$$\tilde{A}_s(\delta) = \int_{-\infty}^{\infty} dt A_s e^{i\delta t} = e^{i\varphi(z)+it_s[\delta-\delta_s]}\tilde{\psi}(\delta - \delta_s),$$

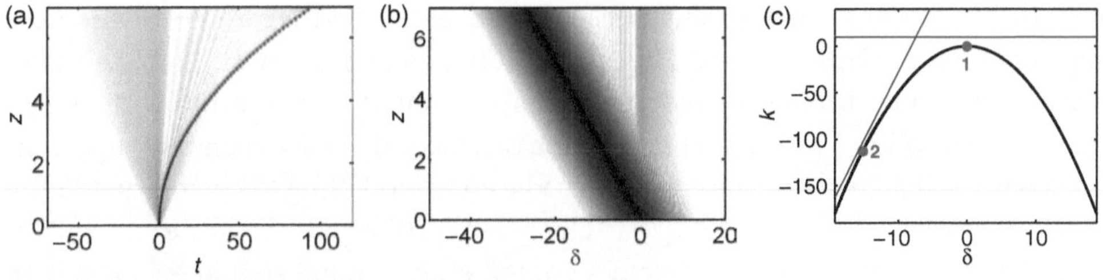

Fig. 9.3. Soliton evolution in (z,t) (a) and (z,δ) (b) planes in presence of the Raman effect. $k(\delta)=\frac{1}{2}k''\delta^2$. Note that the dispersive waves in (a) are not reflected by the soliton. (c) shows the phase matching diagrams between the soliton and dispersive waves. Soliton spectra k_s are shown with straight lines. Point 1 indicates the initial soliton frequency and point 2 is the frequency at some distance down the fibre (as predicted by Eq. (9.6)). Curved lines show the dispersion of linear waves $k(\delta)$. All the units are dimensionless.

where $\tilde{\psi}(\delta)=\int_{-\infty}^{\infty}\psi(t)e^{it\delta}dt$ is a real function. The wavenumber associated with $\tilde{A}_s(\delta)$ is

$$k_s(\delta)=\partial_z[\varphi(z)+[\delta-\delta_s]t_s]=q+\partial_\delta k|_{\delta=\delta_s}[\delta-\delta_s]+k(\delta_s). \tag{9.6}$$

$k_s(\delta)$ is linear in δ, expressing the fact that the soliton is immune to GVD and hence $\partial_\delta^2 k_s$ should be zero. Also it can be seen that $k_s(\delta)$ is actually a tangent to the dispersion of linear waves $k(\delta)=\frac{1}{2}k''\delta^2$ taken at $\delta=\delta_s$ and shifted away from it by the offset q. If the GVD is anomalous for all the frequencies ($k''<0$) the soliton spectrum does not touch the spectrum of linear dispersive waves, see Fig. 9.3(c).

ψ, as in Eq. (9.3), is of course an approximate solution, and the Raman effect forces the soliton to shake off some radiation, see Fig. 9.1(b), Fig. 9.3(a). This radiation can be approximated by Airy functions (Akhmediev et al., 1996; Gorbach and Skryabin, 2008). The Airy waves emitted in a typical supercontinuum setting are very weak and have a broad frequency spectrum almost exclusively belonging to the anomalous GVD range. The latter is the main reason why they practically do not interact with solitons, and their impact on the supercontinuum spectrum is secondary in importance.

9.4 Radiation emission by solitons and supercontinuum

Solitons dominate the long-wavelength edge of the supercontinuum, while the so-called resonant (or 'Cherenkov') radiation (Wai et al., 1990; Karpman, 1993; Akhmediev and Karlsson, 1995; Skryabin et al., 2003; Biancalana et al., 2004) comes to mind as one of the reasons for the spectrum created in the normal GVD range (Herrmann et al., 2002; Gu et al., 2003; Cristiani et al., 2004). Despite the importance of the resonant radiation identified in the first efforts to model fibre supercontinuum (Husakou and Herrmann, 2001; Herrmann et al., 2002), it later became clear that it is only when the initially emitted radiation has a chance to interact with solitons over long propagation distances that the expanding supercontinua observed in experiments can be reproduced in modelling (Skryabin and Yulin, 2005; Gorbach and Skryabin, 2007b,c; Genty et al., 2004b, 2005). The Raman effect is the key factor ensuring such interaction, therefore caution should be taken with interpretation of numerical results where the Raman term has been disregarded (Husakou and Herrmann, 2001; Herrmann et al., 2002).

To model Cherenkov radiation, it is often sufficient to include third order dispersion only, so that the single zero of the GVD is taken into account:

$$i\partial_z A - \left(\frac{1}{2!}\beta_2\partial_t^2 + \frac{i}{3!}\beta_3\partial_t^3\right)A = -\gamma|A|^2A - T\gamma A\partial_t|A|^2. \quad (9.7)$$

The linear dispersive wave solution for Eq. (9.7) is $A = e^{ikz-i\delta t}$ with $k(\delta) = \frac{1}{2}\beta_2\delta^2 + \frac{1}{6}\beta_3\delta^3$. The zero GVD point $\partial_\delta^2 k = 0$ is located at $\delta_0 = -\beta_2/\beta_3$. The soliton existence condition is $\beta_2 < 0$ and hence if, e.g., $\beta_3 > 0$, then the GVD is normal ($\partial_\delta^2 k > 0$) towards higher frequencies $\delta > \delta_0$, see Fig. 9.4(c).

Resonant Cherenkov radiation (unlike the above-mentioned Airy waves) depends on the phase matching conditions and therefore it is narrow band. If F_{Ch} is the resonant radiation, then

$$\left(i\partial_z - \frac{1}{2!}\beta_2\partial_t^2 - \frac{i}{3!}\beta_3\partial_t^3\right)F_{\mathrm{Ch}} \propto \partial_t^3 A_{\mathrm{s}}. \quad (9.8)$$

Equating the wavenumbers of the left- and right-hand sides of Eq. (9.8) we find (Wai et al., 1990; Karpman, 1993; Akhmediev and Karlsson, 1995; Skryabin et al., 2003; Biancalana et al., 2004)

$$k(\delta) = k_{\mathrm{s}}(\delta). \quad (9.9)$$

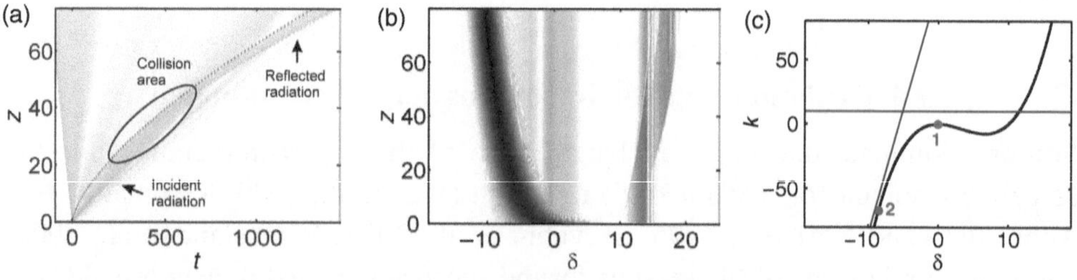

Fig. 9.4. The same as Fig. 9.3, but with $\beta_3 > 0$.

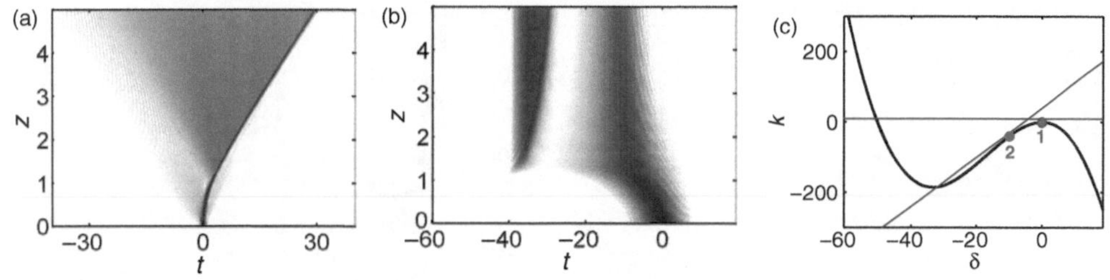

Fig. 9.5. The same as Fig. 9.3, but with $\beta_3 < 0$.

The roots of the above equation are the resonance frequencies δ_{Ch}. Thus $\beta_3 > 0$ ($\beta_3 < 0$) gives the blue (red) shifted dispersive waves, see Figs. 9.4 and 9.5. Accounting for higher order dispersions is effortless and may lead to several roots of Eq. (9.9). The energy of the emitted wave (or waves) is drawn from the entire soliton spectrum resulting in the adiabatic decay of the soliton (Biancalana et al., 2004).

In a typical supercontinuum setting with $\beta_3 > 0$ the Raman effect increases frequency detuning ($\Delta = \delta_{Ch} - \delta_s$) between the soliton ($\delta_s$) and the resonance radiation (δ_{Ch}), see Fig. 9.4(c). For $\beta_3 < 0$ the situation is the opposite, see Fig. 9.5(c). The amplitude of the emitted radiation is proportional to $e^{-a^2|\Delta|}$ (a is a constant), where $\Delta \propto z$ (Wai et al., 1990; Karpman, 1993; Akhmediev and Karlsson, 1995; Skryabin et al., 2003; Biancalana et al., 2004). Thus for $\beta_3 > 0$ the soliton self-frequency shift induces exponential decay of the radiation amplitude with propagation (Biancalana et al., 2004). Practically this implies that the soliton emits significant radiation only at the initial stage of the supercontinuum generation. The radiation emission quickly becomes unnoticeable when solitons are shifted away from the zero GVD point. It can be verified (comparing numerical results and predictions of Eq. (9.9)), that the radiation band continuously changing its wavelength towards shorter values, see Fig. 9.2, does not match the resonant frequencies of the solitons. The roots of Eq. (9.9) are located further towards shorter wavelengths. This observation is illustrated by Fig. 9.4 (see also Fig. 4 of Gorbach et al., 2006).

So, the question is: what causes the continuous blue shift of the short-wavelength edge of a supercontinuum? The answer is – interaction between radiation and solitons (Gorbach and Skryabin, 2007b,c; Genty et al., 2004b, 2005)! If $\beta_3 > 0$, the resonant radiation emitted by a soliton has group velocity less than the soliton itself, so that the radiation appears behind the soliton (Fig. 9.4(a)). However, after some propagation the radiation catches up with the soliton, which is continuously decelerated by the Raman effect, and the two collide (Fig. 9.4(a)). During this collision the radiation is reflected from the soliton backwards, and so the next collision is unavoidable, through the same mechanism. The process can be repeated many times, see Fig. 9.6. Note that the radiation colliding with a soliton is not necessarily

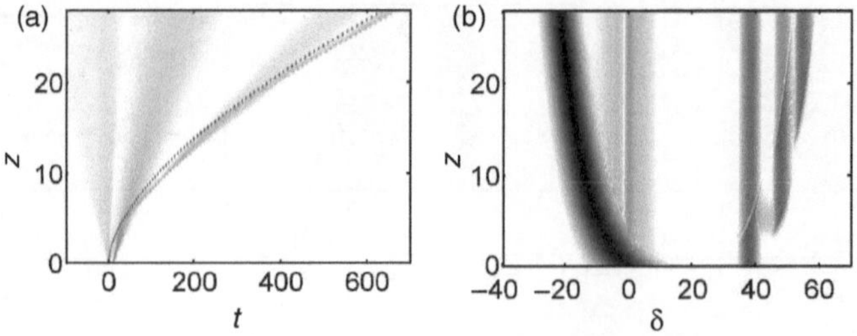

Fig. 9.6. A weak gaussian pulse in normal GVD range initially placed behind a soliton experiences multiple reflections from the latter. After each collision some light gets through the soliton and some is reflected back. The reflected light is localized close to the soliton and almost indistinguishable from it in the (z,t)-plot (a). However, the reflected light is clearly visible in the spectral (z,δ)-plot (b), because its frequency is blue-shifted after each collision.

emitted by the soliton itself, it can be initiated via other mechanisms as well, e.g., via SPM or emitted by other solitons (Fig. 9.6). An important condition though is that the radiation frequency belongs to the normal GVD range.

Reflection of the radiation backwards (towards higher t) from the soliton, implies that the radiation group velocity is further reduced, i.e. the group index $\partial_\delta k$ for the radiation increases. For the normal GVD ($\partial_\delta^2 k > 0$) an increase in the group index happens together with the increase in frequency δ. Therefore backward reflections have to be accompanied by the blue shifts of the radiation frequency, see Fig. 9.6(b). The claim is (Gorbach et al., 2006; Gorbach and Skryabin, 2007b,c) that this is exactly the type of process happening between the solitons and the short-wavelength radiation in the expanding supercontinuum shown in Fig. 9.2.

For $\beta_3 < 0$, the detuning between the radiation and soliton gets smaller with propagation (Fig. 9.5(c)), therefore the radiation is exponentially amplified in z. This amplification leads to the strong spectral recoil on the soliton followed by the compensation of the soliton self-frequency shift (Skryabin et al., 2003; Biancalana et al., 2004; Tsoy and deSterke, 2006) (see Fig. 9.5(b)). In this case the radiation is emitted ahead of the soliton and has no chance of interacting with it, see Fig. 9.5(a). The radiation emitted by one soliton can, however, interact with other solitons present in the fibre (Efimov et al., 2004; Gorbach and Skryabin, 2007a). The issue of supercontinuum generation with two zero GVD points can be found, e.g., in Harbold et al. (2002); Genty et al. (2004a); Falk et al. (2005).

9.5 Scattering of radiation from solitons and supercontinuum

The frequency conversion, crucial for supercontinuum, happens as the result of interaction of solitons and dispersive waves. Scattering of a dispersive wave from a soliton is the four-wave mixing process, which depends on the phase matching conditions (Yulin et al., 2004; Skryabin and Yulin, 2005; Efimov et al., 2004, 2005, 2006; Gorbach et al., 2006). An important difference of the soliton–radiation interaction with the textbook four-wave mixing of dispersive waves only (Agrawal, 2001), is that one of the participating fields (soliton) has a straight-line dispersion, see Eq. (9.6).

Let's assume that $F_{\rm p}$ and $\delta_{\rm p}$ are the amplitude and frequency of the dispersive wave interacting with the soliton $A_{\rm s}$, and F is the generated signal field with the frequency δ. The equation for the weak field F is (Skryabin and Yulin, 2005):

$$\left(i\partial_z - \frac{1}{2!}\beta_2\partial_t^2 - \frac{i}{3!}\beta_3\partial_t^3\right)F = -\gamma|A_{\rm s}|^2F_{\rm p} \tag{9.10}$$
$$= -\gamma A_{\rm s}^2F_{\rm p}^*.$$

Thus F is excited by either $|A_{\rm s}|^2F_{\rm p}$ or $A_{\rm s}^2F_{\rm p}^*$ terms depending on the phase matching. Assuming $F_{\rm p} = \epsilon_{\rm p}e^{ik(\delta_{\rm p})-i\delta_{\rm p}t}$ and using Eqs. (9.2)–(9.6), we work out the spectral content of the FWM terms:

$$|A_{\rm s}|^2F_{\rm p} = \epsilon_{\rm p}e^{ik(\delta_{\rm p})z}\int_{-\infty}^{\infty} f(\delta-\delta_{\rm p})e^{it_{\rm s}[\delta-\delta_{\rm p}]-i\delta t}d\delta\,, \tag{9.11}$$

$$A_{\rm s}^2F_{\rm p}^* = \epsilon_{\rm p}^*e^{2i\varphi(z)-ik(\delta_{\rm p})z}\int_{-\infty}^{\infty} f(\delta+\delta_{\rm p}-2\delta_{\rm s})e^{it_{\rm s}[\delta+\delta_{\rm p}-2\delta_{\rm s}]-i\delta t}d\delta\,, \tag{9.12}$$

where $f(\delta) = \int_{-\infty}^{\infty}\psi^2(t)e^{i\delta t}dt$ and φ is given by Eq. (9.3).

Using the Fourier expansion $F = \int_{-\infty}^{\infty}\epsilon(\delta)e^{ik(\delta)z-i\delta t}d\delta$ and taking z derivatives of the phases involved (cf. Eq. (9.6)) we equate the wavenumber of F to the wavenumbers of (9.11) and (9.12). The result is (Yulin et al., 2004; Skryabin and Yulin, 2005; Efimov et al., 2004, 2005, 2006; Gorbach et al., 2006)

$$k(\delta) = k_{\rm s}(\delta) - [k_{\rm s}(\delta_{\rm p}) - k(\delta_{\rm p})], \tag{9.13}$$
$$k(\delta) = k_{\rm s}(\delta) + [k_{\rm s}(\delta_{\rm p}) - k(\delta_{\rm p})]. \tag{9.14}$$

Equations (9.13) and (9.14) correspond to the first and second terms in the r.h.s. of Eq. (9.10), respectively. Here k is the wavenumber of the generated wave; $k_{\rm s}(\delta)$ and $k_{\rm s}(\delta_{\rm p})$ are the soliton wavenumbers at the generated frequency and at the frequency of the wave incident onto the soliton. If the soliton self-frequency shift is disregarded ($t_{\rm s} = \delta_{\rm s} \equiv 0$), then $k_{\rm s}(\delta) \equiv q$ and Eqs. (9.13) and (9.14) are replaced with $k(\delta) = k(\delta_{\rm p})$, $k(\delta) = 2q - k(\delta_{\rm p})$.

Solving Eqs. (9.13) and (9.14) graphically we find that they predict up to four resonances, see Fig. 9.7(a and c) and Skryabin and Yulin (2005). One of these

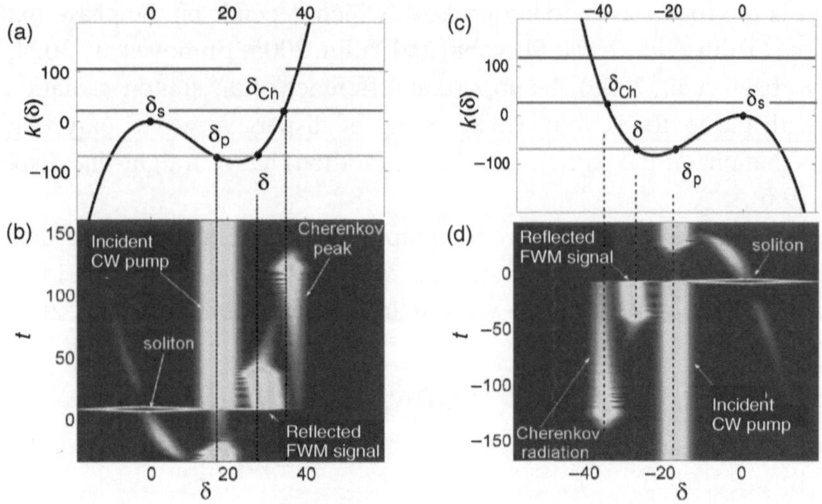

Fig. 9.7. Wavenumber matching diagrams (a,c) and XFROG spectrograms (b,d) for the fibre pumped with the soliton and cw (Raman effect is disregarded). (a,b) correspond to $\beta_3 > 0$ and (c,d) to $\beta_3 < 0$. The black curved lines in (a,c) are $k = \beta_2\delta^2/2 + \beta_3\delta^3/6$. The upper and lower horizontal lines in (a,c) correspond to the FWM resonances given by Eq. (9.13) and (9.14), respectively. The middle horizontal lines in (a,c) give Cherenkov resonances, Eq. (9.9).

resonances always coincides with the frequency of the cw pump: $\delta = \delta_p$. Fig. 9.7(a and b) shows the case when the incident (pump) wave is reflected back from the soliton with simultaneous up-shift of the reflected wave frequency. Fig. 9.7(c and d) shows the case when the radiation–soliton collision happens at the front edge of the soliton and the wave is reflected forward. The frequency is down-shifted in the latter example. One can also see from Fig. 9.7 that in both examples the efficiency of the FWM process due to the $A_s^2 F_p^*$ term in Eq. (9.10) is remarkably low, as compared to the FWM process due to the $|A_s|^2 F_p$ term, though radiation emission due to the $A_s^2 F_p^*$ was observed in other cases (see Fig. 9.8(b)).

Eqs. (9.13) and (9.14) apply without a change if the soliton and dispersive waves are orthogonally polarized, which has been used in the experimental measurements of the soliton–radiation interaction (Efimov et al., 2005, 2006). These measurements fully confirmed the validity of both Eq. (9.13) (Efimov et al., 2005) and Eq. (9.14) (Efimov et al., 2006). Two examples of the experimental XFROG spectrograms are shown in Fig. 9.8. In the recent work (Philbin et al., 2008) the spectral peaks generated in the soliton–radiation interaction have been interpreted as a fibre analogue of the event-horizon and a step towards observation of optical Hawking radiation.

It has been verified that the frequency up-shift of the radiation, resulting from the cascaded back reflection of the radiation due to $\sim |A_s|^2$ term, is the mechanism

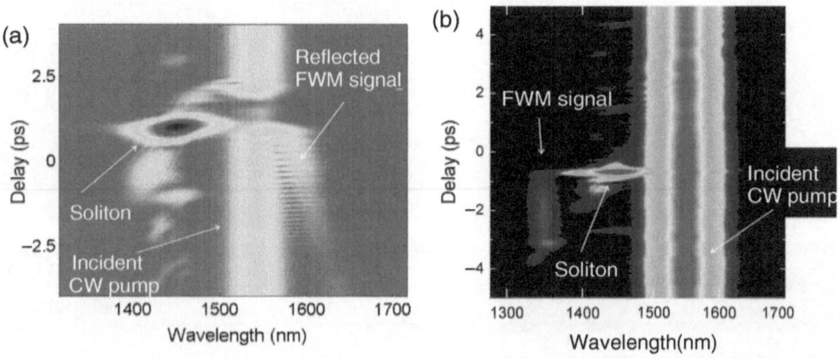

Fig. 9.8. Experimentally measured XFROG spectrograms showing radiation–soliton interaction. (a) corresponds to the resonance given by Eq. (9.13) and (b) to Eq. (9.14). (a) shows the case of the forward reflected wave, as in Fig. 9.7(d). Both (a) and (b) measurements have been taken for $\beta_3 < 0$. Figure (a) is reprinted with permission from Efimov et al. (2005); copyright (2005) by the American Physical Society. Figure (b) is reprinted with permission from Efimov et al. (2006); copyright (2006) by the Optical Society of America.

ensuring blue shift of the short-wavelength edge of the supercontinuum (Gorbach et al., 2006). The phase matching condition (9.13) means that with every reflection the frequencies of the incident and reflected waves both tend towards the limit point, where the group velocity of the dispersive wave coincides with the soliton group velocity (Skryabin and Yulin, 2005; Gorbach et al., 2006). This is well illustrated by Fig. 9.7(a). Indeed, if the instantaneous soliton frequency is assumed to be zero, then the $k(\delta)$ minimum corresponds to the frequency with the group velocity matched to the soliton group velocity. If the frequency of the incident wave (δ_p) is exactly at this minimum or very close to it, then the resonance predicted by Eq. (9.13) is either degenerated or is close to the degeneracy with δ_p. Under such conditions the process of frequency conversion practically happens within a single wave packet and can be termed as the intrapulse four-wave mixing (Gorbach et al., 2006).

Soliton–radiation mixing considered above should be contrasted with the well-known XPM-induced spectral broadening of a weak signal (Islam et al., 1987; Agrawal, 2001). The latter is not sensitive to the phase matching conditions, while the former is. The nonlinear term involved here $|A_s|^2 F_p$ is of course the classical XPM term, but in our case its interplay with dispersion and soliton acceleration is critical. Similarly, the SPM term can, on the one hand, induce spectral broadening, but, on the other (when GVD is accounted for), is responsible for the soliton formation (Agrawal, 2001).

In the above picture of the soliton–radiation interaction, it remains unclear what sustains the efficiency of the process over long propagation distances. Strong normal GVD acting on the dispersive wave packet, should lead to its sufficiently

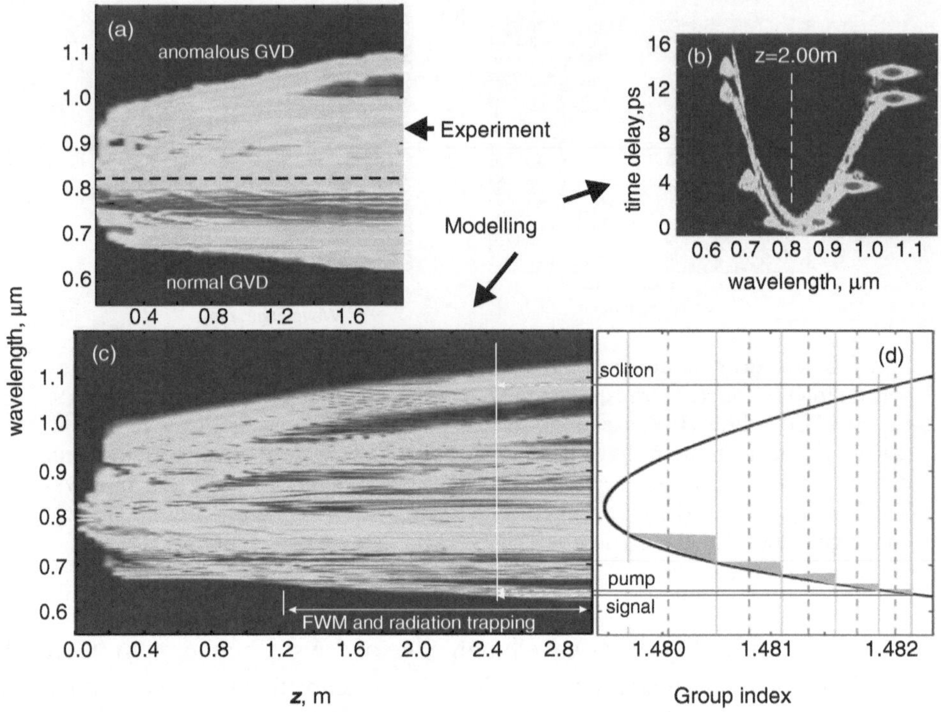

Fig. 9.9. Spectral evolution along the fibre length. (a)-experiment and (c)-modeling. (b) XFROG diagram computed for the propagation distance $z = 2$ m. (d) Group index as a function of wavelength. The decrease in height of the shaded triangles shows the decreasing frequency difference between the incident and reflected waves (Gorbach et al., 2006). Figures are reprinted with permission from Gorbach et al. (2006); copyright (2006) by the Optical Society of America.

fast spreading and drop in the peak intensity, terminating any nonlinear interaction. Instead, as originally measured in Hori et al. (2004) and as can be seen from the modelling results (Figs. 9.2(b) and 9.9(b)), the wavepackets on the short-wavelength edge of the continuum remain localized on a femtosecond time scale and propagate in the soliton-like regime albeit in the normal GVD range. This problem is addressed in the next section.

9.6 Radiation trapping and supercontinuum

So, what does stop dispersive spreading of the radiation wavepackets at the short-wavelength of a supercontinuum and keeps them localized on the femtosecond time scale? The short answer is – the refractive index change created by the decelerated soliton exerts a special type of inertial force and ensures dispersionless propagation (trapping) of the radiation.

Reflection of the radiation from a soliton plays the major role in the process of radiation trapping. Mathematically this is described by Eq. (9.10) with the first nonlinear term in the right-hand side, see the previous section. Switching into the reference frame moving together with the soliton reveals two distinct propagation regimes. If the offset of the group velocities of the soliton and radiation is sufficiently large, then the reflection from the soliton of course happens (Skryabin and Yulin, 2005), but recurrent collisions and radiation trapping can be disregarded for relatively short propagation lengths. The trapping phenomenon becomes the dominant feature of the propagation, when group velocities of the soliton and radiation become sufficiently close (Gorbach and Skryabin, 2007a, 2007b, 2007c).

The formal consideration of the problem starts from Eq. (9.10) complemented by the equation for the soliton field. We choose the reference frequencies of the soliton (δ_1) and radiation (δ_2) so that the group velocities are matched ($k'(\delta_1) = k'(\delta_2)$) across the zero GVD point and assume

$$E = A_s \exp[ik(\delta_1)z - i\delta_1 t] + F \exp[ik(\delta_2)z - i\delta_2 t]. \tag{9.15}$$

For the amplitudes $A_s(z,t)$ and $F(z,t)$ we make substitutions (Gorbach and Skryabin, 2007a, 2007c)

$$A_s = \psi(z,\xi)\exp\left[-it\frac{gz}{k_1''} + iqz + \frac{1}{3k_1''}g^2z^3\right], \tag{9.16}$$

$$F = \phi(z,\xi)\exp\left[-it\frac{gz}{k_2''} + i\lambda z + \frac{1}{3k_2''}g^2z^3\right], \tag{9.17}$$

$$t_s = gz^2/2\,, \xi = t - t_s\,, k_{1,2}'' = \partial_\delta^2 k(\delta_{1,2})\,.$$

We have introduced the parameter g to indicate that the deceleration rate for the soliton interacting with radiation can be different from g_0 for the pure soliton, see Eq. (9.4). GVD for the soliton and radiation are anomalous ($k_1'' < 0$) and normal ($k_2'' > 0$), respectively, and $\delta_1 < \delta_2$ is assumed. Thus the directions of the frequency shifts acquired by the two waves are opposite, see the first terms in the exponential factors of Eqs. (9.16) and (9.17). The group indices felt by the soliton and radiation ($\partial_z t_s = gz$) increase with the same rate (see also discussion in the previous section) and hence they experience an equal negative acceleration. Note also that if for any given z one takes the soliton frequency to be zero, then the radiation frequency coincides with the minimum of $k(\delta)$ in Fig. 9.7, where the phase matching condition (9.13) is satisfied.

Still the critical question is how does the radiation wave survive the dispersive spreading, i.e. is there a solution for ϕ retaining its localized form? Neglecting some non-essential terms, we find evolution equations for ψ and ϕ (Gorbach and

Skryabin, 2007c):

$$i\partial_z\psi - \frac{1}{2}k_1''\partial_\xi^2\psi - q\psi + \gamma|\psi|^2\psi = 2\gamma|\phi|^2\psi + T\gamma\psi\partial_\xi|\psi|^2 - \frac{g\xi}{k_1''}\psi, \quad (9.18)$$

$$i\partial_z\phi - \frac{1}{2}k_2''\partial_\xi^2\phi - \lambda\phi + 2\gamma|\psi|^2\phi + \frac{g\xi}{k_2''}\phi = -\gamma|\phi|^2\phi. \quad (9.19)$$

It is a good approximation to assume that the radiation is weak, so that all the terms $\propto |\phi|^2$ can be neglected. Another reasonable assumption, linked to the previous one, is that the soliton profile $|\psi|$ does not change with propagation. Then z-independent, i.e. shape preserving, radiation waves have to satisfy

$$-\frac{1}{2}k_2''\partial_\xi^2\phi_n + 2\gamma|\psi|^2\phi_n + \frac{g\xi}{k_2''}\phi_n = \lambda_n\phi_n,\ k_2'' > 0. \quad (9.20)$$

The above is the linear Schrödinger equation with the effective potential energy $V = 2\gamma|\psi|^2 + \frac{g\xi}{k_2''}$. The first term in V is the purely repelling potential created by the refractive index change induced by the soliton. The dispersive wave reflects from it in the way described in the previous section. The second term is the potential linearly increasing in ξ, which exists only due to the fact that we have switched into the non-inertial frame of reference, accelerating together with the soliton. Hence this term represents a type of inertial force acting on photons, which is analogous to the gravity force acting on massive particles. It is known from classical mechanics that inertial forces act as the usual ones, but show up in the equations of motion only when the appropriate non-inertial frame of reference is introduced.

Overall the potential $V(\xi)$ has a well-defined minimum and therefore supports (quasi-)localized modes (bound states), see Fig. 9.10. The index n in Eq. (9.20)

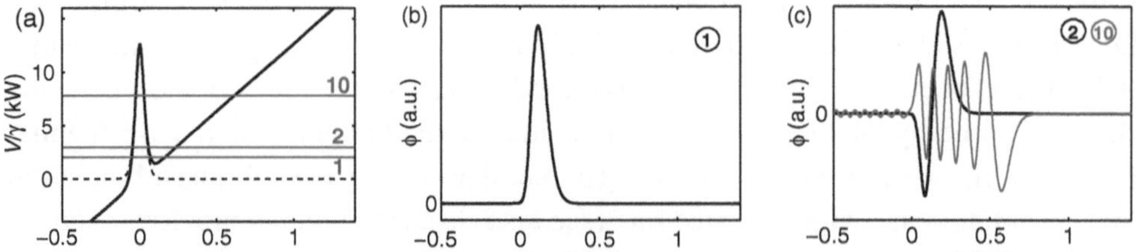

Fig. 9.10. (a) Full line shows the potential V. The dashed line shows the potential with the contribution from the inertial force disregarded, which removes the minimum and makes radiation trapping impossible. Level lines indicate ground state, 2nd and 10th modes. Parameters are taken for a typical 100 fs soliton created during supercontinuum formation process illustrated in Figs. 9.1(a) and 9.2. (b) shows the ground state mode of the potential V. (c) shows 2nd and 10th modes.

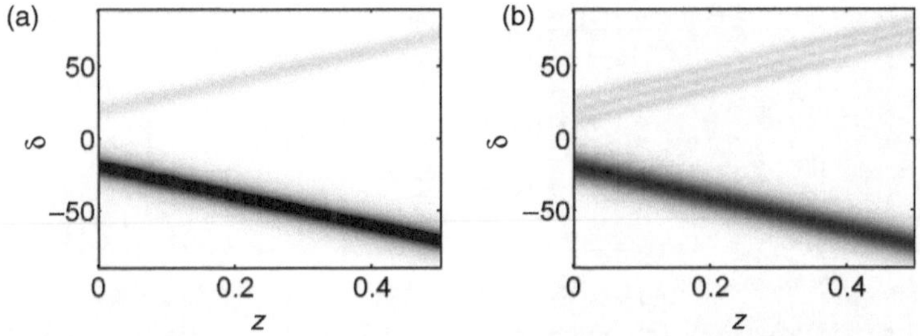

Fig. 9.11. Spectral evolution of the soliton–radiation bound states with $n = 1$ (a) and $n = 3$ (b) given by Eqs. (9.18) and (9.19).

numbers the localized modes, a 100-femtosecond soliton emerging in a typical supercontinuum setup supports one or two dozens of such modes (Gorbach and Skryabin, 2007b). These modes can be found either numerically or using a variational approach (Gorbach and Skryabin, 2007c). Taking the soliton plus one of these bound states results in the spectral evolution shown in Fig. 9.11. The soliton spectrum moves continuously to the smaller frequencies and the radiation spectrum moves towards higher frequencies. One can notice that taking higher order modes leads not only to temporal (Fig. 9.10(c)), but also to spectral broadening of the radiation (Fig. 9.11). Higher order modes have the typical multi-peak spectral structure, see Fig. 9.11(b), which explains fine spectral features often observed at the short-wavelength edge of super continua, see Fig. 9.13.

Spectral trajectories in Fig. 9.11 follow straight lines because Eqs. (9.18) and (9.19) assume frequency independent GVD. In a real fibre the soliton and radiation moving away from the zero GVD point encounter increasing absolute values of the GVD. This leads to the adiabatic broadening of the soliton and hence to the reduced Raman shift (Fig. 9.12(b)). The result is the gradually slowing spectral divergence of the soliton and radiation (Fig. 9.12(a)). The nonlinear mechanism behind the soliton self-frequency shift is the usual intrapulse Raman scattering. While the frequency conversion of the radiation wavepacket at the elementary level is due to the intrapulse four-wave mixing (see previous section), the latter is made possible by the sustained overlap of the radiation and soliton pulses.

Each of the solitons inside the supercontinuum shown in Figs. 9.1(b) and 9.9 has its own radiation pulse continuously drifting towards shorter wavelengths. We have found that the strongest soliton on the long-wavelength edge of the supercontinuum spectrum in Fig. 9.1(b) creates the potential V trapping around 20 modes on the short-wavelength edge. The radiation captured by the soliton can be represented as a superposition of these modes. Adiabatic transformation of the soliton power and width with propagation, caused by the increasing dispersion, induces weak adiabatic evolution of the mode parameters, but apart from this the modes are

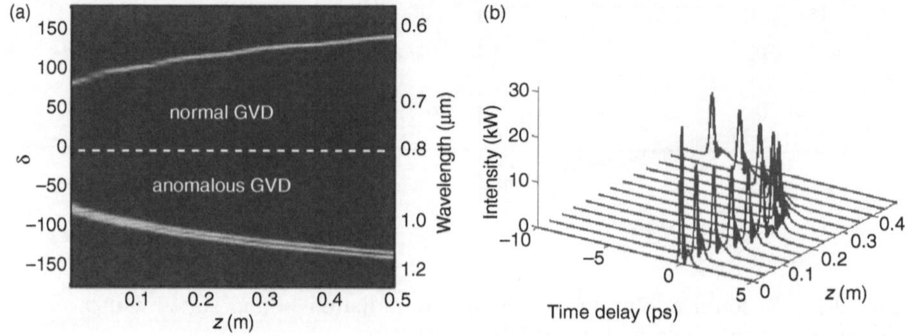

Fig. 9.12. Spectral (a) and time-domain (b) evolution of the soliton and trapped radiation modelled using the realistic frequency dependence of GVD (cf. (a) with Fig. 9.11). (b) is plotted in the reference frame accelerating with g_0 calculated for the initial soliton width, which explains the difference of the soliton trajectory from the one in Fig. 9.4(a).

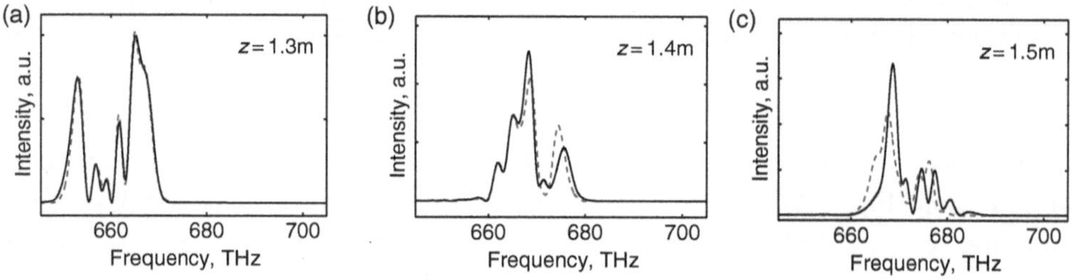

Fig. 9.13. Comparison of the spectral evolution of the short-wavelength edge of the supercontinuum obtained from the modelling of Eq. (9.1) (full lines) and from the modal expansion (9.21) (dashed lines). The spectrum plotted with full lines is taken directly from the edge of supercontinuum in Fig. 9.2 at propagation distances $z = 1.3$ m (a), $z = 1.4$ m (b) and $z = 1.5$ m (c).

stationary solutions and hence their temporal and spectral dynamics are suppressed. Assuming

$$F(z,\xi) = e^{-itgz/k_2''+g^2z^3/(3k_2'')} \sum_n \phi_n(\xi) e^{i\lambda_n z} \tag{9.21}$$

and comparing it with numerically computed spectral evolution of the short-wavelength edge of the supercontinuum (taken from Fig. 9.2) we have found an excellent agreement between the two approaches in predicting both the position of the short-wavelength edge of the continuum and its spectral shape, see Fig. 9.13. Small discrepancies in the shape of the spectral peaks become more noticeable for larger z and are the result of adiabatic changes of the soliton parameters not accounted in ϕ_n.

Finally we'd like to draw your attention to three observations and results. Firstly, the potential barrier on the soliton side is high, but still finite, so that some light leaks through the soliton barrier, creating oscillating tails. The leaked radiation is especially noticeable in higher-order modes, Figs. 9.1(b) and 9.10(c). Secondly, from numerical modelling of supercontinuum it can be found that the rate of the soliton self-frequency shift on the red edge of the supercontinuum is actually higher than that for the bare soliton. Analytically this effect can be captured if g is calculated from Eqs. (9.18) and (9.19) with the nonlinear in ϕ terms accounted for (Gorbach and Skryabin, 2007c). The resulting expression is $g \approx g_0(1 + Pb^2)$. Here P is the peak power of the radiation and b^2 is a constant. A cumbersome analytical expression for b^2 can be found in Gorbach and Skryabin (2007c). Thirdly, to generate the spectra shown in Figs. 9.2 and 9.9(a and c) we have used the same dispersion profiles and input powers, but different input wavelengths. For Fig. 9.9(a and c) the pump was only 10 nm away from the zero GVD point: this has led to formation of less powerful solitons. Hence their frequency shift and associated frequency shift of the shortwavelength radiation were substantially smaller leading to much narrower continua.

It is important to note that the simultaneous and opposite soliton–radiation frequency conversion and radiation trapping have been observed in few experiments, not related to the mainstream of the fibre supercontinuum research. The first paper known to us is the 1987 experiment by Beaud et al. (Beaud et al., 1987). Then there was a gap until 2001, when Nishizawa and Goto reported a series of spectral and time domain measurements of the effect of pulse trapping by a soliton across the zero GVD point (Nishizawa and Goto, 2001, 2002).

9.7 Summary

9.7.1 Soliton–radiation interaction

- The interaction of a soliton with dispersive radiation leads to the phase-matching sensitive generation of new frequencies. The most pronounced, out of a few possible interaction channels, is the reflection of the radiation from the refractive index change created by the soliton intensity. The reflection happens provided the radiation frequency belongs to the normal GVD range. Depending on the sign of the third order dispersion and the incident radiation frequency, the frequency of the reflected wave gets either up- or down-shifted.
- The Raman effect decelerates solitons and down-shifts their frequency. Such solitons can interact with radiation repeatedly, trap it on time scales of 100 fs and continuously up-shift the radiation frequency.

9.7.2 Supercontinuum

Our results (Gorbach et al., 2006; Gorbach and Skryabin, 2007b, 2007c) unambiguously support prevalence of the following scenario of the supercontinuum generation in photonic crystal fibres pumped by femtosecond pulses with the input wavelength around the zero GVD point:

- The spectrum of the input signal is distributed in some proportion between the frequency ranges with normal and anomalous GVD. This happens through the combination of nonlinear processes. SPM dominates during the first few centimetres of propagation. Then the soliton formation accompanied by the Cherenkov radiation and four-wave mixing of the dispersive waves with solitons leads to further spectral broadening.
- The next stage is when the Raman solitons on the long-wavelength edge of the supercontinuum enter into the regime of the cascaded interaction with dispersive radiation. This quickly leads to formation of the bound soliton–radiation states. The trapping mechanism cancels the dispersive spreading of the radiation and sustains stable time-domain overlap of the tails of the radiation and soliton pulses. The resulting four-wave mixing leads to the continuous up-shift of the radiation frequency. The necessary condition for the radiation trapping is a near matching of the group velocities at the supercontinuum edges across the zero GVD point.

Using a nanosecond or cw pump leads to modulational instability and subsequent creation of soliton trains. The latter trap radiation, and the above scenario is realized again albeit with a greater number of solitons (Travers, 2009), see Chapter 8.

9.8 Perspectives

Interaction between solitons and radiation, optical rogue (freak) waves (Solli et al., 2007), development of ideas around the gravity-like force exerted on light by solitons (Gorbach and Skryabin, 2007a), optical turbulence (Barviau et al., 2008), and optical black-holes (Philbin et al., 2008) can all be mentioned on the list of problems stimulated by the supercontinuum research and still holding a significant research potential.

References

Agrawal G. (2001). *Nonlinear Fiber Optics*. Academic Press.

Akhmediev N., Karlsson M. (1995). Cherenkov radiation emitted by solitons in optical fibers. *Phys. Rev. A* **51**, 2602–2607.

Akhmediev N., Krolikowski W., Lowery A. (1996). Influence of the Raman-effect on solitons in optical fibers. *Opt. Commun.* **131**, 260–266.

Barviau B., Kibler B., Coen S., Picozzi A. (2008). Toward a thermodynamic description of supercontinuum generation. *Opt. Lett.* **33**, 2833–2835.

Beaud P., Hodel W., Zysset B., Weber H. (1987). Ultrashort pulse-propagation, pulse breakup, and fundamental soliton formation in a single-mode optical fiber. *IEEE J. Quantum Electron.* **23**, 1938–1946.

Biancalana F., Skryabin D. V., Yulin A. V. (2004). Theory of the soliton self-frequency shift compensation by the resonant radiation in photonic crystal fibers. *Phys. Rev. E* **70**, 016615.

Corwin K. L., Newbury N. R., Dudley J. M. et al. (2003). Fundamental noise limitations to supercontinuum generation in microstructure fiber. *Phys. Rev. Lett.* **90**, 113904.

Cristiani I., Tediosi R., Tartara L., Degiorgio V. (2004). Dispersive wave generation by solitons in microstructured optical fibers. *Opt. Express* **12**, 124–135.

Dudley J. M., Coen S. (2002). Coherence properties of supercontinuum spectra generated in photonic crystal and tapered optical fibers. *Opt. Lett.* **27**, 1180–1182.

Dudley J. M., Genty G., Coen S. (2006). Supercontinuum generation in photonic crystal fiber. *Rev. Mod. Phys.* **78**, 1135–1184.

Efimov A., Taylor A. J., Omenetto F. G. et al. (2004). Time-spectrally-resolved ultrafast nonlinear dynamics in small-core photonic crystal fibers: Experiment and modelling. *Opt. Express* **12**, 6498–6507.

Efimov A., Taylor A. J., Yulin A. V., Skryabin D. V., Knight J. C. (2006). Phase-sensitive scattering of a continuous wave on a soliton. *Opt. Lett.* **31**, 1624–1626.

Efimov A., Yulin A. V., Skryabin D. V. et al. (2005). Interaction of an optical soliton with a dispersive wave. *Phys. Rev. Lett.* **95**, 213902.

Falk P., Frosz M. H., Bang O. (2005). Supercontinuum generation in a photonic crystal fiber with two zero-dispersion wavelengths tapered to normal dispersion at all wavelengths. *Opt. Express* **13**, 7535–7540.

Gagnon L., Belanger P. A. (1990). Soliton self-frequency shift versus galilean-like symmetry. *Opt. Lett.* **15**, 466–469.

Genty G., Lehtonen M., Ludvigsen H. (2004a). Effect of cross-phase modulation on supercontinuum generated in microstructured fibers with sub-30 fs pulses. *Opt. Express* **12**, 4614–4624.

Genty G., Lehtonen M., Ludvigsen H. (2005). Route to broadband blue-light generation in microstructured fibers. *Opt. Lett.* **30**, 756–758.

Genty G., Lehtonen M., Ludvigsen H., Kaivola M. (2004b). Enhanced bandwidth of supercontinuum generated in microstructured fibers. *Opt. Express* **12**, 3471–3480.

Gorbach A. V., Skryabin D. V. (2007a). Bouncing of a dispersive pulse on an accelerating soliton and stepwise frequency conversion in optical fibers. *Opt. Express* **15**, 14560–14565.

Gorbach A. V., Skryabin D. V. (2007b). Light trapping in gravity-like potentials and expansion of supercontinuum spectra in photonic-crystal fibres. *Nature Photonics* **1**, 653–657.

Gorbach A. V., Skryabin D. V. (2007c). Theory of radiation trapping by the accelerating solitons in optical fibers. *Phys. Rev. A* **76**, 053803.

Gorbach A. V., Skryabin D. V. (2008). Soliton self-frequency shift, non-solitonic radiation and self-induced transparency in air-core fibers. *Opt. Express* **16**, 4858–4865.

Gorbach A., Skryabin D., Stone J., Knight J. (2006). Four-wave mixing of solitons with radiation and quasi-nondispersive wave packets at the short-wavelength edge of a supercontinuum. *Opt. Express* **14**, 9854–9863.

Gu X., Kimmel M., Shreenath A. P. et al. (2003). Experimental studies of the coherence of microstructure-fiber supercontinuum. *Opt. Express* **11**, 2697–2703.

Harbold J. M., Ilday F. O., Wise F. W., Birks T. A., Wadsworth W. J., Chen Z. (2002). Long-wavelength continuum generation about the second dispersion zero of a tapered fiber. *Opt. Lett.* **27**, 1558–1560.

Herrmann J., Griebner U., Zhavoronkov N. et al. (2002). Experimental evidence for supercontinuum generation by fission of higher-order solitons in photonic fibers. *Phys. Rev. Lett.* **88**, 173901.

Hori T., Nishizawa N., Goto T., Yoshida M. (2004). Experimental and numerical analysis of widely broadened supercontinuum generation in highly nonlinear dispersion-shifted fiber with a femtosecond pulse. *J. Opt. Soc. Am. B* **21**, 1969.

Husakou A. V., Herrmann J. (2001). Supercontinuum generation of higher-order solitons by fission in photonic crystal fibers. *Phys. Rev. Lett.* **8720**, 203901.

Islam M. N., Mollenauer L. F., Stolen R. H., Simpson J. R., Shang H. T. (1987). Cross-phase modulation in optical fibers. *Opt. Lett.* **12**, 625–627.

Karpman V. I. (1993). Radiation by solitons due to higher-order dispersion. *Phys. Rev. E* **47**, 2073–2082.

Knight J. C., Skryabin D. V. (2007). Nonlinear waveguide optics and photonic crystal fibers. *Opt. Express* **15**, 15365–15376.

Luan F., Skryabin D., Yulin A., Knight J. (2006). Energy exchange between colliding solitons in photonic crystal fibers. *Opt. Express* **14**, 9844–9853.

Mitschke F., Mollenauer L. (1986). Discovery of the soliton self-frequency shift. *Opt. Lett.* **11**, 659–661.

Nishizawa N., Goto T. (2001). Experimental analysis of ultrashort pulse propagation in optical fibers around zero-dispersion region using cross-correlation frequency resolved optical gating. *Opt. Express* **8**, 328–334.

Nishizawa N., Goto T. (2002). Characteristics of pulse trapping by use of ultrashort soliton pulses in optical fibers across the zero-dispersion wavelength. *Opt. Express* **10**, 1151–1159.

Philbin T. G., Kuklewicz C., Robertson S., Hill S., Konig F., Leonhardt U. (2008). Fiber-optical analog of the event horizon. *Science* **319**, 1367–1370.

Ranka J. K., Windeler R. S., Stentz A. J. (2000). Visible continuum generation in air–silica microstructure optical fibers with anomalous dispersion at 800 nm. *Opt. Lett.* **25**, 25–27.

Russell P. (2003). Photonic crystal fibers. *Science* **299**, 358–362.

Skryabin D., Wadsworth W. (2009). *Nonlinear Periodic Structures*. Springer.

Skryabin D. V., Yulin A. V. (2005). Theory of generation of new frequencies by mixing of solitons and dispersive waves in optical fibers. *Phys. Rev. E* **72**, 016619.

Skryabin D. V., Luan F., Knight J. C., Russell P. S. (2003). Soliton self-frequency shift cancellation in photonic crystal fibers. *Science* **301**, 1705–1708.

Smirnov S., Ania-Castanon J. D., Ellingham T. J., Kobtsev S. M., Kukarin S., Turitsyn S. K. (2006). Optical spectral broadening and supercontinuum generation in telecom applications. *Opt. Fiber Tech.* **12**, 122–147.

Solli D. R., Ropers C., Koonath P., Jalali B. (2007). Optical rogue waves. *Nature* **450**, 1054–1057.

Stone J. M., Knight J. C. (2008). Visibly "white" light generation in uniform photonic crystal fiber using a microchip laser. *Opt. Express* **16**, 2670–2675.

Travers J. (2009). Blue solitary waves from infrared continuous wave pumping in optical fibers. *Opt. Express* **17**, 1502–1507.

Tsoy E., deSterke C. M. (2006). Dynamics of ultrashort pulses near zero dispersion wavelength. *J. Opt. Soc. Am. B* **23**, 2425–2433.

Wadsworth W. J., Ortigosa-Blanch A., Knight J. C., Birks T. A., Man T. P. M., Russell P. S. (2002). Supercontinuum generation in photonic crystal fibers and optical fiber tapers: a novel light source. *J. Opt. Soc. Am. B* **19**, 2148–2155.

Wai P. K. A., Chen H. H., Lee Y. C. (1990). Radiations by solitons at the zero group-dispersion wavelength of single-mode optical fibers. *Phys. Rev. A* **41**, 426–439.

Yulin A. V., Skryabin D. V., Russell P. S. J. (2004). Four-wave mixing of linear waves and solitons in fibers with higher-order dispersion. *Opt. Lett.* **29**, 2411–2413.

10

Interaction of four-wave mixing and stimulated Raman scattering in optical fibers

Stéphane Coen, Stuart G. Murdoch and Frédérique Vanholsbeeck

10.1 Introduction

Parametric four-wave mixing (FWM) and stimulated Raman scattering (SRS) are two fundamental processes of third-order nonlinear optical materials, and silica optical fibers in particular (Agrawal, 2006). As will be clear from many of the preceding chapters in this book, they also play a central role in supercontinuum generation. On the one hand, FWM involves the elastic interaction between two pump photons as well as a Stokes and an anti-Stokes photon symmetrically located around the pump frequency (Stolen, 1975; Stolen and Bjorkholm, 1982). It is of tremendous interest for large bandwidth optical amplification (Hansryd et al., 2002), including ultra-low-noise phase-sensitive amplification (McKinstrie and Radic, 2004; Oo-Kaw et al., 2008), but also for parametric oscillators (Coen and Haelterman, 2001; Sharping, 2008), wavelength converters (Islam and Boyraz, 2002) including noise-free schemes (Gnauck et al., 2006; Méchin et al., 2006) or optical signal processing in general (Salem et al., 2008). Its efficiency critically depends on a phase-matching relation, which is related to momentum conservation. On the other hand, SRS is characterized by the stimulated inelastic conversion of photons from an intense optical pump wave into lower frequency Stokes photons through the resonant excitation of a vibration in the transmission medium (i.e., optical phonons) (Bloembergen, 1967; Stolen et al., 1972). Here phase-matching is automatically satisfied. The SRS gain, in contrast to the FWM gain, exhibits an asymmetric profile (Stolen and Ippen, 1973): only the down-shifted Stokes waves are amplified while the up-shifted anti-Stokes waves are exponentially absorbed. Aside from supercontinuum generation (Dudley et al., 2006), other applications include high-power large-bandwidth optical amplifiers and Raman fiber lasers (Islam, 2002).

While FWM and SRS are generally considered as distinct phenomena, it has been known since the early days of laser physics and nonlinear optics that their interaction leads to non-trivial dynamics. A full analysis of parametric or Raman amplification must therefore take into account both phenomena. This was first revealed in the seminal theoretical work of Bloembergen and Shen (1964). These authors predicted that at the peak Raman gain detuning (13.2 THz for a fused silica fiber) the available gain is a strong function of the linear phase-mismatch $\Delta\beta_L$ of the FWM process associated with the optical waves involved in the Raman interaction. In particular, they revealed that the vibrational excitation and the SRS gain are completely suppressed when $\Delta\beta_L$ is zero. Also, Bloembergen and Shen showed that, even under non-phase-matched conditions, the coupling between SRS and FWM can lead to the exponential amplification of the anti-Stokes wave. Accordingly, although a simple theory of SRS predicts that the amplified wave has a pure Stokes character, a slight admixture at the anti-Stokes is in practice always present. Roughly speaking, all these phenomena can be understood by noting that the beat note between the pump wave and an anti-Stokes wave generated by FWM within the Raman bandwidth matches the frequency of the vibrational wave. In this way, the anti-Stokes wave is coupled to the material excitation, which in turn affects the dynamics of the SRS process.

Even though some of these phenomena had been reported qualitatively earlier than 1964, rigorous experimental evidence has been slow to come. The suppression of the Raman gain at zero linear phase-mismatch was only first reported in 1986 (Duncan et al.) in high pressure H_2 and D_2. In addition Duncan et al. (1986) were able to show that the residual gain at this point was not exponential but grew quadratically with distance. Following the extension of Bloembergen and Shen's theoretical treatment to the domain of optical fibers (Blow and Wood, 1989; Trillo and Wabnitz, 1992; Golovchenko and Pilipetskii, 1994), observations of parametric Raman gain suppression were then reported in silica fibers (Golovchenko et al., 1989, 1990) but only for positive values of the linear phase-mismatch. It took another decade for a complete experimental characterization of the gain at the peak Raman gain detuning as a function of the phase-mismatch to be reported (Vanholsbeeck et al., 2003). These latter observations have been made possible thanks to the remarkable dispersive and nonlinear properties of photonic crystal fibers. At the same period, the non-phase-matched amplification of the anti-Stokes wave has also been carefully studied experimentally (Coen et al., 2002). This phenomenon was subsequently shown to be more generally caused by the interaction of FWM gain with dispersive loss (Tanemura et al., 2004).

Following these developments which were found to be in excellent agreement with the predictions of Bloembergen and Shen (1964), more extensive studies have recently been performed. In particular, they include observations made for a broad range of Stokes/anti-Stokes detunings and not just at the peak Raman gain detuning of 13.2 THz. First, it was reported that the phase-matched gain of a parametric amplifier is strongly affected by the *real* part of the Raman susceptibility (Hsieh et al., 2007a,b). This has important consequences for the gain flatness of optical parametric amplifiers which are foreseen as key components in the multi-Tb/s telecommunication networks of the future. Second, intricate destructive interferences of the two gain modes of a mixed parametric-Raman amplifier were also shown to lead to a strong signal suppression (Wang et al., 2008). Finally, let us point out that a quantum treatment of FWM and SRS has also been presented recently (Brainis et al., 2007).

Clearly, all these results reveal the fundamental and practical importance of the interplay between parametric FWM and SRS in optical fibers, and more generally of the frequency dependence of both the real and imaginary parts of the nonlinear susceptibility. The scope of this chapter is therefore to present all these different dynamical behaviors in a unified context. To this end, we start by a detailed presentation of the underlying theory, and then we review recent experimental results obtained in optical fibers and covering all the above-mentioned phenomena. First, we discuss the experimental observations of the dependence of the Raman/parametric gain versus phase-mismatch (at 13.2 THz detuning) (Vanholsbeeck et al., 2003) and versus detuning (at perfect phase-matching) (Hsieh et al., 2007a,b). We then present the experimental results of Wang et al. (2008) demonstrating strong signal suppression. Finally, the non-phase-matched amplification of the anti-Stokes wave is considered (Coen et al., 2002) together with some recent results.

10.2 Theory

10.2.1 The model

To ensure a good understanding of all the experimental observations presented below, we first present the theory of the combined influence of SRS and FWM. In the following, we consider single-mode silica fibers as the nonlinear medium. Note that the use of single-mode waveguides enables one to make observations unhampered by the diffraction of the beams. We also assume that all the waves are collinearly polarized [see Trillo and Wabnitz (1992) and Golovchenko and Pilipetskii (1994) for a generalization of this theory to the case of birefringent fibers]. Silica fibers exhibit a $\chi^{(3)}$ Kerr nonlinearity that mainly stems from the instantaneous anharmonic motion of bound electrons under the influence of an applied field (Agrawal, 2006). SRS arises due to the additional excitation of internal vibrational modes of the silica molecules and can be represented as a delayed contribution to the nonlinearity (Stolen et al., 1989). In these conditions, the combined effect of FWM and SRS can be described by a generalized nonlinear Schrödinger equation

(GNLSE) that reads (Agrawal, 2006),

$$\frac{\partial A(z,t)}{\partial z} = \sum_{k\geq 2} i^{k+1} \frac{\beta_k}{k!} \frac{\partial^k A(z,t)}{\partial t^k} + i\gamma \Big[(1-f)|A(z,t)|^2 + f \int_{-\infty}^{t} \chi_{\mathrm{R}}^{(3)}(t-t')|A(z,t')|^2 dt' \Big] A(z,t). \tag{10.1}$$

Here $A(z,t)$ is the envelope of the electric field, z is the longitudinal coordinate along the fiber axis, while t is the time in a reference frame traveling at the group velocity of the pump light in the fiber. The β_k are the second- and higher-order dispersion coefficients associated with the Taylor series expansion of the propagation constant $\beta(\omega)$ about the carrier frequency ω_0 while γ is the nonlinearity coefficient of the fiber, also evaluated at the carrier frequency. $\chi_{\mathrm{R}}^{(3)}(t)$ is the delayed Raman response function of silica, i.e., the Fourier transform of the complex Raman susceptibility $\widetilde{\chi}_{\mathrm{R}}^{(3)}(\Omega)$, which is plotted in Figure 10.1 [the normalization is chosen such that $\widetilde{\chi}_{\mathrm{R}}^{(3)}(0) = 1$] (Stolen et al., 1989). The small parameter f measures the fractional contribution of the Raman susceptibility to the instantaneous Kerr effect. f is commonly estimated at 0.18 for silica fibers (Stolen et al., 1989) but its value is not well known and may vary from fiber to fiber as will be highlighted by some of our observations. Note that setting $f = 0$ in Eq. (10.1) cancels the contribution of SRS and reduces the model to the case where only parametric FWM is taken into account.

We restrict ourselves to considering the combined effect of the two nonlinearities on a single strong undepleted pump wave and two weak waves symmetrically detuned from the pump by an angular frequency $\pm\Omega$ ($\Omega > 0$). This corresponds to the situation typically found in single-pump parametric amplifiers (Hansryd et al., 2002), parametric oscillators (Sharping, 2008), wavelength converters (Islam and

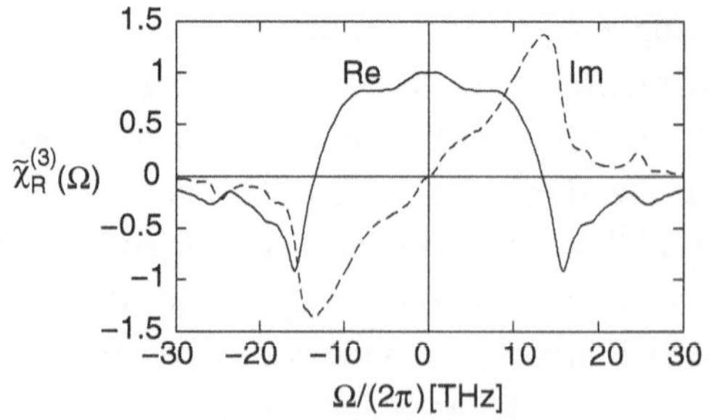

Fig. 10.1. Complex Raman susceptibility of silica, $\widetilde{\chi}_{\mathrm{R}}^{(3)}(\Omega)$. The solid line is the real part, while the dashed line is the imaginary part.

Boyraz, 2002), and standard modulation instability type experiments (Tai et al., 1986). In this case, the wave dynamics can be investigated by introducing into the GNLSE (10.1) the ansatz

$$A(z,t) = \left[\sqrt{P_p} + A_s(z)\exp(i\Omega t) + A_a(z)\exp(-i\Omega t)\right]\exp(i\gamma P_p z). \qquad (10.2)$$

In this expression, P_p is the pump power while $A_s(z)$ and $A_a(z)$ are the amplitudes of the Stokes and anti-Stokes waves, respectively. All three waves are assumed to be monochromatic. For the undepleted pump approximation to be valid, the Stokes and anti-Stokes powers must stay small with respect to the pump, $P_p \gg |A_{s,a}(z)|^2$. In these conditions, we can neglect all direct couplings between the Stokes and the anti-Stokes waves, and consider only first order terms in A_s and A_a, which leads to the linear problem (* denotes the complex conjugate),

$$\frac{d}{dz}\begin{bmatrix} A_s \\ A_a^* \end{bmatrix} = i\gamma P_p \begin{bmatrix} q-K & q \\ -q & -q+K \end{bmatrix}\begin{bmatrix} A_s \\ A_a^* \end{bmatrix}, \qquad (10.3)$$

for the evolution of the Stokes and anti-Stokes fields along the fiber length. In the above equation, the complex variable q gives the normalized strength of the real and imaginary parts of the $\chi^{(3)}$ susceptibility. It is frequency dependent and reads,

$$q = q(\Omega) = 1 - f + f\tilde{\chi}_R^{(3)}(-\Omega) = 1 - f + f\tilde{\chi}_R^{(3)}(\Omega)^*. \qquad (10.4)$$

The second part of the equality is due to the antisymmetry of the Raman response (see Figure 10.1). Note that $\mathrm{Im}(q) < 0$ for all detunings considered here ($\Omega > 0$) and that q is reduced to 1 when one neglects SRS [which occurs, e.g., when $\Omega/(2\pi) \ll 13.2$ THz]. K is the linear phase-mismatch normalized to the nonlinear contribution to the mismatch, $K = -\Delta\beta_L/(2\gamma P_p)$ where $\Delta\beta_L = \beta_a + \beta_s - 2\beta_p = \beta_2\Omega^2 + \beta_4\Omega^4/12 + \cdots$. Here $\beta_a = \beta(\omega_0 + \Omega)$, $\beta_s = \beta(\omega_0 - \Omega)$, and $\beta_p = \beta(\omega_0)$ are the wavenumbers of the anti-Stokes, Stokes, and pump waves, respectively. The standard condition of perfect phase-matching, $\Delta\beta_L + 2\gamma P_p = 0$, corresponds to $K = 1$ (Agrawal, 2006). When higher-order dispersion is negligible, we observe that $K = 0$ when the pump wavelength is set to the zero-dispersion wavelength (ZDW) of the fiber. K can therefore also be interpreted as the normalized detuning of the pump wavelength with respect to the ZDW. It is positive (negative) in the anomalous (normal) dispersion regime, respectively.

10.2.2 The mixed Stokes/anti-Stokes modes

The solution of the linear system of differential equations (10.3) can be written in terms of the eigenvalues $\lambda_\pm$ and eigenvectors $\mathbf{v}_\pm$ of the 2×2 matrix of the system

$$\begin{bmatrix} A_s(z) \\ A_a^*(z) \end{bmatrix} = C_+ \mathbf{v}_+ \exp(\lambda_+ z) + C_- \mathbf{v}_- \exp(\lambda_- z) \tag{10.5}$$

where $C_\pm$ are constants of integration whose values must be determined from the initial conditions (see Section 10.2.4). The eigenvalues are given by $\lambda_\pm = \pm\gamma P_p R$ where

$$R = \sqrt{K(2q - K)}. \tag{10.6}$$

By convention, we will take the real part of R to be positive, so that λ_+ is associated with an exponentially growing mode, while λ_- is associated with an exponentially decaying mode. We will define the gain coefficient g of the modes of the amplification process such that the *power* of the growing (decaying) mode evolves along the fiber as $\exp(gz)$ [respectively $\exp(-gz)$].

$$g = 2\mathrm{Re}(\lambda_+) = 2\gamma P_p \mathrm{Re}(R). \tag{10.7}$$

Note that a special case occurs in the limit where SRS is negligible and only FWM is considered, $q = 1$. When $K < 0$ or $K > 2$, we find that R is purely imaginary. Neither mode is either growing or decaying but rather exhibits an oscillatory behavior along the fiber length, which is typical of a non-phase-matched parametric process. By contrast, $0 \leq K \leq 2$ corresponds to the standard amplifying bandwidth of a pure parametric amplifier. Here R is purely real. Under perfect phase-matching conditions, $K = 1$, the gain coefficient is maximum, $g_{\max} = 2\gamma P_p$ for all detunings Ω. All these results match the textbook results of single pump parametric FWM amplifiers (Agrawal, 2006).

To gain a better insight into the coupled Stokes/anti-Stokes dynamics, let us focus now on the eigenmodes of the amplification process. In the general case, the modes are neither purely Stokes nor anti-Stokes but exhibit a mixed Stokes/anti-Stokes character. It is convenient to write them in the following form:

$$\mathbf{v}_+ = \begin{bmatrix} 1 \\ \frac{1}{q}(K - q - iR) \end{bmatrix}, \quad \mathbf{v}_- = \begin{bmatrix} \frac{1}{q}(K - q - iR) \\ 1 \end{bmatrix}. \tag{10.8}$$

Calculations reveal that the properties of the Raman susceptibility $\widetilde{\chi}_R^{(3)}(\Omega)$, and in particular the fact that $\mathrm{Im}(q) < 0$, implies that the common fraction in the expressions of the two eigenvectors above is such that $|(K - q - iR)/q| \leq 1$ for all values of K and Ω. This reveals that the growing mode $\mathbf{v}_+$ has a dominant Stokes character (i.e., the Stokes wave has a larger amplitude than the anti-Stokes wave) while the decaying mode $\mathbf{v}_-$ has a dominant anti-Stokes character. This is to be expected as SRS always favors the Stokes wave.

Asymptotically, for a large gain or a long propagation distance, $gz \gg 1$, the amplitude of the decaying mode $\mathbf{v}_-$ eventually vanishes and the output of the system is entirely determined by the growing mode $\mathbf{v}_+$, independent of the initial conditions. Despite the fact that this mode is associated with the growth of the Stokes wave, its mixed Stokes/anti-Stokes nature implies that the anti-Stokes wave is always present and grows proportionally to the Stokes wave, a result which contradicts the simple theory of SRS. The asymptotic ratio of power between the anti-Stokes and Stokes sidebands is given by the square of the ratio between the two components of the $\mathbf{v}_+$ eigenvector, and is independent of the propagation length and of the initial conditions,

$$\frac{P_\mathrm{a}}{P_\mathrm{s}} = \left|\frac{A_\mathrm{a}^*}{A_\mathrm{s}}\right|^2 = \left|\frac{K - q - iR}{q}\right|^2 . \tag{10.9}$$

This equation is illustrated in Figure 10.2 as a function of the signal detuning Ω and the normalized mismatch K. The calculation uses the complex Raman susceptibility curve of Stolen et al. (1989) and a value for f of 0.18. In this figure, we can observe that, in the limit of large phase-mismatches, $|K| \gg 1$, the anti-Stokes wave is almost negligible. As a matter of fact, it is only in this limit that the Stokes and anti-Stokes waves decouple and that the modes exhibit a pure Stokes and anti-Stokes character respectively, $\mathbf{v}_+ \to [1;0]$, $\mathbf{v}_- \to [0;1]$. This is to be expected as in this regime the

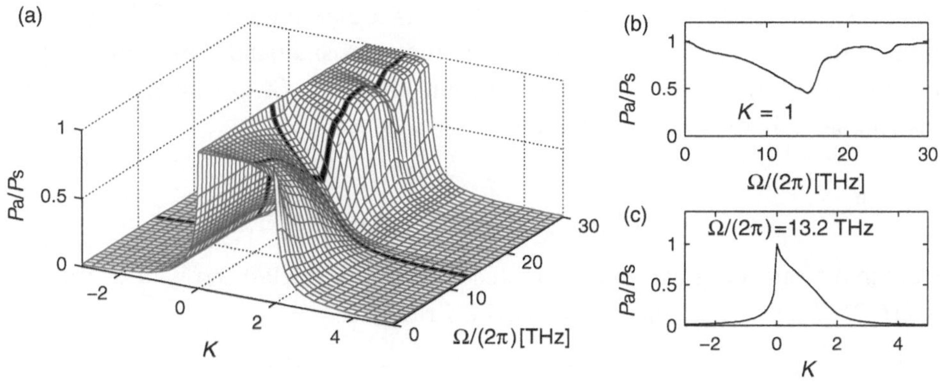

Fig. 10.2. (a) Plot of the asymptotic anti-Stokes to Stokes power ratio, $P_\mathrm{a}/P_\mathrm{s}$, versus the frequency detuning of the Stokes and anti-Stokes sidebands Ω and the normalized phase-mismatch K. Insets show cross-sections (b) at perfect phase-matching $K = 1$ and (c) at the peak Raman gain detuning $\Omega/(2\pi) = 13.2$ THz. The black lines in (a) correspond to the insets.

system is acting primarily as a Raman amplifier. More precisely, at second order in the small parameter q/K, we find that $R \to i\left[q - K + q^2/(2K)\right]$ and that the anti-Stokes amplitude scales inversely proportional to the linear phase-mismatch,

$$\frac{A_\mathrm{a}^*}{A_\mathrm{s}} \xrightarrow[|K|\gg 1]{} \frac{q}{2K} = \frac{-\gamma q P_\mathrm{p}}{\Delta\beta_\mathrm{L}}. \tag{10.10}$$

This equation is well known (Bloembergen and Shen, 1964; Golovchenko et al., 1990; Trillo and Wabnitz, 1992; Coen et al., 2002) and clearly highlights the fact that the amplification of the anti-Stokes wave is due to parametric FWM.

For a phase-matched amplifier ($K = 1$), we can obtain a simplified expression of the anti-Stokes to Stokes power ratio by considering f as a small parameter. Starting from $R = \sqrt{2q - 1}$ [Eq. (10.6) with $K = 1$] and expanding it as a binomial expansion in terms of f, we can then express Eq. (10.9) at first order in the parameter f as

$$\left.\frac{P_\mathrm{a}}{P_\mathrm{s}}\right|_{K=1} \simeq 1 + 2f\,\mathrm{Im}\big[\widetilde{\chi}_\mathrm{R}^{(3)}(-\Omega)\big]. \tag{10.11}$$

In this form, the anti-Stokes to Stokes power ratio appears as a simple function of the imaginary part of $\chi^{(3)}$ with only a weak (second order) dependence on the real part of the Raman susceptibility. This dependence can clearly be observed by comparing Figure 10.2(b) with the imaginary part of $\widetilde{\chi}_\mathrm{R}^{(3)}(\Omega)$ plotted in Figure 10.1. At detunings where the imaginary part of $\chi^{(3)}$ is low ($\Omega \ll$ or $\gg$ 13.2 THz), the anti-Stokes to Stokes power ratio is close to unity over the entire parametric gain bandwidth ($0 < K < 2$), as is expected in the absence of SRS. At detunings where the imaginary part of $\chi^{(3)}$ is large the ratio is naturally weighted in favor of the Stokes sideband. The imaginary part of $\chi^{(3)}$ is a maximum at 13.2 THz, at this detuning the ratio $P_\mathrm{a}/P_\mathrm{s} = 0.5$ for a phase-matched sideband. For completeness, we have also plotted in Figure 10.2(c) the anti-Stokes to Stokes power ratio at the peak Raman gain detuning of 13.2 THz versus K.

10.2.3 The gain coefficient

To complete our analysis, let us now discuss in more detail the gain coefficient, Eq. (10.7). We first note that it can also be written as

$$g = 2\,\mathrm{Re}\sqrt{(\gamma q P_\mathrm{p})^2 - \left(\frac{\Delta\beta_\mathrm{L} + 2\gamma q P_\mathrm{p}}{2}\right)^2}. \tag{10.12}$$

In this form, this expression is a generalization of the standard expression for pure parametric gain (Agrawal, 2006), which is recovered by setting $q = 1$. The above

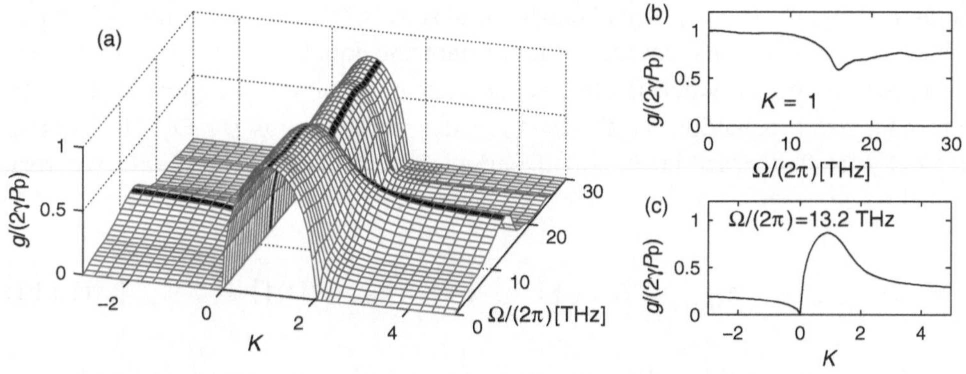

Fig. 10.3. (a) Plot of the normalized gain coefficient, $g/(2\gamma P_{\mathrm{p}}) = \mathrm{Re}(R)$, versus the frequency detuning of the Stokes and anti-Stokes sidebands Ω and the normalized phase-mismatch K. Insets show cross-sections (b) at perfect phase-matching $K = 1$ and (c) at the peak Raman gain detuning $\Omega/(2\pi) = 13.2$ THz. The black lines in (a) correspond to the insets.

expression is plotted in Figure 10.3 in normalized units, $g/(2\gamma P_{\mathrm{p}}) = \mathrm{Re}(R)$, as a function of the signal detuning Ω and the normalized mismatch K. Again we have set $f = 0.18$. This figure demonstrates several key features of the combined interaction of FWM and SRS and of the role of the real and imaginary parts of the third-order susceptibility. At large phase-mismatches, $|K| \gg 1$, we have $R \to i(q - K)$ and $\mathrm{Re}(R) \to -\mathrm{Im}(q)$ so that the gain coefficient g reduces to

$$g \xrightarrow[|K|\gg 1]{} -2\gamma P_{\mathrm{p}}\, f\, \mathrm{Im}\big[\widetilde{\chi}_{\mathrm{R}}^{(3)}(-\Omega)\big]. \tag{10.13}$$

As expected, the gain spectrum is that of a pure Raman amplifier, i.e., proportional to the imaginary part of the susceptibility. Another feature of the combined Raman-parametric interaction clearly shown in Figure 10.3 is that when the normalized mismatch parameter K is zero, the gain coefficient g of the amplifier falls to zero. An analysis of the coupled Eqs. (10.3) shows that at this mismatch a signal does not experience any exponential gain and rather undergoes only a weak quadratic amplification. This result, which was first observed experimentally by Duncan et al. (1986), is somewhat surprising as it shows that the parametric effects at this detuning act to suppress Raman scattering, a process which is usually thought of as self-phase-matched, and so not subject to any phase-matching conditions. Note that other more complicated schemes for suppressing SRS with parametric interactions have also been demonstrated using multiple pump waves (Tchofo Dinda et al., 1998, 1999; Fan and Migdall, 2006). The final important feature of Figure 10.3 can be seen in the variation of the phase-matched gain coefficient ($K = 1$) as a function of detuning [Figure 10.3(b)]. Recall that in the absence of Raman scattering, the phase-matched parametric gain coefficient is constant and equal to $2\gamma P_p$ for all detunings as discussed above, which is clearly not what is observed in Figure 10.3(b). To understand this behavior, we follow the same procedure as for Eq. (10.11) and expand g at first order in the small parameter f. This gives a simple yet very revealing expression for the peak phase-matched gain

$$g\big|_{K=1} \simeq 2\gamma P_p \left(1 - f + f\,\mathrm{Re}[\tilde{\chi}_R^{(3)}(-\Omega)]\right). \tag{10.14}$$

Clearly, the introduction of Raman scattering results in a significant modification of the phase-matched gain coefficient which now explicitly depends on the *real* part of the Raman susceptibility. The real part of the Raman susceptibility of silica is a strong function of detuning (see Figure 10.1), and so large variations in the phase-matched parametric gain with detuning are to be expected. As shown in Figure 10.1 the real part of the normalized Raman susceptibility of silica varies from $+1$ to -1 over the range from 0 to 30 THz. This implies that the phase-matched parametric gain coefficient will vary by a factor $2f$, or almost 40%, over this range. The most prominent feature of the real part of the Raman susceptibility is the strong dip at 15.5 THz detuning. The corresponding reduction in the phase-matched parametric gain at this detuning can clearly be seen in Figure 10.3(b). This dispersion in phase-matched parametric gain with detuning has important implications for the design of ultra-broadband parametric amplifiers with gain bandwidths in excess of 10 THz. For detunings below 10 THz the real part of the Raman susceptibility is relatively flat and so does not have a large effect on the gain flatness of the amplifier. However, for detunings above 10 THz the real part of the Raman susceptibility varies strongly with frequency and so any parametric gain calculations for an amplifier operating in this region must include the full frequency dependence of both the real and imaginary parts of $\chi^{(3)}$.

For completeness, we have also plotted in Figure 10.3(c) the normalized gain coefficient at the peak Raman gain detuning of 13.2 THz versus K. It varies from the peak pure Raman gain coefficient, $g/(2\gamma P_p) \simeq 0.25$, for large phase-mismatches to about $g/(2\gamma P_p) \simeq 0.85$ slightly below the peak pure parametric gain coefficient.

10.2.4 The small-signal gain

We now look at the general solution for the evolution of the Stokes and anti-Stokes sidebands, Eq. (10.5). From the initial values of the sideband amplitudes at the fiber input ($z = 0$), $A_s(0)$ and $A_a(0)$, it is simple to obtain expressions for the small signal evolution along the fiber length,

$$A_s(z) = \frac{1}{2iR}\Big[\big[(K-q+iR)A_s(0) - qA_a^*(0)\big]e^{\gamma RP_p z} + \big[-(K-q-iR)A_s(0) + qA_a^*(0)\big]e^{-\gamma RP_p z}\Big] \quad (10.15)$$

$$A_a^*(z) = \frac{1}{2iR}\Big[\big[qA_s(0) - (K-q-iR)A_a^*(0)\big]e^{\gamma RP_p z} + \big[-qA_s(0) + (K-q+iR)A_a^*(0)\big]e^{-\gamma RP_p z}\Big]. \quad (10.16)$$

These equations, initially derived by Golovchenko et al. (1990), fully describe the evolution of a continuous-wave (cw) pump and two sidebands in the presence of both the real and imaginary parts of the $\chi^{(3)}$ nonlinearity. They can be used to calculate the gain per pass of singly (Sharping, 2008), doubly, or triply resonant (Coen and Haelterman, 2001) $\chi^{(3)}$ optical parametric oscillators, as well as the gain of phase-sensitive (McKinstrie and Radic, 2004) and phase-insensitive optical parametric amplifiers (Hansryd et al., 2002).

Here we concentrate on the standard phase-insensitive single-pump optical parametric amplifier to illustrate the combined interaction of the Raman and Kerr nonlinearities. The small input signal injected into such a parametric amplifier is either a Stokes wave (negatively detuned, $\alpha = \mathrm{s}$) or an anti-Stokes wave (positively detuned, $\alpha = \mathrm{a}$), while the other sideband amplitude is initially zero. The small-signal intensity gain of the signal, $G = P_\alpha(L)/P_\alpha(0)$, where L is the amplifier length and P_α is the signal power, is easily derived from Eqs. (10.15)–(10.16) and reads

$$G = \left|\cosh(\gamma P_p RL) \pm \frac{i(K-q)}{R}\sinh(\gamma P_p RL)\right|^2, \quad (10.17)$$

with the coefficient of the sinh term positive when the signal is positively detuned with respect to the pump and negative when the signal is negatively detuned with respect to the pump.

The small-signal gain is not simply equal to the power gain of the growing mode $\exp(gL)$ because with the initial conditions considered here both the growing and decaying mixed Stokes/anti-Stokes modes of the amplifier are initially excited, which leads to interference between them. It is only in the limit of large phase-mismatches, $|K| \gg 1$, and for a negatively detuned signal (Stokes), that $G = \exp(gL)$. Here the Stokes and anti-Stokes wave are decoupled and the initial condition only excites the growing mode. For the same large mismatches, a positively detuned signal (anti-Stokes) naturally exhibits a small-signal gain

$G = \exp(-gL)$ because only the decaying mode is propagating. This latter result is due to the natural tendency of SRS to transfer all the anti-Stokes power towards the Stokes side over long distances. More generally, in the asymptotic limit of a large gain or long amplifier, $gL \gg 1$, the small-signal gain always becomes proportional to $\exp(gL)$ irrespective of K and whether the signal is positively or negatively detuned,

$$G\big|_{gL\gg 1} = \frac{1}{4}\left|1 \pm i\frac{K-q}{R}\right|^2 \exp(gL)\,, \tag{10.18}$$

but with a multiplicative coefficient that depends on which mixture of the two modes is initially excited. Because of the interferences, the small-signal gain can be larger or smaller than $\exp(gL)$.

In the limit where SRS is negligible ($q = 1$) and within the parametric amplification bandwidth ($0 < K < 2$), we have $g = 2\gamma P_p R$, and Eq. (10.17) reduces to the well-known expression of the small-signal gain of a parametric amplifier (Stolen and Bjorkholm, 1982; Hansryd et al., 2002),

$$G = 1 + \frac{1}{R^2}\sinh^2\left(\frac{gL}{2}\right) = 1 + \left(\frac{2\gamma P_\mathrm{p}}{g}\right)^2 \sinh^2\left(\frac{gL}{2}\right). \tag{10.19}$$

This expression is identical for positively and negatively detuned signals. At perfect phase-matching ($K = 1$), the gain is maximum, $g = 2\gamma P_\mathrm{p}$, and the above expression reduces to $G = \cosh^2(\gamma P_\mathrm{p} L)$ (Hansryd et al., 2002), and in the large gain limit, $G|_{gL\gg 1} = (1/4)\exp(gL)$.

The interferences between the two mixed modes of a parametric amplifier are not purely of academic interest but can have practical implications. As a matter of fact, there exists a regime whereby a strong pump wave and a phase-mismatched (but not strongly phase-mismatched) signal can interact via the $\chi^{(3)}$ nonlinearity to completely suppress the input signal (Wang et al., 2008). This suppression is not due to an exponential attenuation of the signal but rather to a destructive interference between the amplifier's two gain modes. As a result, at least in theory, total signal suppression is possible. To understand the origins of this effect, consider the amplification of a positively detuned (anti-Stokes) signal. Expressing the initial condition in the basis of the amplifier modes, we find that it is of the form

$$\begin{bmatrix} A_\mathrm{s}(0) \\ A_\mathrm{a}^*(0) \end{bmatrix} \propto \mathbf{v}_- - \kappa \mathbf{v}_+\,. \tag{10.20}$$

The parameter κ is set so that the initial Stokes field is zero, $A_\mathrm{s}(0) = 0$. It is clear that $|\kappa| < 1$ because the Stokes component of the growing mode $\mathbf{v}_+$ is larger than that of the decaying mode $\mathbf{v}_-$ [see Eq. (10.8)]. Correspondingly, we note that the anti-Stokes component of the decaying mode is initially larger than that of the growing mode. As the modes evolve along the fiber, we can therefore reach a condition where, for a certain propagation distance, the anti-Stokes component of both modes exactly cancel due to destructive interference. At that point, the small-signal gain is exactly zero, $G = 0$. Note that it must be clear from the above discussion that this phenomenon can only occur for a positively detuned (anti-Stokes) signal; for a negatively detuned signal there is no combination of parameters for which the two modes interfere destructively. Also, it is a clear manifestation of the combined influence of FWM and SRS. In the absence of Raman scattering, $|\kappa| = 1$ and there is no position along the fiber where the mode components can cancel each other. As a result, the small-signal gain of an injected signal is unity or greater at all detunings as confirmed by Eq. (10.19).

Total signal suppression requires a correct combination of signal detuning, phase-mismatch, and pump power which occurs whenever the following condition is satisfied

$$\tanh(\gamma R P_\text{p} L) = \frac{iR}{K - q}. \tag{10.21}$$

Eq. (10.21) follows directly from Eq. (10.17) where we have set $G = 0$ and is a function of three independent variables: K, Ω, and the product $\gamma P_\text{p} L$. As will be shown later, this effect has a low pump power threshold and multiple points of complete suppression can appear in the parameter space of the amplifier as the pump power increases.

10.3 Experiments

In the second part of this chapter, we will review the results of four different experiments in which key features of the above theoretical developments have been observed. A common characteristic of these experiments is their ability to resolve the fine details of the mixed Raman-parametric gain structure (Figure 10.3). In particular, one needs to be able to precisely control the value of the normalized phase-mismatch K within a range going from, say, -2 to $+2$. This amounts to designing a parametric amplifier whose bandwidth largely overlaps the Raman gain bandwidth, i.e., that is phase-matched around the peak Raman gain detuning of 13.2 THz. This condition is not always that easy to satisfy. It typically requires large pump powers, to work very close to the zero-dispersion wavelength (ZDW) of the fiber, and/or to have a significant amount of fourth- or higher-order dispersion (Hansryd et al., 2002; Harvey et al., 2003; Agrawal, 2006). To keep things simple and to get some general insights, we will, however, not consider dispersion orders higher than 3 in the following discussion. K then reads

$$K = -\frac{\beta_2 \Omega^2}{2\gamma P_p}. \tag{10.22}$$

β_2 is the second-order dispersion coefficient *at the pump wavelength.* It can be expressed as $\beta_2 = 2\pi c \beta_3 \left[1/\lambda_p - 1/\lambda_{ZDW}\right]$, where λ_{ZDW} is the ZDW. Clearly, the value of K is tied to the difference between the pump wavelength and the ZDW and this sets important limitations. First, the spectral bandwidth of the pump needs to be much narrower than this difference. Second, the longitudinal dispersion fluctuations due to the non-uniformity of the fiber are particularly relevant close to the ZDW. They must be sufficiently small not to hamper measurements as they lead to longitudinal fluctuations of K and to a blurring of the gain characteristics. This effect also limits the usable fiber length, which has to be compensated for by an increase in pump power. Additionally, considerations must be given to the fiber length to limit pump depletion and reduce the walk-off between pump and signal.

As an example, consider phase-matching, $K = 1$, at the peak Raman gain detuning, $\Omega/(2\pi) = 13.2$ THz, with a pump power of 1 W in a standard single-mode fiber, $\beta_3 \simeq 0.05$ ps^3/km with a 1.3 μm ZDW and $\gamma = 3$ (W km)$^{-1}$. We find that the pump wavelength must only be 0.015 nm away from the ZDW. To make observations possible, Golovchenko et al. (1989, 1990) working with such standard fibers and a 1064 nm pump had to use very high peak pump powers of $>$ 100 kW. K was varied by varying the pump power, which restricted K to negative values. More recently, the observations of the dependence of the gain versus K at the peak Raman gain detuning (Vanholsbeeck et al., 2003) and the dependence of the phase-matched gain on the real part of the $\chi^{(3)}$ susceptibility (Hsieh et al., 2007a) were based on the use of highly nonlinear photonic crystal fibers (PCFs) with nonlinear coefficients γ in excess of 100 (W km)$^{-1}$. To complement this, Vanholsbeeck et al. (2003) used a picosecond Ti:Sapphire laser with up to 180 W peak power. Phase-matching at the peak of the Raman gain was then obtained with a comfortable pump-ZDW separation of about 40 nm. In contrast, in the first experiment of Hsieh et al. (2007a), the peak Raman gain was phase-matched with a 1.3 W pump only 0.4 nm away from the ZDW. Excellent measurements were still obtained thanks to the use of a pulsed nanosecond dye laser with a very narrow spectral bandwidth. In both cases, K could then be precisely adjusted by tuning the pump wavelength. Subsequent experiments were performed in the 1.5 μm wavelength range in highly uniform telecom-grade dispersion shifted fibers (Hsieh et al., 2007b; Wang et al., 2008). Observations were made possible through the use of a highly coherent tunable external cavity laser and a powerful erbium-doped fiber amplifier with up to 200 W peak power. In those conditions, the peak Raman gain was phase-matched with a pump 1.5 nm away from the ZDW.

From here it must be clear that all these experiments have been performed in the pulsed regime, first to reach the necessary peak power, but also to avoid the detrimental effect of stimulated Brillouin scattering (SBS) (Agrawal, 2006). SBS is typically 100 times more efficient than SRS in the cw regime but is completely eliminated when using pulses shorter than 10 ns because the pump spectrum is then larger than the Brillouin gain bandwidth ($\sim$80 MHz). The Brillouin gain is then reduced by a factor equal to the ratio between the pump and the Brillouin gain spectral bandwidth (Lichtman et al., 1987). Observing interactions between SRS and FWM in the cw regime therefore requires appropriate phase dithering of the pump as well as sufficiently large powers.

This discussion clearly highlights the difficulty in observing the interactions between Raman and FWM. They also point out that the phenomena we discuss here are not usually observed in standard Raman or parametric amplifiers. Still, with the development of Tbit/s networks and of ultra-broadband parametric amplifiers with gain bandwidths in excess of 10 THz, they will become increasingly relevant. The work of Boggio et al. (2009) constitutes a very recent example of a cw-pumped parametric amplifier with 155 nm bandwidth in which SRS had to be taken into account.

10.3.1 Frequency dependence of the gain at the peak Raman gain detuning

The first experimental results presented here are those of Vanholsbeeck et al. (2003). These authors measured for the first time the full dependence of the gain at the peak Raman gain detuning of 13.2 THz both for positive and negative values of K. To this end, they measured the on/off small-signal gain of a fiber Raman-parametric amplifier at various pump wavelengths in the vicinity of the ZDW.

As highlighted above, the fiber used was a highly-nonlinear PCF with a nonlinear coefficient $\gamma \simeq 100\ (\mathrm{W\,km})^{-1}$, more than 10 times the value found in standard fibers, and a ZDW of 860 nm ($\beta_3 \simeq 0.056\ \mathrm{ps^3/km}$). A mode-locked Ti:Sapphire laser (Ti:S) tunable between 840 and 900 nm was used to pump the fiber in the vicinity of the ZDW. The laser generated 2.6 ps pulses with an 82 MHz repetition rate. The signal to be amplified was obtained through SRS by propagating half the Ti:S laser power into a Ge-doped conventional step-index fiber so as to generate a Stokes pulse down-shifted by 13.2 THz from the pump wave. The output of the Ge-doped fiber was filtered so that only the Stokes components of the signal were coupled into the PCF amplifier together with the other half of the Ti:S output acting as the pump. A peak pump power of up to $P_\mathrm{p} = 180$ W was coupled into the 0.8 m-long PCF whilst the power density of the input signal was typically kept 35–40 dB below that of the pump beam. This low signal power together with the short amplifier length guaranteed that the measurements were performed in the small-signal gain limit with minimal walk-off between pump and signal (≤ 0.6 ps/m). Additionally,

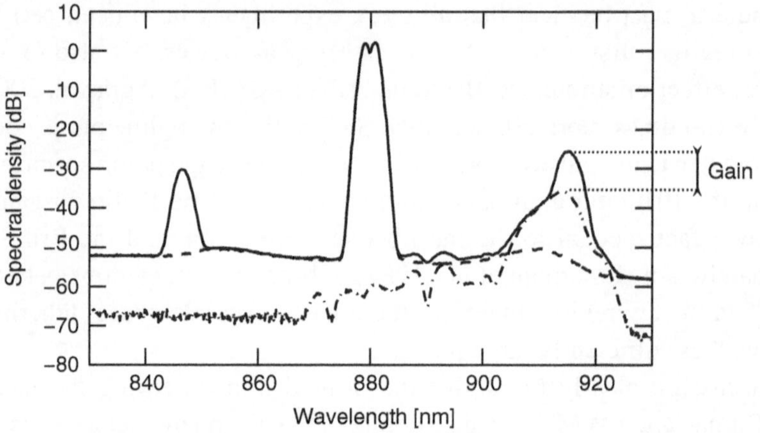

Fig. 10.4. Spectra at the PCF output for a pump power $P_p = 86$ W and a pump wavelength of 880 nm ($K \simeq 0.97$ at the peak Raman gain detuning). Signal only (dash-dotted), pump only (dashed), and amplified signal (solid).

because the Raman and parametric gains are polarization sensitive, polarization optics were used to carefully align the signal and pump polarization at the fiber input.

Figure 10.4 illustrates the amplification process for a pump power of $P_p = 86$ W and a pump wavelength of 880 nm. These conditions set the amplifier close to phase-matching since here $K \simeq 0.97$ at the peak Raman gain detuning of 13.2 THz. The solid line is the output spectrum of the amplifier. A large anti-Stokes signal is present, which is the expected signature of a strong coupling between SRS and FWM. In the same figure, we have also plotted the output spectrum when the pump was injected by itself (dashed line). It is very well superimposed with the residual pump component of the amplified signal, highlighting the fact that pump depletion is negligible. Additionally, we can note that the pump spectrum is slightly broadened by self-phase-modulation and that it is powerful enough that both Stokes and anti-Stokes waves are generated even without an input signal. Finally, the dash-dotted line in Figure 10.4 shows the spectrum of the signal at the fiber output when the pump is absent.

The amplifier gain was simply obtained by monitoring the signal output power on an optical spectrum analyzer while turning the pump on and off. All measurements were performed at a frequency shift of $\Omega/(2\pi) = 13.2$ THz. The pump wavelength was varied by 5 nm steps and at each wavelength the gain was measured for four or five different pump power levels. To obtain the small-signal gain G, these gross gain measurements were corrected to take into account the wavelength-dependent absorption of the PCF as well as the difference between the duration of the pump and signal pulses. As a matter of fact, the signal pulse is largely broadened in

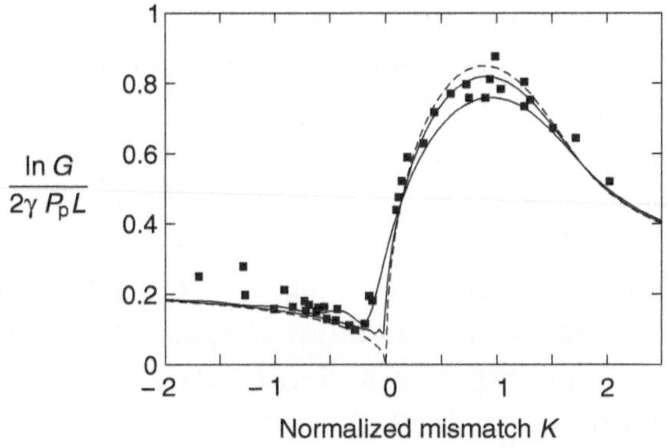

Fig. 10.5. Experimental measurements of the gain of the PCF amplifier at the peak Raman gain detuning of 13.2 THz (squares). The pump power and the pump wavelength were varied in the 60–180 W and 840–890 nm range, respectively. The solid lines are the theoretical predictions [Eq. (10.17)] calculated for $P_p = 60$ and 180 W. The dashed line is calculated in the asymptotic limit of a large gain amplifier, $gL \gg 1$, and corresponds to the normalized gain coefficient of the growing mode, $g/(2\gamma P_p) = \mathrm{Re}(R)$ [Eq. (10.7)].

the Ge-doped fiber and is about 200 times longer than the pump pulses, i.e., only ~0.5% of the signal is actually amplified in the PCF. The results are presented as solid squares in Figure 10.5. Here we have plotted the normalized small signal-gain on a logarithmic scale $(\ln G)/(2\gamma P_p L)$ versus the normalized mismatch K. The normalization has been chosen because $(\ln G)/L$ tends towards the gain coefficient of the growing mode g [Eq. (10.7)] in the asymptotic large gain limit. The vertical axis of Figure 10.5 is therefore immediately comparable to that of Figure 10.3(c).

Comparison between theory and experiment is, however, not straightforward because the amplifier is not always operated in the asymptotic limit. This point has actually been overlooked in the original analysis of Vanholsbeeck et al. (2003) where it was simply assumed that the small-signal gain G was equal to that of the growing mode, $\exp(gL)$. The theory presented in Section 10.2.4 makes, however, clear that interferences between the two mixed Stokes/anti-Stokes modes can sometimes lead to important differences. We will therefore present here a new more thorough analysis of the experimental data of Vanholsbeeck et al. (2003). In doing so, we have only kept the data for pump power levels between 60 and 180 W and eliminated all the data points obtained for lower pump powers. We estimate that the gain here was too small to be reliably obtained given that the optical spectrum analyzer measures the time-integrated power of the signal pulses out of which only 0.5% is actually amplified.

The data analysis is complicated by the fact that the experimental measurements were not taken at the exact same power levels for all pump wavelengths. It was deemed unimportant at the time since the asymptotic limit assumed in the original analysis made possible a complete renormalization. However, when the amplifier is *not* in the asymptotic limit, the small-signal gain G [Eq. (10.17)] cannot be normalized to a universal curve independent of the pump power. Hence, the experimental data do not all fit on a single curve. To circumvent this problem, we have plotted as solid lines in Figure 10.5 the theoretical small-signal gain for the two extreme pump power levels we consider in the data, $P_{\mathrm{p}} = 60$ and 180 W. Here we have used $f = 0.18$ and the Raman susceptibility of Stolen et al. (1989). All the other parameters are those of the experiment. Remarkably, nearly all the experimental points fit in between these two lines. Note that the three data points that agree the least have all been measured with the lowest pump power level of 60 W. The peak of the gain at phase-matching is particularly well reproduced and is clearly below the pure parametric gain (normalized to a value of 1 in the figure). Considering that the two extreme theoretical curves are not too different from each other, the agreement is actually excellent. For comparison, we have also plotted as a dashed line in Figure 10.5 the gain in the asymptotic limit. In particular, we can observe that the gain cancellation expected at $K = 0$ is not fully developed yet for the pump powers considered in the experiment, a fact which is nicely reproduced by the experimental data. Note that this excellent agreement between theory and experiment is obtained without any fit parameter. Finally, let us point out that the theoretical curves obtained for $P_{\mathrm{p}} = 60$ and 180 W (solid) as well as in the large gain limit (dashed) are all reasonably close to each other, which explains the good agreement originally presented by Vanholsbeeck et al. (2003). We believe, however, that the more rigorous analysis we have presented here provides an even better fit to the experimental data and further reinforces the validity of the theoretical model.

10.3.2 Dependence of the phase-matched gain on the real part of the susceptibility

Complementing the above results, the mixed Raman parametric gain was measured at perfect phase-matching ($K = 1$) versus the detuning Ω by Hsieh et al. (2007a,b). This corresponds to Figure 10.3(b) and highlights the role played by the real part of the $\chi^{(3)}$ susceptibility of silica, Eq. (10.14).

The initial observations on this topic were performed with 20 m of PCF pumped by 1.3 W peak power 13 ns pulses from a tunable dye laser (Hsieh et al., 2007a). The gain was measured by observing the amplification of a cw signal generated by a second dye laser and copropagating with the pump pulses. Measurements were performed for four different pump wavelengths, and for a range of pump-signal detunings. The results clearly highlighted the presence of a strong dip in the small-signal gain for a 15.5 THz detuning as predicted by the theory and due to the corresponding dip in the real part of the Raman susceptibility.

More thorough measurements were then performed in 40 m of standard telecommunications DSF with a zero dispersion wavelength of 1556.5 nm that was used as a single-pump parametric amplifier (Hsieh et al., 2007b). Here the pump was a C-band cw tunable external cavity laser (ECL) followed by an intensity modulator and a high gain, high power erbium-doped fiber amplifier. The modulation was necessary to avoid the detrimental effect of SBS as well as to reach sufficiently high peak power. The resulting optical pulse train consisted of 1.5 ns rectangular pulses with 82 W peak power tunable between 1520 and 1560 nm. Tuning the pump wavelength around the ZDW allowed the phase-matched frequency of the amplifier to be varied between 7 and 22 THz (Marhic et al., 2004) as observed from the spontaneous generation of Stokes and anti-Stokes sidebands. At each of these phase-matched detunings the small-signal gain G [Eq. (10.17)] of the amplifier was measured by copropagating an anti-Stokes (positively detuned) signal generated by a second tunable ECL. The results are shown in Figure 10.6. The open circles are the experimentally measured phase-matched gains at each detuning, the solid line is the prediction of Eq. (10.17). The agreement between the experiment and theory is good with the phase-matched gain reproducing exactly the frequency dependence of the real part of the Raman susceptibility (see Figure 10.1). The strong dip in the parametric gain around 15.5 THz is clearly evident. The 40% ($2f$) reduction in the combined Raman-parametric gain coefficient at this detuning results in a 30 dB drop in the amplifier's gain. This demonstrates the importance of including the full frequency dependence of both the real and imaginary parts of the $\chi^{(3)}$ nonlinearity in parametric systems that operate at detunings larger than 10 THz.

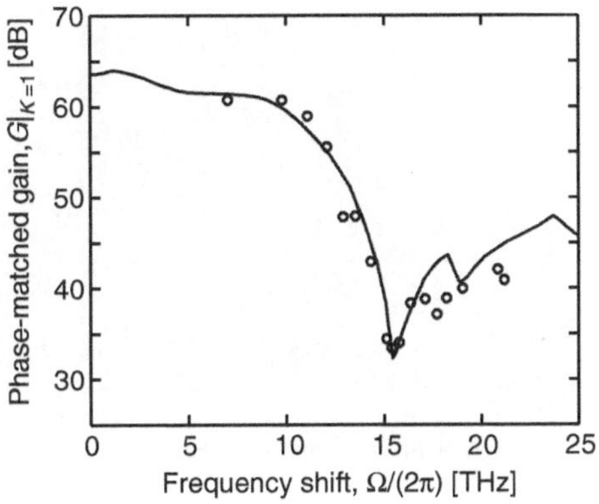

Fig. 10.6. Measured phase-matched gain for an anti-Stokes signal as a function of detuning Ω (circles). The solid line is the prediction of Eq. (10.17) with $\gamma = 2.53\ \mathrm{W}^{-1}\,\mathrm{km}^{-1}$ and $f = 0.18$. The input pump peak power is 82 W.

10.3.3 Complete signal suppression due to the combined effects of Raman and parametric gain

We now turn our attention to the complete signal suppression ($G = 0$) that can occur when amplifying a positively detuned (anti-Stokes) signal. As explained in the theoretical section, the signal suppression is due to destructive interferences between the two mixed Stokes/anti-Stokes modes of the combined Raman and FWM dynamics. This phenomenon occurs whenever the following condition is satisfied [Eq. (10.21)]

$$\tanh(\gamma R P_{\mathrm{p}} L) = \frac{iR}{K-q},$$

i.e., for a correct combination of signal detuning Ω, phase-mismatch K, and pump power P_{p}. Using the experimentally measured Raman susceptibility, Figure 10.1, and a value for f of 0.25 taken to match the experiments which follow we find that no signal suppression can occur while the product $\gamma P_{\mathrm{p}} L < 1.95$. This sets a low power threshold for the effect. At $\gamma P_{\mathrm{p}} L = 1.95$ the first point at which complete signal suppression occurs appears at $\Omega/(2\pi) = 12.9$ THz, $K = 2.12$. At higher values of $\gamma P_{\mathrm{p}} L$ multiple points of complete suppression appear in the parameter space of the amplifier, but always at positive signal detunings. In Figure 10.7 we plot a false color image of the small-signal gain G of a parametric amplifier as a function of K and Ω for a positively detuned signal. The parameter $\gamma P_{\mathrm{p}} L$ is set to 4. Under these conditions eight points of total signal suppression are present (marked in Figure 10.7 by white arrows). All these eight points occur at values of K greater than the phase-matched mismatch ($K = 1$). For the dispersion parameters of the

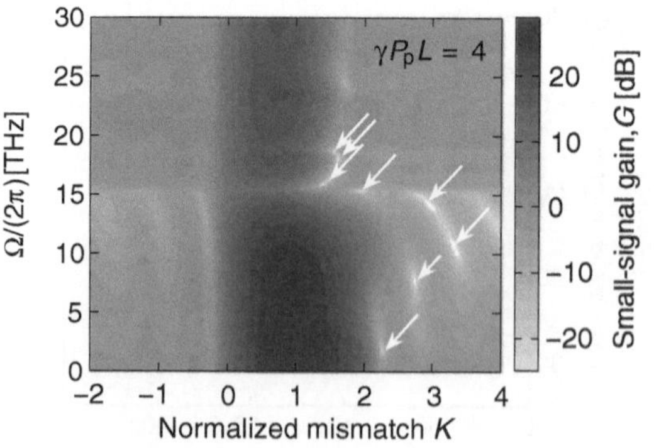

Fig. 10.7. Signal gain in decibels as a function of normalized mismatch K and detuning Ω for $\gamma P_{\mathrm{p}} L = 4$. The regions shaded in white correspond to a signal attenuation of more than 25 dB. The points at which complete signal suppression occurs are indicated by white arrows.

fiber used in the experiments that follow this implies that the signal detuning at which the suppression occurs will be larger than the phase-matched detuning. At higher values of $\gamma P_p L$ points of suppression can also occur on the other side of the phase-matching curve, $K < 1$.

The strong suppression of an optical signal as described above has been experimentally observed in Wang et al. (2008). The parametric amplifier used was the same as that of the experiment presented in the previous section except that a shorter 10 m long piece of the DSF was used ($\lambda_{ZDW} = 1556.5$ nm). A positively detuned signal was injected into the amplifier and the gain measured as a function of detuning. The pump peak power was set to 85 W so that the parameter $\gamma P_p L = 2.15$, just above the low power threshold, so only one point of complete suppression is present. The results of this measurement at a pump wavelength of 1556.1 nm are shown in Figure 10.8. The experimentally measured gains are plotted as open circles, the solid line is the prediction of Eq. (10.17) with the parameter f set to 0.25, the dashed line is the same curve with the parameter f set to 0.18. A strong dip (13 dB) in the parametric gain due to the destructive interference of the two parametric gain modes is clearly evident at a detuning of 11.1 THz. Note that the fiber used is not polarization maintaining and that any depolarization between the signal and the pump will result in a component of the signal orthogonally polarized with respect to the pump that does not experience any suppression. We believe that this small depolarized unsuppressed component of the signal is the reason we can observe only a 95% signal suppression instead of the complete suppression predicted by Eq. (10.21).

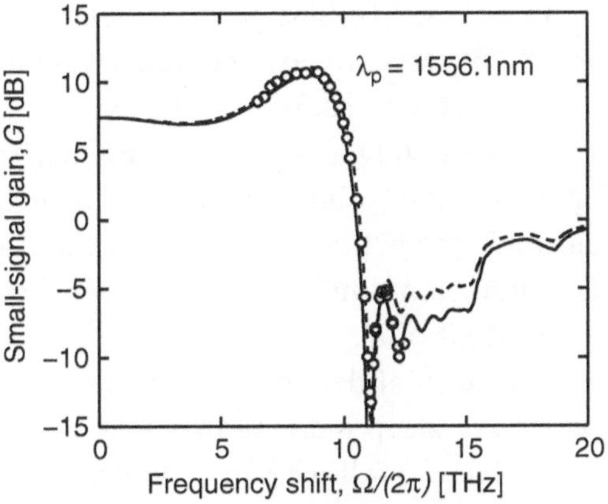

Fig. 10.8. Experimentally measured signal gain as a function of detuning for a pump wavelength of 1556.1 nm and a pump power of 85 W (circles). The solid curve is the prediction of Eq. (10.17) with $f = 0.25$; the dashed curve is the prediction of Eq. (10.17) with $f = 0.18$.

While both values of f give a reasonable fit to the parametric gain around the phase-matched peak (~ 9 THz detuning), $f = 0.25$ appears to provide a better fit to the experimentally measured gains on the high frequency side of the dip. Another recent measurement (Lin and Agrawal, 2006) also suggests a value for f of close to 0.25. Careful measurements of amplifier gain as a function of detuning, especially around a point of destructive interference between the gain modes, can be used as a sensitive measure of the parameter f. Any such measurement will implicitly rely on an accurate knowledge of the shape of the normalized complex Raman susceptibility, which is itself a function of the material composition of the specific fiber under test.

10.4 Anti-Stokes to Stokes power ratio

In the three previous experimental subsections, we have presented various measurements of the mixed Raman/FWM gain and shown how all experiments very nicely matched theoretical predictions. The other cornerstone of the theory is of course the presence and the amplification of the anti-Stokes wave. The modes of the amplification process in the presence of both FWM and SRS have a mixed Stokes/anti-Stokes character which implies that the anti-Stokes wave will always be present and amplified proportionally to the Stokes wave even in the asymptotic large gain limit. The signal suppression discussed above is actually a direct consequence of the mixed character of the modes. To complement this, we now look at some direct experimental measurements of the anti-Stokes to Stokes power ratio.

Figure 10.9 presents such measurements. These unpublished results have been obtained by Hsieh et al. with the same experimental setup as that described in Subsection 10.3.2 (Hsieh et al., 2007b). The pump wavelength and the pump power were set at 1555.7 nm and 82 W, respectively. The signal was tuned in the anti-Stokes band with detunings in the range 7.5–12.5 THz. The experimentally obtained anti-Stokes to Stokes power ratios are shown as crosses in Figure 10.9 and are compared with the exact analytical prediction (solid line) obtained from Eqs. (10.15)–(10.16) and with the large gain limit prediction [dashed line, Eq. (10.9)]. The amplifier is operating close to the large gain limit so both theoretical curves nearly overlap except for small negative values of K.

As can be seen, the experimental data neatly reproduces the shape of the theoretical curve. We have a very steep increase of the anti-Stokes to Stokes power ratio and a maximum around $K = 0$ followed by a slower decrease for larger values of the normalized mismatch K. The experimental measurements are, however, too small by about 20% in comparison with the theoretical predictions. The squares in Figure 10.9 correspond to the experimental data points multiplied by 1.2. We do not know for certain what causes this discrepancy but it is most likely due to an

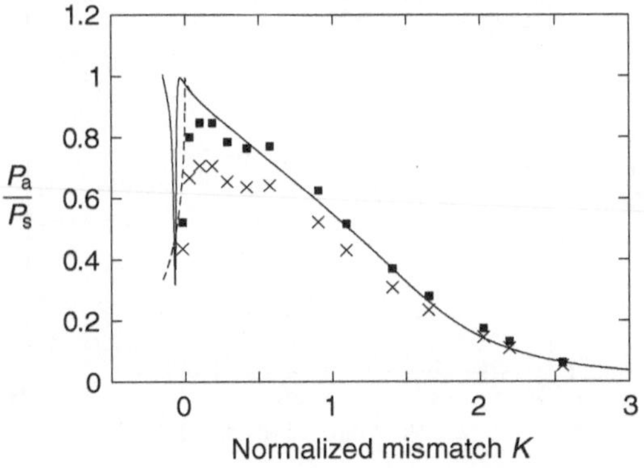

Fig. 10.9. Experimental measurements of the anti-Stokes to Stokes power ratio (crosses) compared with the asymptotic theoretical prediction [dashed, Eq. (10.9)] and with the exact theoretical prediction [solid, Eqs. (10.15)–(10.16)] with $f = 0.25$. The squares correspond to the experimental points multiplied by 1.2. The pump wavelength was set at 1555.7 nm while the peak pump power was 82 W.

insufficient accuracy in the power calibration of the optical spectrum analyzer used to make this measurement.

The amplification of the anti-Stokes wave at 13.2 THz detuning has also been studied by Coen et al. (2002) in the large phase-mismatch limit, $K = -24$ to -16. In this work, the experiment has been performed with 60 ps pulses of up to 460 W peak power from a 647 nm krypton ion laser launched into a 2.85 m long single mode polarization maintaining step-index silica fiber [$\beta_2 \simeq 50$ ps^2/km and $\gamma \simeq 23$ (W km)$^{-1}$]. Measurements of the Stokes and anti-Stokes waves intensities were performed with a calibrated photomultiplier tube as a function of the pump power both in the spontaneous and in the stimulated regime and the transition between the two regimes is particularly well observed by the authors (Coen et al., 2002). Note that there is no seed in this experiment. Both sidebands grow from noise and, as a result, the stimulated regime is definitely in the asymptotic large gain limit. In Figure 10.10, we only present the measurements obtained in the stimulated regime for which the theory presented in this chapter holds. These correspond to peak powers above 310 W. The experimental results (solid) are compared with the theoretical prediction, Eq. (10.9) [dashed]. As can be seen, even though the two curves do not quite overlap, the order of magnitude of the anti-Stokes to Stokes power ratio, 0.5–1×10^{-3}, matches between theory and experiment. Also, the measured ratio increases with K in good agreement with theory. Again, there is no fit parameter. The discrepancies could simply be explained by some uncertainties in some of the experimental parameters. Note that a quantum treatment of SRS and FWM has been shown to match these results as well (Brainis et al., 2007).

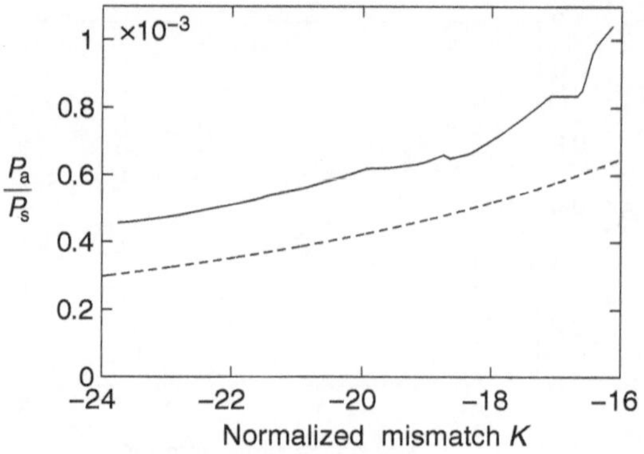

Fig. 10.10. Experimental measurements of the anti-Stokes to Stokes power ratio (solid) compared with the theoretical prediction in the large gain limit (dashed) for large phase-mismatches. The measurements have all been obtained at 13.2 THz detuning.

10.5 Conclusion

In conclusion, we have presented in this chapter a thorough study of the interaction between SRS and FWM. Our analysis builds upon the seminal theoretical work of Bloembergen and Shen (1964) and includes results from several recent experimental publications (Coen et al., 2002; Vanholsbeeck et al., 2003; Hsieh et al., 2007a,b; Wang et al., 2008). Whereas these publications sometimes used different notations and only presented parts of the underlying theory, we have provided here a unified theoretical framework to discuss all the experimental results in the same context. In particular, we have shown that all the phenomena described here can be understood in terms of the properties of the mixed Stokes/anti-Stokes modes of the amplification process and of the interferences between these modes. We believe we have also clearly highlighted the important and sometimes confusing difference between the asymptotic gain coefficient g of the individual modes and the small-signal gain G obtained for specific initial conditions.

The experimental results we have presented demonstrate all the key features of the theory for a broad range of parameters including large detunings, phase-matched and non-phase-matched conditions, as well as small and large gains. This complete theoretical description, fitted extremely well by experimental results, fully reveals the importance of the combined influence of SRS and FWM from gain suppression and enhancement, to the crucial role of the real part of the Raman susceptibility, and the amplification of non-phase-matched anti-Stokes waves. It is worth noting that most of the experimental results on this topic have been obtained only recently, mostly thanks to technological advances in laser and fiber technologies, even though the theoretical background was laid more than half a century ago.

We anticipate that the interaction of FWM and SRS will become increasingly relevant in practical applications as the move towards multi-Tbit/s telecommunication systems intensifies. In particular, the phenomena discussed here have important consequences for the gain flatness of ultra-broadband fiber optical parametric amplifiers as recently exemplified by the work of Boggio et al. (2009). Supercontinuum generation is another important practical area where the combined influence of FWM and SRS needs to be taken into account.

References

Agrawal G. P. (2006). *Nonlinear Fiber Optics*, 4th edn. Optics and Photonics Series, San Diego: Academic Press.

Bloembergen N. (1967). The stimulated Raman effect. *Am. J. Phys.* **35**, 989–1023.

Bloembergen N., Shen Y. R. (1964). Coupling between vibrations and light waves in Raman laser media. *Phys. Rev. Lett.* **12**, 504–507.

Blow K. J., Wood D. (1989). Theoretical description of transient stimulated Raman scattering in optical fibers. *IEEE J. Quantum Electron.* **25**, 2665–2673.

Boggio J., Moro S., Myslivets E., Windmiller J., Alic N., Radic S. (2009). 155 nm continuous wave two-pump parametric amplification. *IEEE Photon. Technol. Lett.* **21**, 612–614.

Brainis E., Clemmen S., Massar S. (2007). Spontaneous growth of Raman Stokes and anti-Stokes waves in fibers. *Opt. Lett.* **32**, 2819–2821.

Coen S., Haelterman M. (2001). Continuous-wave ultrahigh-repetition-rate pulse train generation through modulational instability in a passive fiber cavity. *Opt. Lett.* **26**, 39–41.

Coen S., Wardle D. A., Harvey J. D. (2002). Observation of non-phase-matched parametric amplification in resonant nonlinear optics. *Phys. Rev. Lett.* **89**, 273901/1–4.

Dudley J. M., Genty G., Coen S. (2006). Supercontinuum generation in photonic crystal fiber. *Rev. Mod. Phys.* **78**, 1135–1184.

Duncan M. D., Mahon R., Reintjes J., Tankersley L. L. (1986). Parametric Raman gain suppression in D_2 and H_2. *Opt. Lett.* **11**, 803–805.

Fan J., Migdall A. (2006). Phase-sensitive four-wave mixing and Raman suppression in a microstructure fiber with dual laser pumps. *Opt. Lett.* **31**, 2771–2773.

Gnauck A. H., Jopson R. M., McKinstrie C. J., Centanni J. C., Radic S. (2006). Demonstration of low-noise frequency conversion by Bragg scattering in a fiber. *Opt. Express* **14**, 8989–8994.

Golovchenko E. A., Pilipetskii A. N. (1994). Unified analysis of four-photon mixing, modulational instability, and stimulated Raman scattering under various polarization conditions in fibers. *J. Opt. Soc. Am. B* **11**, 92–101.

Golovchenko E., Mamyshev P. V., Pilipetskii A. N., Dianov E. M. (1990). Mutual influence of the parametric effects and stimulated Raman scattering in optical fibers. *IEEE J. Quantum Electron.* **26**, 1815–1820.

Golovchenko E. A., Dianov E. M., Mamyshev P. V., Pilipetskiĭ A. N. (1989). Parametric suppression of stimulated Raman scattering. *Pis'ma Zh. Eksp. Teor. Fiz.* **50**, 170–172. [JETP Lett. **50**, 190–193 (1989)].

Hansryd J., Andrekson P. A., Westlund M., Li J., Hedekvist P.-O. (2002). Fiber-based optical parametric amplifiers and their applications. *IEEE J. Sel. Top. Quantum Electron.* **8**, 506–520.

Harvey J. D., Leonhardt R., Coen S., Wong G. K. L., Knight J. C., Wadsworth W. J., Russell P. St. J. (2003). Scalar modulation instability in the normal dispersion regime by use of a photonic crystal fibre. *Opt. Lett.* **28**, 2225–2227.

Hsieh A. S. Y., Murdoch S. G., Coen S., Leonhardt R., Harvey J. D. (2007a). Influence of Raman susceptibility on optical parametric amplification in optical fibers. *Opt. Lett.* **32**, 521–523.

Hsieh A. S. Y., Wong G. K. L., Murdoch S. G., Coen S., Vanholsbeeck F., Leonhardt R., Harvey J. D. (2007b). Combined effect of Raman and parametric gain on single-pump parametric amplifiers. *Opt. Express* **15**, 8104–8114.

Islam M. N. (2002). Raman amplifiers for telecommunications. *IEEE J. Sel. Top. Quantum Electron.* **8**, 548–559.

Islam M. N., Boyraz Ö. (2002). Fiber parametric amplifiers for wavelength band conversion. *IEEE J. Sel. Top. Quantum Electron.* **8**, 527–537.

Lichtman E., Friesem A. A., Waarts R. G., Yaffe H. H. (1987). Stimulated Brillouin scattering excited by two pump waves in single-mode fibers. *J. Opt. Soc. Am. B* **4**, 1397–1403.

Lin Q., Agrawal G. P. (2006). Raman response function for silica fibers. *Opt. Lett.* **31**, 3086–3088.

Marhic M. E., Wong K. K.-Y., Kazovsky L. G. (2004). Wide-band tuning of the gain spectra of one-pump fiber optical parametric amplifiers. *IEEE J. Sel. Top. Quantum Electron.* **10**, 1133–1141.

McKinstrie C. J., Radic S. (2004). Phase-sensitive amplification in a fiber. *Opt. Express* **12**, 4973–4979.

Méchin D., Provo R., Harvey J. D., McKinstrie C. J. (2006). 180-nm wavelength conversion based on Bragg scattering in an optical fiber. *Opt. Express* **14**, 8995–8999.

Oo-Kaw L., Grigoryan V., Shin M., Kumar P. (2008). Ultra-low-noise inline fiber-optic phase-sensitive amplifier for analog optical signals. In *2008 Optical Fiber Communication Conference/National Fiber Optic Engineers Conference, OFC/NFOEC 2008*. Piscataway, NJ: IEEE.

Salem R., Foster M. A., Turner A. C., Geraghty D. F., Lipson M., Gaeta A. L. (2008). Optical time lens based on four-wave mixing on a silicon chip. *Opt. Lett.* **33**, 1047–1049.

Sharping J. E. (2008). Microstructure fiber based optical parametric oscillators. *IEEE J. Lightwave Technol.* **26**, 2184–2191.

Stolen R. H. (1975). Phase-matched-stimulated four-photon mixing in silica-fiber waveguides. *IEEE J. Quantum Electron.* **QE-11**, 100–103.

Stolen R. H., Bjorkholm J. E. (1982). Parametric amplification and frequency conversion in optical fibers. *IEEE J. Quantum Electron.* **QE-18**, 1062–1072.

Stolen R. H., Ippen E. P. (1973). Raman gain in glass optical waveguides. *Appl. Phys. Lett.* **22**, 276–278.

Stolen R. H., Gordon J. P., Tomlinson W. J., Haus H. A. (1989). Raman response function of silica-core fibers. *J. Opt. Soc. Am. B* **6**, 1159–1166.

Stolen R. H., Ippen E. P., Tynes A. R. (1972). Raman oscillation in glass optical waveguide. *Appl. Phys. Lett.* **20**, 62–64.

Tai K., Hasegawa A., Tomita A. (1986). Observation of modulational instability in optical fibers. *Phys. Rev. Lett.* **56**, 135–138.

Tanemura T., Ozeki Y., Kikuchi K. (2004). Modulational instability and parametric amplification induced by loss dispersion in optical fibers. *Phys. Rev. Lett.* **93**, 163902/1–4.

Tchofo Dinda P., Millot G., Wabnitz S. (1998). Polarization switching and suppression of stimulated Raman scattering in birefringent optical fibers. *J. Opt. Soc. Am. B* **15**, 1433–1441.

Tchofo Dinda P., Wabnitz S., Coquet E., Sylvestre T., Maillotte H., Lantz E. (1999). Demonstration of stimulated-Raman-scattering suppression in optical fibers in a multifrequency pumping configuration. *J. Opt. Soc. Am. B* **16**, 757–767.

Trillo S., Wabnitz S. (1992). Parametric and Raman amplification in birefringent fibers. *J. Opt. Soc. Am. B* **9**, 1061–1082.

Vanholsbeeck F., Emplit Ph., Coen S. (2003). Complete experimental characterization of the influence of parametric four-wave mixing on stimulated Raman gain. *Opt. Lett.* **28**, 1960–1962.

Wang J. C. C., Murdoch S. G., Leonhardt R., Harvey J. D. (2008). Strong signal suppression in single-pump optical parametric amplifiers. *Opt. Lett.* **33**, 935–937.

11

Nonlinear optics in emerging waveguides: revised fundamentals and implications

Shahraam Afshar V., Mark Turner and Tanya M. Monro

11.1 Introduction

Guided-wave nonlinear optics has attracted significant interest because of the unique environment that waveguides provide for nonlinear interactions, including tight confinement (high intensity), long interaction lengths (especially for fibres), control of propagation constants, and the possibility to incorporate them with integrated circuits (mainly for planar waveguides) [see (Lin et al., 2007; Knight and Skryabin, 2007; Foster et al., 2008; Afshar and Monro, 2009) and references therein]. Recent and rapid progress in design and manufacturing of complex structured microstructured optical fibres and planar waveguides with subwavelength features (including both subwavelength inclusions and voids) has further extended the opportunities to develop nonlinear devices by enabling extreme nonlinearity to combine with tailorable chromatic dispersion (Lin et al., 2007; Knight and Skryabin, 2007; Foster et al., 2008; Koos et al., 2007).

The nonlinear optical phenomena that occur in waveguides are determined through two main factors; the linear and nonlinear properties of the constituent bulk materials, and the optical properties of the waveguide. Recent advances in the design and fabrication of complex structured waveguides with high contrast linear refractive indices, inhomogeneous cross-sections, and subwavelength features have provided great potential to accelerate the field of guided-wave nonlinear optics. We define a new class of optical waveguides, "emerging waveguides", as waveguides with a combination of the following features:

(i) High index materials
(ii) Inhomogeneous and complex structure
(iii) Subwavelength features such as voids or material inclusions.

Despite the importance and growing interest, applications, and publications in the field of nonlinear processes in these emerging waveguides, nonlinear pulse propagation models for these structures still mainly rely on the well-known scalar Helmholtz equation (Lin et al., 2007; Foster et al., 2005, 2007; Magi et al., 2007; Karasawa et al., 2001; Boyraz et al., 2004a, 2004b; Zheltikov, 2005; Kibler et al., 2005; Leong et al., 2006; Genty et al., 2007; Zhang et al., 2007; Agrawal, 2007). This equation is based on the weak guidance approximation, and assumes that the waveguide cross-section is homogeneous. However, even a cursory inspection of the key characteristics of these emerging waveguides (i.e. inhomogeneous, high index contrast transverse structure incorporating subwavelength features) reveals that these waveguides operate far from the weak guidance regime. Indeed, one can argue that the emerging waveguides considered here exhibit strong guidance. In particular, it is observed that no Helmholtz wave equation can be obtained for the emerging waveguides since $\nabla.D = 0$ in Maxwell's equation does not result in $\nabla.E = 0$ because of the inhomogeneous nature of the susceptibility tensor $\epsilon(x,y)$ (Koos et al., 2007). Also, it has been pointed out that for subwavelength structures, such as optical nanowires, pulse propagation based on scalar theory does not give a good approximation (Zheltikov, 2005).

There have been some reports of new models of nonlinear pulse propagation that take into account the inhomogeneous and vectorial solutions of Maxwell's equations (Koos et al., 2007; Kolesik and Moloney, 2004; Kolesik et al., 2004; Chen et al., 2006; Laegsgaard, 2007; Dadap et al., 2008). The papers by Koos et al. (2007), Kolesik and Moloney (2004), Kolesik et al. (2004), Chen et al. (2006) and Dadap et al. (2008) ignored the potential for coupling between different modes including the two polarisations of one mode (in the case of single mode waveguides), which as we will show later, is a key characteristic of linear and nonlinear pulse propagation when the full vectorial solutions of Maxwell's equations are considered. In fact, it is shown (Afshar and Monro, 2009) that there are parameter regimes for which this modal coupling makes a significant contribution to both the predicted nonlinear and dispersive effects. In some reports (Kolesik and Moloney, 2004; Kolesik et al., 2004; Laegsgaard, 2007), no consideration has been given to the possibility of using an inhomogeneous cross-section which, as mentioned above, is a key feature of emerging waveguides. In addition, one vital aspect of the vectorial formulation of nonlinear pulse propagation, which is usually not considered in the literature, is the correct vectorially-based form of the effective nonlinear coefficient γ, the modal Raman gain g and the impact of the longitudinal component of the modal fields [the component along the propagation direction, (z)] on the dispersion, nonlinear, and modal/polarisation coupling behaviour of a pulse propagating through an emerging waveguide. It should be noted that the paper by Dadap et al. (2008) recognises the contribution of the longitudinal component of the electric field to the effective nonlinearity of a waveguide γ. The same authors also recently have discussed the significance of the z-component of the propagating modes in silicon waveguides (Driscoll et al., 2009).

Here a general vectorially-based nonlinear Schrödinger equation (VNSE) (Afshar and Monro 2008, 2009), is derived for pulse propagation through waveguides with complex transverse structures including inhomogeneous refractive index profiles and subwavelength features. We consider both Kerr (Section 11.4) and Raman (Section 11.5) nonlinear effects. We demonstrate that in the strong guidance regime, the propagating modes have a significant electric field component along the direction of propagation, which causes the propagating modes to be non-transverse. As a result, our formalism predicts a range of new tempo-spatial effects within emerging waveguides, such as dispersion-induced depolarisation.

Based on this VNSE, this chapter presents a new and generalised equation for the A_{eff}, the parameter that defines the effective mode area, γ, the parameter commonly used to describe the effective nonlinearity of an optical fibre (Agrawal, 2007) and g, which is used to describe the modal Raman gain of an optical waveguide. These new definitions are generalised versions of the standard definitions that take into account both an inhomogeneous refractive index profile and subwavelength features. We apply these definitions to optical nanowires, and show that in some parameter regimes, the value of γ can be a factor of two higher than that obtained using the standard definitions. We provide an analysis of the value of γ predicted by this new generalised model. Similarly, increases in the calculated modal Raman gain are observed using the VNSE over the standard model, and the shape of the modal Raman gain spectrum is shown to be significantly altered from that of the bulk material within these emerging waveguides. The new model also predicts new coupling terms between different modes or polarisations of propagating modes, which are due to the non-transverse nature of the modes.

The structure of this chapter is as follows. Recent advances in the field of emerging waveguides both in terms of their constituent materials and their structures are reviewed in Sections 11.2 and 11.3. Section 11.4 is devoted to the development of the theory of the vectorially-based nonlinear Schrödinger equation (VNSE) for Kerr nonlinearity. In this section, the new model is applied to step index cylindrical waveguides and the results of the new model for Kerr nonlinearity are analysed and compared with those of the standard model. Experimental confirmation of the new vectorially-based definition of γ is provided. In Section 11.5 we develop the VNSE to include the stimulated Raman scattering between modes of the optical waveguides and redefine the modal Raman gain. The impact of the z-component of the electric field vector on the modal Raman gain, using the VNSE, as well as the possibility for tailoring the shape of the modal Raman gain spectrum through design of the optical waveguide are then investigated.

11.2 Host material

The characterisation of the nonlinear properties of bulk materials is indeed a rich and established field and, depending on the nonlinear process, is usually expressed in terms of different coefficients such as nonlinear refractive index, Raman and Brillouin gains, frequency shifts, bandwidths and response times, etc. However, the possibility of using highly nonlinear materials (mainly non-silica glasses for fibres) or semiconductors to fabricate fibres and waveguides has opened up new opportunities for nonlinear photonic devices. The advent of microstructured optical fibres (MOFs) in the 1970s (Kaiser et al., 1973) and the quick growth of this field in the 1990s [see for example (Knight et al., 1996)], was a major step towards optical fibres made of a single material (Monro and Ebendorff-Heidepriem, 2006). This overcame one of the major restrictions in fabricating optical fibres. Conventional fibre designs require the use of two glasses for the fibre's core and cladding regions that are not only thermally and chemically compatible but also have appropriate optical characteristics for guidance. MOF technology can thus be seen as a tool for realising optical fibres from an extremely broad range of soft glasses because only a single material is required. Apart from different optical loss and UV and IR cutoffs of these glasses, the range of linear and nonlinear refractive indices and Raman and Brillouin coefficients of these glasses provide great flexibility in tailoring the linear and nonlinear behaviour of optical waveguides.

For example, a quick look at the linear and nonlinear (defined in Section 11.4) refractive indices of the families of glasses shows the great range of options in choosing the optical host material for waveguides. Fig. 11.1 (Monro and Ebendorff-Heidepriem, 2006) shows the empirical relation, usually known as the Miller rule, between the linear and nonlinear refractive indices of different families of glasses ranging from pure silica to chalcogenide.

In general and in terms of Kerr nonlinearity, waveguides fabricated from glasses with higher linear and nonlinear refractive indices can provide higher nonlinearity since tighter mode confinement, higher intensity, and higher nonlinear material lead to overall higher effective Kerr nonlinearity of a waveguide.

The bulk Raman gain coefficients for soft glasses are also typically larger than that of silica glass. For example for pumping in the near infrared, the bulk Raman gain coefficient of silica glass is 5.4×10^{-14} m/W (Abedin, 2006), whereas for bismuth, tellurite and chalcogenide based glasses the bulk Raman gain coefficients are approximately 4.2×10^{-13} m/W (Cantini et al., 2005), 2.2×10^{-12} m/W (Stegeman et al., 2003) and 5.1×10^{-11} m/W (Slusher et al., 2004) respectively. The exact bulk Raman gain coefficient is dependent on the pump wavelength and composition of the glass as we shall discuss later in Section 11.5.

Over recent years many soft glass-based MOFs or planar waveguides have been reported in the literature, examples of which include chalcogenide MOFs

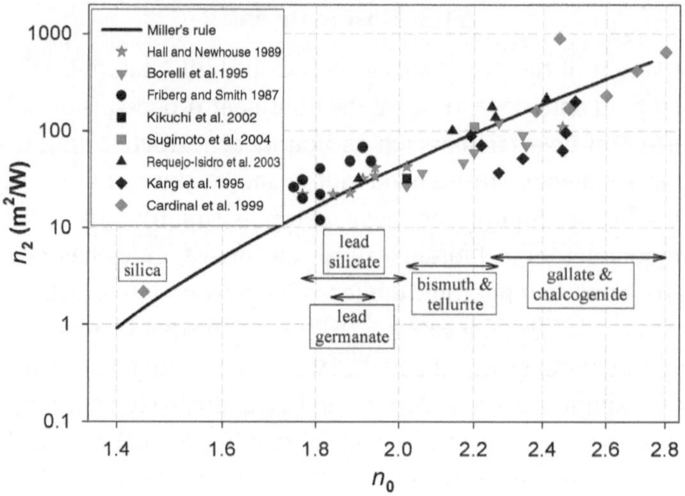

Fig. 11.1. Measured values of the nonlinear index (n_2) versus linear index (n_0) for a range of glasses (coloured symbols) and Miller's rule (straight black line). Data for the range of glasses are from Hall and Newhouse (1989), Borelli et al. (1995), Friberg and Smith (1987), Kikuchi et al. (2002), Sugimoto et al. (2004), Requejo-Isidro et al. (2003), Kang et al. (1995) and Cardinal et al. (1999).

(Monro et al., 2000; Abedin, 2005; Song et al., 2006; Brilland et al., 2006; Désévédavy et al., 2008), chalcogenide planar waveguides (Madden et al., 2007; Prasad et al., 2008), bismuth MOFs (Ebendorff-Heidepriem et al., 2004b; Lee et al., 2005a, 2005b), tellurite MOFs (Kumar et al., 2003; Feng et al., 2005, 2008; Abedin 2006), and SF57 MOFs (Leong et al., 2006).

Another feature of emerging waveguides that can be classified under the host material is the possibility to use postprocessing techniques to fill or coat complex structured waveguides with highly nonlinear materials. This also provides extended flexibility in tailoring the nonlinear effects in waveguides and hence has opened new horizons for applications of nonlinear optical phenomena. The use of postprocessing techniques provides a method to have waveguides with inhomogeneous cross-sections. The propagation of guided modes in waveguides with inhomogeneous cross-sections is affected by their structure both directly, through the linear part of the refractive index which determines the modal characteristics, and indirectly, since the waveguide modes, under propagation through the structure, experience different losses, nonlinearity, material dispersion etc. Examples of such structures include liquid-filled (Fuerbach et al., 2005; Yiou et al., 2005; Cox et al., 2006; Zhang et al., 2006), gas-filled (Perrone et al., 1997; Tomasi, 2001; Meng et al., 2002; Benabid et al., 2004, 2005a), silicon nanocrystals-filled (Barrios, 2004), atomic vapour-filled (Ghosh et al., 2005, 2006; Benabid et al., 2005b; Light et al., 2007), or surface-functionalised structures (Debs et al., 2008).

11.3 Structure

With the advancement of fabrication techniques, many MOFs with different cross-sectional structures have been successfully fabricated over the last 15 years. In particular, using extrusion techniques it has been possible to fabricate complex-structured, soft glass (non-silica glass) MOFs (Ebendorff-Heidepriem and Monro, 2007). The variety of cross-sectional structures in these MOFs indicates the great flexibility that these fibres provide in tailoring the optical properties, see Fig. 11.2. The cross-section of these MOFs, in general, can be considered as inhomogeneous refractive index distributions consisting of glass and air. More complex and specific optical properties can be obtained by filling these structures with liquids, or gases as mentioned in Section 11.2.

Recently, waveguides with subwavelength features have attracted significant interest, both in planar waveguides (Foster et al., 2004, 2005, 2007, 2008; Xu et al., 2004; Almeida et al., 2004a, 2004b, 2004c; Mullner and Hainberger, 2006; Magi et al., 2007; Salem et al., 2007) and fibres (Nagel et al., 2006; Wiederhecker et al., 2007; Afshar et al., 2007; Atakaramians et al., 2008). The key concept of subwavelength structures is the possibility to confine the light beyond the diffraction limit and in a subwavelength region (Wiederhecker et al., 2007). The concept can be explained by considering the continuity of the tangent and normal components of the electric, E_{t}, and displacement, D_{n}, fields, respectively, at the interface of any two media. The continuity of the normal components of the displacement fields implies that the normal component of electric fields are discontinuous at the interface and

$$E_{1\mathrm{n}}/E_{2\mathrm{n}} = \varepsilon_2/\varepsilon_1.$$

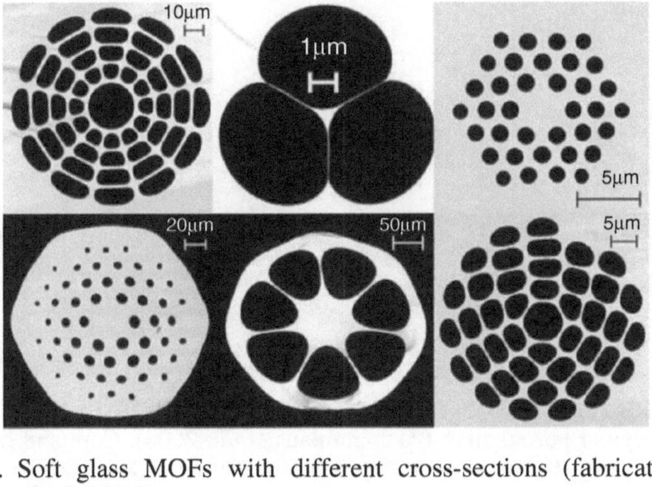

Fig. 11.2. Soft glass MOFs with different cross-sections (fabricated at the University of Adelaide).

As a result, the normal component of the electric field shows a discontinuous enhancement at the interface of two media and in the medium with lower refractive index. This has been shown in Fig. 11.3, where the normal component of the electric and displacement fields are plotted as a function of distance from the centre of a circular step index fibre with a subwavelength bore hole in the centre.

The normal component of the electric field, shown as the solid line in Fig. 11.3, shows an enhancement of $\varepsilon_2/\varepsilon_1$ at both core–cladding and bore hole interfaces,

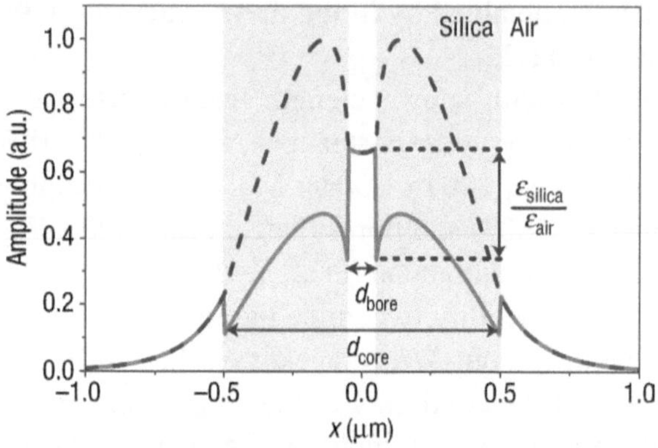

Fig. 11.3. Normal component of the displacement (dashed) and electric (solid) fields at different positions from the centre of an air–silica fibre with a subwavelength hole in the centre (Wiederhecker et al., 2007). Adapted by permission from Macmillan Publishers Ltd: [Nature Photonics], copyright (2007).

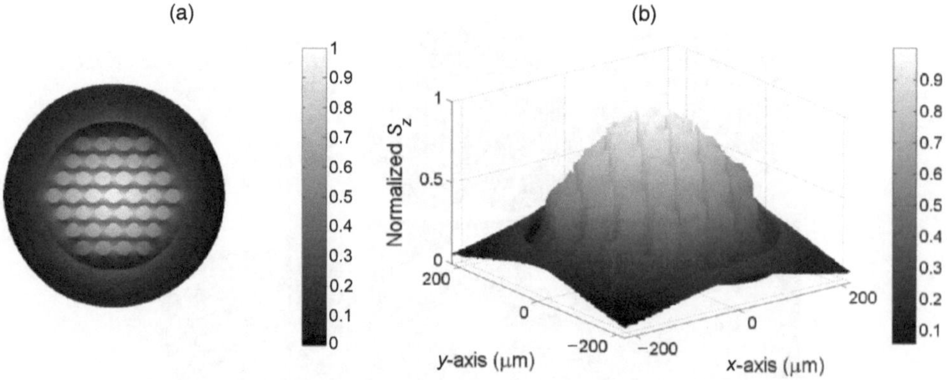

Fig. 11.4. 2D (a) and 3D (b) profiles of the z-component of the Poynting vector S_z, for a polymer porous fibre (Atakaramians et al., 2008). Core and hole diameters are 400 and 40 μm, respectively. The wavelength is $\lambda = 600$ μm. Intensity enhancement within subwavelength holes is observed.

which results in a high intensity layer at the low refractive index side of the interface between the two dielectric media. For the case of subwavelength voids in dielectrics, the enhanced intensity region forms at the surface of the void in the air side and extends over the whole void region since the evanescent decay of the field is negligible within a subwavelength void. This key characteristic of subwavelength features, when deployed within high-intensity regions within the mode cross-section, can be used to achieve arbitrary distributions of high-intensity regions within waveguides, see Fig. 11.4 (Atakaramians et al., 2008). The figure shows the profile of the z-component of the Poynting vector, S_z, across a porous fibre at THz wavelength (Atakaramians et al., 2008).

11.4 Kerr nonlinearity

Kerr nonlinearity is a third order nonlinear phenomena through which the refractive index of the material is modified by the intensity of the light itself, i.e., $n = n_0 + n_2 I$. Here, n_0 and n_2 are the linear and nonlinear refractive indices of the material and I is the intensity of the light. Kerr nonlinearity in a waveguide is considered through a nonlinear polarisation field. Using Maxwell's equations, assuming weak guidance approximation, for which the propagating (along the longitudinal direction of the waveguide, namely z) modes of the waveguide are described by transverse scalar fields, and writing the electric field as $E(r,t) = F(x,y)A(z,t)\exp(i\beta_0 z)$, where β_0 is the propagation constant, $F(x,y)$ is the transverse field distribution, and $A(z,t)$ is the field amplitude, two differential equations are obtained for $F(x,y)$ and $A(z,t)$ as (Agrawal, 2007)

$$\frac{\partial^2 F}{\partial x^2} + \frac{\partial^2 F}{\partial y^2} + (\varepsilon(\omega)k_0^2 - \tilde{\beta})F = 0, \tag{11.1}$$

$$\frac{\partial A}{\partial z} + \beta^{(1)}\frac{\partial A}{\partial t} + \frac{i\beta^{(2)}}{2}\frac{\partial^2 A}{\partial t^2} + \frac{\alpha}{2}A = i\gamma|A|^2 A. \tag{11.2}$$

Here, $\beta^{(n)} = \partial^n \beta/\partial\omega$ are higher order dispersion terms, α is the attenuation, and γ is the effective nonlinear coefficient of the waveguide and is given by

$$\gamma^{(\mathrm{S})} = (2\pi/\lambda)n_2/A_{\mathrm{eff}}, \tag{11.3}$$

where we use the superscript S to indicate the standard definition of γ.

In the next section, we develop a vectorially-based nonlinear Schrödinger equation (VNSE) for emerging waveguides. In this model, unlike other models, we consider waveguides with inhomogeneous cross-sections and explore the potential for coupling between different modes including different polarisations. Most importantly, we have considered the impact of the z-component of the modal fields on the dispersion, nonlinear, and modal/polarisation coupling behaviour of a pulse propagating through emerging waveguides. We also provide a correct vectorially-based definition for γ, namely γ^{V}, and explore the behaviour and significance of γ in different regimes, especially in the regime of subwavelength structure and high index contrast when the vectorial solutions of Maxwell's equations become important.

11.4.1 Theory

We start with Maxwell's equations for electric, magnetic and induced polarisation fields, $\tilde{\boldsymbol{E}}$, $\tilde{\mathbf{H}}$, and $\tilde{\mathbf{P}}$ in the Fourier domain as

$$\nabla \times \tilde{\mathbf{E}}(\mathbf{r},\omega) = i\mu_0\omega\tilde{\mathbf{H}}(\mathbf{r},\omega) \tag{11.4}$$

$$\nabla \times \tilde{\mathbf{H}}(\mathbf{r},\omega) = -i\varepsilon_0\omega\tilde{\mathbf{E}}(\mathbf{r},\omega) - i\omega\tilde{\mathbf{P}}(\mathbf{r},\omega), \tag{11.5}$$

where the Fourier transformation is given by

$$F(\mathbf{r},t) = \frac{1}{2\pi}\int \tilde{F}(\mathbf{r},\omega)e^{-i\omega t}d\omega, \tag{11.6}$$

and $F = \mathbf{E}$, $\mathbf{H}$, or $\mathbf{P}$. By considering a perturbative expansion

$$\tilde{\mathbf{P}}(\mathbf{r},\omega) = \sum_{n=1}^{\infty}\tilde{\mathbf{P}}^{(n)}(\mathbf{r},\omega),$$

where (n) represents the order of induced polarisation, and

$$\tilde{\mathbf{P}}^{(1)}(\mathbf{r},\omega) = \varepsilon_0\chi^{(1)}(-\omega;\omega)\tilde{\mathbf{E}}(\mathbf{r},\omega),$$

$$\tilde{\mathbf{P}}_{\mathrm{NL}}(\mathbf{r},\omega) = \sum_{n=2}^{\infty}\tilde{\mathbf{P}}^{(n)}(\mathbf{r},\omega),$$

in which the second rank tensor $\chi^{(1)}(-\omega;\omega)$ is assumed to be a scalar and related to the refractive index n through $n^2(\mathbf{r},\omega) = 1 + \chi^{(1)}(-\omega;\omega)$; we find

$$\nabla \times \tilde{\mathbf{E}}(\mathbf{r},\omega) = i\mu_0\omega\tilde{\mathbf{H}}(\mathbf{r},\omega), \tag{11.7}$$

$$\nabla \times \tilde{\mathbf{H}}(\mathbf{r},\omega) = -i\varepsilon_0 n^2(\mathbf{r},\omega)\tilde{\mathbf{E}}(\mathbf{r},\omega) - i\omega\tilde{\mathbf{P}}_{\mathrm{NL}}(\mathbf{r},\omega). \tag{11.8}$$

Next we consider Eqs. (11.7) and (11.8) for two sets of fields: unperturbed fields $\tilde{\boldsymbol{E}}_0(\mathbf{r},\omega_0)$ and $\tilde{\boldsymbol{H}}_0(\mathbf{r},\omega_0)$, which represent the electromagnetic fields of narrowband pulses at frequency ω_0 for which the dispersion, loss, and nonlinearity terms are zero, and perturbed fields $\tilde{\boldsymbol{E}}(\mathbf{r},\omega)$ and $\tilde{\boldsymbol{H}}(\mathbf{r},\omega)$, representing electromagnetic fields of frequency ω associated with wideband pulses centred at ω_0, where the dispersion, loss and nonlinearity terms are nonzero. Vectorial solutions of Maxwell's equation for the unperturbed fields result in a complete orthonormal set of forward, backward and radiation propagating modes (labelled μ) with propagating constants of β_μ and forward modal fields of (Snyder and Love, 1995)

$$\widehat{\mathbf{e}}_\nu = \frac{\mathbf{e}_\nu(x,y,\omega_0)}{\sqrt{N_\nu}} e^{i\beta_\nu z} \tag{11.9}$$

$$\widehat{\mathbf{h}}_\nu = \frac{\mathbf{h}_\nu(x,y,\omega_0)}{\sqrt{N_\nu}} e^{i\beta_\nu z}, \tag{11.10}$$

where

$$\int \mathbf{e}_\mu(x,y,\omega) \times \mathbf{h}^*_\nu(x,y,\omega).\hat{z} dA = N_\mu \delta_{\mu\nu} \tag{11.11}$$

$$N_\mu = \frac{1}{2}\left| \int \mathbf{e}_\mu(x,y,\omega) \times \mathbf{h}^*_\mu(x,y,\omega).\hat{z}\, dA \right| .$$

The propagation constant β_ν and modal field distributions, $\mathbf{e}_\nu(x,y,\omega_0)$ and $\mathbf{h}_\nu(x,y,\omega_0)$, of propagating modes of a waveguide, in general, can be obtained through various numerical methods including finite element methods (Okamoto, 2006), the multipole method (White et al., 2002; Kuhlmey et al., 2002a), etc. By constructing a function $\mathbf{F}_\mathrm{C}$ defined as

$$\mathbf{F}_\mathrm{C} = \tilde{\mathbf{E}}_0 \times \tilde{\mathbf{H}}^* + \tilde{\mathbf{E}}^* \times \tilde{\mathbf{H}}_0$$

and using Eqs. (11.7) and (11.8) for perturbed and unperturbed fields, we find (Snyder and Love, 1995)

$$\begin{aligned}\nabla \cdot \mathbf{F}_\mathrm{C} = &-i\mu_0(\omega-\omega_0)\tilde{\mathbf{H}}^*.\tilde{\mathbf{H}}_0 - i\varepsilon_0[\omega n^2(\mathbf{r},\omega) - \omega_0 n^2(\mathbf{r},\omega_0)]\tilde{\mathbf{E}}^*.\tilde{\mathbf{E}}_0 \\ &+ i\omega\tilde{\mathbf{E}}_0.\tilde{\mathbf{P}}^*_\mathrm{NL}(\mathbf{r},\omega).\end{aligned} \tag{11.12}$$

Next we expand the perturbed fields $\tilde{\mathbf{E}}$ and $\tilde{\mathbf{H}}$ according to the orthonormal and complete modal set of forward, backward, and radiation modes of the unperturbed field as (Snyder and Love, 1995)

$$\begin{aligned}\tilde{\mathbf{E}}(\mathbf{r},\omega) = &\sum_\mu \tilde{a}'_\mu(z,\omega)\frac{\mathbf{e}_\mu(x,y,\omega_0)}{\sqrt{N_\mu}} e^{i\beta_\mu z} + \tilde{a}'_{-\mu}(z,\omega)\frac{\mathbf{e}_{-\mu}(x,y,\omega_0)}{\sqrt{N_{-\mu}}} e^{-i\beta_\mu z} \\ &+ \text{radiation modes,}\end{aligned} \tag{11.13}$$

$$\tilde{\mathbf{H}}(\mathbf{r},\omega) = \sum_{\mu} \tilde{a}'_{\mu}(z,\omega)\frac{\mathbf{h}_{\mu}(x,y,\omega_0)}{\sqrt{N_{\mu}}}e^{i\beta_{\mu}z} + \tilde{a}'_{-\mu}(z,\omega)\frac{\mathbf{h}_{-\mu}(x,y,\omega_0)}{\sqrt{N_{-\mu}}}e^{-i\beta_{\mu}z}$$
$$+ \text{ radiation modes.} \tag{11.14}$$

Here index $-\mu$ refers to backward propagating modes and both forward and backward modes are orthogonal to radiation modes. Here, we only consider unidirectional pulse propagation for which we neglect the backscattering of a forward propagating laser beam and the nonlinearity associated with it (Kolesik and Moloney, 2004; Kolesik et al., 2002). This is not strictly true, especially for nonlinear and coupling processes where counter-propagating fields exist in the fibre. It can be shown that the backscattered field affects the overall nonlinearity for the forward modes but we leave the full investigation of this effect to future publications. Therefore, we only consider the first term in Eqs. (11.13) and (11.14) for expanding perturbed fields and, hence, ignore the coupling between the unperturbed field with the backward and radiation modes of the perturbed field. This will be discussed further later in this section. Unlike other reports that consider the modal expansion only for the nonlinear term (Koos et al., 2007; Kolesik and Moloney, 2004; Chen et al., 2006), we consider the modal expansion in Eqs. (11.13) and (11.14) for both dispersion and nonlinear effects. It should also be noted that the frequency dependence of perturbed fields is totally contained within the coefficients $\tilde{a}'_{\mu}(z,\omega)$. Assuming that the unperturbed fields $\tilde{\boldsymbol{E}}_0(\mathbf{r},\omega)$ and $\tilde{\mathbf{H}}_0(\mathbf{r},\omega)$ are one of the propagating modes (e.g., mode ν) of the unperturbed case i.e.,

$$\tilde{\mathbf{E}}_0(\mathbf{r},\omega_0) = \frac{\mathbf{e}_{\nu}(x,y,\omega_0)}{\sqrt{N_{\nu}}}e^{i\beta_{\nu}z} \tag{11.15}$$

$$\tilde{\mathbf{H}}_0(\mathbf{r},\omega_0) = \frac{\mathbf{h}_{\nu}(x,y,\omega_0)}{\sqrt{N_{\nu}}}e^{i\beta_{\nu}z}, \tag{11.16}$$

and using the reciprocal theorem (Snyder and Love, 1995)

$$\frac{\partial}{\partial z}\int \mathbf{F}_{\mathrm{C}}.\hat{z}\, dA = \int \nabla.\mathbf{F}_{\mathrm{C}}\, dA, \tag{11.17}$$

results in

$$\frac{\partial}{\partial z}\tilde{a}'_{\nu}(z,\omega) = \frac{1}{4}\sum_{\mu}[A_{\nu\mu} + B_{\nu\mu}]\tilde{a}'_{\mu} - \frac{i\omega e^{-i\beta_{\nu}z}}{4\sqrt{N_{\nu}}}\int \mathbf{e}^*_{\nu}.\tilde{\mathbf{P}}_{\mathrm{NL}}(\mathbf{r},\omega)\, dA. \tag{11.18}$$

Here,

$$A_{\nu\mu} = \frac{i\mu_0 e^{-i(\beta_\nu-\beta_\mu)z}}{\sqrt{N_\nu N_\mu}}(\omega-\omega_0)\int \mathbf{h}_\mu.\mathbf{h}_\nu^*\, dA \tag{11.19}$$

$$B_{\nu\mu} = \frac{-i\varepsilon_0 e^{-i(\beta_\nu-\beta_\mu)z}}{\sqrt{N_\nu N_\mu}}\int [\omega n^2(x,y,\omega)-\omega_0 n^2(x,y,\omega_0)]\mathbf{e}_\mu.\mathbf{e}_\nu^*\, dA. \tag{11.20}$$

Eq. (11.18) is a general first order differential equation that describes the propagation of amplitudes of the coupling coefficient of the perturbed field based on the unperturbed one. This equation is similar to those reported in Kolesik and Moloney (2004) and Laegsgaard (2007), except that the dispersion terms in Eq. (11.18) include the coupling between different modes.

Although $\tilde{a}'_\nu$ are the coefficients of the perturbed fields, Eq. (11.18) is exact in the sense that no perturbation has been considered for dispersion and nonlinearity and hence this equation can be applied to describe, in general, any nonlinear or dispersion-based processes in an optical waveguide. The first and the second terms on the right-hand side of the equation represent the dispersion and nonlinearity, respectively. Next, we perform a Taylor series expansion for the dispersion term in Eq. (11.18), around ω_0. Depending on the bandwidth of the pulse around ω_0, higher orders in the Taylor series can be considered to achieve better approximation for dispersion terms. For some cases, such as supercontinuum generation where extra wideband pulses propagate along the waveguide, however, it may be more appropriate to work with Eq. (11.18) directly. We separate the sum over the modes in Eq. (11.18) into self and cross terms to find

$$\frac{\partial}{\partial z}\tilde{a}'_\nu(z,\omega) = i\sum_{n=1}^{\infty}\frac{(\Delta\omega)^n}{n!}\beta_\nu^{(n)}\tilde{a}'_\nu + i\sum_{\mu\neq\nu}\sum_n\frac{(\Delta\omega)^n}{n!}\beta_{\nu\mu}^{(n)}\tilde{a}'_\mu \tag{11.21}$$

$$-\frac{i\omega e^{-i\beta_\nu z}}{4\sqrt{N_\nu}}\int \mathbf{e}_\nu^*.\tilde{\mathbf{P}}'_{\mathrm{NL}}(\mathbf{r},\omega)\, dA$$

where

$$\beta_\nu^{(1)} = \frac{1}{4N_\nu}\int\left[\mu_0\,|\mathbf{h}_\nu|^2 + \varepsilon_0\frac{\partial}{\partial\omega}(\omega n^2)|_{\omega=\omega_0}\,|\mathbf{e}_\nu|^2\right] dA \tag{11.22}$$

$$\beta_\nu^{(n)} = \frac{\partial^n}{\partial\omega^n}\beta_\nu^1 \tag{11.23}$$

$$\beta^{(1)}_{\nu\mu} = \frac{e^{-i(\beta_\nu - \beta_\mu)z}}{4\sqrt{N_\nu N_\mu}} \int \left[\mu_0 \mathbf{h}_\mu . \mathbf{h}^*_\nu + \varepsilon_0 \frac{\partial}{\partial \omega}(\omega n^2)|_{\omega=\omega_0} \mathbf{e}_\mu . \mathbf{e}^*_\nu \right] dA \tag{11.24}$$

$$\beta^{(n)}_{\nu\mu} = \frac{\partial^n}{\partial \omega^n} \beta^1_{\nu\mu}. \tag{11.25}$$

Here, to avoid confusion, superscripts (1) and (n) correspond to the first and higher order dispersion of the propagating modes, respectively, and subscripts μ label the mode number. It is straightforward to show that $\beta^{(1)}_\nu$ in Eq. (11.22) is in fact $\beta^{(1)}_\nu = 1/V_g$ where V_g is the group velocity as given in Snyder and Love (1995). One important aspect of Eq. (11.21), which has not been reported before, to the best of our knowledge, is the existence of cross dispersive terms $\beta^{(1)}_{\nu\mu}$, Eq. (11.24), and their derivatives, Eq. (11.25). Such terms can only become significant if they are phase matched, i.e., $\beta_\nu = \beta_\mu$, otherwise fast oscillations of the $e^{-i(\beta_\nu - \beta_\mu)z}$ average to a negligible value.

Eqs. (11.24) and (11.25) result in a new process when the μ and ν refer to the two polarisations, 1 and 2, of one mode. It is well known that waveguides with three or higher-fold symmetries are not birefringent (Steel et al., 2001) i.e., for these waveguides $\beta_1 = \beta_2$. In this case, the phase terms in Eqs. (11.24) and (11.25) are equal to unity and hence do not average to a negligible value as in the non-phase-matched case. The cross dispersive terms, $\beta^{(1)}_{12}$ and $\beta^{(n)}_{12}$ in Eqs. (11.24) and (11.25) are basically modifications to the group velocity $\beta^{(1)}_1$ and higher order dispersion terms $\beta^{(n)}_1$ of polarisation 1. They have nonzero values which, as will be shown later, is mainly due to the fact that in the strong guidance regime the dot product of the two polarisations of one mode i.e., $\mathbf{e}_1.\mathbf{e}^*_2$ and $\mathbf{h}_1.\mathbf{h}^*_2$ are nonzero because of the strong z-component of the fields which results in non-transversality of the modes. This key finding is discussed in more detail later in this section. The physical consequence of this is dispersion-induced depolarisation of the guided mode, i.e., a polarised guided mode depolarises even if the incident beam is initially coupled perfectly to one of the polarisation axes of the waveguide. For instance, assuming that the incident beam is perfectly launched along polarisation 1, then it can be deduced from Eq. (11.21) that the amplitude of the field along polarisation 2, i.e., $\tilde{a}'_2$, inside the fibre grows through $\sum_n \frac{(\Delta\omega)^n}{n!} \beta^{(n)}_{21} \tilde{a}'_1$.

The next step in this derivation is to derive the time domain equivalent of Eq. (11.21). This is done by multiplying both sides of Eq. (11.21) by $e^{-i(\omega-\omega_0)t}$ and integrating with respect to ω. Consider the following definitions

$$a'_\nu(z,t) \equiv 1/2[a_\nu(z,t)e^{-i\omega_0 t} + c.c]$$

$$\mathbf{P}'_{\mathrm{NL}}(\mathbf{r},t) = 1/2[\mathbf{P}_{\mathrm{NL}}(z,t)e^{-i\omega_0 t} + c.c],$$

where $a'_\nu(z,t)$ and $\mathbf{P}'_{\mathrm{NL}}(\mathbf{r},t)$ are the inverse Fourier transforms of $\tilde{a}'_\nu(z,\omega)$ and $\tilde{\mathbf{P}}'_{\mathrm{NL}}(\mathbf{r},\omega)$, respectively, to find

$$\begin{aligned}\frac{\partial}{\partial z}a_\nu(z,t) = i\sum_{n=1}^{\infty}\frac{(i\partial/\partial t)^n}{n!}\beta_\nu^{(n)}a_\nu + i\sum_{\mu\neq\nu}\sum_{n=1}^{\infty}\frac{(i\partial/\partial t)^n}{n!}\beta_{\nu\mu}^{(n)}a_\mu \\ - i\omega_0\frac{e^{-i\beta_\nu z}}{4\sqrt{N_\nu}}(1+\tau_{\mathrm{shock}}\partial/\partial t)\int \mathbf{e}_\nu^*.\mathbf{P}_{\mathrm{NL}}(\mathbf{r},t)\,dA,\end{aligned} \quad (11.26)$$

where $\tau_{\mathrm{shock}} = i/\omega_0$. Eq. (11.26) is a general equation that describes the nonlinear pulse propagation in the time domain. The first two terms on the right-hand side of this equation describe the dispersion of a pulse propagating through a waveguide. The last term includes all the nonlinear effects and considers a shock term of $(1+\tau_{\mathrm{shock}}\partial/\partial t)$ which is responsible for self phase modulation and self steepening of the pulse. This term, in various forms, has been considered in many publications (Karasawa et al., 2001; Kibler et al., 2005; Genty et al., 2007; Blow and Wood, 1989; Mamyshev and Chernikov, 1990; Brabec and Krausz, 1997; Gaeta, 2000; Dudley and Coen, 2002; Biancalana et al., 2003; Chang et al., 2003).

A few points should be noted here about the shock term: (1) Similar to Kolesik and Moloney (2004), it is naturally derived through the $(\partial/\partial t)\mathbf{P}_{\mathrm{NL}}$ of Maxwell's equations without any approximation of a second order time derivative in the scalar wave equation (Helmholtz equation) which is usually used to describe the nonlinear pulse propagation (Karasawa et al., 2001; Blow and Wood, 1989; Mamyshev and Chernikov, 1990; Brabec and Krausz, 1997). (2) The whole frequency dependence of the perturbed field is inherently included through the expansions Eqs. (11.13) and (11.14) and is contained completely within the $\tilde{a}'_2$ coefficients. Thus there is no need to include the frequency dependence of the propagating modes in the shock term, through A_{eff}, as has been done in Karasawa et al. (2001), Kibler et al. (2005), Agrawal (2007), Blow and Wood (1989) and Mamyshev and Chernikov (1990). (3) There is no dispersive term associated with $\chi^{(3)}$, i.e., $(\partial/\partial\omega)\chi^{(3)} = 0$ since assuming a delta function form for the response function of Kerr nonlinearity results in frequency independence of $\chi^{(3)}$. For the nonlinear term in Eq. (11.26), since $\chi^{(2)} = 0$ for isotropic media such as glasses, we only consider the third order Kerr nonlinearity for which we approximate $\mathbf{P}_{\mathrm{NL}}(\mathbf{r},t) \approx \mathbf{P}^{(3)}(\mathbf{r},t)$ and assume that the nonlinear response function can be expressed in terms of delta functions and hence $\mathbf{P}^{(3)}(\mathbf{r},t)$, for Kerr nonlinearity, can be written as (Butcher and Cotter, 1990)

$$\mathbf{P}^{(3)}(\mathbf{r},t) = (3/4)\varepsilon_0\chi^{(3)}(-\omega_0;\omega_0,\omega_0,-\omega_0)|\mathbf{E}(\mathbf{r},t)\mathbf{E}(\mathbf{r},t)\mathbf{E}^*(\mathbf{r},t), \quad (11.27)$$

where $\chi^{(3)}$ is a rank four tensor and | indicates tensorial multiplication. Other third order nonlinear effects such as Raman scattering, for which the response function is not an instantaneous function of time will be considered in Section 11.5. Also it is assumed that $\mathbf{P}_{\mathrm{NL}}(\mathbf{r},t) \approx \mathbf{P}^{(3)}(\mathbf{r},t)$ is a small perturbation compared to the linear induced polarisation $\mathbf{P}_{\mathrm{L}} = \varepsilon_0 \chi^{(1)}(-\omega;\omega)\tilde{\mathbf{E}}(\mathbf{r},\omega)$, and higher order nonlinear effects are negligible. This is usually a valid approximation at low intensity fields and for typical optical glasses due to their relatively low nonlinear properties. However, for some materials such as semiconductor-doped glasses (Roussignol et al., 1987; Stegeman and Stolen, 1989; Gatz and Herrmann, 1991; Coutaz and Kull, 1991; Konar et al., 2005) and some organic materials such as paratoluene sulphonate (PTS) (Stegeman and Stolen, 1989; Konar et al., 2005) optical processes based on higher order nonlinearity can be observed, due to their higher order nonlinear susceptibilities, at moderate pulse intensity. For such materials, higher order terms must be considered in the nonlinear polarisation field $\tilde{\mathbf{P}}_{\mathrm{NL}}(\mathbf{r},\omega) = \sum_{n=2}^{\infty} \tilde{\mathbf{P}}^{(n)}(\mathbf{r},\omega)$.

The components of $\chi^{(3)}$ depend on the class symmetry of the crystal. Silica glasses have an isotropic crystal structure (Agrawal, 2007) and silicon crystal, which is usually used in waveguides, have m3m point-group symmetry (Lin et al., 2007). For isotropic materials, it can be shown that among 81 elements of $\chi^{(3)}_{ijkl}$ $(i,j,k,l = x,y,z)$ only 21 are nonzero, which depend on only three independent quantities (Boyd, 2003) i.e.

$$\chi^{(3)}_{ijkl} = \chi^{(3)}_{xxyy}\delta_{ij}\delta_{kl} + \chi^{(3)}_{xyxy}\delta_{ik}\delta_{jl} + \chi^{(3)}_{xyyx}\delta_{il}\delta_{jk}, \tag{11.28}$$

where

$$\chi_{xxxx} = \chi_{yyyy} = \chi_{zzzz} = \chi^{(3)}_{xxyy} + \chi^{(3)}_{xyxy} + \chi^{(3)}_{xyyx}. \tag{11.29}$$

Considering Eq. (11.28), Eq. (11.27) can be written as

$$P_i^{(3)}(\mathbf{r},t) = (3/4)\varepsilon_0 \left[\sum_j \chi^{(3)}_{xxyy} \left|E_j\right|^2 E_i + \sum_j \chi^{(3)}_{xyxy} \left|E_j\right|^2 E_i + \sum_j \chi^{(3)}_{xyyx} (E_j)^2 E_i^* \right], \tag{11.30}$$

where i and j refer to x, y, z. For Kerr nonlinearity with the choice of frequencies in Eq. (11.27), i.e., $\chi^{(3)}(-\omega_0;\omega_0,\omega_0,-\omega_0)$, the condition of permutation symmetry requires that $\chi^{(3)}_{xxyy} = \chi^{(3)}_{xyxy}$. The magnitude of the terms in the right-hand side of Eq. (11.29) depends on the origin of the nonlinear term. In the case of silica and other glasses they have mainly nonresonant electronic origins for which $\chi^{(3)}_{xyxy} \approx \chi^{(3)}_{xyyx}$ (Agrawal, 2007; Boyd, 2003) and hence Eq. (11.30) can be simplified to

$$\mathbf{P}^{(3)}(\mathbf{r},t) = (1/2)\varepsilon_0 \chi^{(3)}_{xxxx}[(\mathbf{E}.\mathbf{E}^*)\mathbf{E} + (1/2)(\mathbf{E}.\mathbf{E})\mathbf{E}^*]. \tag{11.31}$$

For silicon, however, the third order nonlinearity can be expressed based on four independent values as

$$\chi^{(3)}_{ijkl} = \chi^{(3)}_{xxyy}\delta_{ij}\delta_{kl} + \chi^{(3)}_{xyxy}\delta_{ik}\delta_{jl} + \chi^{(3)}_{xyyx}\delta_{il}\delta_{jk} + \chi_d\delta_{ijkl}, \tag{11.32}$$

where $\chi_d \equiv \chi_{xxxx} - \chi_{xxyy} - \chi_{xyxy} - \chi_{xyyx}$ represent the nonlinearity isotropy. Similar to silica, for the choice of frequencies in the third order susceptibility and photon energies $\hbar\omega$ well above E_g (Lin et al., 2007)

$$\chi_{xxyy}(-\omega;\omega,-\omega,\omega) = \chi_{xyyx}(-\omega;\omega,-\omega,\omega) \approx \chi_{xyxy}(-\omega;\omega,-\omega,\omega).$$

As a result, χ_{ijkl} becomes

$$\chi^{(3)}_{ijkl} = \chi_{xxxx}\left[\frac{\rho}{3}(\delta_{ij}\delta_{kl} + \delta_{ik}\delta_{jl} + \delta_{il}\delta_{jk}) + (1-\rho)\delta_{ijkl}\right],$$

where $\rho \equiv 3\chi_{xxyy}/\chi_{xxxx}$ characterises the nonlinear anisotropy and its value in the telecom band is real and close to 1.27 (Lin et al., 2007). Using this, Eq. (11.27) can be written for silicon as

$$\begin{aligned}\mathbf{P}^{(3)}(\mathbf{r},t) &= (\rho/2)\varepsilon_0\chi^{(3)}_{xxxx}[(\mathbf{E}.\mathbf{E}^*)\mathbf{E} + (1/2)(\mathbf{E}.\mathbf{E})\mathbf{E}^*] \\ &\quad + (3/4)\varepsilon_0(1-\rho)\chi^{(3)}_{xxxx}\mathbf{E}.\mathbf{E}.\mathbf{E}^*,\end{aligned} \tag{11.33}$$

where $\mathbf{E}.\mathbf{E}.\mathbf{E}^* \equiv \sum_i E_iE_iE_i^*\mathbf{v}_i$ ($\mathbf{v}_i$ is a Cartesian unit vector). Within the rest of this chapter we ignore the last term of Eq. (11.33), which in fact affects the polarisation dependence of nonlinear phenomena inside silicon waveguides, and thus Eq. (11.33) is the same as Eq. (11.31) except the factor ρ. Considering the expansion in Eq. (11.13) and using Eq. (11.31) we can evaluate the integrand in the last term of Eq. (11.26), i.e., $(1/\sqrt{N_\nu})e^{-i\beta_\nu z}\mathbf{e}^*_\nu.\mathbf{P}_{\mathrm{NL}}(\mathbf{r},t)$ as

$$\begin{aligned}(1/\sqrt{N_\nu})e^{-i\beta_\nu z}\mathbf{e}^*_\nu.\mathbf{P}_{\mathrm{NL}}(\mathbf{r},t) &= (1/2)\varepsilon_0\chi^{(3)}_{xxxx} \\ \times \sum_{\mu,\eta,\zeta}[(1/\sqrt{N_\mu N_\eta N_\zeta N_\nu})a_\mu a^*_\eta a_\zeta(\mathbf{e}_\mu.\mathbf{e}^*_\eta)(\mathbf{e}^*_\nu.\mathbf{e}_\zeta)e^{-i(\beta_\nu-\beta_\mu+\beta_\eta-\beta_\zeta)z} \\ + (1/2\sqrt{N_\mu N_\eta N_\zeta N_\nu})a_\mu a_\eta a^*_\zeta(\mathbf{e}_\mu.\mathbf{e}_\eta)(\mathbf{e}^*_\nu.\mathbf{e}^*_\zeta)e^{-i(\beta_\nu-\beta_\mu-\beta_\eta+\beta_\zeta)z}]\end{aligned} \tag{11.34}$$

where Greek indices μ, ν, η, ζ represent different modes of the waveguide. The terms on the right-hand side of this equation, once integrated over the waveguide cross-section, are overlap integrals representing how different propagating modes of the fibre couple to each other through the nonlinearity. Eq. (11.34) can be expanded as a sum of terms with and without phase terms as

$$
\begin{aligned}
&(1/\sqrt{N_\nu})e^{-i\beta_\nu z}\mathbf{e}_\nu^*.\mathbf{P}_{\mathrm{NL}}(\mathbf{r},t) \\
&= (3/4)\varepsilon_0\chi_{xxxx}^{(3)}\left\{\left(\frac{|a_\nu|^2 a_\nu}{3N_\nu^2}\right)\left[2\,|\mathbf{e}_\nu|^4 + \left|\mathbf{e}_\nu^2\right|^2\right]\right. \\
&+\sum_{\mu\neq\nu}\left(\frac{2a_\nu\,|a_\mu|^2}{3\sqrt{N_\nu^2N_\mu^2}}\right)\left[\left|\mathbf{e}_\nu.\mathbf{e}_\mu^*\right|^2 + \left|\mathbf{e}_\nu.\mathbf{e}_\mu\right|^2 + |\mathbf{e}_\nu|^2\left|\mathbf{e}_\mu\right|^2\right] \\
&+\sum_{\mu\neq\nu}\left(\frac{a_\mu^* a_\nu^2}{3\sqrt{N_\nu^3N_\mu}}\right)\left[2\,|\mathbf{e}_\nu|^2(\mathbf{e}_\mu^*.\mathbf{e}_\nu) + (\mathbf{e}_\nu)^2(\mathbf{e}_\mu^*.\mathbf{e}_\nu^*)\right]e^{-i(\beta_\mu-\beta_\nu)z} \\
&+\sum_{\mu\neq\nu}\left(\frac{2a_\mu\,|a_\nu|^2}{3\sqrt{N_\nu^3N_\mu}}\right)\left[2\,|\mathbf{e}_\nu|^2(\mathbf{e}_\mu.\mathbf{e}_\nu^*) + (\mathbf{e}_\nu^*)^2(\mathbf{e}_\mu.\mathbf{e}_\nu)\right]e^{-i(\beta_\nu-\beta_\mu)z} \\
&+\sum_{\mu\neq\nu}\left(\frac{\left|a_\mu\right|^2 a_\mu}{3\sqrt{N_\mu^3N_\nu}}\right)\left[2\left|\mathbf{e}_\mu\right|^2(\mathbf{e}_\mu.\mathbf{e}_\nu^*) + (\mathbf{e}_\mu)^2(\mathbf{e}_\mu^*.\mathbf{e}_\nu^*)\right]e^{-i(\beta_\nu-\beta_\mu)z} \\
&+\sum_{\mu\neq\nu}\left(\frac{a_\mu^2 a_\nu^*}{3\sqrt{N_\nu^2N_\mu^2}}\right)\left[2(\mathbf{e}_\mu.\mathbf{e}_\nu^*)^2 + (\mathbf{e}_\mu)^2(\mathbf{e}_\nu)^2\right]e^{-2i(\beta_\nu-\beta_\mu)z} \\
&\left.+\sum_{\mu\neq\eta\neq\zeta\neq\nu}\text{otherphaseterms}\right\}.
\end{aligned}
\tag{11.35}
$$

The first two terms on the right-hand side of Eq. (11.35) are automatically phase matched while the rest of the terms require phase matching in order to make significant contributions. The phase terms are responsible for nonlinear-induced depolarisation or four-wave mixing (Agrawal, 2007). They can be phased matched, depending on $\Delta\beta_{\nu\mu} = \beta_\nu - \beta_\mu$. This can be achieved by employing the flexibility in controlling the dispersion properties of MOFs through structure design and glass choice.

By substituting Eq. (11.35) into Eq. (11.26), a first order differential equation is obtained which describes the nonlinear pulse propagation in a multimode waveguide. An important aspect of our formalism is related to the orthogonality of the waveguide propagating modes. Contrary to the standard formalism (Agrawal, 2007; Boyd, 2003; Lin and Agrawal, 2004), for which $\mathbf{e}_\nu$s are approximated to transverse modes and $\int \mathbf{e}_\nu^*.\mathbf{e}_\mu dA = \int \mathbf{e}_{\nu t}^*.\mathbf{e}_{\mu t} dA = 0$, or in the case of different polarisations $\mathbf{e}_\nu.\mathbf{e}_\mu = 0$, in our formalism $\int \mathbf{e}_\mu^*.\mathbf{e}_\nu dA \neq 0$ (or $\mathbf{e}_\nu.\mathbf{e}_\mu \neq 0$ if μ and ν are the two polarisations of the same mode) because the modes are non-transverse, i.e., they have a nonzero z-component. The generalised orthogonality condition, which is valid even in the strong guidance regime, is $\int (\widehat{\mathbf{e}}_\nu \times \widehat{\mathbf{h}}_\mu^*).\hat{z} dA = \delta_{\nu\mu}$ and inherently includes the z-component of the fields. Considering that

$$(\widehat{\mathbf{e}} \times \widehat{\mathbf{h}}^*).\hat{z} = (\widehat{\mathbf{e}}_{\mathrm{t}} \times \widehat{\mathbf{h}}_{\mathrm{t}}^*).\hat{z}, \tag{11.36}$$

and (Snyder and Love, 1995)

$$\widehat{\mathbf{h}}_t = \left(\frac{\varepsilon_0}{\mu_0}\right)^{1/2} \frac{1}{k}\hat{z} \times [\beta \widehat{\mathbf{e}}_{\mathrm{t}} + i\nabla_{\mathrm{t}}\widehat{e}_z], \tag{11.37}$$

one can show that

$$\begin{aligned} \int (\widehat{\mathbf{e}}_\nu \times \widehat{\mathbf{h}}_\mu^*).\hat{z}\, dA &= \left(\frac{\varepsilon_0}{\mu_0}\right)^{1/2} \frac{1}{k} \int (\widehat{\mathbf{e}}_{\nu\mathrm{t}} \times \{\hat{z} \times [\beta \widehat{\mathbf{e}}_{\mu\mathrm{t}}^* - i\nabla_{\mathrm{t}} e_{\mu z}^*]\}).\hat{z}\, dA, \qquad (11.38) \\ &= \left(\frac{\varepsilon_0}{\mu_0}\right)^{1/2} \frac{1}{k} \int (\beta \widehat{\mathbf{e}}_{\nu\mathrm{t}} \cdot \widehat{\mathbf{e}}_{\mu\mathrm{t}}^* - i\widehat{\mathbf{e}}_{\nu\mathrm{t}} \cdot \nabla_{\mathrm{t}} \widehat{e}_{\mu z}^*)\, dA \end{aligned}$$

which considering the general orthogonality relation $\int (\widehat{\mathbf{e}}_\nu \times \widehat{\mathbf{h}}_\mu^*).\hat{z}\; dA = \delta_{\nu\mu}$ results in

$$\int \widehat{\mathbf{e}}_{\nu\mathrm{t}} \cdot \widehat{\mathbf{e}}_{\mu\mathrm{t}}^* dA = \left(\frac{\mu_0}{\varepsilon_0}\right)^{1/2} \frac{k}{\beta}\delta_{\nu\mu} + \left(\frac{i}{\beta}\right) \int (\widehat{\mathbf{e}}_{\nu\mathrm{t}} \cdot \nabla_{\mathrm{t}} \widehat{e}_{\mu z}^*)\, dA, \tag{11.39}$$

where subscript t refers to the transverse component of fields and operators. Eq. (11.39) clearly shows that $\int \widehat{\mathbf{e}}_{\nu\mathrm{t}} \cdot \widehat{\mathbf{e}}_{\mu\mathrm{t}}^* dA \neq 0$ in the parameter regime where the z-component of electromagnetic fields are nonzero. It should also be noted that Eq. (11.35) has been obtained by ignoring the backward and radiation terms in Eqs. (11.13) and (11.14). Considering these two terms in the expansion Eqs. (11.13) and (11.14) results in coupling between forward-backward and forward-radiation modes, which are represented by dot products of forward modes with backward and radiation modes in Eq. (11.35). These terms describe the power coupling between a forward propagating mode and backward and radiation modes due to nonlinearity.

In the case of single mode fibres, where two independent polarisations exist in the waveguide, μ and ν refer to the two polarisations 1 and 2 and $\Delta\beta_{\nu\mu}$ is the linear birefringence of the waveguide. For waveguides with strong birefringence, the beat length $L_{\mathrm{B}} = 2\pi/\Delta\beta$ is short, and hence for fibre lengths $L \gg L_{\mathrm{B}}$ the phase terms oscillate very fast and so have negligible contributions. However, for waveguides with weak birefringence, for which $L < L_B$, the phase terms are not negligible and should be taken into account. In the following sections we develop a model for nonlinear pulse propagation in single mode waveguides for both weak and strong birefringence.

11.4.1.1 Single mode highly birefringent waveguides

In the case of single mode waveguides with high birefringence, where only the first two terms in Eq. (11.35) are significant, substituting Eq. (11.35) into Eq. (11.26), considering (Lin et al., 2007; Agrawal, 2007) $\varepsilon_0 c n^2 n_2 = (3/4)\mathrm{Re}\chi^{(3)}_{xxxx}$, where n_2 is the nonlinear refractive index of the material measured in m²/W, and ignoring nonlinear two-photon absorption we find:

$$\frac{\partial}{\partial z}a_\nu(z,t) = i\sum_n \frac{(i\partial/\partial t)^n}{n!}\beta_\nu^{(n)}a_\nu + \frac{-ik}{4}\left(\frac{\varepsilon_0}{\mu_0}\right)(1+\tau_{\text{shock}}\partial/\partial t) \tag{11.40}$$
$$\times\left\{\frac{1}{3N_\nu^2}|a_\nu|^2 a_\nu\int n^2(x,y)n_2(x,y)\left[2|\mathbf{e}_\nu|^4 + \left|\mathbf{e}_\nu^2\right|^2\right]dA\right.$$
$$+\frac{2}{3N_\nu N_\mu}|a_\mu|^2 a_\nu\int n^2(x,y)n_2(x,y)\left[\left|\mathbf{e}_\nu.\mathbf{e}_\mu^*\right|^2 + |\mathbf{e}_\nu.\mathbf{e}_\mu|^2\right.$$
$$\left.\left.+|\mathbf{e}_\nu|^2|\mathbf{e}_\mu|^2\right]dA\right\},$$

where μ, $\nu = 1,2$ and $\mu \neq \nu$ refer to the two polarisations of the fundamental mode. This equation can finally be written in a simple form:

$$\frac{\partial}{\partial z}a_\nu(z,t) = i\sum_n \frac{(i\partial/\partial t)^n}{n!}\beta_\nu^{(n)}a_\nu \tag{11.41}$$
$$-(1+\tau_{\text{shock}}\partial/\partial t)[i\gamma_\nu|a_\nu|^2 a_\nu + i\gamma_{\mu\nu}|a_\mu|^2 a_\nu],$$

where

$$\gamma_\nu = k\left(\frac{\varepsilon_0}{\mu_0}\right)\frac{\int n^2(x,y)n_2(x,y)[2|\mathbf{e}_\nu|^4 + |\mathbf{e}_\nu^2|^2]\,dA}{3\left|\int(\mathbf{e}_\nu\times\mathbf{h}_\nu^*).\hat{z}\,dA\right|^2}, \tag{11.42}$$

$$\gamma_{\mu\nu} = \gamma_{\mu\nu}^{(1)} + \gamma_{\mu\nu}^{(2)} = k\left(\frac{\varepsilon_0}{\mu_0}\right)\left[\left(\frac{2\int n^2(x,y)n_2(x,y)\left[\left|\mathbf{e}_\nu.\mathbf{e}_\mu^*\right|^2 + |\mathbf{e}_\nu.\mathbf{e}_\mu|^2\right]dA}{3\left|\int(\mathbf{e}_\mu\times\mathbf{h}_\mu^*).\hat{z}\,dA\right|\left|\int(\mathbf{e}_\nu\times\mathbf{h}_\nu^*).\hat{z}\,dA\right|}\right.\right.$$
$$\left.+\frac{2\int n^2(x,y)n_2(x,y)|\mathbf{e}_\mu|^2|\mathbf{e}_\nu|^2 dA}{3\left|\int(\mathbf{e}_\mu\times\mathbf{h}_\mu^*).\hat{z}\,dA\right|\left|\int(\mathbf{e}_\nu\times\mathbf{h}_\nu^*).\hat{z}\,dA\right|}\right]. \tag{11.43}$$

Eq. (11.41) is the final form of nonlinear pulse propagation inside a single mode birefringent waveguide, which in form is similar to the commonly used equation (see Agrawal, 2007) but the effective nonlinear coefficients of the waveguide are now given by the generalised forms as in Eqs. (11.42) and (11.43). By generalising the definition of the A_{eff} as

$$A_{\text{eff}} = \frac{\left|\int (\mathbf{e}_\nu \times \mathbf{h}_\nu^*).\hat{z}\, dA\right|^2}{\int \left|(\mathbf{e}_\nu \times \mathbf{h}_\nu^*).\hat{z}\right|^2 dA} \tag{11.44}$$

the nonlinear coefficient γ_ν can be rewritten as

$$\gamma_\nu = \frac{2\pi}{\lambda}\frac{\overline{n_2}}{A_{\text{eff}}} \tag{11.45}$$

$$\overline{n_2} = \left(\frac{\varepsilon_0}{\mu_0}\right) \frac{\int n^2(x,y) n_2(x,y)[2\,|\mathbf{e}_\nu|^4 + \left|\mathbf{e}_\nu^2\right|^2]\, dA}{\int \left|(\mathbf{e}_\nu \times \mathbf{h}_\nu^*).\hat{z}\right|^2 dA},$$

where $\overline{n_2}$ can be viewed as the nonlinear refractive index averaged over an inhomogeneous cross-section weighted with respect to field distribution. The advantages of writing γ as in Eq. (11.45) over the other reported form (Koos et al., 2007) is that it allows the analysis of γ to be separated into parts describing linear (geometry and $n(x,y) \Longrightarrow A_{\text{eff}}$) and nonlinear (mode profile and $n_2(x,y) \Longrightarrow \overline{n_2}$) characteristics, providing a more intuitive analysis of γ. In our formalism, A_{eff} has its standard interpretation as the effective area of the propagating modes, which can be determined purely based on the geometry and the linear refractive index of the waveguide $n(x,y)$. This avoids the need to define an "effective nonlinear interaction area" as in Koos et al. (2007). Considering Eqs. (11.36) and (11.38), we find that Eq. (11.44) can be written as

$$A_{\text{eff}} = \frac{\left|\int [\beta\,|\mathbf{e}_t|^2 + i(\mathbf{e}_t.\nabla_t e_z)]\, dA\right|^2}{\int \left|\beta\,|\mathbf{e}_t|^2 + i(\mathbf{e}_t.\nabla_t e_z)\right|^2 dA}. \tag{11.46}$$

This equation in the limit of small z-component of electric field, where $\nabla_t e_z$ can be ignored in comparison with $\beta \mathbf{e}_t$, simplifies as

$$A_{\text{eff}} = \frac{\left|\int |\mathbf{e}_t|^2\, dA\right|^2}{\int |\mathbf{e}_t|^4\, dA}, \tag{11.47}$$

which is the standard definition of A_{eff} (Agrawal, 2007).

Comparing Eqs. (11.41)–(11.45) with the expression commonly used for nonlinear birefringent terms (Agrawal, 2007), shows (1) the vectorial-based γ developed here (referred to hereafter as γ^{V}) in Eq. (11.45) includes both the inhomogeneous waveguide structure and vectorial nature of electromagnetic fields and (2) an extra term that induces the depolarisation of an initially polarised beam through a nonzero coupling term

$$\gamma_{\mu\nu}^{(1)} \propto \int n^2(x,y)n_2(x,y)\left(\left|\mathbf{e}_\nu.\mathbf{e}_\mu^*\right|^2+\left|\mathbf{e}_\nu.\mathbf{e}_\mu\right|^2\right) dA \neq 0.$$

The nonzero nature of this term, especially in the strong guidance regime as will be shown in Section 11.4.2, is the direct result of two facts: (1) the two different polarisations are not perpendicular to each other in the common sense, i.e., $\mathbf{e}_\mu.\mathbf{e}_\nu = 0$ or $\int \mathbf{e}_\mu.\mathbf{e}_\nu^* dA = 0$ because of the strong z-component of the fields and (2) the transverse integral, due to transverse dependence of $n^2(x,y)$ and $n_2(x,y)$, can be evaluated over different regions with different n and n_2.

11.4.1.2 Single mode non-birefringent waveguides

For waveguides with perfect three or higher fold symmetry, the fundamental mode of the waveguide is degenerate or non-birefringent (Steel et al., 2001), i.e., $\beta_1 = \beta_2$ where 1 and 2 refer to the two polarisations. Therefore, the phase factors in Eq. (11.35) are equal to 1 for the pair of fundamental modes of these waveguides and hence Eq. (11.35) and Eq. (11.26), result in

$$\begin{aligned}
&\frac{\partial}{\partial z}a_\nu(z,t)\\
&= i\sum_{n=1}^{\infty}\frac{(i\partial/\partial t)^n}{n!}\beta_\nu^{(n)}a_\nu + i\sum_{n=1}^{\infty}\frac{(i\partial/\partial t)^n}{n!}\beta_{\nu\mu}^{(n)}a_\mu - \frac{ik}{4}\left(\frac{\varepsilon_0}{\mu_0}\right)(1+\tau_{\text{shock}}\partial/\partial t)\\
&\times\left\{\frac{1}{3N_\nu^2}|a_\nu|^2 a_\nu\int n^2(x,y)n_2(x,y)\left[2\left|\mathbf{e}_\nu\right|^4+\left|\mathbf{e}_\nu^2\right|^2\right]dA\right.\\
&+\frac{2}{3N_\nu N_\mu}\left|a_\mu\right|^2 a_\nu\int n^2(x,y)n_2(x,y)\left[\left|\mathbf{e}_\nu.\mathbf{e}_\mu^*\right|^2+\left|\mathbf{e}_\nu.\mathbf{e}_\mu\right|^2+\left|\mathbf{e}_\nu\right|^2\left|\mathbf{e}_\mu\right|^2\right]dA\\
&+\frac{1}{3\sqrt{N_\nu^3 N_\mu}}a_\mu^* a_\nu^2\int n^2(x,y)n_2(x,y)\left[2\left|\mathbf{e}_\nu\right|^2(\mathbf{e}_\mu^*.\mathbf{e}_\nu)+(\mathbf{e}_\nu)^2(\mathbf{e}_\mu^*.\mathbf{e}_\nu^*)\right]dA\\
&+\frac{2}{3\sqrt{N_\nu^3 N_\mu}}a_\mu\left|a_\nu\right|^2\int n^2(x,y)n_2(x,y)\left[2\left|\mathbf{e}_\nu\right|^2(\mathbf{e}_\mu.\mathbf{e}_\nu^*)+(\mathbf{e}_\nu^*)^2(\mathbf{e}_\mu.\mathbf{e}_\nu)\right]dA
\end{aligned}$$

$$+\frac{1}{3\sqrt{N_\mu^3 N_\nu}}|a_\mu|^2 a_\mu \int n^2(x,y)n_2(x,y)\left[2\left|\mathbf{e}_\mu\right|^2(\mathbf{e}_\mu.\mathbf{e}_\nu^*)+(\mathbf{e}_\mu)^2(\mathbf{e}_\mu^*.\mathbf{e}_\nu^*)\right]dA$$

$$\left.+\frac{1}{3\sqrt{N_\nu^2 N_\mu^2}}a_\mu^2 a_\nu^* \int n^2(x,y)n_2(x,y)\left[2(\mathbf{e}_\mu.\mathbf{e}_\nu^*)^2+(\mathbf{e}_\mu)^2(\mathbf{e}_\nu)^2\right]dA\right\},\tag{11.48}$$

where μ, $\nu = 1,2$ and $\mu \neq \nu$ refer to the two polarisations of the fundamental mode. Eq. (11.48) is the vectorial generalisation of the common nonlinear pulse propagation for the two polarisations of single mode fibre (Agrawal, 2007). The main differences are the extra contributions from the different combinations of $\mathbf{e}_\mu.\mathbf{e}_\nu$. These terms, as indicated in the previous section, have nonzero values in the regime of high index and subwavelength core diameters.

11.4.2 Results and discussion

The formalism developed in the previous section is general, and can be applied to an arbitrary waveguide. However, here the above formalism is applied to a simple step index rod waveguide (i.e. a nanowire) and demonstrates that its nonlinear behaviour is predicted to be significantly different in the regime of high index contrast and subwavelength features than the predictions made using the standard formalism (Agrawal, 2007). It should be pointed out that, throughout, we use the unit of $\mathrm{W^{-1}m^{-1}}$ for γ instead of the commonly used unit $\mathrm{W^{-1}km^{-1}}$ (Agrawal, 2007). This is justified considering the fact that small core structures in high index silicon or glasses can now provide access to waveguides with extremely large γ values (Koos et al., 2007; Magi et al., 2007; Leong et al., 2006; Ebendorff-Heidepriem et al., 2004b).

A comparison between γ^{V} developed here and the standard definition given by Agrawal (2007):

$$\gamma^{\mathrm{A}} = (2\pi n_2/\lambda)\frac{\int |F|^4\, dA}{\left(\int |F|^2\, dA\right)^2}\tag{11.49}$$

where $F(x,y) = e_{\mathrm{t}}(x,y)$ is the scalar transverse electric field, indicates that γ^{V} accounts for both the inhomogeneous waveguide cross-section and full vectorial nature of the propagating modes of the waveguide, especially the z-component of the modes. Fig. 11.5 shows the effective nonlinearity for the two definitions of γ as a function of core diameter for three step index rods with different host materials

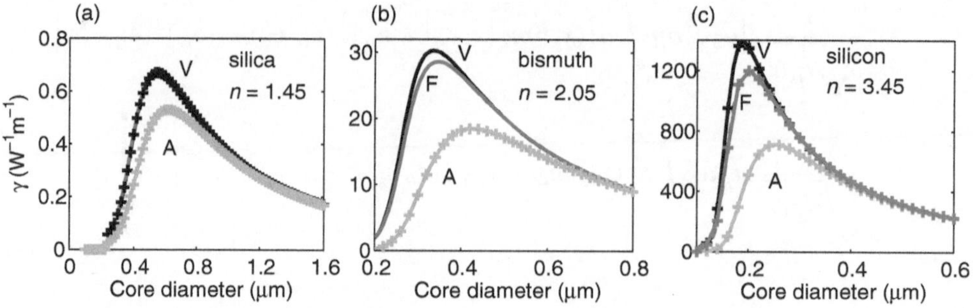

Fig. 11.5. Three definitions of γ, label V is γ^{V} based on VNSE, label F is γ^{F} reported by Foster et al., and label A is γ^{A} reported by Agrawal, as a function of core diameter and for three different materials: silica ($n=1.45, n_2=2.6\times10^{-20}$ m^2/W), bismuth ($n=2.05, n_2=3.2\times10^{-19}$ m^2/W), and silicon ($n=3.45, n_2=4.5\times10^{-18}$ m^2/W). The wavelength is $\lambda=800$ nm, and the cladding is air with $n=1.0$ in (a) and (b) and is silica, $n=1.45$ in (c). Plus signs and solid lines show the actual calculated data and the lines of best fit, respectively.

of silica (Agrawal, 2007) ($n=1.45$, $n_2=2.6\times10^{-20}$ m^2/W), bismuth (Ebendorff-Heidepriem et al., 2004a) ($n=2.05$, $n_2=3.2\times10^{-19}$ m^2/W), and silicon (Dinu et al., 2003) ($n=3.45$, $n_2=4.5\times10^{-18}$ m^2/W). While the γ^{V} based on VNSE approaches γ^{A} in the limit of large core diameter, it is significantly higher for small core diameters. For example, in Fig. 11.5(c) γ^{V} is a factor of 2 higher than the γ^{A} for silicon at the core diameter of $D=0.19$ μm. Fig. 11.5 also indicates that the difference between the γ values increases as the index contrast of the core and cladding increases. Here, we have also considered another definition of γ, given by Foster et al. (2004), as

$$\gamma^{\mathrm{F}}=(2\pi/\lambda)\frac{\int n_2[(e_\nu\times h_\nu^*).\hat{z}]^2dA}{[\int(e_\nu\times h_\nu^*).\hat{z}\,dA]^2}. \tag{11.50}$$

Foster et al. (2004) refer this equation to Agrawal (2007). It seems that Foster et al. have just simply replaced $|F(x,y)|^2=|e_{\mathrm{t}}(x,y)|^2$ in Eq. (11.49) with $S_z=(e_\nu\times h_\nu^*).\hat{z}$, arguing that S_z is the intensity of light propagating down the waveguide. However, this replacement is only valid for the transverse mode approximation where $|F(x,y)|^2=|e_{\mathrm{t}}(x,y)|^2$ is proportional to $(e_\nu\times h_\nu^*).\hat{z}$ and interpreted as the intensity of the light, see Eq. (11.38). In general, for full vectorial formalism, the nonlinear-induced polarisation is expanded in terms of different powers of electric field strength, see theory section, which for the $\chi^{(3)}$ nonlinearity has the form given in Eq. (11.33), which in turn results in the new definition of γ^{V} as in Eq. (11.42).

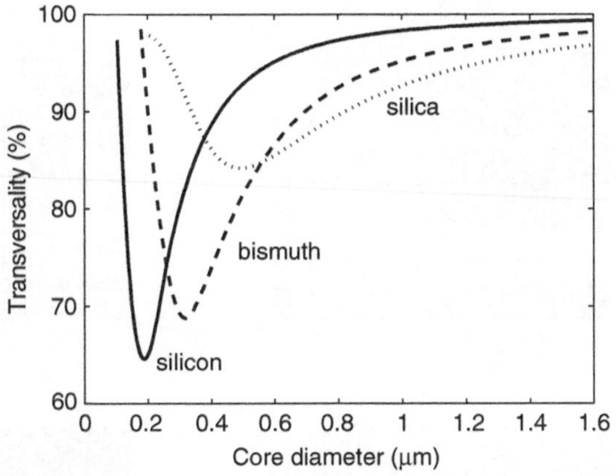

Fig. 11.6. Transversality versus core diameter for two glasses: silica ($n = 1.45$) and bismuth ($n = 2.05$), and silicon ($n = 3.45$). The structure is a simple rod in the air for the glasses and a rod within a substrate of silica for silicon.

In Fig. 11.5, the values of γ^{F} as a function of core diameter for different glass materials are plotted. While for silica, with the lowest index, the γ^{V} and γ^{F} curves are on the top of each other, for silicon with the highest index, there is a maximum difference of 40% between the two curves at $D = 0.15\ \mu$m. Also there is a difference of about 14% between the maximum values of γ^{V} and γ^{F}.

The differences between γ^{V} developed here and γ^{A} and γ^{F} are attributed to the fact that propagating modes of a waveguide are not transverse in a strong guidance regime. In order to demonstrate this, we have defined the transversality (Steel, 2004) of a mode as $T_\nu = 1 - \int \left|e_{\nu z}^2\right| dA / \int \left|\mathbf{e}_\nu^2\right| dA$ and plotted it as a function of core diameter for different materials, as shown in Fig. 11.6. It indicates that in the regime of large core diameter, the transversality approaches 100%, i.e., the modes become essentially transverse as expected. However, in the limit of small cores, the transversality is reduced, indicating that a large fraction of the electric field power is contained within the z-component of the field. This effect is more profound for the waveguides made from high index glass than the low index ones.

Fig. 11.7 shows a 2D plot of the z-component, e_z, and transverse electric field $\sqrt{e_r^2 + e_\theta^2}$, normalised to the power as in Eq. (11.9) for two different core diameters $D = 0.4\ \mu$m and $D = 1.8\ \mu$m at the wavelength of $\lambda = 1550$ nm. It is evident from Fig. 11.7(a and c) that for large and small cores, the e_z field is strongly localised at the edge of the fibre. For the core diameter $D = 0.4\ \mu$m, however, the value of e_z is one order of magnitude larger than that of the core diameter $D = 1.8\ \mu$m. Contrary to this, the distribution of the transverse field changes widely from a completely confined beam at large diameter, Fig. 11.7(d), to a beam with high intensity regions at the fibre interface in Fig. 11.7(b).

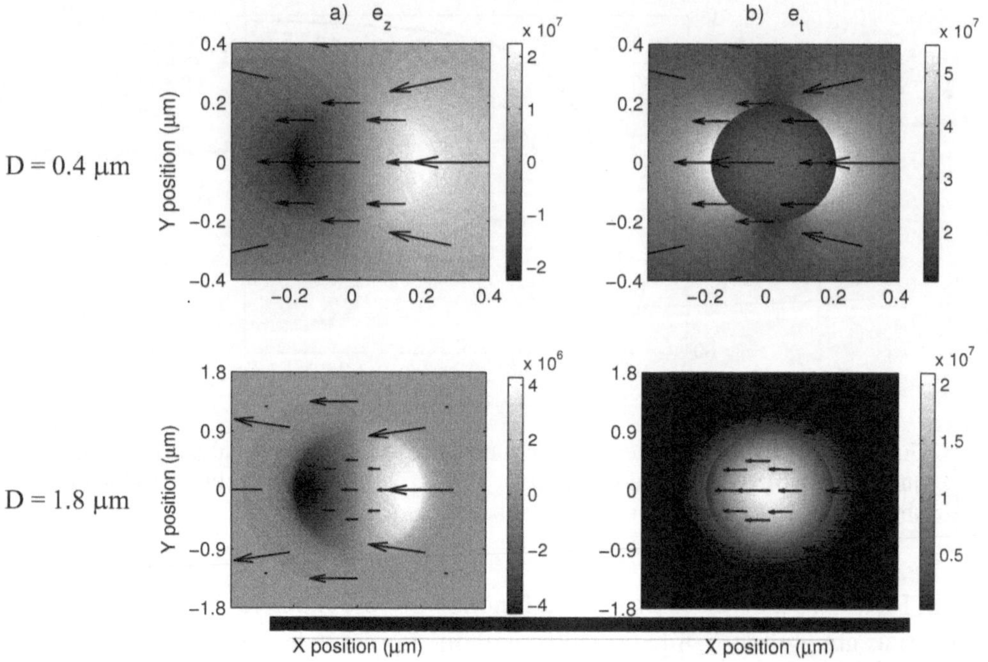

Fig. 11.7. 2D plot of e_z (a,c) and $\sqrt{e_r^2 + e_\theta^2}$ (b,d) for two step index rods with core diameters 0.4 μm (a,b) and 1.8 μm (c,d) at the wavelength 1550 nm. The material is bismuth with refractive index $n = 2.05$.

The first term in Eq. (11.43), $\gamma_{\mu\nu}^{(1)}$, is proportional to

$$\int n^2(x,y) n_2(x,y) \left(\left| \mathbf{e}_\nu . \mathbf{e}_\mu^* \right|^2 + \left| \mathbf{e}_\nu . \mathbf{e}_\mu \right|^2 \right) dA,$$

which contributes to the overall nonlinearity of mode ν, and is due to the overlap of the two different modes (two different polarisations in the case of a single mode waveguide). This term does not appear in formalisms (Agrawal, 2007) where fully-transverse propagating modes and a homogenous cross-section are assumed since either $\int \mathbf{e}_\nu . \mathbf{e}_\mu^* dA = 0$, due to the orthogonality of transverse modes or $\mathbf{e}_\nu . \mathbf{e}_\mu = 0$ if the modes are the two polarisations of a single fundamental mode.

In the formalism developed here, however, this term appears because of the nonzero z-component of the fields which results in $\int \left| \mathbf{e}_\nu . \mathbf{e}_\mu^* \right|^2 dA \neq 0$ for any two propagating modes of a waveguide or $\mathbf{e}_\nu . \mathbf{e}_\mu \neq 0$ even when the two modes are the polarisations of the fundamental mode of a single mode waveguide. Fig. 11.8 shows the magnitude of $\gamma_{\mu\nu}^{(1)}$ relative to γ_ν, i.e., $\gamma_{\mu\nu}^{(1)}/\gamma_\nu$ where μ and ν are the two polarisations of a simple step index rod, as a function of core diameter. It demonstrates that indeed in the limit of large core, the relative value of $\gamma_{\mu\nu}^{(1)}$ approaches

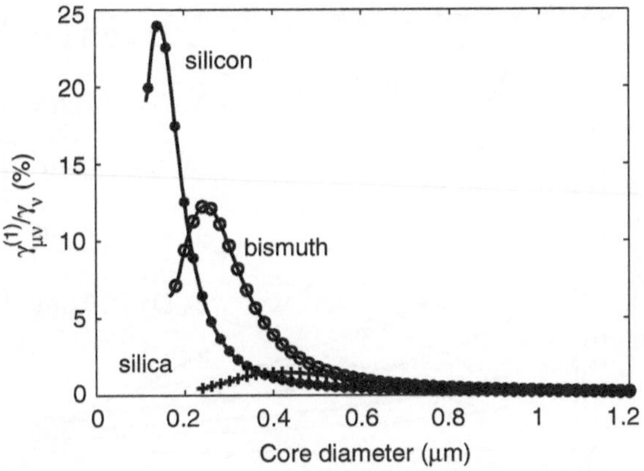

Fig. 11.8. Ratio of nonlinear coefficients $\gamma^{(1)}_{\mu\nu}$ and γ_ν, see Eqs. (11.43) and (11.42), as a function of core diameter for step index rods with host materials silica ($n = 1.45$), bismuth ($n = 2.05$), and silicon ($n = 3.45$). The cladding material for glass is air and for silicon is silica. Signs and the solid lines are the calculated data and lines of best fit, respectively.

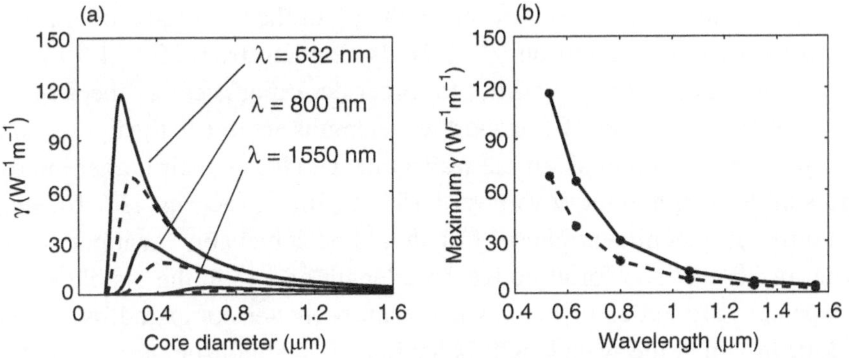

Fig. 11.9. (a) γ as a function of core diameter of a step index rod for different wavelengths. The solid lines are γ^{V} and the dashed lines are γ^{A}. In (b) the maximum of γ in (a) have been plotted as a function of wavelength. The host material is bismuth ($n = 2.05$).

zero but for small core diameters its value is enhanced significantly. Comparing the ratio $\gamma^{(1)}_{\mu\nu}/\gamma_\nu$ for three different materials: silica ($n = 1.45$), bismuth ($n = 2.05$) and silicon ($n = 3.45$) demonstrates that the value of $\gamma^{(1)}_{\mu\nu}$ is more significant in the subwavelength regime and for large index contrast host materials.

The behaviour of the nonlinear coefficient γ^{V} as a function of wavelength also shows a significant difference between the usual definitions γ^{A} and the one based on our VNSE. Fig. 11.9(a) shows the behaviour of γ as a function of core diameter for different wavelengths: $\lambda = 532$, 800, and 1550 nm. As also evident from

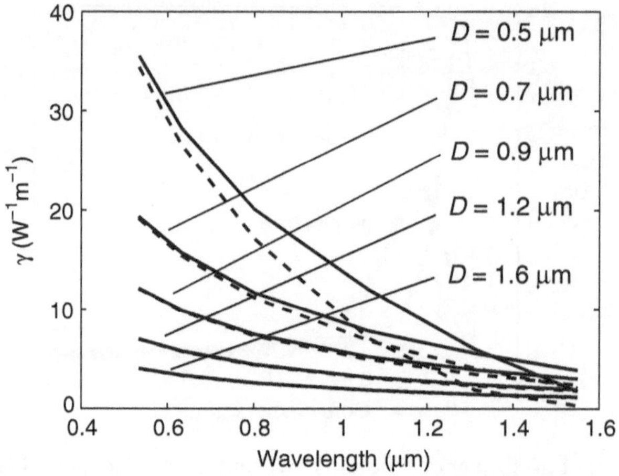

Fig. 11.10. γ^V (solid lines) and γ^A (dashed lines) vs wavelength for different core diameters for a step index rod with host material of bismuth glass ($n = 2.05$).

Fig. 11.9(b), the γ values decrease as the wavelength increases, but the decrease in the maximum value of γ is much faster for the γ^V of the vectorial model compared to that of the common definition γ^A. It is also evident from Fig. 11.9(a) that the position of the maximum of γ shifts to larger core diameters as the wavelength increases. Very high values of γ at short wavelengths are due to the tighter confinement of the propagating mode of the waveguide and hence their higher intensities. The possibility of achieving γ values of 150 $W^{-1}m^{-1}$, see Fig. 11.9(b), suggests that an order of π nonlinear phase shift should be achievable for input powers of the order of 20 mW and for effective fibre length of 1 m in the visible spectrum. The dispersion properties of the γ values can be better compared by examining Fig. 11.10 in which the wavelength behaviour of γ at constant core diameters is shown. For large core diameter, e.g. $D = 1.6\ \mu$m, the γ value at $\lambda = 1550$ nm is higher than the value at $\lambda = 532$ nm by a factor of 3.5 and the two definitions of γ are very close. However, for small core diameter, $D = 0.5\mu$m, the difference between the γ values at $\lambda = 1550$ nm and $\lambda = 532$ nm is the order of 20 times and a large difference between the two definitions of γ is observed. This indicates higher dispersion of γ at small core diameters than that of large core diameters.

11.4.3 Experimental confirmation

Eqs. (11.45) and (11.49), as indicated by Fig. 11.5, predict (Afshar and Monro, 2009) a significant difference between the standard γ^S and the new vectorial-based γ^V even for simple step index rod waveguides with high index glasses and in the regime of subwavelength dimensions. For bismuth and silicon step index rod waveguides (nanowires), the difference can be more than a factor of 2 (Afshar and Monro, 2009). The important feature of the comparison between γ^{S} and γ^{V} of a step index rod waveguide is that, as shown by Fig. 11.5, unsurprisingly the two values approach each other for large core diameters, i.e., $D \approx> \lambda$ but differ significantly for $D \ll \lambda$, around the peak of γ vs D curve. Continued research on nonlinear pulse propagation and the design and development of emerging waveguide devices demand the use of an accurate definition of γ. Hence, it is a matter of great importance to experimentally confirm the validity of this vectorially-based definition of γ.

In this section, we present the experimental confirmation of the validity of the new vectorially-based definition of γ, i.e., γ^{V}, in the regime of waveguides with subwavelength structures and high index materials. We first compare previously reported experimental values for the highest γ in different waveguides with the calculated values of γ i.e., γ^{S} and γ^{V}. It is shown that for the parameter regime of these waveguides, the core dimensions are not small enough to lead to experimentally distinguishable values of γ^{S} and γ^{V}. We then calculate both γ^{S} and γ^{V} for our recent suspended core fibre, which is a breakthrough in fabricating a fibre with the smallest core diameter reported to date, $D = 530$ nm, and demonstrate a good agreement between the experimental value of γ and γ^{V}.

Fig. 11.11 shows the record experimental values of γ reported in the literature for some bismuth (Ebendorff-Heidepriem et al., 2004b) and SF57 (Ebendorff-Heidepriem et al., 2005; Leong et al., 2006) suspended core microstructured optical fibres (MOF) and chalcogenide nanowires (Magi et al., 2007; Yeom et al., 2008), respectively, together with the numerical modelling results of γ^{S} and γ^{V}. For suspended-core MOFs, we have accessed the original SEM image of one of the fibres (insets in Fig. 11.11(a and b) corresponding to a core diameter of $D = 1.4\ \mu$m and $D = 1.0\ \mu$m, respectively) and scaled it to reduce or increase the core diameter. For each diameter, we have used a finite element package, COMSOL 3.4, to calculate the mode fields distributions from which we have calculated γ^{S} and γ^{V} values.

Measured γ values for different bismuth suspended core fibres ($n = 2.02$ and $n_2 = 3.2 \times 10^{-19}$ m^2/W) with different core diameters are shown in Fig. 11.11(a) (Ebendorff-Heidepriem et al., 2004b). Errors in γ values are due to uncertainty in measuring the input power and fibre effective length and errors in core diameters are related to the non-uniformity of the core diameters of the MOFs along their length. No errors are reported for the measured values in Fig. 11.11(c). Comparing the measured γ values with the theoretical calculations according to γ^{S} and γ^{V} definitions indicates that for the corresponding core diameters there are no distinguishable differences between the two theoretical models. This is also clearly true for suspended core SF57 ($n = 1.81$ and $n_2 = 4.1 \times 10^{-19}$ m^2/W) fibres with different core diameters (Ebendorff-Heidepriem et al., 2005; Leong et al., 2006)

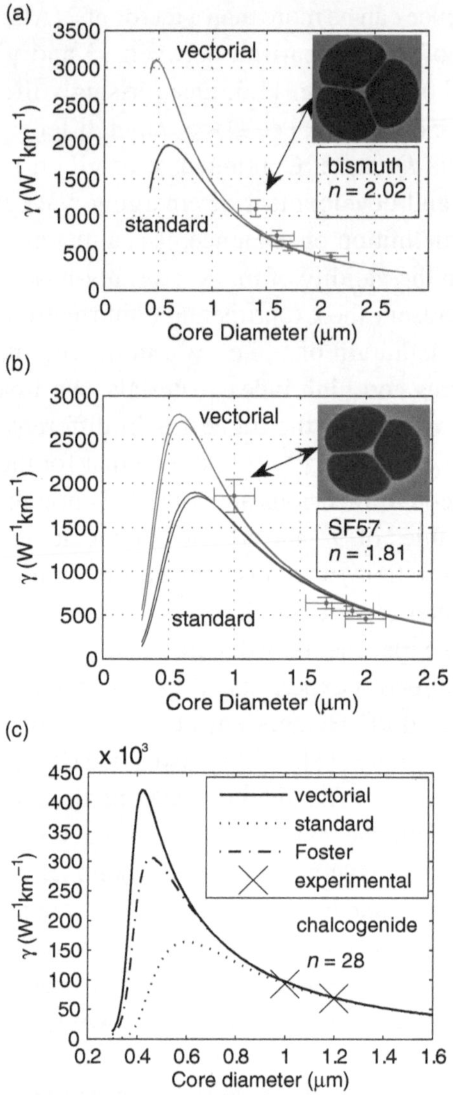

Fig. 11.11. (a) Experimental (data points) (Ebendorff-Heidepriem et al., 2004b, 2005; Leong et al., 2006) and numerical results of γ vs core diameter for bismuth (a) and SF57 (b) suspended core and chalcogenide (c) (Magi et al., 2007; Yeom et al., 2008) tapered (nanowire) fibres. For numerical calculations, different diameters have been obtained by scaling the whole structure.

in Fig. 11.11(b). The measured γ value at the core diameter of 1 μm, although showing some indication of matching with the new model, does not support a clear conclusion considering the experimental errors of measuring γ. In Fig. 11.11(c) we have shown recent experimental values of γ for two chalcogenide tapered fibres (nanowire) (Magi et al., 2007; Yeom et al., 2008) together with the theoretical curves of γ vs core diameter based on three definitions, our new vectorial, γ^{V}, standard, γ^{S}, and a definition reported originally by Foster et al. (2008), γ^{F}, and used in Magi et al. (2007). Like Fig. 11.11(a and b), the experimental values correspond to core diameters that do not support any distinguishable differences between γ^{V} and γ^{S}.

Zhang et al. have recently fabricated the world's smallest core bismuth suspended core fibre with a core diameter of only $D \approx 530$ nm (Zhang et al., 2008). They have measured the effective nonlinear coefficient γ of this fibre (Zhang et al., 2008) using the Boskovic (Boskovic et al., 1996) method, in which two slightly separated CW laser beams ($\lambda_1 = 1551.6$ and $\lambda_2 = 1552.4$ nm) are coupled into the fibre and the ratio of the output side peaks to the main peaks are used to evaluate the nonlinear phase shift and hence γ. The γ value is the slope of the linear fit to the experimental results of the phase shift vs input power and is $\gamma = 5400 \pm 200\ \mathrm{W^{-1}km^{-1}}$ (see Fig. 11.13(a)) (Zhang et al., 2008). The overall experimental error $\pm 200\ \mathrm{W^{-1}km^{-1}}$ is due to error in measuring the fibre loss and input power, which is shown by error bars in Fig. 11.13(a). Fig. 11.12(a) shows the SEM image of this fibre and the z-component of the Poynting vector for the two polarisations of the fundamental mode of this fibre. The fibre is birefringent but single mode at $\lambda = 1550$ nm.

The bismuth glass used in the fibre (code AB061) is different from the one reported in Ebendorff-Heidepriem et al. (2004b) (code A1310) and is provided by Asahi Glass Co., Ltd. It has a lower refractive index of $n = 1.9771$ at 1550 nm. To compare the experimental value of γ with the calculated γ^{S} and γ^{V}, the n_2 of a $17 \times 10 \times 2$ mm polished sample of bismuth glass was measured using a z-scan experiment (Prasad et al., 2008). An n_2 of $6.0 \times 10^{-19}\ \mathrm{m^2/W} \pm 10\%$ at 1550 nm was measured using the n_2 of a similar sample of SF57 glass as a reference. The $\pm 10\%$ accounts for the experimental uncertainty associated with the z-scan experiment (Prasad et al., 2008). We have calculated the field distributions of the real fibre (using the SEM image in Fig. 11.12(a) and the finite element

(a) (b) (c)

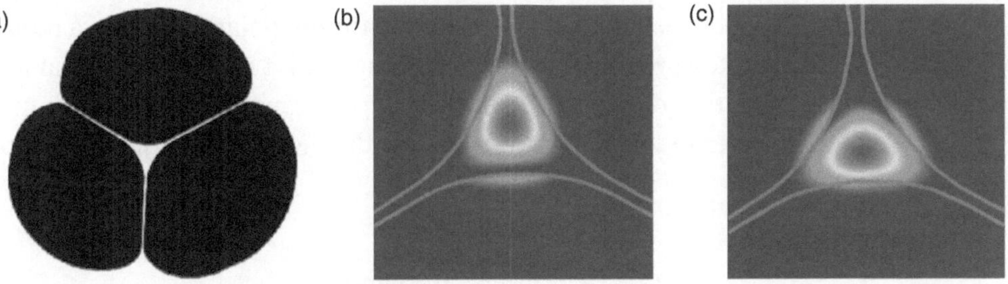

Fig. 11.12. SEM image (a) and calculated z-component of Poynting vector of the two polarisations of the fundamental mode (b) and (c) of a bismuth fibre with the core diameter of 530 nm.

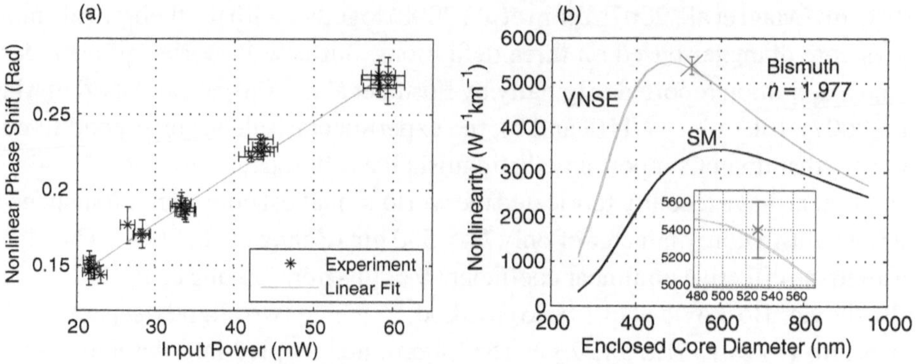

Fig. 11.13. Experimental (a) measurement of γ for the fibre shown in Fig. 11.12(a). The slope of the linear fit is $\gamma = 5400 \pm 200\,\mathrm{W}^{-1}\mathrm{km}^{-1}$. Numerical calculation (b) of γ vs core diameter for the same fibre as in Fig. 11.12(a). Curves labelled VNSE and SM are calculated according to the new vectorially-based and standard definitions of γ respectively.

method) from which the γ^{S}, γ^{V}, $A^{\mathrm{S}}_{\mathrm{eff}}$ and $A^{\mathrm{V}}_{\mathrm{eff}}$ values are $\gamma^{\mathrm{S}} = (3400, 3400) \pm 300$ $\mathrm{W}^{-1}\mathrm{km}^{-1}$, $\gamma^{\mathrm{V}} = (5200, 5300) \pm 500$ $\mathrm{W}^{-1}\mathrm{km}^{-1}$, $A^{\mathrm{S}}_{\mathrm{eff}} = (0.640, 6393)$ $\mu\mathrm{m}^2$, and $A^{\mathrm{V}}_{\mathrm{eff}} = (0.496, 0.498)$ $\mu\mathrm{m}^2$, respectively. Comparing γ^{S} and γ^{V} with the experimental value of $\gamma = 5400 \pm 200$ $\mathrm{W}^{-1}\mathrm{km}^{-1}$, a clear agreement between the experimental value and γ^{V} is demonstrated within the experimental errors (Afshar *et al.*, 2009). The uncertainties in the γ^{S} and γ^{V} are due to the 10% uncertainties in the experimental value of n_2.

The SEM image of the fibre has been scaled in order to achieve different core structures. Fig. 11.13(b) then shows the plot of γ^{S} and γ^{V} as a function of core diameter, achieved by scaling the fibre. The diameter of the real structure is $D = 530$ nm, corresponding to a scale of 1, and the experimental value of $\gamma = 5400 \pm 200$ $\mathrm{W}^{-1}\mathrm{km}^{-1}$. Fig. 11.13(b) shows that the γ value is very close to the highest value of γ that can be achieved in this structure.

The experimental investigation carried out here only relates to the SPM term i.e., Eq. (11.45). However, as Eq. (11.42) indicates there are other XPM terms, originating from coupling between different modes or different polarisations of a mode, which also contribute towards the overall nonlinearity of a mode. Numerical and experimental investigation of the effects of these terms, especially for birefringent waveguides, will further reveal the new aspects of the vectorially-based definition of γ^{V}.

11.5 Raman nonlinearity

Stimulated Raman scattering (SRS) is a third order nonlinear optical process where the delayed nonlinear response of the medium leads to optical amplification of a long wavelength pulse at the expense of a shorter wavelength pulse. SRS within optical waveguides is an area of interest for its potential for creating lasers at new wavelengths, broadband Raman amplifiers, supercontinuum generation as well as other applications.

In this section the VNSE will be further extended to consider the effects of SRS by considering two pulses operating at different wavelengths and their interactions through a time dependent Raman response.

The standard model (SM) (Agrawal, 2007; Headley and Agrawal, 1996) used in literature for pulse propagation in the presence of SRS (ignoring waveguide losses, and assuming pulses much longer than the Raman response), is given by the two coupled differential equations (Agrawal, 2007):

$$\frac{\partial A_{\mathrm{p}}(z,t)}{\partial z} = -\frac{g(\mathrm{m}^{-1}\mathrm{W}^{-1})}{2} |A_{\mathrm{s}}(z,t)|^2 A_{\mathrm{p}}(z,t), \tag{11.51}$$

$$\frac{\partial A_{\mathrm{s}}(z,t)}{\partial z} = +\frac{g(\mathrm{m}^{-1}\mathrm{W}^{-1})}{2} \left|A_{\mathrm{p}}(z,t)\right|^2 A_{\mathrm{s}}(z,t). \tag{11.52}$$

Where $A_{\mathrm{p}}(z,t)$ and $A_{\mathrm{s}}(z,t)$ are the propagating pulse envelopes of the pump and Stokes pulses respectively and $g(\mathrm{m}^{-1}\mathrm{W}^{-1})$ is the modal Raman gain defined as

$$g(\mathrm{m}^{-1}\mathrm{W}^{-1}) = \frac{g_{\mathrm{R}}(\Delta\omega)}{A_{\mathrm{eff}}}, \tag{11.53}$$

$$A_{\mathrm{eff}} = \frac{\left(\int |F(x,y)|^2\, dxdy\right)^2}{\int |F(x,y)|^4\, dxdy}. \tag{11.54}$$

A_{eff} is the effective area of the modal field distribution $F(x,y)$ and $g_{\mathrm{R}}(\Delta\omega)$ is the bulk Raman gain coefficient of the host material of the optical waveguide measured in (m/W). This model fails to consider certain features of optical waveguides by making the following assumptions:

(i) The spatial dependence of the bulk Raman gain coefficient is negligible, i.e. $g_{\mathrm{R}}(\Delta\omega,x,y) = g_{\mathrm{R}}(\Delta\omega)$.
(ii) The differences in the modal field distributions of the pump and Stokes modes is negligible, i.e. $F_{\mathrm{p}}(x,y) = F_{\mathrm{s}}(x,y)$.
(iii) The vectorial form of the modal field distributions can be reduced to the scalar form assuming polarisation in one direction, i.e. $\mathbf{E}(x,y) = E_x(x,y)\hat{\mathbf{x}}$.

Whilst these assumptions made by the SM are valid for optical fibres in the weak guidance regime, these assumptions break down when one considers emerging waveguides with high refractive index contrast, subwavelength features and complex transverse structure.

In the next section, a VNSE for SRS is developed, redefining the modal Raman gain and effective area for SRS. In Section 11.5.2 the VNSE is used to calculate the modal Raman gain of an optical nanowire to compare with the SM to observe in which regimes the SM fails to accurately calculate the modal Raman gain, and what physical processes cause these differences. Lastly, Section 11.5.2.1 investigates the possibility of changing the shape of the modal Raman gain spectrum, through the strong wavelength dependence of the modal field distributions in emerging waveguides.

11.5.1 Theory

To develop a VNSE for SRS, a propagation equation for the pump and Stokes pulses will be derived using perturbation theory, as done for Kerr nonlinearity. To begin, Maxwell's equations will be defined with and without the effects of dispersion and SRS for both pump and Stokes fields. Then the unperturbed fields will be defined in terms of a particular eigenmode of the waveguide without dispersion and nonlinearity. The perturbed fields are then created from a linear superposition of the basis set of unperturbed eigenmodes. The evolution of the amplitudes of these modes will describe the pulse propagation of each mode.

First, let us construct the unperturbed and perturbed systems through Maxwell's equations in the absence of sources:

$$\nabla \times \mathbf{E} = -\mu_0 \frac{\partial}{\partial t}\mathbf{H},$$

$$\nabla \times \mathbf{H} = \varepsilon_0 \frac{\partial}{\partial t}\mathbf{E} + \frac{\partial}{\partial t}\mathbf{P},$$

where $\mathbf{P}$ is the polarisation vector field of the medium containing both linear and nonlinear components $\mathbf{P} = \mathbf{P}^{\mathrm{L}} + \mathbf{P}^{\mathrm{NL}}$. Let us assume that the set of full vectorial eigenmodes for the unperturbed system has been calculated, based entirely on the linear refractive index geometry of the waveguide. Here let us define the unperturbed fields as a single eigenmode of the optical waveguide, i.e. the μth eigenmode. We shall develop a full vectorial nonlinear pulse propagation equation that describes the evolution of this particular mode as it interacts with the other modes through nonlinearity. The unperturbed fields $\mathbf{E}_{\mathrm{op}}(\mathbf{r},t)$ and $\mathbf{E}_{\mathrm{os}}(\mathbf{r},t)$ have been calculated independently at two different narrowband frequencies ω_{p} and ω_{s} where p, s denote pump and Stokes respectively and these can be obtained numerically using techniques such as the finite element method (Okamoto, 2006), the multipole method (Kuhlmey et al., 2002b; White et al., 2002) etc. The unperturbed fields thus have the form (Snyder and Love, 1995):

$$\mathbf{E}_{\mathrm{os}} = \frac{1}{2}\mathbf{E}_{\mathrm{s}\mu} e^{-i\omega_{\mathrm{s}}t} + \mathrm{c.c.}, \tag{11.55}$$

$$\mathbf{H}_{\mathrm{os}} = \frac{1}{2}\mathbf{H}_{\mathrm{s}\mu} e^{-i\omega_{\mathrm{s}}t} + \mathrm{c.c.}, \tag{11.56}$$

$$\mathbf{E}_{\mathrm{op}} = \frac{1}{2}\mathbf{E}_{\mathrm{p}\mu} e^{-i\omega_{\mathrm{p}}t} + \mathrm{c.c.}, \tag{11.57}$$

$$\mathbf{H}_{\mathrm{op}} = \frac{1}{2}\mathbf{H}_{\mathrm{p}\mu} e^{-i\omega_{\mathrm{p}}t} + \mathrm{c.c.}, \tag{11.58}$$

$$\mathbf{E}_{\mathrm{p}\mu} = \frac{\mathbf{e}_{\mathrm{p}\mu}(x,y)e^{i\beta_{\mathrm{p}\mu}z}}{\sqrt{N_{\mathrm{p}\mu}}}, \qquad \mathbf{H}_{\mathrm{p}\mu} = \frac{\mathbf{h}_{\mathrm{p}\mu}(x,y)e^{i\beta_{\mathrm{p}\mu}z}}{\sqrt{N_{\mathrm{p}\mu}}}, \tag{11.59}$$

$$\mathbf{E}_{\mathrm{s}\mu} = \frac{\mathbf{e}_{\mathrm{s}\mu}(x,y)e^{i\beta_{\mathrm{s}\mu}z}}{\sqrt{N_{\mathrm{s}\mu}}}, \qquad \mathbf{H}_{\mathrm{s}\mu} = \frac{\mathbf{h}_{\mathrm{s}\mu}(x,y)e^{i\beta_{\mathrm{s}\mu}z}}{\sqrt{N_{\mathrm{s}\mu}}}. \tag{11.60}$$

Here $\beta_{\mathrm{p}\mu}$ and $\beta_{\mathrm{s}\mu}$ are the propagation constants for the μth mode of the unperturbed eigenmodes of the pump and Stokes fields and c.c. is the complex conjugate which will be left out from here on for simplicity. The eigenmodes of the optical waveguide form an orthonormal basis set:

$$\frac{1}{2}\int(\mathbf{e}^*_{\kappa\mu} \times \mathbf{h}_{\kappa\eta}) \cdot \hat{z}\, dA = N_{\kappa\mu\eta}\delta_{\mu\eta} = N_{\kappa\mu}, \qquad N_{\kappa\mu} = \frac{1}{2}\int(\mathbf{e}^*_{\kappa\mu} \times \mathbf{h}_{\kappa\mu}) \cdot \hat{z}\, dA, \tag{11.61}$$

where $\kappa = \mathrm{p, s}$. Let us now define the unperturbed and perturbed systems through Maxwell's equations. We assume here that since each of Maxwell's equations will contain electric, magnetic and polarisation fields oscillating at either ω_{p} or ω_{s}, Maxwell's equations can be separated into two sets, one each for the pump and Stokes fields. This is valid as long as the two quasi-monochromatic pump and Stokes pulses do not spectrally overlap. Here the derivation for the Stokes pulse propagation equation is presented. The exact approach can be used to find the pump pulse propagation equation. Thus the unperturbed set of Maxwell's equations assuming no dispersion or nonlinearity for the Stokes pulse is

$$\nabla \times \mathbf{E}_{\mathrm{os}} = -\mu_0 \frac{\partial}{\partial t}\mathbf{H}_{\mathrm{os}}, \tag{11.62}$$

$$\nabla \times \mathbf{H}_{\mathrm{os}} = \varepsilon_0 \frac{\partial}{\partial t}\mathbf{E}_{\mathrm{os}} + \frac{\partial}{\partial t}\mathbf{P}^{\mathrm{L}}_{\mathrm{s}}. \tag{11.63}$$

Similarly the perturbed system is defined through Maxwell's equations which now include both nonlinearity and dispersion, by now considering two quasi-monochromatic electric and magnetic fields $\mathbf{E}_{\mathrm{p,s}}$ and $\mathbf{H}_{\mathrm{p,s}}$ which will be explicitly defined later on as the perturbed fields:

$$\nabla \times \mathbf{E}_\mathrm{s} = -\mu_0 \frac{\partial}{\partial t}\mathbf{H}_\mathrm{s}, \tag{11.64}$$

$$\nabla \times \mathbf{H}_\mathrm{s} = \varepsilon_0 \frac{\partial}{\partial t}\mathbf{E}_\mathrm{s} + \frac{\partial}{\partial t}\left(\mathbf{P}_\mathrm{s}^\mathrm{L} + \mathbf{P}_\mathrm{s}^\mathrm{NL}\right). \tag{11.65}$$

Let us now continue our analysis in the Fourier domain by using a tilde above the fields to represent analysis done in the Fourier domain.

$$F(\mathbf{r},t) = \frac{1}{2\pi}\int_{-\infty}^{\infty} \tilde{F}(\mathbf{r},\omega)e^{-i\omega t}d\omega. \tag{11.66}$$

The unperturbed fields in the Fourier domain then become

$$\tilde{\mathbf{E}}_\mathrm{os} = \frac{1}{2}\mathbf{E}_{\mathrm{s}\mu}\delta(\omega - \omega_\mathrm{s}), \tag{11.67}$$

$$\tilde{\mathbf{H}}_\mathrm{os} = \frac{1}{2}\mathbf{H}_{\mathrm{s}\mu}\delta(\omega - \omega_\mathrm{s}). \tag{11.68}$$

In the Fourier domain Maxwell's equations for the Stokes fields become

$$\nabla \times \tilde{\mathbf{E}}_\mathrm{os} = \frac{i\mu_0}{2}\omega_\mathrm{s}\mathbf{H}_{\mathrm{s}\mu}\delta(\omega - \omega_\mathrm{s}), \tag{11.69}$$

$$\nabla \times \tilde{\mathbf{H}}_\mathrm{os} = -i\omega_\mathrm{s}\left[\frac{\varepsilon_0}{2}\mathbf{E}_{\mathrm{s}\mu} + \tilde{\mathbf{P}}_\mathrm{s}^\mathrm{L}(\mathbf{r},\omega)\right]\delta(\omega - \omega_\mathrm{s}), \tag{11.70}$$

where the unperturbed linear polarisation is given by

$$\mathbf{P}_\mathrm{s}^\mathrm{L}(\mathbf{r},t) = \varepsilon_0 \mathbf{R}^{(1)}(\mathbf{r},\tau) \otimes \mathbf{E}_\mathrm{os},$$

$$\tilde{\mathbf{P}}_\mathrm{s}^\mathrm{L}(\mathbf{r},\omega) = \varepsilon_0 \chi^{(1)}(-\omega_\mathrm{s};\omega_\mathrm{s})\tilde{\mathbf{E}}_\mathrm{os}, \tag{11.71}$$

$$n^2(\mathbf{r},\omega_\mathrm{p,s}) = 1 + \chi^{(1)}(-\omega_\mathrm{s};\omega_\mathrm{s}), \tag{11.72}$$

where the delta function in Eq. (11.67) forces the susceptibility to be evaluated at ω_s. Using Eqs. (11.71) and (11.72) in Eq. (11.70):

$$\nabla \times \tilde{\mathbf{H}}_\mathrm{os} = \frac{-i\varepsilon_0}{2}\omega_\mathrm{s} n^2(\mathbf{r},\omega_\mathrm{s})\mathbf{E}_{\mathrm{s}\mu}\delta(\omega - \omega_\mathrm{s}). \tag{11.73}$$

The perturbed fields $\mathbf{E}_\mathrm{p,s}$ and $\mathbf{H}_\mathrm{p,s}$ are the fields in the presence of both dispersion and nonlinear Raman effects. Unlike previous models that consider just the interactions between two modes of the optical waveguide (Agrawal, 2007; Chen et al., 2006; Dadap et al., 2008), here a complete model is to be developed, which requires the use of a basis set of eigenmodes. Hence the perturbed fields will be constructed through the linear combination of the unperturbed modes:

$$\mathbf{E}_{\mathrm{s}} = \frac{1}{2}\sum_{\eta} a_{\mathrm{s}\eta}(z,t)\mathbf{E}_{\mathrm{s}\eta}e^{-i\omega_{\mathrm{s}}t}, \tag{11.74}$$

$$\mathbf{H}_{\mathrm{s}} = \frac{1}{2}\sum_{\eta} a_{\mathrm{s}\eta}(z,t)\mathbf{H}_{\mathrm{s}\eta}e^{-i\omega_{\mathrm{s}}t}, \tag{11.75}$$

$$\mathbf{E}_{\mathrm{p}} = \frac{1}{2}\sum_{\eta} a_{\mathrm{p}\eta}(z,t)\mathbf{E}_{\mathrm{p}\eta}e^{-i\omega_{\mathrm{p}}t}, \tag{11.76}$$

$$\mathbf{H}_{\mathrm{p}} = \frac{1}{2}\sum_{\eta} a_{\mathrm{p}\eta}(z,t)\mathbf{H}_{\mathrm{p}\eta}e^{-i\omega_{\mathrm{p}}t}. \tag{11.77}$$

The coefficients $a_{\kappa\eta}(z,t)$ $(\kappa = \mathrm{p,s})$ are the modal amplitudes and $|a_{\kappa\eta}(z,t)|^2$ is the optical power for the ηth (pump/Stokes) mode. Physically, these modal amplitudes describe how the modes of the optical fibre evolve as they propagate along the waveguide with the effects of both dispersion and nonlinearity where $|a_{\kappa\eta}(0,t)|^2$ is the initial pulse of the ηth mode at the start of the waveguide. As the Stokes pulse propagates along the fibre, the Stokes modal amplitudes will experience Raman gain at the expense of the pump modal amplitudes.

The discrete summation of modes includes both the bound and leaky modes propagating in both the forward and backward directions. However, the space wave radiation modes which do not propagate along the fibre due to their rapid attenuation away from the source (Snyder and Love, 1995) have been left out. Therefore, whilst our basis set is not strictly complete, it spans the useful set of modes for numerical modelling of nonlinear effects. The extra completeness achieved by introducing these space wave radiation fields is left to further investigation.

Again, analysis shall be carried out in the Fourier domain where the perturbed Stokes fields become:

$$\tilde{\mathbf{E}}_{\mathrm{s}} = \frac{1}{2}\sum_{\eta} \tilde{a}_{\mathrm{s}\eta}(z,\omega-\omega_{\mathrm{s}})\mathbf{E}_{\mathrm{s}\eta}, \tag{11.78}$$

$$\tilde{\mathbf{H}}_{\mathrm{s}} = \frac{1}{2}\sum_{\eta} \tilde{a}_{\mathrm{s}\eta}(z,\omega-\omega_{\mathrm{s}})\mathbf{H}_{\mathrm{s}\eta}. \tag{11.79}$$

Maxwell's equations for the perturbed Stokes system in the Fourier domain are given by

$$\nabla \times \tilde{\mathbf{E}}_{\mathrm{s}} = \frac{i\mu_0\omega}{2}\sum_{\eta} \tilde{a}_{\mathrm{s}\eta}\mathbf{H}_{sm}, \tag{11.80}$$

$$\nabla \times \tilde{\mathbf{H}}_s = \frac{-i\varepsilon_0 \omega n^2(\mathbf{r},\omega)}{2} \left[\sum_\eta \tilde{a}_{s\eta} \mathbf{E}_{s\eta} - i\omega \mathbf{P}_s^{NL} \right]. \tag{11.81}$$

Since the perturbed fields now have finite bandwidth due to the time dependence of the modal amplitudes, the dispersion of the linear susceptibility is now considered in the perturbed system:

$$\mathbf{P}_s^L(\mathbf{r},t) = \varepsilon_0 \mathbf{R}^{(1)}(\mathbf{r},\tau) \otimes \mathbf{E}_s \tag{11.82}$$

$$= \varepsilon_0 \chi^{(1)}(-\omega;\omega) \tilde{\mathbf{E}}_s$$

$$n^2(\mathbf{r},\omega) = 1 + \chi^{(1)}(-\omega;\omega). \tag{11.83}$$

Let's apply a Taylor series expansion on $\omega n^2(\mathbf{r},\omega)$ about the Stokes frequency ω_s:

$$\omega n^2(\mathbf{r},\omega) = \omega_s n^2(\mathbf{r},\omega_s) + \Delta\omega_s \left[\frac{\partial}{\partial \omega} \left(\omega n^2(\mathbf{r},\omega) \right) \right]_{\omega=\omega_s} + O(\Delta\omega_s^2),$$

where $\Delta\omega_s = \omega - \omega_s$.

Having now explicitly defined Maxwell's equations and the fields for the unperturbed and perturbed systems, let us now derive a pulse propagation equation of a particular mode propagating under the effects of dispersion and nonlinearity.

The standard model is derived using the wave equation, which makes the assumption that $\nabla \cdot \mathbf{E} = 0$, by assuming that the permittivity of the optical waveguide has negligible spatial dependence. However, emerging waveguides that in general have permittivities with strong spatial dependence cannot use this assumption, thus this technique cannot be used. Instead let us invoke the reciprocal theorem (Snyder and Love, 1995), using the unperturbed and perturbed Stokes fields:

$$\frac{\partial}{\partial z} \int \mathbf{F}_{cs} \cdot \hat{z}\, dA = \int \nabla \cdot \mathbf{F}_{cs}\, dA, \tag{11.84}$$

where the vector field $\mathbf{F}_{cs}$ is given by

$$\mathbf{F}_{cs} = \tilde{\mathbf{E}}_{os}^* \times \tilde{\mathbf{H}}_s + \tilde{\mathbf{E}}_s \times \tilde{\mathbf{H}}_{os}^*. \tag{11.85}$$

Eq. (11.84) allows the unperturbed fields to be related to the perturbed fields and create a propagation equation for Stokes pulses under dispersion and nonlinear effects such as SRS. The right-hand side of Eq. (11.84) can be expanded using Eqs. (11.69), (11.73), (11.80), (11.81), Maxwell's equations for the unperturbed and perturbed system. Eq. (11.61), the orthonormality condition, is then applied to the left-hand side of Eq. (11.84). This results in Eq. (11.86): the general nonlinear pulse propagation equation.

$$\frac{\partial \tilde{a}_{s\mu}(z,\omega-\omega_s)}{\partial z} = i\Delta\omega_s \beta^1_{s\mu}\tilde{a}_{s\mu}(z,\omega-\omega_s) + i\Delta\omega_s \sum_{\eta\neq\mu} \beta^1_{s\eta}\tilde{a}_{s\eta}(z,\omega-\omega_s) + O(\Delta\omega_s^2)$$
$$+ \frac{i}{2}\int \mathbf{E}^*_{s\mu}\cdot \omega\tilde{\mathbf{P}}^{NL}_s(\mathbf{r},\omega)\, dA \tag{11.86}$$

$$\beta^1_{s\mu} = \frac{1}{4N_{s\mu}}\int \left[\mu_0 \left|\mathbf{h}_{s\mu}\right|^2 + \varepsilon_0 \left[\frac{\partial}{\partial\omega}\left(\omega n^2(\mathbf{r},\omega)\right)\right]|_{\omega=\omega_s} \left|\mathbf{e}_{s\mu}\right|^2\right] dA$$
$$\beta^1_{s\eta} = \frac{e^{i(\beta_{s\eta}-\beta_{s\mu})z}}{4\sqrt{N_{s\eta}N_{s\mu}}}\int \left[\mu_0 \mathbf{h}_{s\eta}\cdot\mathbf{h}^*_{s\mu} + \varepsilon_0 \left[\frac{\partial}{\partial\omega}\left(\omega n^2(\mathbf{r},\omega)\right)\right]|_{\omega=\omega_s}\mathbf{e}_{s\eta}\cdot\mathbf{e}^*_{s\mu}\right] dA$$

Eq. (11.86) is very general and is of the same form as in Chen et al. (2006). However, here a basis set of modes is used to define the perturbed fields, which is shown later to lead to a series of interactions in the pulse propagation model in both dispersion and nonlinearity. In the time domain this general nonlinear pulse propagation equation has the form:

$$\frac{\partial a_{s\mu}(z,t)}{\partial z} = \hat{D}a_{s\mu}(z,t) - e^{i\omega_s t}\frac{\partial}{\partial t}\frac{e^{-i\beta_{s\mu}z}}{2\sqrt{N_{s\mu}}}\int \mathbf{e}^*_{s\mu} \bullet \mathbf{P}^{\mathrm{NL}}_s(\mathbf{r},t)\, dA \tag{11.87}$$
$$\hat{D} = i\beta^1_{s\mu}\frac{\partial}{\partial t} + i\sum_{\eta\neq\mu}\beta^1_{s\eta}\frac{\partial}{\partial t} + O\left(\frac{\partial^2}{\partial t^2}\right).$$

To derive the nonlinear Raman polarisation the formalism in Butcher and Cotter (1990) will be used. The nonlinear Stokes polarisation can be split into its fast and slow time varying components:

$$\mathbf{P}^{\mathrm{NL}}_s(\mathbf{r},t) = \frac{1}{2}\mathbf{P}_{\omega_s}(\mathbf{r},t)e^{-i\omega_s t} + \text{c.c.}$$

where $\mathbf{P}_{\omega_s}(\mathbf{r},t)$ is given by the convolution of the electric fields with the third order nonlinear response function $\Phi^{(3)}_{-\omega_s;\omega_p,-\omega_p,\omega_s}$ (Butcher and Cotter, 1990):

$$\mathbf{P}^{(3)}_{\omega_s}(t) = \varepsilon_0 \int\int\int \Phi^{(3)}_{-\omega_s;\omega_p,-\omega_p,\omega_s} \vdots \mathbf{E}_{\omega_p}(\mathbf{r},t_1)\mathbf{E}_{-\omega_p}(\mathbf{r},t_2)\mathbf{E}_{\omega_s}(\mathbf{r},t_3)\, dt_1 dt_2 dt_3$$
$$= \frac{3\varepsilon_0}{2}\int\int\int \mathbf{R}^{(3)}(t-t_1,t-t_2,t-t_3) \vdots \mathbf{E}_{\omega_p}(\mathbf{r},t_1)\mathbf{E}^*_{\omega_p}(\mathbf{r},t_2)\mathbf{E}_{\omega_s}(\mathbf{r},t_3)$$
$$\times \exp\left(i\sum_{r=1}^{3}\omega_r(t-t_r)\right) dt_1 dt_2 dt_3, \tag{11.88}$$

where $\mathbf{E}_{-\omega'} = \mathbf{E}^*_{\omega'}$, $\omega_{1,2,3} = \omega_{\mathrm{p},-\mathrm{p},\mathrm{s}}$ and $\mathbf{R}^{(3)}$ is the rank-4 Raman response tensor that is defined as (Agrawal, 2007; Lin et al., 2007):

$$\mathbf{R}^{(3)}(t-t_1, t-t_2, t-t_3) = \mathbf{R}(x,y,t-t_2)\delta(t-t_1)\delta(t_2-t_3). \tag{11.89}$$

It is important to note that the spatial dependence of the Raman response tensor for waveguides must not be ignored in the case of optical waveguides containing complex geometry in the areas where the modal field distributions have overlaps. Using Eq. (11.89) in Eq. (11.88) and expanding the electric field vectors using Eqs. (11.74) and (11.76), the nonlinear Stokes polarisation becomes:

$$\begin{aligned}
\mathbf{P}^{(3)}_{\omega_\mathrm{s}}(t) &= \frac{3\varepsilon_0}{2}\mathbf{E}_{\omega_\mathrm{p}} \int \mathbf{R}(t-t_2) \vdots \mathbf{E}_{-\omega_\mathrm{p}}(\mathbf{r},t_2)\mathbf{E}_{\omega_\mathrm{s}}(\mathbf{r},t_2)\exp(-i(\omega_\mathrm{p}-\omega_\mathrm{s})(t-t_2))\,dt_2 \\
&= \frac{3\varepsilon_0}{2}\sum_{\eta,\sigma,\xi}\frac{e^{i(\beta_{p\eta}-\beta_{\mathrm{p}\sigma}+\beta_{\mathrm{s}\xi})}}{\sqrt{N_{\mathrm{p}\eta}N_{\mathrm{p}\sigma}N_{\mathrm{s}\xi}}}a_{\mathrm{p}\eta}(z,t) \\
&\quad \times \int \mathbf{R}(t-t_2) \vdots \mathbf{e}_{\mathrm{p}\eta}\mathbf{e}^*_{\mathrm{p}\sigma}\mathbf{e}_{\mathrm{s}\xi}\, a^*_{\mathrm{p}\sigma}(z,t_2)a_{\mathrm{s}\xi}(z,t_2) \\
&\quad \times \exp(-i(\omega_\mathrm{p}-\omega_\mathrm{s})(t-t_2))\,dt_2.
\end{aligned} \tag{11.90}$$

Using Eq. (11.90) in Eq. (11.87) one arrives at the pulse propagation equation for SRS:

$$\begin{aligned}
\frac{\partial a_{\mathrm{s}\mu}(z,t)}{\partial z} &= \hat{D}a_{\mathrm{s}\mu}(z,t) + i\omega_\mathrm{s}\left(1+\frac{i}{\omega_\mathrm{s}}\frac{\partial}{\partial t}\right)\frac{e^{-i\beta_{\mathrm{s}\mu}z}}{4\sqrt{N_{\mathrm{s}\mu}}}\int \mathbf{e}^*_{\mathrm{s}\mu}\cdot\mathbf{P}^{(3)}_{-\omega_\mathrm{s}}(t)\,dA \\
&= \hat{D}a_{\mathrm{s}\mu}(z,t) + i\frac{3\varepsilon_0\omega_\mathrm{s}}{8}\left(1+\frac{i}{\omega_\mathrm{s}}\frac{\partial}{\partial t}\right)\sum_{\eta,\sigma,\xi}\frac{e^{i(\beta_{\mathrm{p}\eta}-\beta_{\mathrm{p}\sigma}+\beta_{\mathrm{s}\xi}-i\beta_{\mathrm{s}\mu}z)}}{\sqrt{N_{\mathrm{p}\eta}N_{\mathrm{p}\sigma}N_{\mathrm{s}\xi}N_{\mathrm{s}\mu}}}a_{\mathrm{p}\eta}(z,t) \\
&\quad \times \int \mathbf{e}^*_{\mathrm{s}\mu}\cdot\int \mathbf{R}(t-t_2) \vdots \mathbf{e}_{\mathrm{p}\eta}\mathbf{e}^*_{\mathrm{p}\sigma}\mathbf{e}_{\mathrm{s}\xi}\, a^*_{\mathrm{p}\sigma}(z,t_2)a_{\mathrm{s}\xi}(z,t_2) \\
&\quad \times \exp(-i(\omega_\mathrm{p}-\omega_\mathrm{s})(t-t_2))\,dt_2 dA.
\end{aligned} \tag{11.91}$$

Eq. (11.91) is very general and can be used for both short and long pulses, and for any optical waveguide structure. For silica glass, the Raman response function is defined by Eq. (11.92):

$$\mathbf{R}(\tau) = \frac{\chi^{(3)}_{xxxx}}{2}\left(f_\mathrm{a}h_\mathrm{a}(\tau)\delta_{ij}\delta_{kl} + \frac{1}{2}f_\mathrm{b}h_\mathrm{b}(\tau)\left(\delta_{ik}\delta_{jl}+\delta_{il}\delta_{jk}\right)\right), \tag{11.92}$$

where $i, j, k, l = x, y, z$. Eq. (11.92) has two components corresponding to the isotropic and anisotropic contributions to SRS (Agrawal, 2007; Hellwarth, 1977) which are denoted by a and b respectively. However, due to the amorphous nature of glasses such as silica the isotropic contribution to SRS dominates throughout most of the Raman spectrum as discussed in Agrawal (2007) and Hellwarth (1977). The anisotropic component will be included here, however, for completeness. Let us now specialise the pulse propagation equation to MOFs that contain amorphous materials such as silica glass, or liquids and gases. Thus let's assume that all materials within the optical fibre have the same form of the Raman response as in Eq. (11.92), if this is not valid the more general Eq. (11.91) must be used. The third order susceptibility tensor component, $\chi^{(3)}_{xxxx}$, is given by (Agrawal, 2007)

$$\chi^{(3)}_{xxxx}(x,y) = \frac{4\varepsilon_0 c n^2(x,y;\omega_\mathrm{s}) n_2(x,y;\omega_\mathrm{s})}{3}. \tag{11.93}$$

Again it is important to note the spatial dependence of $n^2(x,y;\omega_\mathrm{s})$, $n_2(x,y;\omega_\mathrm{s})$, $f_{\mathrm{a,b}}(x,y)$ and $h_{\mathrm{a,b}}(x,y,\tau)$, however, for simplicity the spatial dependence of these terms will be implied. Using Eqs. (11.92) and (11.93) in Eq. (11.91) the general nonlinear pulse propagation due to dispersion and SRS for silica glass becomes

$$\begin{aligned}
\frac{\partial a_{\mathrm{s}\mu}(z,t)}{\partial z} &= \hat{D} a_{\mathrm{s}\mu}(z,t) + i\frac{\varepsilon_0^2 c\omega_\mathrm{s}}{4}\left(1+\frac{i}{\omega_\mathrm{s}}\frac{\partial}{\partial t}\right)\sum_{\eta,\sigma,\xi}\frac{e^{i(\beta_{\mathrm{p}\eta}-\beta_{\mathrm{p}\sigma}+\beta_{\mathrm{s}\xi}-i\beta_{\mathrm{s}\mu}z)}}{\sqrt{N_{\mathrm{p}\eta}N_{\mathrm{p}\sigma}N_{\mathrm{s}\xi}N_{\mathrm{s}\mu}}} a_{\mathrm{p}\eta}(z,t)\\
&\times\left\{\int f_\mathrm{a} n^2 n_2\left(\mathbf{e}^*_{\mathrm{s}\mu}\cdot\mathbf{e}_{\mathrm{p}\eta}\right)\left(\mathbf{e}^*_{\mathrm{p}\sigma}\cdot\mathbf{e}_{\mathrm{s}\xi}\right) dA\right.\\
&\times\int h_\mathrm{a}(t-t_2)a^*_{\mathrm{p}\sigma}(z,t_2)a_{\mathrm{s}\xi}(z,t_2)\exp(-i(\omega_\mathrm{p}-\omega_\mathrm{s})(t-t_2))\,dt_2\\
&\times+\frac{1}{2}\int f_\mathrm{b} n^2 n_2\left[\left(\mathbf{e}^*_{\mathrm{p}\sigma}\cdot\mathbf{e}_{\mathrm{p}\eta}\right)\left(\mathbf{e}^*_{\mathrm{s}\mu}\cdot\mathbf{e}_{\mathrm{s}\xi}\right)+\left(\mathbf{e}^*_{\mathrm{s}\mu}\cdot\mathbf{e}^*_{\mathrm{p}\sigma}\right)\left(\mathbf{e}_{\mathrm{p}\eta}\cdot\mathbf{e}_{\mathrm{s}\xi}\right)\right] dA\\
&\times\left.\int h_\mathrm{b}(t-t_2)a^*_{\mathrm{p}\sigma}(z,t_2)a_{\mathrm{s}\xi}(z,t_2)\exp(-i(\omega_\mathrm{p}-\omega_\mathrm{s})(t-t_2))\,dt_2\right\}.
\end{aligned} \tag{11.94}$$

Eq. (11.94) is a general pulse propagation equation for dispersion and SRS within optical fibres as long as the materials have the same Raman response form as given by Eq. (11.89). Eq. (11.94) does not use the adiabatic approximation and thus can be used for ultrashort pulses. Due to the full vectorial formulation as well as the use of a basis set of modes the form of Eq. (11.94) is now different to that in Agrawal (2007). The standard definition of the effective area does not match with this general form. This is further discussed in Section 11.5.2, where numerical calculations show significant differences between these two definitions of effective area. Let us now assume that the pulse widths of the pump and Stokes pulses are much longer than the Raman response time, which leads to the adiabatic approximation:

$$\int h_{\rm a,b}(t-t_2)a^*_{\rm p\sigma}(z,t_2)a_{\rm s\xi}(z,t_2)\exp(-i\Delta\omega(t-t_2))$$
$$= a^*_{\rm p\sigma}a_{\rm s\xi}\int h_{\rm a,b}(t-t_2)\exp(-i\Delta\omega(t-t_2))$$
$$= a^*_{\rm p\sigma}a_{\rm s\xi}\tilde{h}_{\rm a,b}(-\Delta\omega),$$

where $\Delta\omega = \omega_{\rm p} - \omega_{\rm s}$. In this adiabatic limit the pulse propagation equation becomes:

$$\frac{\partial a_{\rm s\mu}(z,t)}{\partial z} = \hat{D}a_{\rm s\mu}(z,t) + i\frac{\varepsilon_0^2 c\omega_{\rm s}}{4}\left(1+\frac{i}{\omega_{\rm s}}\frac{\partial}{\partial t}\right)\sum_{\eta,\sigma,\xi}\frac{e^{i(\beta_{\rm p\eta}-\beta_{\rm p\sigma}+\beta_{\rm s\xi}-i\beta_{\rm s\mu}z)}}{\sqrt{N_{\rm p\eta}N_{\rm p\sigma}N_{\rm s\xi}N_{\rm s\mu}}}$$
$$\times\, a_{\rm p\eta}(z,t)a^*_{\rm p\sigma}(z,t)a_{\rm s\xi}(z,t)$$
$$\times\left\{\int f_{\rm a}\tilde{h}_{\rm a}(-\Delta\omega)n^2(x,y;\omega_{\rm s})n_2(x,y;\omega_{\rm s})\left(\mathbf{e}^*_{\rm s\mu}\cdot\mathbf{e}_{\rm p\eta}\right)\left(\mathbf{e}^*_{\rm p\sigma}\cdot\mathbf{e}_{\rm s\xi}\right)dA\right.$$
$$+\frac{1}{2}\int f_{\rm b}\tilde{h}_{\rm b}(-\Delta\omega)n^2(x,y;\omega_{\rm s})n_2(x,y;\omega_{\rm s})$$
$$\left.\times\left[\left(\mathbf{e}^*_{\rm p\sigma}\cdot\mathbf{e}_{\rm p\eta}\right)\left(\mathbf{e}^*_{\rm s\mu}\cdot\mathbf{e}_{\rm s\xi}\right)+\left(\mathbf{e}^*_{\rm s\mu}\cdot\mathbf{e}^*_{\rm p\sigma}\right)\left(\mathbf{e}_{\rm p\eta}\cdot\mathbf{e}_{\rm s\xi}\right)\right]dA\right\}. \qquad (11.95)$$

The (isotropic/anisotropic) bulk Raman coefficient measured in m/W is given by (Agrawal, 2007)

$$g_{\rm a,b} = g_{\rm a,b}(\Delta\omega,x,y) = \frac{2f_{\rm a,b}n_2\omega_{\rm s}{\rm Im}(\tilde{h}_{\rm a,b}(\Delta\omega))}{c}.$$

From here on, only the imaginary part of $\tilde{h}_{\rm a,b}(\Delta\omega)$ will be considered for simplicity, as this is the term that leads to Raman gain. The real part of $\tilde{h}_{\rm a,b}(\Delta\omega)$ can be trivially included if required. Also, due to ${\rm Im}\left[\tilde{h}_{\rm a,b}(\Delta\omega)\right]$ being an odd function

$\mathrm{Im}(\tilde{h}_{\mathrm{a,b}}(-\Delta\omega)) = -\mathrm{Im}(\tilde{h}_{\mathrm{a,b}}(\Delta\omega))$.

$$\frac{\partial a_{\mathrm{s}\mu}(z,t)}{\partial z} = \hat{D}a_{\mathrm{s}\mu}(z,t) + \frac{\varepsilon_0^2 c^2}{8}\left(1 + \frac{i}{\omega_{\mathrm{s}}}\frac{\partial}{\partial t}\right)\sum_{\eta,\sigma,\xi}\frac{e^{i(\beta_{\mathrm{p}\eta} - \beta_{\mathrm{p}\sigma} + \beta_{\mathrm{s}\xi} - i\beta_{\mathrm{s}\mu} z)}}{\sqrt{N_{\mathrm{p}\eta}N_{\mathrm{p}\sigma}N_{\mathrm{s}\xi}N_{\mathrm{s}\mu}}}$$
$$\times\, a_{\mathrm{p}\eta}(z,t)a^*_{\mathrm{p}\sigma}(z,t)a_{\mathrm{s}\xi}(z,t)\int g_{\mathrm{a}}n^2\left(\mathbf{e}^*_{\mathrm{s}\mu}\cdot\mathbf{e}_{\mathrm{p}\eta}\right)\left(\mathbf{e}^*_{\mathrm{p}\sigma}\cdot\mathbf{e}_{\mathrm{s}\xi}\right)$$
$$+\,\frac{1}{2}g_{\mathrm{b}}n^2\left[\left(\mathbf{e}^*_{\mathrm{p}\sigma}\cdot\mathbf{e}_{\mathrm{p}\eta}\right)\left(\mathbf{e}^*_{\mathrm{s}\mu}\cdot\mathbf{e}_{\mathrm{s}\xi}\right) + \left(\mathbf{e}^*_{\mathrm{s}\mu}\cdot\mathbf{e}^*_{\mathrm{p}\sigma}\right)\left(\mathbf{e}_{\mathrm{p}\eta}\cdot\mathbf{e}_{\mathrm{s}\xi}\right)\right]dA. \tag{11.96}$$

It is important to note that in a standard optical fibre in the weak guidance regime, the spatial dependence of $n^2(x,y;\omega_{\mathrm{s}})$ and $g_{\mathrm{a,b}}(x,y,\Delta\omega)$ can be ignored since material interfaces are usually located where the fields have a small value, and thus these terms can be taken outside the integral in Eq. (11.96). Also in the weak guidance regime the modes are purely transverse thus the x-polarised and y-polarised modes are completely orthogonal (i.e. $\mathbf{e}^*_{\mathrm{s}x}\cdot\mathbf{e}_{\mathrm{p}y} = 0$). Hence any interference terms between x and y polarised modes will have zero isotropic Raman gain. However, contributions coming from the anisotropic component of the Raman gain can occur, as discussed later.

Emerging waveguides are interesting since the nonlinear susceptibility can have strong spatial dependence where the electric fields are at high intensity (Atakaramians et al., 2008; Afshar and Monro, 2009), and, as well, the x and y polarised modes can have significant overlap with each other due to their non-transverse nature, as discussed in Section 11.4. This is quite fascinating as it leads to a new contribution to Raman gain not evident in previous analysis.

From here on, only the contributions from the terms that do not require any phase matching, i.e. $\beta_{\mathrm{p}\eta} - \beta_{\mathrm{p}\sigma} + \beta_{\mathrm{s}\xi} - \beta_{\mathrm{s}\mu} = 0$, for any arbitrary optical waveguide is explicitly shown. This contracts the summation of modes to $\sigma = \eta$ and $\xi = \mu$. For $\sigma \neq \eta$ and/or $\xi \neq \mu$, the phase term $\beta_{\mathrm{p}\eta} - \beta_{\mathrm{p}\sigma} + \beta_{\mathrm{s}\xi} - \beta_{\mathrm{s}\mu}$ in general will be nonzero and typically very large. Thus the beat length of these interactions will be of the order of μm, thus will not contribute to any physically observed Raman effects within typical optical waveguides. It is, however, possible to phase match these modes such that $\beta_{\mathrm{p}\eta} - \beta_{\mathrm{p}\sigma} + \beta_{\mathrm{s}\xi} - \beta_{\mathrm{s}\mu} = 0$ even for $\sigma \neq \eta$ and/or $\xi \neq \mu$, by correct tailoring of the modes of the MOF such that these phase terms cancel (Efimov et al., 2003; Dimitropoulos et al., 2004).

$$\frac{\partial a_{\mathrm{s}\mu}(z,t)}{\partial z} = \hat{D}a_{\mathrm{s}\mu} + \frac{g_{\mu\mu}(\mathrm{m}^{-1}\mathrm{W}^{-1})}{2}\left[1 + \frac{i}{\omega_{\mathrm{s}}}\frac{\partial}{\partial t}\right]\left|a_{\mathrm{p}\mu}\right|^2 a_{\mathrm{s}\mu} \tag{11.97}$$
$$+\sum_{\eta\neq\mu}\frac{g_{\mu\eta}(\mathrm{m}^{-1}\mathrm{W}^{-1})}{2}\left[1 + \frac{i}{\omega_{\mathrm{s}}}\frac{\partial}{\partial t}\right]\left|a_{\mathrm{p}\eta}\right|^2 a_{\mathrm{s}\mu} + \text{phase terms},$$

where $g_{\mu\eta}(\mathrm{m}^{-1}\mathrm{W}^{-1})$ is the modal Raman gain experienced by the μth Stokes mode due to the presence of the ηth pump mode, which is now defined as

$$g_{\mu\eta}(\mathrm{m}^{-1}\mathrm{W}^{-1}) = \frac{\varepsilon_0^2 c^2}{4N_{\mathrm{p}\eta}N_{\mathrm{s}\mu}} \int g_{\mathrm{a}} n^2 \left|\mathbf{e}_{\mathrm{p}\eta}\cdot\mathbf{e}_{\mathrm{s}\mu}^*\right|^2 + g_{\mathrm{b}} n^2 \left(\left|\mathbf{e}_{\mathrm{p}\eta}\cdot\mathbf{e}_{\mathrm{s}\mu}\right|^2 + \left|\mathbf{e}_{\mathrm{p}\eta}\right|^2\left|\mathbf{e}_{\mathrm{s}\mu}\right|^2\right) dA. \tag{11.98}$$

The modal Raman gain has now become a normalised weighted integral of the spatially dependent, bulk Raman gain coefficients, and refractive index profiles of the MOF, with the pump and Stokes fields. Since the modal Raman gain contribution from the isotropic component is dependent on the dot product of the pump and Stokes fields, this component is very small for two orthogonally polarised modes, but not necessarily zero as will be shown later; the z-component of the electric field vector can create significant contributions to this term. The VNSE can be rewritten in similar notation to the SM where one now defines $\bar{g}_{\mathrm{R}}$ as the effective Raman gain coefficient and $\bar{A}_{\mathrm{eff}}$ to be the new effective area for SRS:

$$g(\mathrm{m}^{-1}\mathrm{W}^{-1}) = \frac{\bar{g}_{\mathrm{R}}}{\bar{A}_{\mathrm{eff}}}, \tag{11.99}$$

$$\bar{A}_{\mathrm{eff}} = \frac{\left(\int \mathbf{e}_{\mathrm{p}\eta}^*\times\mathbf{h}_{\mathrm{p}\eta}\cdot\hat{z}\, dA\right)\left(\int \mathbf{e}_{\mathrm{s}\mu}^*\times\mathbf{h}_{\mathrm{s}\mu}\cdot\hat{z}\, dA\right)}{\int \left(\mathbf{e}_{\mathrm{p}\eta}^*\times\mathbf{h}_{\mathrm{p}\eta}\cdot\hat{z}\right)\left(\mathbf{e}_{\mathrm{s}\mu}^*\times\mathbf{h}_{\mathrm{s}\mu}\cdot\hat{z}\right) dA}, \tag{11.100}$$

$$\bar{g}_{\mathrm{R}} = \frac{\varepsilon_0^2 c^2 \int g_{\mathrm{a}} n^2 \left|\mathbf{e}_{\mathrm{p}\eta}\cdot\mathbf{e}_{\mathrm{s}\mu}^*\right|^2 + g_{\mathrm{b}} n^2 \left(\left|\mathbf{e}_{\mathrm{p}\eta}\cdot\mathbf{e}_{\mathrm{s}\mu}\right|^2 + \left|\mathbf{e}_{\mathrm{p}\eta}\right|^2\left|\mathbf{e}_{\mathrm{s}\mu}\right|^2\right) dA}{\int \left(\mathbf{e}_{\mathrm{p}\eta}^*\times\mathbf{h}_{\mathrm{p}\eta}\cdot\hat{z}\right)\left(\mathbf{e}_{\mathrm{s}\mu}^*\times\mathbf{h}_{\mathrm{s}\mu}\cdot\hat{z}\right) dA}. \tag{11.101}$$

The definition of the effective area in Eq. (11.100) is very different to that in the SM which defines the effective area purely by the scalar transverse electric field distribution which is assumed to be equal at both the pump and Stokes wavelengths. However, it can be shown that Eq. (11.100) reduces to the standard definition of the effective area (Agrawal, 2007) in the regime of weak guidance, which is discussed further in Section 11.5.2. The use of the Poynting vector to define the effective area in Eq. (11.100) means that the effective area now describes the area of the energy density distribution rather than the area of the electric field distribution. Eq. (11.100) also considers the differences in the pump and Stokes energy density distributions, such that $\bar{A}_{\mathrm{eff}}$ is an "average" of the confinement of both the pump and Stokes energy field distributions. In the regime of weakly guiding optical waveguides, it can be shown that $\bar{A}_{\mathrm{eff}}$ converges to that in the SM, as discussed in Section 11.4.

The advantages in defining the modal Raman gain as in Eq. (11.99) is that it allows analysis to be separated into the linear properties of the optical waveguide that define $\bar{A}_{\mathrm{eff}}$ and the nonlinear properties that are determined by $g_{\mathrm{a,b}}(\Delta\omega,x,y)$.

If one assumes that the anisotropic component of the Raman susceptibility is negligible, then $f_{\mathrm{b}} = 0$ and $f_{\mathrm{R}} = f_{\mathrm{a}}$, and the analysis further simplifies to:

$$g(\mathrm{m}^{-1}\mathrm{W}^{-1}) = \frac{\varepsilon_0^2 c^2}{4N_{p\eta}N_{\mathrm{s}\mu}} \int g_{\mathrm{R}} n^2 \left|\mathbf{e}_{\mathrm{p}\eta} \cdot \mathbf{e}_{\mathrm{s}\mu}^*\right|^2 dA, \tag{11.102}$$

$$\bar{g}_{\mathrm{R}} = \frac{\varepsilon_0^2 c^2 \int g_{\mathrm{R}} n^2 \left|\mathbf{e}_{\mathrm{p}\eta} \cdot \mathbf{e}_{\mathrm{s}\mu}^*\right|^2 dA}{\int \left(\mathbf{e}_{\mathrm{p}\eta}^* \times \mathbf{h}_{\mathrm{p}\eta} \cdot \hat{z}\right) \left(\mathbf{e}_{\mathrm{s}\mu}^* \times \mathbf{h}_{\mathrm{s}\mu} \cdot \hat{z}\right) dA}.$$

The full form of the nonlinear pulse propagation equation as given by Eq. (11.96) is quite complicated and one must analyse all eigenmodes and their propagation constants to see if the phase can be matched. For example in a single mode non-birefringent fibre $\beta_{\mathrm{p}\eta} - \beta_{\mathrm{p}\sigma} = 0$ where η and σ are the x and y polarised modes respectively. In a standard optical fibre in the weak guidance regime, these modes have negligible overlap since the z-component of the electric field vector is negligible, but in the strong guidance regime where the z-component of the electric field vector becomes significant, the overlap can become significant. One may also need to consider the possibility of phase matching of higher order modes with lower order modes, the forward and backward propagating modes for both the pump and Stokes fields.

Having developed the VNSE and formulated new definitions of the effective area and modal Raman gain for SRS, an investigation is required to see how these new definitions compare with the SM. Section 11.5.2 explores which regimes show large differences between the two models and discusses the physical effects that cause these differences.

11.5.2 Results

The VNSE allows us to investigate where the SM fails and what physical effects cause this. In the regime of weak guidance one can show that Eq. (11.102) contracts to Eq. (11.53) as would be expected.

The definition of the modal Raman gain by the VNSE now contains the contributions from the z-component of the electric field vector, which is neglected in the SM. In Section 11.4 it was shown that the z-component of the electric field vector (e_z) can become significant in the presence of subwavelength features with high refractive index contrast. For example in Section 11.4 it was shown for a high refractive index nanowire, as the size of the core decreases to subwavelength dimensions, the transverse fields become less confined to the core and spread out into the air increasing the effective area of the mode as well localising the electric fields in air where there is no nonlinearity.

Interestingly, within this subwavelength nanowire regime where the transverse fields are weakly confined to the core, the e_z fields remain strongly confined to the core. Also, as the magnitudes of the e_z fields become larger in this subwavelength nanowire regime, the overall mode becomes significantly more confined to the core. The implications of this extra confinement will cause an increase in nonlinear effects such as Kerr nonlinearity and Raman gain.

To show the significance of this effect on Raman gain, the effective area, $\bar{g}_R$ and $g\,(\mathrm{m}^{-1}\mathrm{W}^{-1})$ of a chalcogenide nanowire are calculated for varying core diameter to observe in which regimes differences between the VNSE and SM can be observed. Here the refractive index of chalcogenide is equal to 2.4 (Ta'eed et al., 2007) and due to the amorphous nature of chalcogenide glass, the bulk Raman gain coefficient is assumed to be completely isotropic at the peak of the gain spectrum as found for silica glass (Agrawal, 2007; Hellwarth, 1977). For all our numerical calculations only the fundamental mode is considered, and the pump and Stokes fields are copolarised in all cases, except for VNSE ORTH where the pump and Stokes fields are orthogonally polarised. The pump wavelength is set to 1550 nm and only one value of the Raman gain spectrum is considered, i.e. the peak which is located at 230 cm^{-1} (Slusher et al., 2004) which corresponds to a Stokes wavelength of 1608 nm. At this Raman shift $g_R = 5.1 \times 10^{-11}$m/W (Slusher et al., 2004).

Fig. 11.14(a) shows the effective area of a chalcogenide nanowire for varying core diameter using the VNSE (solid curve) as defined by Eq. (11.100), and using the SM (Agrawal, 2007) (dashed curve). For large core diameters the two models are equal, which is where the modal field distributions have very small e_z fields, and thus the scalar approximation of the fields used by the SM is valid. However, at subwavelength core diameters the VNSE predicts smaller effective areas with a minimum of 0.233 $\mu\mathrm{m}^2$ compared with the SM that has a minimum of 0.36 $\mu\mathrm{m}^2$.

Thus the energy distribution of the mode can be confined more tightly than the transverse electric field distribution at these subwavelength nanowires near the minimum effective area. This can perhaps be attributed to the z-component of the electric field vector as it was shown in Section 11.4 that the time averaged z-component of the Poynting vector is dependent on the e_z fields by

$$S_z \propto \beta \left|\hat{e}_t\right|^2 - i\hat{e}_t \cdot \nabla_t e_z^*.$$

Thus when e_z becomes significant it can contribute to the effective area. Also, the e_z fields maintain tighter confinement than the transverse electric fields in the subwavelength nanowire regimes. Thus the e_z fields help to keep the mode tightly confined to the subwavelength nanowire, further reducing the effective area.

Fig. 11.14(b) contains a plot of the effective Raman gain coefficient of a chalcogenide nanowire for varying core diameter with the VNSE (solid curve) and bulk

Raman gain coefficient (dashed line) which is $g_R = 5.1 \times 10^{-11}$m/W. For large core diameters the effective Raman gain coefficient is approximately equal to the bulk Raman gain coefficient. As the size of the core becomes highly subwavelength, the effective Raman gain coefficient increases above the bulk Raman gain coefficient, showing that the contributions from the e_z fields can create a significant enhancement to the effective Raman gain coefficient. For smaller core diameters the effective Raman gain coefficient decreases sharply from the peak, which is due to the fields starting to spread out into the air decreasing the overlap of the fields with the Raman-active core.

The z-component of the electric field vector can contribute significantly to both reducing the effective area and enhancing the effective Raman gain coefficient at subwavelength core diameters. These two effects will both create an increase in the modal Raman gain. Fig. 11.15 shows a numerical comparison of the modal Raman gain of a chalcogenide nanowire for varying core diameter for all three models considered here. Again, the pump wavelength is 1550 nm and the Stokes wavelength is 1608 nm.

The convergence of the VNSE, SM, and ASM for core sizes of the order of a wavelength and larger indicates that in this regime where both pump and Stokes fields have high overlap, are contained completely within the core, and have negligible e_z component, the three models are equivalent. This is the regime of standard optical fibres where the weak guidance approximation is valid.

However, in the subwavelength core size regime significant differences can be seen between the three models with a factor of over 2.5 observed between the VNSE and SM at a core diameter of 0.5 μm. Both the VNSE and ASM show a sharp decrease in modal Raman gain for core sizes smaller than 0.5 μm which is due to the increase in effective area as well as a decrease in the overlap with the Raman active core as seen in Fig. 11.14.

The modal Raman gain for orthogonally polarised pump and Stokes fields is shown by plus symbols in Fig. 11.15. In the regime of weak guidance where the core size is not subwavelength this contribution to Raman gain is negligible as assumed in Agrawal (2007). However, at the subwavelength core sizes around 0.5 μm this contribution to Raman gain becomes significant and can be as large as 10% of the modal Raman gain due to copolarised Stokes and pump fields at the same core diameter. This effect is attributed to the z-component of the electric fields that now creates a nonzero dot-product between the x and y polarised pump and Stokes modes as discussed in Section 11.4.

11.5.2.1 Tailoring the modal Raman gain

Changing the shape of the Raman gain spectrum can be very useful for applications such as telecommunications, where gain flattening techniques are applied to

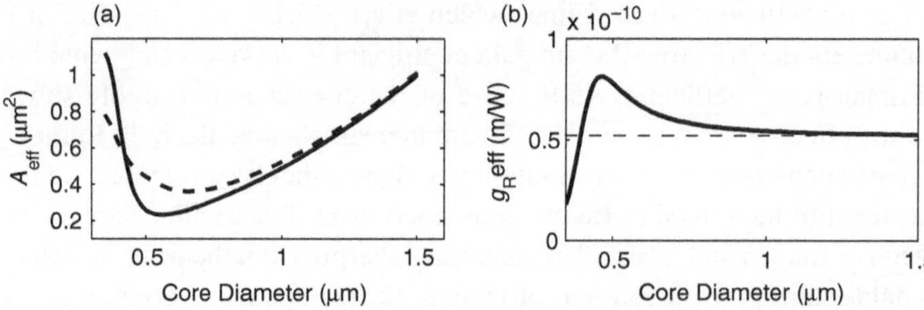

Fig. 11.14. (a) Effective area of the fundamental mode of a chalcogenide nanowire versus core diameter, using the VNSE (solid curve) and SM (dashed curve). (b) Effective Raman gain coefficient of a chalcogenide nanowire versus core diameter using VNSE (solid curve) and the bulk Raman gain coefficient is shown as a dashed line.

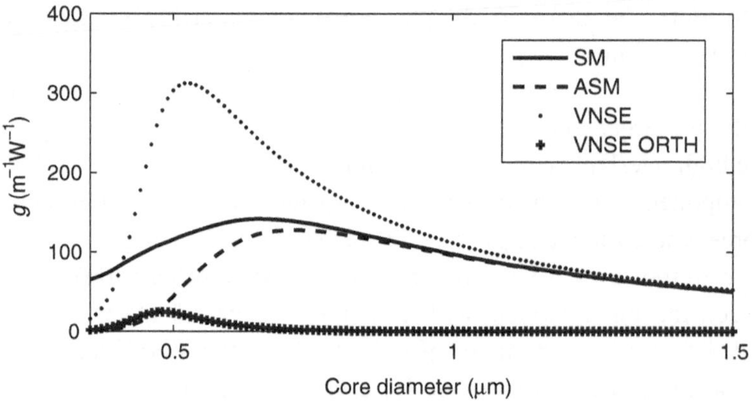

Fig. 11.15. Modal Raman gain of a chalcogenide nanowire for varying core diameter. SM (solid curve), VNSE (dotted curve), ASM (dashed curve) and VNSE ORTH (plus symbol). Pump and Stokes wavelengths are 1550 and 1608 nm, respectively.

obtain an ultra-flat gain spectrum. Typical approaches to gain flattening of a Raman amplifier include using multiple pump sources (Perlin and Winful, 2002; Cui et al., 2004), designing the losses of the optical fibre (Kakkar and Thyagarajan, 2005), and tailoring the composition of the glass (Jose and Ohishi, 2007), etc.

However, these techniques do not attempt to flatten the modal Raman gain spectrum of the optical waveguide. Tailoring the modal field distributions and the optical characteristics of a MOF can be achieved via complex fibre geometry, choice of the host glass material, as well as post fabrication processes such as filling the MOF with gases, liquids and atomic vapour. Thus MOFs have great flexibility to engineer the modal Raman gain to achieve gain enhancement, gain suppression or even change the shape of the modal Raman gain spectrum.

In Section 11.5.2, significant contributions from the z-component of the electric field vector using the VNSE led to an enhanced modal Raman gain than that predicted by the SM. The VNSE also has two other significant differences from the SM. The bulk Raman gain coefficient and refractive index of the optical waveguide in general contain spatial dependence and the pump and Stokes fields have different modal field distributions dependent on their wavelength of operation. These two attributes of the VNSE suggest that the wavelength dependence of the Stokes modal field distributions $\mathbf{e}_{s\mu}$ will create a wavelength dependent overlap of the Stokes fields with both the pump fields and the Raman active materials. Therefore, the shape of the modal Raman gain spectrum will not only depend on the bulk Raman gain coefficient spectrum, $g_R(\Delta\omega)$, but also on the wavelength dependence of the waveguide's modal field distributions over the entire Raman spectrum.

In contrast, the SM assumes the effective area to be constant and hence the shape of the modal Raman gain to be governed entirely by the bulk Raman gain coefficient. In Section 11.4 it was observed that at these subwavelength dimensions optical nanowires can exhibit strong wavelength dependence on the confinement of the modal fields to the core. Here an investigation is given in the use of this effect to tailor the shape of the modal Raman gain spectrum and its use for gain flattening in Raman amplifiers.

Tellurite glass has a broad bandwidth bulk Raman gain coefficient (Stegeman et al., 2003; Jose and Ohishi, 2007; Guanshi et al., 2007; Mori et al., 2003), but requires significant gain flattening to be used in a Raman amplifier. Fig. 11.16(a) shows an approximation of the tellurite bulk Raman gain coefficient spectrum based upon that in (Stegeman et al., 2003) using a super-Gaussian for the peak at 10 THz and a Gaussian at 22 THz. The exact magnitude and shape of the bulk Raman gain spectra are dependent on the composition of the tellurite glass (Jose and Ohishi, 2007; Stegeman et al., 2003). Let us define the height of the peak at 22 THz divided by the height of the peak at 10 THz as R, then here $R = 2.0$ for the bulk Raman gain coefficient. The value of R for the modal Raman gain is dependent on the design of the tellurite MOF. As discussed previously, in the regime of weak guidance, the effective area of the optical fibre is assumed to be constant over the Raman spectrum. This is shown in Fig. 11.16(b) as the solid curve where the modal Raman gain spectrum for a tellurite nanowire of non-subwavelength core diameter 1.5 μm is shown. At this core size, both pump and Stokes fields are confined tightly to the Raman active core over all the Stokes wavelengths within the spectrum and the shape of the modal Raman gain looks identical to that of the bulk Raman gain coefficient with $R \approx 2.0$.

However, if one considers a tellurite nanowire at a highly subwavelength core diameter of 0.5 μm as shown in Fig. 11.16(b) (dashed curve), a significant change in the shape of the modal Raman gain spectrum is observed with a decrease of the ratio

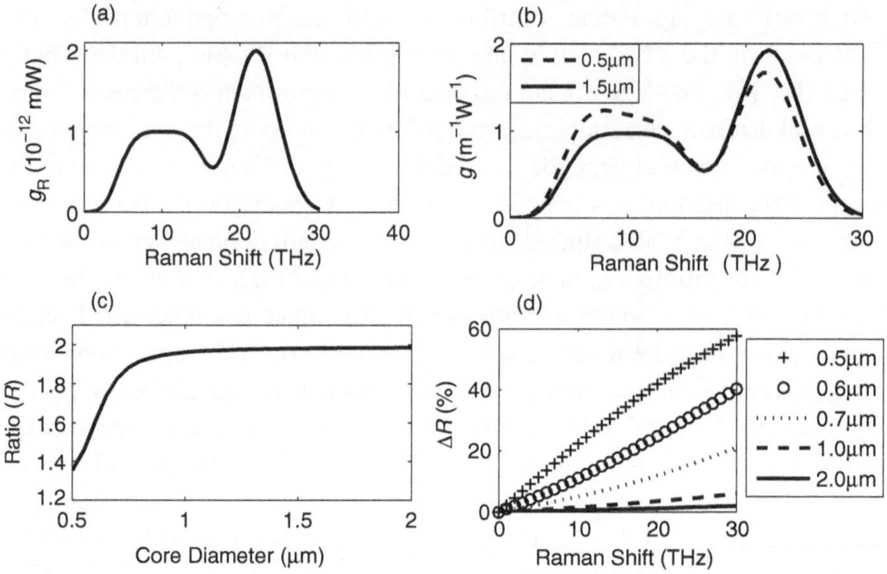

Fig. 11.16. (a) Toy model of bulk Raman gain coefficient of tellurite glass. (b) Modal Raman gain spectrum of a tellurite nanowire for 1.5 μm core diameter (solid) and 0.5 μm core diameter (dashed) calculated with the VNSE. (c) Ratio (R) of the two peaks for varying core diameter. (d) Plot of the amount of decrease in modal Raman gain due to finite overlap of the Stokes fields with the pump fields and the Raman active core.

of the peaks to $R = 1.35$. At this highly subwavelength core diameter, an increase in the Stokes wavelength causes the modal field distributions to further spread out from the core into the air, increasing the effective area as well as decreasing the overlap of the Stokes fields with both the Raman active core and pump modal field distribution. The net effect at this core diameter is a decrease in the confinement of the Stokes modal field distributions for increasing Raman shift (longer wavelength). The resultant modal Raman gain spectrum is a significantly flattened gain spectrum making this optical fibre more suitable in a Raman amplifier system.

To see how this flattening effect varies with core diameter, in Fig. 11.16(c) a plot of R is given for varying core diameter of the optical nanowire. For larger core diameters R is close to 2.0 showing the convergence of the modal Raman gain to the SM in the regime of weak guidance. As the core diameter decreases to subwavelength dimensions, R reduces rapidly as the Stokes fields start to diverge out into the air. In order to observe the strength of this flattening effect across the spectrum, Fig. 11.16(d) contains a plot of ΔR, the amount of decrease in the modal Raman gain relative to the bulk Raman gain spectrum for a number of core diameters. As expected the larger Raman shifts are affected more by this effect and these effects become more significant as we decrease the core size to subwavelength dimensions. Decreases in the modal Raman gain of up to nearly 60% are observed demonstrating the importance of considering these effects in the regime of strong guidance.

To conclude, this section has shown that significant alteration to the shape of the Raman gain spectrum can be achieved through design of the MOF. The analysis here in an optical nanowire has only one free parameter to tailor the tellurite modal Raman gain spectrum, i.e. the core size of the nanowire. However, microstructured optical fibres can have an arbitrary number of free parameters, such as the use of holes and inclusions of arbitrary shape and size within the MOF. Thus MOFs have more potential to further tailor the modal Raman gain by engineering the fibre geometry to achieve greater flattening, which is yet to be explored.

11.6 Discussion, conclusion and future work

A new frontier in the field of optical waveguides is the design and fabrication of waveguides, referred to here as "emerging waveguides" with three main characteristics: (1) complex and inhomogeneous structure, (2) high index contrast, and (3) subwavelength features. A direct consequence of these features is that the common nonlinear Schrödinger equation which is based on weak guidance approximation does not provide an accurate description of the nonlinear processes in these waveguides. Here, a vectorial-based nonlinear Schrödinger equation (VNSE) was developed, without relying on weak guidance approximation, which can be applied to these emerging waveguides.

An important feature of these waveguides is the fact that their propagating modes have a much larger z-component (z: direction of propagation) in comparison with those waveguides for which the weak guidance approximation is valid. As a result, the modes of these waveguides are not fully transverse and hence different orthogonality conditions govern them. Nonlinear and dispersion processes can be associated to this third "direction of propagation polarisation", which also couples the transverse polarisations through dispersion and nonlinear processes. The model provides a platform for generalising nonlinear processes such as XPM, modulation instability, soliton formation and propagation, four-wave mixing, parametric processes and Raman and Brillouin scattering for emerging waveguides. The model also predicts new tempo-spatial processes such as dispersion-induced depolarisation of the guided modes. Despite the complexity that this new concept brings into guided-wave nonlinear optics, early redevelopment of Kerr and Raman processes indicates the great flexibility for "engineering" nonlinear processes. New effects are expected to be observed in supercontinuum generation in emerging waveguides since these waveguides have extreme nonlinearity and dispersion, and couple different propagating modes, including the two different polarisations of one mode, through their strong z-components.

Based on the model developed here, new vectorially-based definitions for the effective nonlinear coefficient of waveguides (γ^{V}) and modal Raman gain g were developed. Although the developed model is general, it has been applied to a simple step index cylindrical waveguide and shown that even for such a simple structure, γ^{V} and g can be a factor of two higher than the common definition of γ, in the regime of high index contrast and subwavelength dimensions. However, the full extent of the model in terms of exploring the rich physics behind the new pulse propagation model and the new definitions of effective nonlinearity γ, effective mode area A_{eff}, and cross-mode effective nonlinearity $\gamma_{\mu\nu}$, especially for waveguides with inhomogeneous structures, is yet to be explored.

Experimental confirmation of the new vectorially-based effective nonlinear coefficient γ^{V} in a bismuth suspended-core fibre, presented here, is the first step in experimentally exploring the nonlinear effects in emerging waveguides. Experimental investigation of other nonlinear effects such as Raman, FWM, MI, or SC, where the dispersion properties and modal coupling of different frequencies can be quite different, will reveal the rich physics behind the nonlinear phenomena in emerging waveguides.

An active research area that benefits significantly from high index glass MOFs is SC generation in mid-IR. Reports of SC generation in mid-IR, include: SF6 MOF (Omenetto et al., 2006), unstructured ZBLAN fibre (Xia et al., 2006, 2007), tellurite MOF (Ishizawa et al., 2008; Domachuk et al., 2008), sapphire fibre (Kim et al., 2008a, 2009b), and bismuth MOF (Price et al., 2007). The new VNSE model will be valuable in designing highly nonlinear and tailorable dispersion MOFs using high index glasses for SC generation in mid-IR.

It should be noted that there are, however, some interesting practical issues with the applications of emerging waveguides. First of all, the coupling into these waveguides has its own complexity due to their subwavelength structure and high index host materials. New full-vectorial models are required in order to accurately model the coupling efficiency of these waveguides for free-space or waveguide-coupled methods. Practically, one possible method can be coupling using tapered waveguides such as the one reported in Boyraz et al. (2004a).

The second issue is related to the loss in the waveguides in the high index and subwavelength regime. It has been shown that scattering of guided modes from surface roughness results in scattering losses (Roberts et al., 2005a, 2005b; Zhai and Tong, 2007; Ebendorff-Heidepriem et al., 2009). Surface roughness arises from thermally excited surface capillary waves, which become frozen-in during cooling of a glass melt (e.g. during fibre drawing or tapering) at the glass transition temperature (Roberts et al., 2005b). The loss of the small core MOFs or nanowires has been found to increase as the core size decreases (Tong et al., 2003; Ebendorff-Heidepriem et al., 2009) or as the refractive index increases (Zhai and Tong, 2007). This has been attributed to the increased field strength at the glass-air interfaces leading to larger roughness scattering loss component (Roberts et al., 2005a; Ebendorff-Heidepriem et al., 2009). Considering the vectorial nature of propagating modes in emerging waveguides and their large z-component, which is strongly localised at the glass–air interface, see Fig. 11.17(a), it would be valuable to explore the loss of guided modes of emerging waveguides due to surface roughness.

Acknowledgements

Author Shahraam Afshar V. would like to acknowledge the invaluable support of Zari Pirvali, and Shayan and Sheida Afshar.

References

Abedin, K. S. (2005), 'Observation of strong stimulated Brillouin scattering in single-mode As_2Se_3 chalcogenide fiber', *Opt. Express* **13**(25), 10266–10271.

Abedin, K. S. (2006), 'Stimulated Brillouin scattering in single-mode tellurite glass fiber', *Opt. Express* **14**(24), 11766–11772.

Afshar V. S. and Monro, T. M. (2008), Kerr nonlinearity in small core optical fibres and nanowires: A generalised model, and application to microstructured fibres, in 'Proceeding of ACOFT 2008', Australian Optical Society.

Afshar V. S. and Monro, T. M. (2009), 'A full vectorial model for pulse propagation in emerging waveguides with subwavelength structures Part I: Kerr nonlinearity', *Opt. Express* **17**(4), 2298–2318.

Afshar, V. S., Zhang, W. Q., Ebendorff-Heidepriem, H. and Monro, T. M. (2009), 'Small core optical waveguides are more nonlinear than expected: experimental confirmation', *Opt. Lett.* **34**, 3577–3579.

Afshar V. S., Warren-Smith, S. C. and Monro, T. M. (2007), 'Enhancement of fluorescence-based sensing using microstructured optical fibres', *Opt. Express* **15**(26), 17891–17901.

Agrawal, G. P. (2007), *Nonlinear Fiber Optics*, Academic Press.

Almeida, V., Barrios, C., Panepucci, R., Lipson, M., Foster, M., Ouzounov, D. and Gaeta, A. (2004a), 'All-optical switching on a silicon chip', *Opt. Lett.* **29**(24), 2867–2869.

Almeida, V. R., Barrios, C. A., Panepucci, R. R. and Lipson, M. (2004b), 'All-optical control of light on a silicon chip', *Nature* **43**, 1081.

Almeida, V. R., Xu, Q., Barrios, C. A. and Lipson, M. (2004c), 'Guiding and confining light in void nanostructure', *Opt. Lett.* **29**, 1209.

Atakaramians, S., Afshar V., S., Fischer, B. M., Abbott, D. and Monro, T. M. (2008), 'Porous fibers: a novel approach to low loss THz waveguides', *Opt. Express* **16**(12), 8845–8854.

Barrios, C. A. (2004), 'High-performance all-optical silicon microswitch', *Electronics Letters* **40**, 862.

Benabid, F., Bouwmans, G., Knight, J. C. and Russell, P. S. J. (2004), 'Ultrahigh efficiency laser wavelength conversion in a gas-filled hollow core photonic crystal fiber by pure stimulated rotational Raman scattering in molecular hydrogen', *Physical Review Letters* **93**, 123903–1.

Benabid, F., Couny, F., Knight, J. C., Birks, T. A. and Russell, P. S. J. (2005a), 'Compact, stable and efficient all-fibre gas cells using hollow-core photonic crystal fibres', *Nature* **434**, 488.

Benabid, F., Light, P., Couny, F. and Russell, P. (2005b), 'Electromagnetically-induced transparency grid in acetylene-filled hollow-core PCF', *Opt. Express* **13**(15), 5694–5703.

Biancalana, F., Skryabin, D. V., and Russell, P. S. J. (2003), 'Four-wave mixing instabilities in photonic-crystal and tapered fibers', *Phys. Rev. E* **68**, 046, 603–1.

Blow, K. J. and Wood, D. (1989), 'Theoretical description of transient stimulated Raman scattering in optical fibers', *IEEE J. Quantum Electronics* **25**, 2665.

Borelli, N. F., Aitken, B. G. and Newhouse, M. A. (1995), 'Resonant and non-resonant effects in photonics glasses', *J. Non-Cryst. Solids* **185**, 109–122.

Boskovic, A., Chernikov, S. V., Taylor, J. R., Gruner-Nielsen, L. and Levring, O. A. (1996), 'Direct continuous-wave measurement of N_2 in various types of telecommunication fiber at 1.55 μm', *Opt. Lett.* **21**, 1966.

Boyd, R. W. (2003), *Nonlinear Optics*, Academic Press, San Diego, CA.

Boyraz, O., Indukuri, T. and Jalali, B. (2004b), 'Self-phase-modulation induced spectral broadening in silicon waveguides', *Opt. Express* **12**, 829.

Boyraz, O., Koonath, P., Raghunathan, V. and Jalali, B. (2004a), 'All optical switching and continuum generation in silicon waveguides', *Opt. Express* **12**, 4094.

Brabec, T. and Krausz, F. (1997), 'Nonlinear optical pulse propagation in the single-cycle regime', *Physical Review Letters* **78**, 3282.

Brilland, L., Smektala, F., Renversez, G., Chartier, T., Troles, J., Nguyen, T., Traynor, N. and Monteville, A. (2006), 'Fabrication of complex structures of holey fibers in chalcogenide glass', *Opt. Express* **14**(3), 1280–1285.

Butcher, P. N. and Cotter, D. (1990), *The Elements of Nonlinear Optics*, Cambridge University Press, Cambridge.

Cantini C., Abedin, K. S., Sacchi, G., Di Pasquale, F. and Kubota, F. (2005), 'Measurement of Raman gain coefficient in bismuth-based single mode optical fibers', *Technical Digest of the 10th OptoElectronics and Communications Conference (OECC2005)*, Seoul, Korea, pp. 234–235.

Cardinal, T., Richardson, K. A., Shim, J., Schulte, A., Beatty, R. et al. (1999), 'Non-linear optical properties of chalcogenide glasses in the system As-S-Se', *J. Non-Cryst. Solids* **256/257**, 353–360.

Chang, G., Norris, T. B. and Winful, H. G. (2003), 'Optimization of supercontinuum generation in photonic crystal fibers for pulse compression', *Opt. Lett.* **28**, 546.

Chen, X., Panoiu, N. C. and Osgood, R. M. (2006), 'Theory of Raman-mediated pulsed amplification in silicon-wire waveguides', *IEEE J. Quantum Electronics* **42**, 160.

Coutaz, J.-L. and Kull, M. (1991), 'Saturation of the nonlinear index of refraction in semiconductor-doped glass', *J. Opt. Soc. Am. B* **8**(1), 95–98.

Cox, F. M., Argyros, A. and Large, M. C. J. (2006), 'Liquid-filled hollow core microstructured polymer optical fiber', *Opt. Express* **14**(9), 4135–4140.

Cui, S., Liu, J. S. and Ma, X. M. (2004), 'A novel efficient optimal design method for gain-flattened multiwavelength pumped fiber Raman amplifier', *IEEE PhotonTechnol. Lett.* **16**(11), 2451–2453.

Dadap, J. L., Panoiu, N. C., Chen, X., Hsieh, I.-W., Liu, X., Chou, C.-Y., Dulkeith, E., McNab, S. J., Xia, F., Green, W. M. J., Sekaric, L., Vlasov, Y. A. and Osgood Jr., R.

M. (2008), 'Nonlinear-optical phase modification in dispersion-engineered Si photonic wires', *Opt. Express* **16**, 1280.

Debs, J. E., Ebendorff-Heidepriem, H., Quinton, J. S. and Monro, T. M. (2008), 'A fundamental study into the surface functionalization of soft glass microstructured optical fibers via silane coupling agents', *IEEE J. of Lightwave Technol.* **27**(5), 578–582.

Désévédavy, F., Renversez, G., Brilland, L., Houizot, P., Troles, J., Coulombier, Q., Smektala, F., Traynor, N. and Adam, J.-L. (2008), 'Small-core chalcogenide microstructured fibers for the infrared', *Appl. Opt.* **47**(32), 6014–6021.

Dimitropoulos, D., Raghunathan, V., Claps, R. and Jalali, B. (2004), 'Phase-matching and nonlinear optical processes in silicon waveguides', *Opt. Express* **12**(1), 149–160.

Dinu, M., Quochi, F. and Garcia, H. (2003), 'Third-order nonlinearities in silicon at telecom wavelengths', *Appl. Phys. Lett* **82**, 2954.

Domachuk, P., Wolchover, N. A., Cronin-Golomb, M., Wang, A., George, A. K., Cordeiro, C. M. B., Knight, J. C. and Omenetto, F. G. (2008), 'Over 4000 nm bandwidth of mid-IR supercontinuum generation in sub-centimeter segments of highly nonlinear tellurite PCFs', *Opt. Express* **16**(10), 7161–7168.

Driscoll, J. B., Liu, X., Yasseri, S., Hsieh, I., Dadap, J. I. and Osgood Jr., R. M. (2009), 'Large longitudinal electric fields (ez) in silicon nanowire waveguides', *Opt. Express* **17**, 2797.

Dudley, J. M. and Coen, S. (2002), 'Coherence properties of supercontinuum spectra generated in photonic crystal and tapered optical fibers', *Opt. Lett.* **27**, 1180.

Ebendorff-Heidepriem, H. and Monro, T. M. (2007), 'Extrusion of complex preforms for microstructured optical fibers', *Opt. Express* **15**(23), 15086–15092.

Ebendorff-Heidepriem, H., Petropoulos, P., Asimakis, S., Finazzi, V., Moore, R. C., Frampton, K., Koizumi, F., Richardson, D. J. and Monro, T. M. (2004a), 'Bismuth glass holey fibers with high nonlinearity', *Opt. Express* **12**, 5082.

Ebendorff-Heidepriem, H., Petropoulos, P., Asimakis, S., Finazzi, V., Moore, R., Frampton, K., Koizumi, F., Richardson, D. and Monro, T. (2004b), 'Bismuth glass holey fibers with high nonlinearity', *Opt. Express* **12**(21), 5082–5087.

Ebendorff-Heidepriem, H., Petropoulos, P., Finazzi, V., Asimakis, S., Leong, J., Koizumi, F., Frampton, K., Moore, R. C., Richardson, D. J. and Monro, T. M. (2005), Heavy metal oxide glass holey fibers with high nonlinearity, in 'OFC 2005 Proceeding', OSA, p. OThA3.

Ebendorff-Heidepriem, H., Warren-Smith, S. C. and Monro, T. M. (2009), 'Suspended nanowires: fabrication, design and characterization of fibers with nanoscale cores', *Opt. Express* **17**(4), 2646–2657.

Efimov, A., Taylor, A., Omenetto, F., Knight, J., Wadsworth, W. and Russell, P. (2003), 'Phase-matched third harmonic generation in microstructured fibers', *Opt. Express* **11**(20), 2567–2576.

Feng, X., Loh, W. H., Flanagan, J. C., Camerlingo, A., Dasgupta, S., Petropoulos, P., Horak, P., Frampton, K. E., White, N. M., Price, J. H. V., Rutt, H. N. and Richardson, D. J. (2008), 'Single-mode tellurite glass holey fiber with extremely large mode area for infrared nonlinear applications', *Optics Express* **16**, 13651.

Feng, X., Monro, T. M., Finazzi, V., Moore, R. C., Frampton, K., Petropoulos, P. and Richardson, D. J. (2005), 'Extruded single mode, high-nonlinearity, tellurite glass holey fibre', *Electronic Letters* **41**, No. 15.

Foster, M., Dudley, J., Kibler, B., Cao, Q., Lee, D., Trebino, R. and Gaeta, A. (2005), 'Nonlinear pulse propagation and supercontinuum generation in photonic nanowires: experiment and simulation', *Appl. Phys. B* **81**(2-3), 363–367.

Foster, M., Moll, K. and Gaeta, A. (2004), 'Optimal waveguide dimensions for nonlinear interactions', *Opt. Express* **12**(13), 2880–2887.

Foster, M. A., Turner, A. C., Lipson, M. and Gaeta, A. L. (2008), 'Nonlinear optics in photonic nanowires', *Opt. Express* **16**, 1300.

Foster, M. A., Turner, A. C., Salem, R., Lipson, M. and Gaeta, A. L. (2007), 'Broad-band continuous-wave parametric wavelength conversion in silicon nanowaveguides', *Opt. Express* **15**(20), 12949–12958.

Friberg, S. R. and Smith, P. A. (1987), 'Nonlinear optical glasses for ultrafast optical switches', *IEEE J. Quant. Electron.* **23**, 2089–2094.

Fuerbach, A., Steinvurzel, P., Bolger, J. and Eggleton, B. (2005), 'Nonlinear pulse propagation at zero dispersion wavelength in anti-resonant photonic crystal fibers', *Opt. Express* **13**(8), 2977–2987.

Gaeta, A. L. (2000), 'Catastrophic collapse of ultrashort pulses', *Phys. Rev. lett.* **84**, 3582.

Gatz, S. and Herrmann, J. (1991), 'Soliton propagation in materials with saturable nonlinearity', *J. Opt. Soc. Am. B* **8**(11), 2296–2302.

Genty, G., Kinsler, P., Kibler, B. and Dudley, J. M. (2007), 'Nonlinear envelope equation modeling of sub-cycle dynamics and harmonic generation in nonlinear waveguides', *Optics Express* **15**, 5382.

Ghosh, S., Bhagwat, A. R., Renshaw, C. K., Goh, S., Gaeta, A. L. and Kirby, B. J. (2006), 'Low-light-level optical interactions with rubidium vapor in a photonic band-gap fiber', *Physical Review Letters* **97**(2), 023603.

Ghosh, S., Sharping, J. E., Ouzounov, D. G. and Gaeta, A. L. (2005), 'Resonant optical interactions with molecules confined in photonic band-gap fibers', *Physical Review Letters* **94**(9), 093902.

Guanshi, Q., Jose, R. and Ohishi, Y. (2007), 'Design of ultimate gain-flattened O-, E-, and S+ C+ L ultrabroadband fiber amplifiers using a new fiber Raman gain medium', *IEEE J. Lightwave Technol.* **25**(9), 2727–2738.

Hall, D. W. and Newhouse, M. A. (1989), 'Nonlinear optical susceptibilities of high-index glasses', *Appl. Phys. Lett.* **54**, 1293–1295.

Headley, C. and Agrawal, G. P. (1996), 'Unified description of ultrafast stimulated Raman scattering in optical fibers', *J. Opt. Soc. Am. B* **13**(10), 2170–2177.

Hellwarth, R. W. (1977), 'Generation of time-reversed wave fronts by nonlinear refraction', *J. Opt. Soc. Am.* **67**, 1–3.

Ishizawa, A., Nishikawa, T., Aozasa, S., Mori, A., Tadanaga, O., Asobe, M. and Nakano, H. (2008), 'Demonstration of carrier envelope offset locking with low pulse energy', *Opt. Express* **16**(7), 4706–4712.

Jose, R. and Ohishi, Y. (2007), 'Higher nonlinear indices, Raman gain coefficients, and bandwidths in the TeO_2-ZnO-Nb_2O_5-MoO_3 quaternary glass system', *Applied Physics Letters* **90**(21), 211104 1–3.

Kaiser, P., Marcatili, E. A. J. and Miller, S. E. (1973), 'A new optical fiber', *Bell Sys. Tech. J.* **52**, 265–269.

Kakkar, C. and Thyagarajan, K. (2005), 'High gain Raman amplifier with inherent gain flattening and dispersion compensation', *Optics Communications* **250**(1-3), 77–83.

Kang, I., Krauss, T. D., Wise, F. W., Aitken, B. G. and Borrelli, N. F. (1995), 'Femtosecond measurement of enhanced optical nonlinearities of sulphide glasses and heavy-metal-doped oxide glasses', *JOSA B* **12**, 2053–2059.

Karasawa, N., Nakamura, S., Nakagawa, N., Shibata, M., Morita, R., Shigekawa, H. and Yamashita, M. (2001), 'Comparison between theory and experiment of nonlinear propagation for a-few-cycle and ultrabroadband optical pulses in a fused-silica fiber', *IEEE J. Quantum Electronics* **37**, 398.

Kibler, B., Dudley, J. M. and Coen, S. (2005), 'Supercontinuum generation and nonlinear pulse propagation in photonic crystal fiber: Influence of the frequency-dependent effective mode area', *Applied Physics B* **81**, 337–342.

Kikuchi, K., Taira, K. and Sugimoto, N. (2002), 'Highly nonlinear bismuth oxide-based glass fibers for all-optical signal processing', *Electron. Lett.* **38**, 166–167.

Kim, J. H., Chen, M.-K., Yang, C.-E., Lee, J., Yin, S. S., Ruffin, P., Edwards, E., Brantley, C. and Luo, C. (2008a), 'Broadband IR supercontinuum generation using single crystal sapphire fibers', *Opt. Express* **16**(6), 4085–4093.

Kim, J. H., Chen, M.-K., Yang, C.-E., Lee, J., Shi, K., Liu, Z., Yin, S. S., Reichard, K., Ruffin, P., Edwards, E., Brantley, C. and Luo, C. (2008b), 'Broadband supercontinuum generation covering uv to mid-IR region by using three pumping sources in single crystal sapphire fiber', *Opt. Express* **16**(19), 14792–14800.

Knight, J. C. and Skryabin, D. V. (2007), 'Nonlinear waveguide optics and photonic crystal fibers', *Opt. Express* **15**, 15365.

Knight, J. C., Birks, T. A., Russell, P. S. J. and Atkin, D. M. (1996), 'All-silica single-mode optical fiber with photonic crystal cladding', *Opt. Lett.* **21**, 1547–1549.

Kolesik, M. and Moloney, J. V. (2004), 'Nonlinear optical pulse propagation simulation: From Maxwell's to unidirectional equations', *Phys. Rev. E* **70**(3), 036604.

Kolesik, M., Moloney, J. V. and Mlejnek, M. (2002), 'Unidirectional optical pulse propagation equation', *Physical Review Letters* **89**, 283902–1.

Kolesik, M., Wright, E. M. and Moloney, J. V. (2004), 'Simulation of femtosecond pulse propagation in sub-micron diameter tapered fibers', *Appl. Phys. B* **79**, 293.

Konar, S., Jana, S. and Mishra, M. (2005), 'Induced focusing and all optical switching in cubic quintic nonlinear media', *Opt. Commun* **255**(1-3), 114–129.

Koos, C., Jacome, L., Poulton, C., Leuthold, J. and Freude, W. (2007), 'Nonlinear silicon-on-insulator waveguides for all-optical signal processing', *Opt. Express* **15**(10), 5976–5990.

Kuhlmey, B. T., McPhedran, R. C. and Sterke, C. d. (2002a), 'Modal cutoff in microstructured optical fibres', *Opt. Lett.* **27**, 1684–1686.

Kuhlmey, B. T., White, T. P., Renversez, G., Maystre, D., Botten, L. C., de Sterke, C. M. and McPhedran, R. C. (2002b), 'Multipole method for microstructured optical fibers. II. implementation and results', *J. Opt. Soc. Am. B* **19**(10), 2331–2340.

Kumar, V. V. R. K., George, A., Knight, J. and Russell, P. (2003), 'Tellurite photonic crystal fiber', *Opt. Express* **11**(20), 2641–2645.

Laegsgaard, J. (2007), 'Mode profile dispersion in the generalised nonlinear Schrödinger equation', *Opt. Express* **15**(24), 16110–16123.

Lee, J. H., Kikuchi, K., Nagashima, T., Hasegawa, T., Ohara, S. and Sugimoto, N. (2005b), 'All fiber-based 160-Gbit/s add/drop multiplexer incorporating a 1-m-long bismuth oxide-based ultra-high nonlinearity fiber', *Opt. Express* **13**(18), 6864–6869.

Lee, J. H., Nagashima, T., Hasegawa, T., Ohara, S., Sugimoto, N. and Kikuchi, K. (2005a), 'Four-wave-mixing-based wavelength conversion of 40-Gb/s nonreturn-to-zero signal using 40-cm bismuth oxide nonlinear optical fiber', *IEEE Photon. Technol. Lett.* **17**(7), 1474–1476.

Leong, J. Y. Y., Petropoulos, P., Price, J. H. V., Ebendorff-Heidepriem, H., Asimakis, S., Moore, R. C., Frampton, K. E., Finazzi, V., Feng, X., Monro, T. M. and Richardson, D. J. (2006), 'High-nonlinearity dispersion-shifted lead-silicate holey fibers for efficient 1-μm pumped supercontinuum generation', *J. Lightwave Technol.* **24**(1), 183.

Light, P. S., Benabid, F., Couny, F., Maric, M. and Luiten, A. N. (2007), 'Electromagnetically induced transparency in Rb-filled coated hollow-core photonic crystal fiber', *Opt. Lett.* **32**(10), 1323–1325.

Lin, Q. and Agrawal, G. P. (2004), 'Vector theory of cross-phase modulation: Role of nonlinear polarization rotation', *IEEE J. Quantum Electronics* **40**, 958.

Lin, Q., Painter, O. J. and Agrawal, G. P. (2007), 'Nonlinear optical phenomena in silicon waveguides: Modeling and applications', *Opt. Express* **15**, 16604.

Madden, S. J., Choi, D.-Y., Bulla, D. A., Rode, A. V., Luther-Davies, B., Ta'eed, V. G., Pelusi, M. D. and Eggleton, B. J. (2007), 'Long, low loss etched As_2S_3 chalcogenide waveguides for all-optical signal regeneration', *Opt. Express* **15**(22), 14414–14421.

Magi, E. C., Fu, L. B., Nguyen, H. C., Lamont, M. R. E., Yeom, D. I. and Eggleton, B. J. (2007), 'Enhanced Kerr nonlinearity in sub-wavelength diameter As_2Se_3 chalcogenide fiber tapers', *Opt. Express* **15**, 10324.

Mamyshev, P. V. and Chernikov, S. V. (1990), 'Ultrashort-pulse propagation in optical fibers', *Opt. Lett.* **19**, 1076.

Meng, L. S., Roos, P. A. and Carlsten, J. L. (2002), 'Continuous-wave rotational Raman laser in H_2', *Opt. Lett.* **27**, 1226.

Monro, T. M. and Ebendorff-Heidepriem, H. (2006), 'Progress in microstructured optical fibers', *Annu. Rev. Mater. Res.* **36**, 467–495.

Monro, T. M., West, Y. D., Hewak, D. W., Broderick, N. G. R. and Richardson, D. J. (2000), 'Chalcogenide holey fibres', *Elect. Lett.* **36**, 1998–2000.

Mori, A., Masuda, H., Shikano, K. and Shimizu, M. (2003), 'Ultra-wide-band tellurite-based fiber Raman amplifier', *IEEE J. Lightwave Technol.* **21**(5), 1300–1306.

Mullner, P. and Hainberger, R. (2006), 'Structural optimization of silicon-on-insulator slot waveguides', *IEEE Photon. Technol. Lett.* **18**, 2557.

Nagel, M., Marchewka, A. and Kurz, H. (2006), 'Low-index discontinuity terahertz waveguides', *Opt. Express* **14**(21), 9944–9954.

Okamoto, K. (2006), *Fundamentals of Optical Waveguides*, Academic Press.

Omenetto, F. G., Wolchover, N. A., Wehner, M. R., Ross, M., Efimov, A., Taylor, A. J., Kumar, V. V. R. K., George, A. K., Knight, J. C., Joly, N. Y. and Russell, P. S. J. (2006), 'Spectrally smooth supercontinuum from 350 nm to 3 μm in sub-centimeter lengths of soft-glass photonic crystal fibers', *Opt. Express* **14**(11), 4928–4934.

Perlin, V. E. and Winful, H. G. (2002), 'Optimal design of flat-gain wide-band fiber Raman amplifiers', *IEEE J. Lightwave Technol.* **20**(2), 250–254.

Perrone, M. R., Piccinno, V., Nunzio, D. and Nassisi, V. (1997), 'Dependence of rotational and vibrational Raman scattering on focusing geometry', *IEEE. J. Quantum Electron* **33**, 938.

Prasad, A., Zha, C.-J., Wang, R.-P., Smith, A., Madden, S. and Luther-Davies, B. (2008), 'Properties of $Ge_xAs_ySe_{1-x-y}$ glasses for all-optical signal processing', *Opt. Express* **16**, 2804.

Price, J. H. V., Monro, T. M., Ebendorff-Heidepriem, H., Poletti, F., Horak, P., Finazzi, V., Leong, J. Y. Y., Petropoulos, P., Flanagan, J. C., Brambilla, G., Feng, X. and Richardson, D. J. (2007), 'Mid-IR supercontinuum generation from nonsilica microstructured optical fibers', *IEEE J. Selected Topics in Quantum Electronics* **13**, 738.

Requejo-Isidro, J., Mairaj, A. K., Pruneri, V., Hewak, D. W., Netti, M. C. and Baumberg, J. J. (2003), 'Self-refractive non-linearities in chalcogenide based glasses', *J. Non-Cryst. Solids* **317**, 241–246.

Roberts, P., Couny, F., Sabert, H., Mangan, B., Birks, T., Knight, J. and Russell, P. (2005a), 'Loss in solid-core photonic crystal fibers due to interface roughness scattering', *Opt. Express* **13**(20), 7779–7793.

Roberts, P., Couny, F., Sabert, H., Mangan, B., Williams, D., Farr, L., Mason, M., Tomlinson, A., Birks, T., Knight, J. and Russell, P. S. J. (2005b), 'Ultimate low loss of hollow-core photonic crystal fibres', *Opt. Express* **13**(1), 236–244.

Roussignol, P., Ricard, D., Lukasik, J. and Flytzanis, C. (1987), 'New results on optical phase conjugation in semiconductor-doped glasses', *J. Opt. Soc. Am. B* **4**(1), 5–13.

Salem, R., Foster, M. A., Turner, A. C., Geraghty, D. F., Lipson, M. and Gaeta, A. L. (2007), 'All-optical regeneration on silicon chip', *Opt. Express* **15**, 7802.

Slusher, R. E., Lenz, G., Hodelin, J., Sanghera, J., Shaw, L. B. and Aggarwal, I. D. (2004), 'Large Raman gain and nonlinear phase shifts in high-purity As_2Se_3 chalcogenide fibers', *J. Opt. Soc. Am. B* **21**, 1146.

Snyder, A. W. and Love, J. D. (1995), *Optical Waveguide Theory*, Chapman and Hall, London.

Song, K. Y., Abedin, K. S., Hotate, K., Herráez, M. G. and Thévenaz, L. (2006), 'Highly efficient Brillouin slow and fast light using As_2Se_3 chalcogenide fiber', *Opt. Express* **14**(13), 5860–5865.

Steel, M. (2004), 'Reflection symmetry and mode transversality in microstructured fibers', *Opt. Express* **12**(8), 1497–1509.

Steel, M. J., White, T. P., de Sterke, C. M., McPhedran, R. C. and Botten, L. C. (2001), 'Symmetry and degeneracy in microstructured optical fibers', *Opt. Lett.* **26**, 488.

Stegeman, G. I. and Stolen, R. H. (1989), 'Waveguides and fibers for nonlinear optics', *J. Opt. Soc. Am. B* **6**(4), 652–662.

Stegeman, R., Jankovic, L., Hongki, K., Rivero, C., Stegeman, G., Richardson, K., Delfyett, P., Yu, G., Schulte, A. and Cardinal, T. (2003), 'Tellurite glasses with peak absolute Raman gain coefficients up to 30 times that of fused silica', *Optics Letters* **28**(13), 1126–1128.

Sugimoto, N., Nagashima, T., Hasegawa, T., Ohara, S., Taira, K. and Kikuki, K. (2004), Bismuth-based optical fiber with nonlinear coefficient of 1360 w^{-1} km^{-1}, in 'Proc. OFC', Opt. Soc. Am., Anaheim, CA, p. PDP26.

Ta'eed, V., Baker, N. J., Fu, L., Finsterbusch, K., Lamont, M. R. E., Moss, D. J., Nguyen, H. C., Eggleton, B. J., Choi, D. Y., Madden, S. and Luther-Davies, B. (2007), 'Ultrafast all-optical chalcogenide glass photonic circuits', *Opt. Express* **15**, 9205.

Tomasi, F. D. (2001), 'Stimulated rotational and vibrational Raman scattering by elliptical polarized pump radiation', *Physical Review A* **64**, 023812.

Tong, L., Gattass, R. R., Ashcom, J. B., He, S., Lou, J., Shen, M., Maxwell, I. and Mazur, E. (2003), 'Subwavelength-diameter silica wires for low-loss optical wave guiding', *Nature* **426**, 816–819.

White, T. P., Kuhlmey, B. T., McPhedran, R. C., Maystre, D., Renversez, G., de Sterke, C. M. and Botten, L. C. (2002), 'Multipole method for microstructured optical fibers. I. Formulation', *J. Opt. Soc. Am. B* **19**(10), 2322–2330.

Wiederhecker, G. S., Cordeiro, C. M. B., Couny, F., Benabid, F., Maier, S. A., Knight, J. C., Cruz, C. H. B. and Fragnito, H. L. (2007), 'Field enhancement within an optical fibre with a subwavelength air core', *Nature Photon.* **1**, 115.

Xia, C., Kumar, M., Cheng, M.-Y., Hegde, R. S., Islam, M. N., Galvanauskas, A., Winful, H. G., Fred, L., Terry, J., Freeman, M. J., Poulain, M. and Mazé, G. (2007), 'Power scalable mid-infrared supercontinuum generation in ZBLAN fluoride fibers with up to 1.3 watts time-averaged power', *Opt. Express* **15**(3), 865–871.

Xia, C., Kumar, M., Kulkarni, O. P., Islam, M. N., Fred, L., Terry, J., Freeman, M. J., Poulain, M. and Mazé, G. (2006), 'Mid-infrared supercontinuum generation to 4.5 μm in ZBLAN fluoride fibers by nanosecond diode pumping', *Opt. Lett.* **31**(17), 2553–2555.

Xu, Q., Almeida, V. R., Panepucci, R. R. and Lipson, M. (2004), 'Experimental demonstration of guiding and confining light in nanometer-size low-refractive-index material', *Opt. Lett.* **29**, 1626.

Yeom, D.-I., Mägi, E. C., Lamont, M. R., Roelens, M. A., Fu, L. and Eggleton, B. J. (2008), Low-energy threshold supercontinuum generated in highly nonlinear As_2Se_3 chalcogenide submicron tapers, in 'CLEO/QELS 2008 Proceeding', OSA, p. CMDD6.
Yiou, S., Delaye, P., Rouvie, A., Chinaud, J., Frey, R., Roosen, G., Viale, P., Février, S., Roy, P., Auguste, J.-L. and Blondy, J.-M. (2005), 'Stimulated Raman scattering in an ethanol core microstructured optical fiber', *Opt. Express* **13**(12), 4786–4791.
Zhai, G. and Tong, L. (2007), 'Roughness-induced radiation losses in optical micro or nanofibers', *Opt. Express* **15**(21), 13805–13816.
Zhang, J., Lin, Q., Piredda, G., Boyd, R. W., Agrawal, G. P. and Fauchet, P. M. (2007), 'Optical solitons in a silicon waveguide', *Opt. Express* **15**(12), 7682–7688.
Zhang, R., Teipel, J. and Giessen, H. (2006), 'Theoretical design of a liquid-core photonic crystal fiber for supercontinuum generation', *Opt. Express* **14**(15), 6800–6812.
Zhang, W. Q., Afshar V., S., Ebendorff-Heidepriem, H. and Monro, T. M. (2008), 'Record nonlinearity in optical fibre', *Electronics Letters* **44**, 1453–1455.
Zheltikov, A. (2005), 'Gaussian-mode analysis of waveguide-enhanced Kerr-type nonlinearity of optical fibers and photonic wires', *J. Opt. Soc. Am. B* **22**, 1100.

12

Supercontinuum generation in dispersion-varying fibers

G. Genty

12.1 Introduction

Studies of supercontinuum generation in photonic crystal fibers or dispersion-shifted highly nonlinear fibers have led to renewed interest in SC generation in other fiber types. For example, it was shown that fibers with a steadily-decreasing diameter in which the dispersive and nonlinear characteristics change as a function of propagation distance can lead to significant enhancement of the SC bandwidth and allow for an additional degree of control of the SC spectrum. In fact, it was the use of fibers with dispersion-varying profiles that motivated in the mid-1990s the pioneering work on SC generation intended for telecommunication applications [see e.g. Morioka et al., 1994a, 1994b, 1995, 1996; Mori et al., 1995]. Of course, it was already well known that efficient adiabatic pulse compression could be achieved in dispersion-varying fibers (Dianov et al., 1986), but these mid-1990s studies, anterior to the development of PCFs, were the first to specifically suggest that fibers with longitudinally-varying dispersion could be advantageous in generating broad and/or flat SC spectra (Lou et al., 1997; Mori et al., 1997). With the development of the theoretical understanding of SC generation process (see other chapters) renewed interest in dispersion-engineering has therefore occurred naturally. In this context, a key result that was reported is the demonstration of coherent SC generation in millimeter lengths of tapered fibers where the diameter of the fiber is controlled as a function of length to a great degree of accuracy (Lu and Knox, 2004). Alternative methods of controlling the dispersion and nonlinear characteristics vs. propagation distance have subsequently been developed in the form of a discrete dispersion map by concatenating fiber segments with different properties. Interestingly, we note with the development of PCFs, soliton compression has also regained interest and adiabatic or near adiabatic soliton compression

Supercontinuum Generation in Optical Fibers, ed. by J. M. Dudley and J. R. Taylor. Published by Cambridge University Press.

in longitudinally dispersion-varying PCFs or discrete dispersion maps fibers has been investigated both theoretically and experimentally (Tse et al., 2006b, 2008; Hu et al., 2006; Travers et al., 2007).

In this chapter, we review SC generation in the main types of dispersion-varying fibers and analyze in detail the underlying dynamics and specific characteristics of this regime of propagation. The chapter is organized as follows. In Section 12.2 the case of a fiber with a decreasing zero-dispersion wavelength (ZD-DF) is reviewed. Section 12.3 discusses SC generation processes in dispersion-decreasing fibers with a flattened convex dispersion profile (DF-DDF). The use of discrete fiber segments to create a dispersion map is addressed in Section 12.4.

12.2 Fibers with decreasing zero-dispersion wavelength

In a zero-dispersion wavelength decreasing fiber the diameter is steadily reduced along the direction of propagation, producing a longitudinally-varying dispersion profile. Typically, the fiber exhibits a single ZDW at short wavelengths but if the diameter is reduced below 1 μm the dispersion profile will also cross the zero line in the longer wavelengths region. Zero-dispersion wavelength decreasing fibers can be fabricated from a standard single-mode fiber (Birks and Li, 1992) or from a PCF (Nguyen et al., 2005) using a heat and pull technique. The former approach requires a substantial reduction in the fiber diameter which can dramatically alter its mechanical robustness. The latter only necessitates a small reduction factor but careful pressure control is needed to prevent collapsing of the micro-cladding structure (Wadsworth et al., 2005). In a set of experiments Lu et al. have demonstrated that it is possible to generate supercontinuum spectra with enhanced bandwidths in ZD-DFs drawn from PCFs (Lu and Knox, 2004, 2006; Lu et al., 2005). More specifically, launching 100 fs, 20 kW pulses at 880 nm in a PCF whose diameter was reduced from 3.3 μm to 1.45 μm over a distance of only 5 mm, they have generated an SC that expanded down to a 400 nm wavelength. Comparatively, launching the same pulses into a PCF of similar length but with a constant 3.3 μm diameter did not lead to any significant spectral broadening. The key factor in Lu's experiments was the steady decrease in the PCF diameter which induced a gradual longitudinal shift of the zero-dispersion wavelength from 865 nm at the input end to 630 nm at the output end. Although it was then neglected in the analysis, the reduction of the PCF diameter affects the area of the propagating fundamental mode and thereby also leads to a longitudinally-increasing nonlinearity which contributes to the SC bandwidth enhancement. The other important factor in the success of the experiments was the precise control of the launched input power so as to shift the soliton fission distance near the fiber end. This particular feature allowed the generation of a structure-free SC spectrum, although it also prevented further extension of the spectrum on the short wavelengths side through soliton-dispersive wave interaction as is typically observed in long length of PCFs with constant dispersion profile (Genty et al., 2004; Yulin et al., 2004; Gorbach and Skryabin, 2007). Another attractive specificity of this dispersion-management scheme is the high coherence and low noise sensitivity that was reported across the entire SC bandwidth as a result of the very short length of fiber utilized. In another set of experiments, Lu et al. also showed that one can engineer in a similar way the dispersion characteristics of both principal axes of polarization in a polarization-maintaining PCF which allows for an extra degree of freedom in the control of the blue spectral edge of the SC (Lu and Knox, 2006). Another aspect of practical importance is the fact that the holes of a dispersion-managed PCF can be readily collapsed at both ends to improve the environmental stability (Pal and Knox, 2007a).

In order to illustrate the basic dynamics of SC generation in a ZD-DF fiber we consider the case depicted in Fig. 12.1 where the fiber diameter decreases linearly from $D_{in} = 2.5$ μm to $D_{out} = 1.5$ μm over a length of 10 cm. Note that similar dynamics are observed for sub-cm lengths when pumped by pulses with higher peak power than envisaged here. The zero-dispersion wavelength of such a fiber changes continuously from 780 nm at the input to 640 nm at the output. We choose a pump wavelength at 820 nm in the anomalous dispersion region which is known to yield the broadest SC spectrum (Dudley et al., 2006). For this choice of pump wavelength the dispersion increases nearly quadratically by a factor of 6 from the input to the output. Over the same length the nonlinearity is enhanced by a factor of 2.5. When 100 fs, 3 kW peak power pulses are injected into this ZD-DF a coherent SC spanning from 420 to 1100 nm is generated (see Fig. 12.2 top curves). The input pulse parameters correspond to an average power of 25 mW at a repetition

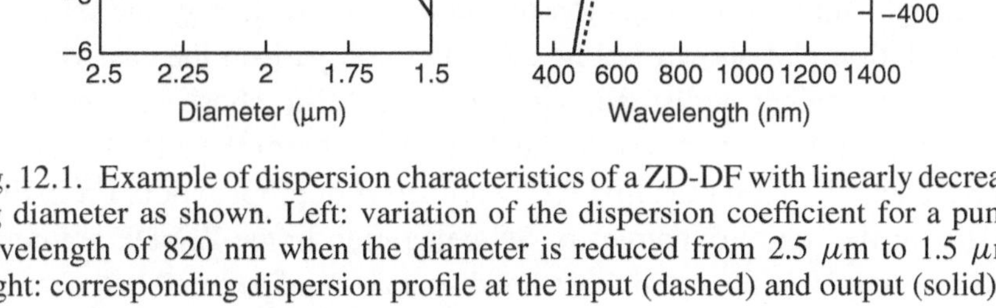

Fig. 12.1. Example of dispersion characteristics of a ZD-DF with linearly decreasing diameter as shown. Left: variation of the dispersion coefficient for a pump wavelength of 820 nm when the diameter is reduced from 2.5 μm to 1.5 μm. Right: corresponding dispersion profile at the input (dashed) and output (solid).

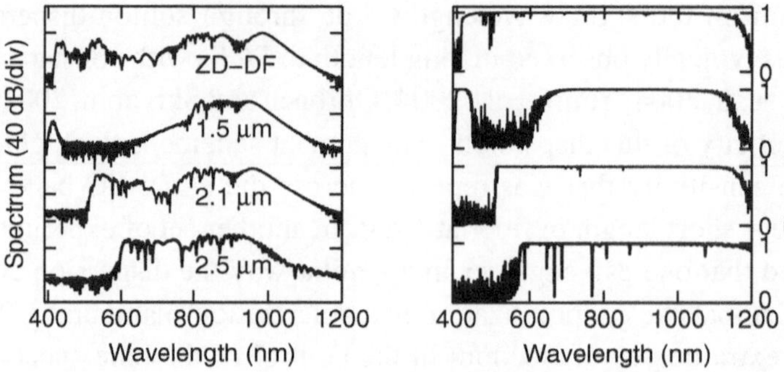

Fig. 12.2. Comparison of SC generated in the ZD-DF of Fig. 12.1 and in fiber with constant diameter. Left: from top to bottom, spectra after 10 cm in the ZD-DF, fiber with 1.5 μm diameter, fiber with 2.1 μm diameter and fiber with 2.5 μm diameter. Right: Corresponding degree of coherence.

rate of 80 MHz, readily produced by unamplified mode-locked Ti:Sapphire lasers. To illustrate the benefit of a zero-dispersion wavelength decreasing fiber the SC spectra generated in similar fiber lengths, but with a constant dispersion profile, are also plotted for various diameter values (and therefore dispersion profiles). It is clear that all the SC are coherent but, most importantly, that the ZD-DF generates the broadest continuous SC spectrum. On the one hand, the long wavelength edge is only slightly affected by the particular dispersion profile but, on the other hand, a dramatic difference is observed on the short wavelength edge. In a constant dispersion fiber the phase-matching condition for DW generation allows us to obtain a large SC bandwidth only for large detuning values of the pump with respect to the ZDW (see Chapter 4). Consequently, obtaining a large bandwidth with a constant dispersion fiber requires a small fiber diameter in order to increase this detuning which will shift the DW location further to the blue. However, although we do get a large bandwidth in this case, it is mitigated by a substantial spectral gap between the blue DW spectral components and the higher-order soliton pump. The ZD-DF scheme allows us to overcome this limitation and the SC generated in the ZD-DF contains more energy in the visible spectral region than SC generated in a constant diameter fiber. These specific characteristics arise from the particular dynamics associated with pulse propagation in a ZD-DF.

More insight into the propagation dynamics is provided by the temporal and spectral evolution plot shown in Fig. 12.3. For comparison the case of a 2.1 μm constant diameter fiber is also shown in Fig. 12.3. We chose this diameter purposefully as it nearly corresponds to the mean diameter of the ZD-DF and yields a similar SC bandwidth after the first few centimeters of propagation. From Fig. 12.3 it is clear that the initial dynamics of SC generation in the ZD-DF are consistent with that observed in a fiber with constant dispersion and pumped in the anomalous

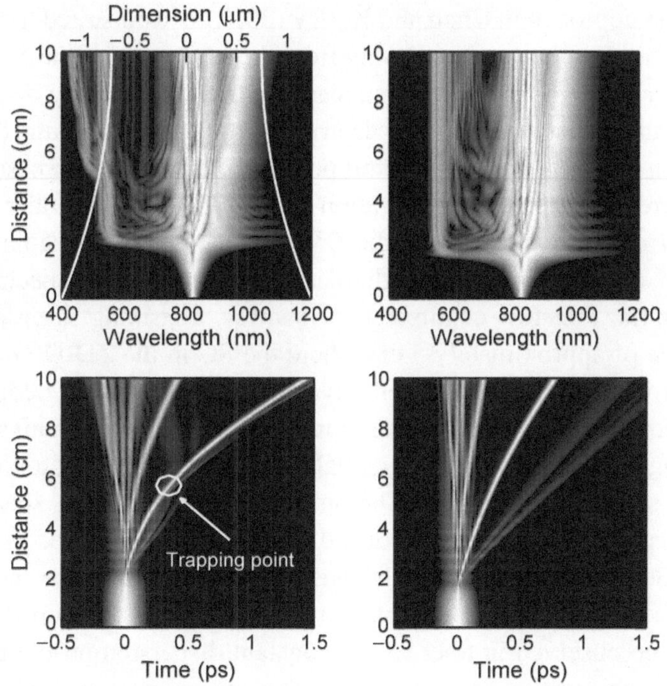

Fig. 12.3. Numerical simulations showing spectral evolution (top) and temporal evolution (bottom) in the ZD-DF (left) and fiber with constant 2.1 μm diameter (right). The solid line illustrates the change in fiber dimension along propagation.

pumping regime, i.e. (i) a short stage of initial higher-order soliton temporal compression and corresponding symmetric spectral broadening and (ii) the generation of a dispersive wave on the blue side as the higher-order soliton spectrum expands in the normal dispersion regime. Yet, the distance of maximum temporal compression or, equivalently, the fission length L_{fiss} at which higher-order soliton fission occurs is slightly longer for the ZD-DF case. In fact, this is not surprising as the increasing dispersion at the pump wavelength in the ZD-DF slows down the compression process. In both cases, the proximity of the ZDW to the pump wavelength at the point of fission results in a large spectral overlap with a the DW and leads to substantial transfer of power on the blue side of the SC spectrum with a minor gap between the DW and pump spectral components. Adapting the scaling laws of self-compression in a constant dispersion fiber (Chen and Kelley, 2002) and for input pulses of the form $P_0\,\mathrm{sech}^2(T/T_0)$ with soliton number $N \gg 1$ the fission distance in the ZD-DF is found to be relatively well approximated by the implicit relation

$$z - \sqrt{\frac{T_0^2}{\beta_2(z)\gamma(z)P_0}} = 0. \qquad (12.1)$$

Note that the approach in Chen and Kelley (2002) is well suited to the case of a ZD-DF since the cumulative error arising from the separation of the contribution of the dispersion and the Kerr effect is reduced due to the increasing dispersion. The fiber diameter at the distance L_{fiss} then determines the phase-matching condition for DW generation and thus the DW spectral position. For the ZD-DF case considered here, L_{fiss} corresponds to the propagation length at which the fiber diameter is equal to 2.1 μm. This explains why the DW is generated at the same wavelength as in the fiber with constant 2.1 μm diameter and why the two spectra are nearly identical after the first few centimeters. However, a striking discrepancy arises after a distance of approximately 5 cm where the SC in the ZD-DF experiences a sudden expansion towards the shorter wavelengths while the blue edge of the SC in the fiber with constant dispersion has already stabilized and remains unaffected. Further propagation in the ZD-DF gradually generates new spectral components towards even shorter wavelengths. The blue extension of the SC spectrum in the ZD-DF is caused by the interaction of red solitons and DW through cross-phase modulation (Genty et al., 2004), which leads to localization of some energy fraction at the short wavelength edge (Gorbach and Skryabin, 2007). A similar mechanism can of course be observed in fiber with a constant dispersion profile but typically requires longer lengths that envisaged here. The energy localization in a fiber with constant dispersion originates from Raman-induced self-frequency shift by which a soliton is constantly slowed-down (see Chapter 9 for more details). A similar effect occurs in the ZD-DF but, significantly, it is caused by the steady change in the group-velocity of the ZD-DF, which effectively simultaneously slows down the red-shifting soliton (in the same way as the soliton self-frequency shift does in a fiber with constant dispersion) and accelerates the DW spectral components. The combination of these two phenomena which are clearly observed in the temporal evolution color plots in Fig. 12.3 leads to very efficient soliton–DW interaction in the ZD-DF and extended blue edge in the SC spectrum (Travers and Taylor, 2009).

A natural question is whether the distance over which the ZD-DF diameter is decreased has a significant impact on the generated continuum. To answer this question Fig. 12.4 shows a false color representation of the SC spectrum vs. the ZD-DF length. In each case shown here the initial and final diameters of the fiber are taken to be 2.5 μm and 1.5 μm, respectively. Several observations can be made when considering the effect of the length parameter. If the length of the taper is too short (less than 1.3 cm in the case illustrated in Fig. 12.4), the soliton fission distance is not reached, no dispersive wave generation process takes place and the propagation resembles that of a higher-order soliton. For increased distances, the point of soliton fission is attained with simultaneous dispersive wave generation on the short wavelength edge. Note that the minimum distance needed to induce fission and consequently SC generation in the ZD-DF can be determined from Eq. 12.1. The

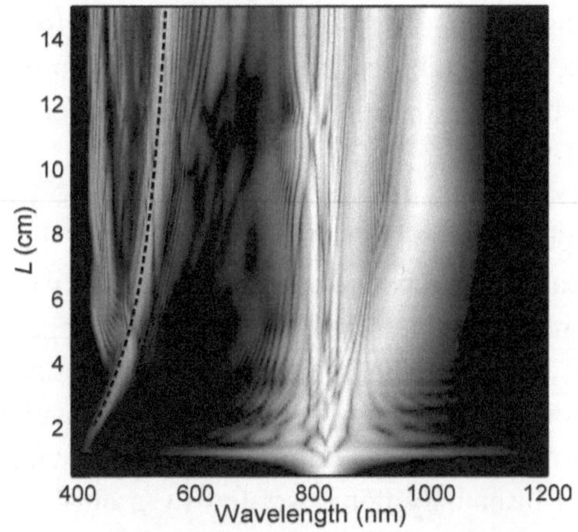

Fig. 12.4. Numerical simulations showing the SC spectrum vs. the ZD-DF length. The dotted line shows the position of the DW as generated.

spectral location of the DW is governed by the phase-matching condition imposed by the dispersion profile and thereby the fiber diameter at the soliton fission point. In the case of short ZD-DF length the soliton fission point is reached at the near-end of the fiber corresponding to smaller diameters resulting in a larger incursion of the dispersive wave into the blue spectral region at the expense of a large gap with the pump spectral components. The late generation of the dispersive wave also prevents any further interaction with the red-shifted solitonic component and further blue-shift. For increased lengths the diameter of the ZD-DF at the point of fission is larger which decreases the spectral separation between the dispersive wave and pump component. Soliton–dispersive wave interaction also occurs leading to further blue extension of the SC spectrum. Ultimately, the SC bandwidth on the short wavelengths side is limited by the group-velocity profile of the fiber at the output end so that increasing further the fiber length does not affect significantly the SC spectrum.

The high sensitivity of the initial nonlinear compression stage to the input pulse parameters leads correspondingly to a strong dependence of the fission point on these. Significantly, the phase-matching condition is thus very sensitive to the input pulse parameters offering the possibility of controlling the spectral location of the DW (Lu and Knox, 2006). Furthermore, the ability of ZD-DF to produce a SC with a large fraction of energy localized in the blue wing over a reduced distance compared to PCF with a constant dispersion is also beneficial to the stability SC characteristics as longer distance is typically associated with coherence degradation. This feature of ZD-DF was indeed demonstrated in an experiment where the generation of a SC in a short length of ZD-DF was found to yield improved stability compared to fibers with constant dispersion (Lu et al., 2005). It is therefore possible to generate a smooth SC with excellent stability properties over a sub-cm length of fiber that can be conveniently applied to the realization of a self-referenced frequency comb (Pal and Knox, 2007b).

The concept of ZD-DF for efficient SC generation has also been applied successfully to the long pulse regime (Kudlinski et al., 2006). In a set of experiments, Kudlinski et al. fabricated a PCF where control parameters were adjusted during the drawing process in order to reduce the diameter while maintaining a constant air-filling fraction in the cladding. This yielded a longitudinally diameter-decreasing fiber with corresponding decreasing zero-dispersion wavelength from 1 μm to 600 nm over a few-meter distance. Pumping the fiber with 0.6 ns pulses from a microchip Nd:YAG laser they demonstrated by a cut-back technique that the longitudinal change in the dispersion profile of the fiber led to the generation of shorter and shorter wavelengths in the SC spectrum. In the same set of experiments but using a PCF with slightly different dispersion-varying characteristics a similar evolution was observed when launching 3 ps pulses with tens of kW peak power from a passively mode-locked Yb^{3+} fiber laser operating at 1064 nm. In the long pulse regime the dynamics behind the extension of the blue SC edge to shorter wavelengths can be understood in the same way as in the short pump pulse regime. Specifically, the change in the fiber group-velocity enhances the interaction between the multiple solitons (resulting from the initial long pulse break-up induced by modulation instability) with the short wavelength spectral components that are either directly generated from the initial modulation instability or from subsequent DW generation processes.

12.3 Fibers with dispersion-decreasing convex profile

In this section we address SC generation in dispersion-flattened fibers with longitudinally-decreasing dispersion (DF-DDF). The specificity of these fibers lies in their convex dispersion profile which exhibits two zero-dispersion wavelengths whose separation reduces with distance (see Fig. 12.5). In this type of fiber, the dispersion at the pump wavelength typically changes from anomalous at the input end to normal at the output end.

Fibers with two ZDWs were demonstrated in the early 1980s using double clad fiber technology (Monerie, 1983; Mamyshev et al., 1993) and it was in fact the use of such "dispersion-flattened" fibers with longitudinally decreasing dispersion that originally led to the extensive studies of SC generation at telecommunications wavelengths.

The initial work on SC generation in DF-DDF was pioneered by Morioka et al. who successfully demonstrated SC generation using a high speed Er^{3+}-based

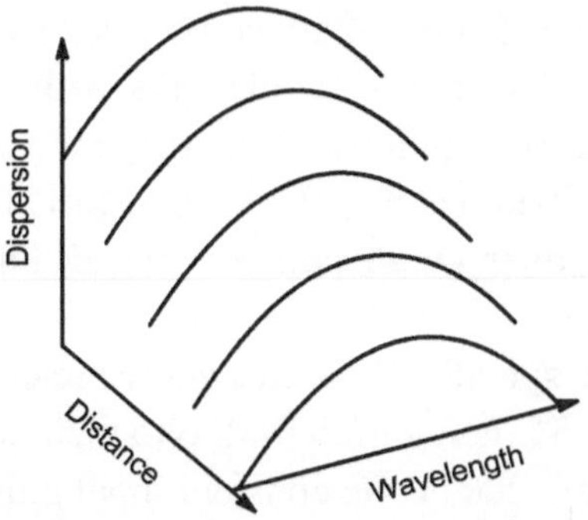

Fig. 12.5. Convex dispersion profile of a dispersion-decreasing fiber.

mode-locked fiber ring laser (Morioka et al., 1994a). In these experiments, 1.7 W peak power pulses with 3.3 ps duration were injected into a 3 km length of fiber with a small anomalous dispersion at the pump wavelength. A spectrally flat 200 nm bandwidth SC extending from 1440 to 1640 nm at the −20 B level was observed, and stable spectral slicing over four wavelength channels was demonstrated. Subsequent experiments using the same SC source quantified the stability of the generated SC by characterizing the time-domain properties of the spectrally-sliced pulses obtained (Morioka et al., 1994b, 1995). In another experiment using 3 km of fiber, an extended SC spanning over 1150–1770 nm was demonstrated using a lower repetition rate with a source generating 0.7 ps pulses with 1.2 kW peak power (Mori et al., 1995). The potential of SC generated in DF-DDF for telecommunications was then shown in a system experiment where a similar setup using a 10 GHz fiber-based source was used to successfully demonstrate 1 Tbit/s transmission over a 40 km fiber link (Morioka et al., 1996). In a systems context where stability properties are important considerations, Nakazawa et al. showed that the use of DF-DDF fiber could lead to improved stability by avoiding higher-order soliton break-up (Nakazawa et al., 1998).

In these early studies, although the detailed dispersion profiles of the particular fibers used to generate the broadband SC were not reported, later research showed that this was, in fact, a critical factor in determining key SC characteristics such as the overall bandwidth and the spectral flatness. Indeed, an important study by Lou et al. suggested that fibers with longitudinally-varying dispersion could be advantageous in generating flatter and broader SC spectra when compared to fibers with uniform longitudinal dispersion (Lou et al., 1997). This was then examined in detail by Mori et al. (1997) for the particular SC fiber that had been used in the previous experiments reported in Morioka et al. (1994a, 1995). A major result of this work was the statement of clear fiber design criteria for the generation of a uniform SC spectra: (i) a dispersion profile presenting decreasing anomalous GVD with propagation and (ii) a dispersion-flattened convex wavelength variation of the GVD exhibiting two zero-dispersion wavelengths at the fiber input and centered near the pump. Detailed investigations of the dispersion-decreasing design were subsequently reported by Okuno et al. (1998) and Mori et al. (2001) who explored SC generation in DF-DDF over a wide parameter space but restricted their analysis to low input soliton number.

In order to illustrate the specific dynamics and characteristics of SC generation in DF-DDFs we examine SC generation with parameters typical of those used in previous studies (Mori et al., 2001). We consider input pulses at 1550 nm with peak power $P_0 = 1$ W and duration (FWHM) of $\Delta\tau = 5$ ps propagating in a 1 km length of DF-DDF. The input pulse parameters correspond to an average power of 50 mW at a repetition rate of 10 GHz, typical of amplified pulses from high-repetition rate harmonically-mode-locked fiber lasers. The dispersion coefficients are calculated from a longitudinally varying dispersion model (Mori et al., 2001) given by

$$D(z) = D_0(1 - z/L_0) + (1/2)D_2(\lambda - \lambda_0)^2 \tag{12.2}$$

with $D_0 = 6$ ps $\times$ nm^{-1} $\times$ km^{-1}, $D_2 = -2 \times 10^{-4}$ ps $\times$ nm^{-3} $\times$ km^{-1}, and $L_0 =$ 900 m. In this model the parameter D_2 sets the convexity of the profile while L_0 represents the distance at which the dispersion changes from anomalous to normal at the pump wavelength. Note that the dispersion parameters used in the simulations are within the realistic range of what might be expected with recent advances in dispersion-flattened and PCF technology around 1550 nm (Okuno et al., 1999; Reeves et al., 2003; Saitoh and Koshiba, 2004; Falk et al., 2005). Fig. 12.6 plots the dispersion parameter D at selected distances to illustrate the longitudinal evolution of the convex dispersion from anomalous to normal values in the vicinity of the

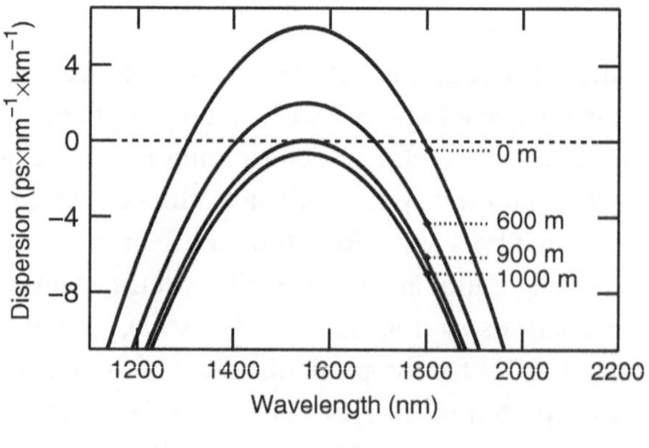

Fig. 12.6. Dispersion profile for the DF-DDF as described in the text at selected fiber lengths.

pump. For values of $z < L_0$, the profile presents two ZDWs whose separation decreases with distance while at distances exceeding L_0 the dispersion is normal at all wavelengths. The pump wavelength is assumed to correspond to λ_0, and the nonlinear coefficient of the fiber is taken to be constant along propagation $\gamma = 5\ \mathrm{W}^{-1} \times \mathrm{km}^{-1}$.

In Fig. 12.7, we plot the spectrum and corresponding temporal pulse profile at selected propagation distances as shown. The SC generation process consists of two distinct stages, an initial stage of slow symmetrical spectral broadening followed

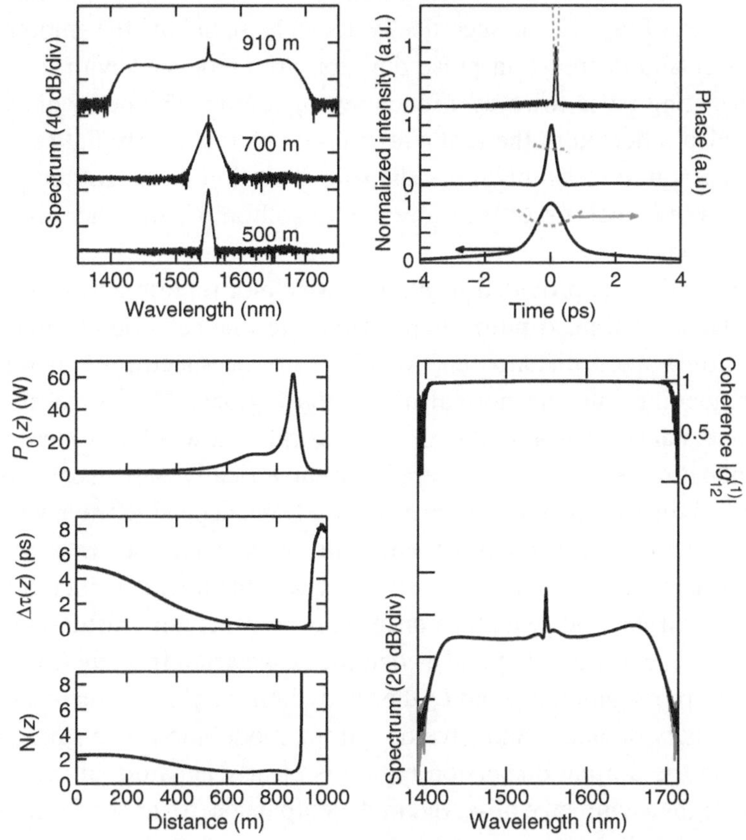

Fig. 12.7. Top: spectra (left) and corresponding temporal profile (right) for the DF-DDF case as described in the text at selected fiber lengths. The dashed lines show the temporal phase. Bottom: longitudinal evolution of the peak power $P_0(z)$, pulse duration (FWHM) $\Delta\tau(z)$ and (for propagation distance $z < L_0$) the associated soliton order (left) and results from 20 simulations (right) showing the output spectra (bottom curves, left axis) and corresponding degree of coherence (top curve, right axis). For the spectral plots, the gray curves show the individual spectra from the ensemble while the solid line shows the calculated mean. Note that the difference between individual shots is apparent only in the spectral wings.

by an abrupt transition into a smooth SC with a perfect degree of coherence. The initial pump pulse here corresponds to a second-order soliton $N = 2.3$, and the adiabatically decreasing dispersion combined with the Kerr nonlinearity induces an initial phase of temporal compression. Such adiabatic pulse compression in dispersion-decreasing fibers has been analyzed originally for injected fundamental solitons (Kuehl, 1988; Chernikov et al., 1993) but higher-order soliton compression in these fibers leads to similar dynamics except for a much reduced optimum compression distance and larger compression ratio (Pelusi and Liu, 1997; Fatemi, 2002). The phase of initial compression can be more conveniently visualized by plotting the evolution of the key pulse parameters along the length of the DF-DDF (see bottom part of Fig. 12.7). Specifically, it can be seen how the smooth increase in peak power and decrease in pulse duration are associated with the evolution of the propagating pulse towards a fundamental soliton. The near-adiabatic compression is also reflected in the temporal phase which is nearly flat as the soliton number reaches unity. It is this near-adiabatic evolution in the soliton parameters that precludes the break-up of the pulse due to soliton fission otherwise observed in SC generation.

Additional insight is provided using a density plot representation as shown in Fig. 12.8. The point of maximum compression is reached at around 850 m, slightly before the propagation distance L_0. At this point the spectrum experiences significant broadening into the normal dispersion regions. This is because after a propagation distance of around 850 m the spectral bandwidth has increased sufficiently to induce dispersive wave generation symmetrically with respect to the pump wavelength. In the case of a fiber with a single ZDW, dispersive wave generation is associated with a modification of the pump spectrum due to spectral recoil, but the simultaneous generation of two dispersive waves in the fiber considered here leads to opposite spectral recoils which balance. This prevents any shift in the spectrum of the pump and contributes to yield a continuous spectrum from the edges towards the center. For propagation beyond L_0 the bandwidth stabilizes somewhat while the temporal intensity broadens and develops strong modulation. The strong temporal modulation in the normal dispersion region is clearly apparent in the false color figure. The slight asymmetry observed in the output spectra arises from stimulated Raman scattering. The extension of the SC spectrum into the normal dispersion region has been sometimes mistakenly attributed in the literature to a four-wave mixing process. To emphasize that this extension is indeed due to a DW generation mechanism the theoretically calculated DW phase-matched wavelengths are superimposed on the spectrum of Fig. 12.8, showing excellent agreement with the spectral evolution.

Besides the symmetrical aspect of the spectrum, another noteworthy difference with SC generation in other types of fibers is the apparent flatness of the SC

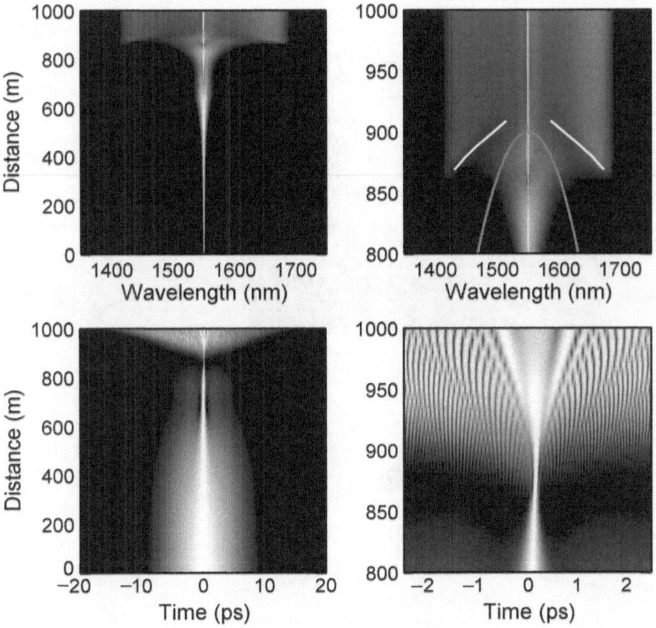

Fig. 12.8. The evolution of spectral (top) and temporal intensity (bottom) for a 5 ps duration, 1 W peak power pulse at 1550 nm propagating in a DF-DDF as described in the text. The right figures show a zoom over the region where the dispersion at the pump wavelength changes from anomalous to normal. The gray and white lines mark the ZDWs and the calculated DW phase-matched wavelengths, respectively.

spectrum. Two factors principally influence the flatness of the spectrum. Firstly, although the phase-matching condition for dispersive wave generation is intrinsically narrowband, the longitudinally varying separation between the pump and the ZDWs results in a continuous and broadband range of phase-matched wavelengths. Secondly, after propagating beyond L_0 when the dispersion is normal everywhere, the residual pump components temporally broaden and overlap with the frequency-shifted dispersive waves, facilitating interaction through cross-phase modulation (Genty et al., 2004; Yulin et al., 2004). However, there is no localization of energy in the normal dispersion regions in this case. The combined result of these dynamics is a broad and flat SC almost free of fine structure. The residual spectral structure at the pump wavelength is a typical feature of higher-order soliton compression, and corresponds to the broad time-domain pedestal. The absence of soliton fission results in a highly coherent propagation regime as is shown explicitly in Fig. 12.7 where one clearly observes negligible pulse-to-pulse intensity noise, and near perfect coherence across the entire SC bandwidth.

Aspects of these specific dynamics can be better visualized in the projected-axis spectrogram plotted in Fig. 12.9. The particular dispersion profile of the fiber

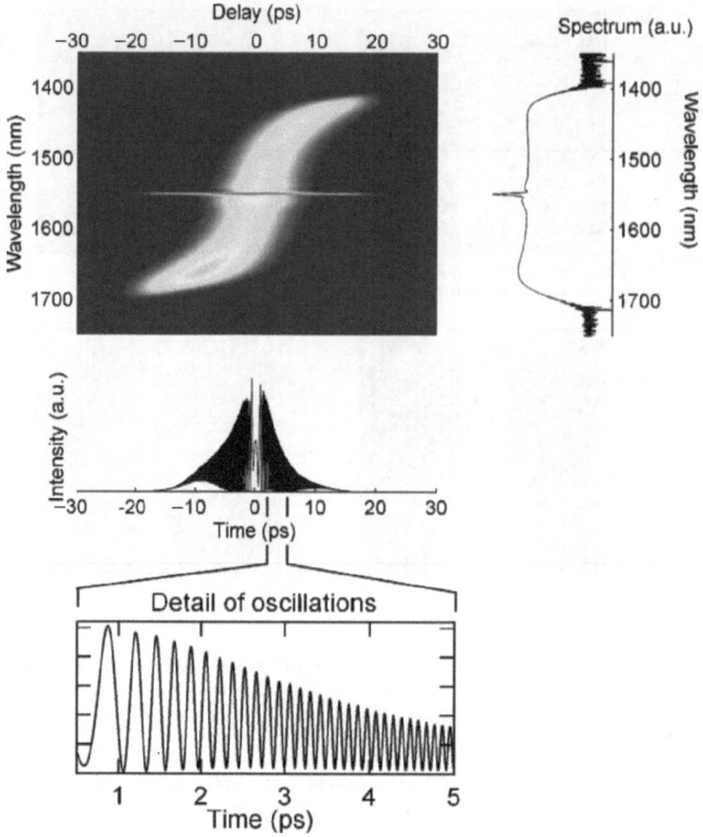

Fig. 12.9. Projected axis spectrogram of the output pulse of Fig. 12.7 illustrating how the flattened spectrum is associated with a complex modulated temporal structure. The gate function used in the spectrogram calculation had duration (FWHM) of $\Delta\tau = 1$ ps, the same as the initial pump pulse. After Genty et al. (2007); reproduced with permission.

manifests itself in the fact that the dispersive waves are spread in time-wavelength space in an S-like structure. A careful inspection of the spectrogram also reveals the physical origin of the strong temporal modulation in the pulse wings as arising from the beating of the unconverted and dispersed pump and the DW components. The spectrogram also clearly reveals how the unconverted pump is sufficiently temporally broadened to overlap with all the dispersive wave components. Of additional interest is the fact that the S-like characteristic of the spectrogram leads to a temporal chirp of this intensity modulation (see the expanded portion of the intensity profile).

Although propagation in DF-DDF was initially studied in the context of low power SC generation for telecommunications applications, suitable scaling of the

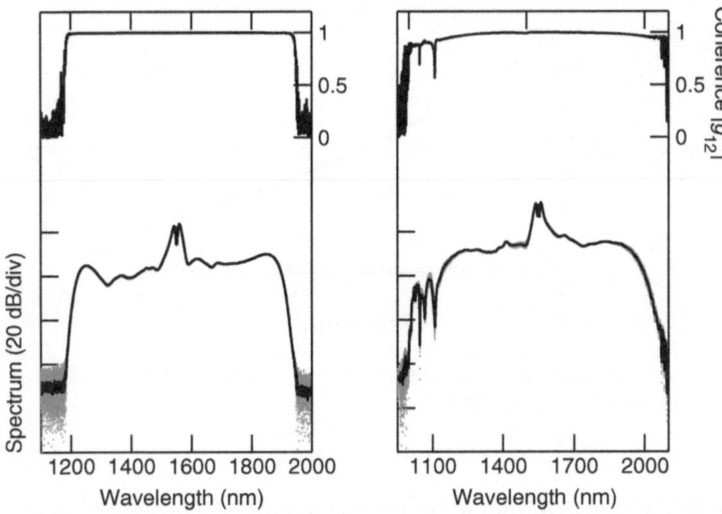

Fig. 12.10. Results showing evolution in DF-DDF of pulses with 1 ps duration and 200 W peak power (left) and 1 ps duration and 1.2 kW peak power (right). The figures present results from 20 simulations showing the output spectra (bottom curves, left axis) and corresponding degree of coherence (top curves, right axis). For the spectral plots, the gray curves show the individual spectra from the ensemble while the solid line shows the calculated mean, although the difference is apparent only in the wings of the spectra.

fiber dispersion can lead to similar generation of broadband flat SC with input pulses over a much wider parameter range. Two examples of such designs yielding coherent SC around 1550 m with significantly higher peak power than considered above are displayed in Fig. 12.10. The first example considers 200 W peak power pulses of duration (FWHM) 1 ps propagating in a 7 m length of DF-DDF with $D_0 = 10$ ps $\times$ nm^{-1} $\times$ km^{-1}, $D_2 = -2 \times 10^{-4}$ ps $\times$ nm^{-3} $\times$ km^{-1}, $L_0 = 6$ m, and $\gamma = 5$ W^{-1} $\times$ km^{-1}. This yields similar spectral characteristics as with a low input soliton order, again with excellent coherence. The input soliton order is in this case $N = 5$. Results for 1.2 kW peak power pulses of duration (FWHM) 1 ps propagating in a 1.4 m length of DF-DDF with $D_0 = 30$ ps $\times$ nm^{-1} $\times$ km^{-1}, $D_2 = -2 \times 10^{-4}$ ps $\times$ nm^{-3} $\times$ km^{-1}, $L_0 = 1.2$ m, and $\gamma = 5$ W^{-1} $\times$ km^{-1} are also shown in Fig. 12.10. Although we do not see the same degree of spectral flatness at this very high power level, we can still observe broadband SC generation. The input soliton order here is $N = 7$.

The trend observed in the SC generation process for increased input soliton order allows us to give guidelines that will typically result in a broadband coherent SC: (i) the length L_0 should be approximately equal to the fission length calculated for the input pulse although slightly longer lengths are needed for low input soliton

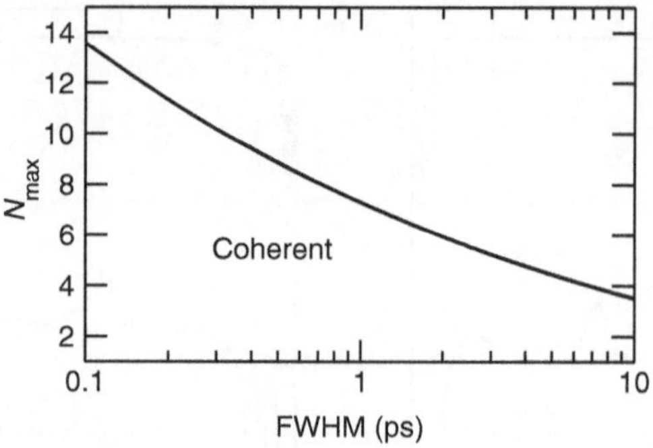

Fig. 12.11. Maximum input soliton number N vs. initial pulse duration to generate a highly coherent SC in a DF-DDF.

numbers; (ii) the fraction of fiber length corresponding to propagation in the normal dispersion regime should be restricted to a few percent of the total length as an increased fraction will simply result in spectral interference causing degradation of the SC smoothness; (iii) a smaller convexity parameter of the dispersion profile generally yields better coherence and smoothness as it prevents soliton, but at the expense of the bandwidth; (iv) a larger initial pulse duration will result in smaller SC bandwidth; (v) in terms of stability characteristics, and as is usually the case with SC generation in the anomalous pumping regime, there is an upper limit to the soliton order that one can launch in the fiber to maintain a high degree of coherence across the entire SC spectrum. In a DF-DDF the maximum soliton order strongly depends on the pump pulse duration as shown in Fig. 12.11, longer pulses requiring smaller N numbers to ensure perfect coherence. The curve of Fig. 12.11 can serve as a general guideline for coherent SC generation in DF-DDF, although it is of course possible to shift the curve slightly up if fibers with very large values of D_2 are used.

A significant potential application of the flattened SC spectra generated in DF-DDF is in the field of pulse compression. The initial nonlinear evolution phase in dispersion-decreasing fibers has previously been studied in detail to directly yield a temporally compressed pulse (Pelusi and Liu, 1997). In this case, the compressed pulse is accompanied by a very broad low amplitude pedestal spanning over several picoseconds. The extension of the initial propagation into the SC generation regime is expected to yield vastly improved compressed pulse quality. Indeed, although the output temporal pulse in this case is complex and could not be exploited directly, the high coherence across the spectrum would, however, be expected to lead to very high quality compressed pulses after appropriate spectral phase compensation (Chang et al., 2003; Dudley and Coen, 2004).

The success of SC generation in DF-DDFs occurred in large part due to the development of highly nonlinear fibers having tailored dispersion characteristics around the operating wavelengths of ultrafast sources at 1550 nm. In parallel, the availability of PCF has lead to further developments in the field of SC generation at other wavelengths. As a matter of fact, it is also possible to fabricate PCFs with two ZDWs that can be tuned from the visible to near-infrared, and experiments using femtosecond pump pulses injected in PCFs with two ZDWs have reported similar nonlinear effects related to the simultaneous transfer of energy across both ZDWs (Hilligsoe et al., 2004; Frosz et al., 2005; Tse et al., 2006a). Exploring tapered PCF design in this wavelength range could therefore provide further possibilities for coherent SC generation for longer pulses with Yb^{3+} based sources. Note that when considering tapered PCF, however, the ideal dynamics described above can be modified by the longitudinal variation of effective area of the fundamental mode (Hu et al., 2006; Tse et al., 2008).

12.4 SC generation in fibers with discrete dispersion map

In earlier sections we have considered fibers with constantly varying dispersion profiles. Nevertheless, it is also possible to create a discrete dispersion map to extend the SC bandwidth by concatenating fibers with different dispersion and nonlinear properties. For instance, a study by Nicholson et al. using 200 fs pulses from a passively mode-locked femtosecond fiber laser at 1550 nm reported improved SC generation spanning over an octave using a series of four 1.5 m segments of highly nonlinear fibers to obtain a discrete map of decreasing anomalous GVD (Nicholson et al., 2003). The fibers were arranged so that the dispersion decreased from the input to output end. The net effect was a compression of the Raman-shifting solitons (generated from the initial fission) leading to enhanced soliton self-frequency shift and extending the SC towards longer wavelengths. The concept of dispersion management for enhancing SC generation was theoretically investigated by Kutz et al. who suggested concatenating two fibers with high and low dispersion values, respectively (Kutz et al., 2005). The main idea was to break up the initial long pump pulse into highly energetic multiple sub-pulses in the first fiber section with a large dispersion coefficient at the pump wavelength. These sub-pulses were subsequently launched into a fiber with a low anomalous dispersion value where they propagate as high-order solitons with associated significant dispersive wave generation and soliton self-frequency shift. A similar idea was demonstrated in a set of experiments where the use of two spliced PCFs with zero-dispersion wavelengths at 1040 nm and 780 nm, respectively, was demonstrated to produce a substantial extension of the SC towards the short wavelengths when pumped with a mode-locked Yb^{3+} laser producing 3 ps pulses (Travers et al., 2005). In this case the extension of the SC resulted from the combination of two characteristics: (i) efficient generation of short wavelength spectral components through modulation instability and

DW generation in the first PCF due to the proximity of the ZDW to the pump and (ii) enhanced dispersive wave trapping through group-velocity matching with red-shifting solitons in the second PCF. The same concept was subsequently applied to the CW regime (Sylvestre et al., 2006) where the combined use of a standard dispersion-shifted fiber and a highly nonlinear dispersion-shifted fiber was shown to produce an increase in the SC bandwidth, both on the red and blue edge of the spectrum. The larger bandwidth in this case was caused by the lower dispersion in the second fiber span which enhanced the soliton self-frequency shift and correspondingly soliton–dispersive wave interactions. Combining up to four different fibers and three pumps the SC was further expanded to span 1000 nm at the −20 dB level. A similar technique was implemented with a set of four standard optical fibers having slightly different ZDWs (Abrardi et al., 2007). The commercial Yb^{3+} CW pump at 1104 nm and a series of highly reflective fiber Bragg gratings were used to construct a nested linear cavity to shift the wavelength of the primary laser to the anomalous dispersion region of the fibers resulting in an SC with a 250 nm bandwidth centered at 1400 nm and 1 mW/nm spectral power density. An alternative technique to extend the bandwidth of SC towards short wavelengths was demonstrated by Xiong et al. (2006) by using a single PCF scheme in combination with a microchip Q-switched laser operating at 1064 nm and emitting 0.6 ns pulses. Using a hole inflation and tapering technique they could manufacture a monolithic PCF consisting of two segments with different dispersion characteristics. In these experiments, the first segment was designed for efficient wavelength conversion of the pump to 742 nm through degenerate four-wave mixing. The second segment was designed to exhibit a ZDW at 700 nm close to the converted pump wavelength at 742 nm. The result was an effective two-pump scheme for the second PCF segment yielding an SC that extended from 400 nm to beyond 1300 nm.

12.5 Conclusion

Fibers with decreasing zero-dispersion wavelength allow us to generate supercontinua with large bandwidths in short lengths of fibers while fibers with dispersion-decreasing and convex profiles will generally result in symmetrical and smooth continua. In the long pulse regime DF-DDF can be perhaps envisaged as a general route towards SC generation of picosecond pulses, with precise fiber designs yielding coherent SC generation at peak power levels in the hundreds of watts range and beyond. A somewhat more straightforward approach relying on a discrete dispersion map created through direct splicing of different fibers has also proved to be an efficient method for extending the short or long wavelength edge, especially in the long pulse regime. Although much work studying SC generation focuses on the fiber nonlinearity characteristics, it is likely that future work in this field will establish more systematic guidelines for dispersion design and optimization. With fiber tapering technologies and the use of discrete-like dispersion profiling both becoming commonplace, it can be anticipated that the study of longitudinal dispersion management strategies will be increasingly important in the development of fiber supercontinuum sources.

References

Abrardi L., Martin-Lopez S., Carrasco-Sanz A., Corredera P., Hemanz M. L., Gonzalez-Herraez M. (2007). Optimized all-fiber supercontinuum source at 1.3 μm generated in a stepwise dispersion-decreasing-fiber arrangement. *J. Lightwave Technol.* **25**, 2098–2102.

Birks T., Li Y. (1992). The shape of fiber tapers. *J. Lightwave Technol.* **10**, 432–438.

Chang G., Norris T., Winful H. (2003). Optimization of supercontinuum generation in photonic crystal fibers for pulse compression. *Opt. Lett.* **28**, 546–548.

Chen C.-M., Kelley P. L. (2002). Nonlinear pulse compression in optical fibers: scaling laws and numerical analysis. *J. Opt. Soc. Am. B* **19**, 1961–1967.

Chernikov S. V., Dianov E. M., Richardson D. J., Payne D. N. (1993). Soliton pulse compression in dispersion-decreasing fiber. *Opt. Lett.* **18**, 476–478.

Dianov E. M., Nikonova Z. S., Prokhorov A. M., Serkin V. N. (1986). Optimal compression of multi-soliton pulses in optical fibers. *Pis'ma Zh. Tekh. Fiz.* **12**, 756–760 [Sov. Tech. Phys. Lett. **12**, 311–313 (1986)].

Dudley J. M., Coen S. (2004). Fundamental limits to few-cycle pulse generation from compression of supercontinuum spectra generated in photonic crystal fiber. *Opt. Express* **12**, 2423–2428.

Dudley J. M., Genty G., Coen S. (2006). Supercontinuum generation in photonic crystal fiber. *Rev. Mod. Phys.* **78**, 1135–1184.

Falk P., Frosz M., Bang O. (2005). Supercontinuum generation in a photonic crystal fiber with two zero-dispersion wavelengths tapered to normal dispersion at all wavelengths. *Opt. Express* **13**, 7535–7540.

Fatemi F. (2002). Analysis of nonadiabatically compressed pulses from dispersion-decreasing fiber. *Opt. Lett.* **27**, 1637–1639.

Frosz M. H., Falk P., Bang O. (2005). The role of the second zero-dispersion wavelength in generation of supercontinua and bright-bright soliton-pairs across the zero-dispersion wavelength. *Opt. Express* **13**, 6181–6192.

Genty G., Coen S., Dudley J. (2007). Fiber supercontinuum sources. *J. Opt. Soc. Am. B* **24**, 1771–1785.

Genty G., Lehtonen M., Ludvigsen H. (2004). Effect of cross-phase modulation on supercontinuum generated in microstructured fibers with sub-30 fs pulses. *Opt. Express* **12**, 4614–4624.

Gorbach A. V., Skryabin D. V. (2007). Light trapping in gravity-like potentials and expansion of supercontinuum spectra in photonic-crystal fibres. *Nature Photon.* **1**, 653–657.

Hilligsoe K., Andersen T., Paulsen H. et al. (2004). Supercontinuum generation in a photonic crystal fiber with two zero dispersion wavelengths. *Opt. Express* **12**, 1045–1054.

Hu J., Marks B. S., Menyuk C. R. et al. (2006). Pulse compression using a tapered microstructure optical fiber. *Opt. Express* **14**, 4026–4036.

Kudlinski A., George A., Knight J. et al. (2006). Zero-dispersion wavelength decreasing photonic crystal fibers for ultraviolet-extended supercontinuum generation. *Opt. Express* **14**, 5715–5722.

Kuehl H. H. (1988). Solitons on an axially nonuniform optical fiber. *J. Opt. Soc. Am. B* **5**, 709–713.

Kutz J., Lynga C., Eggleton B. (2005). Enhanced supercontinuum generation through dispersion-management. *Opt. Express* **13**, 3989–3998.

Lou J. W., Xia T. J., Boyraz O., Shi C.-X., Nowak G. A., Islam M. N. (1997). Broader and flatter supercontinuum spectra in dispersion tailored fibers. *Tech. Digest OFC, Paper TuH6*, 32–34.

Lu F., Knox W. (2004). Generation of a broadband continuum with high spectral coherence in tapered single-mode optical fibers. *Opt. Express* **12**, 347–353.

Lu F., Knox W. (2006). Generation, characterization, and application of broadband coherent femtosecond visible pulses in dispersion micromanaged holey fibers. *J. Opt. Soc. Am. B* **23**, 1221–1227.

Lu F., Deng Y., Knox W. (2005). Generation of broadband femtosecond visible pulses in dispersion-micromanaged holey fibers. *Opt. Lett.* **30**, 1566–1568.

Mamyshev P. V., Wigley P. G. J., Wilson J. et al. (1993). Adiabatic compression of Schrödinger solitons due to the combined perturbations of higher-order dispersion and delayed nonlinear response. *Phys. Rev. Lett.* **71**, 73–76.

Monerie M. (1983). Propagation in doubly-clad single mode fibers. *IEEE J. Quantum Electron.* **18**, 535–542.

Mori K., Morioka T., Saruwatari M. (1995). Ultrawide spectral range group-velocity dispersion measurement utilizing supercontinuum in an optical-fiber pumped by a 1.5 μm compact laser source. *IEEE Trans. Instrum. Meas.* **44**, 712–715.

Mori K., Takara H., Kawanishi S., Saruwatari M., Morioka T. (1997). Flatly broadened supercontinuum spectrum generated in a dispersion decreasing fibre with convex dispersion profile. *Electron. Lett.* **33**, 1806–1808.

Mori K., Takara H., Kawanishi S. (2001). Analysis and design of supercontinuum pulse generation in a single-mode optical fiber. *J. Opt. Soc. Am. B* **18**, 1780–1792.

Morioka T., Kawanishi S., Mori K., Saruwatari M. (1994a). Nearly penalty-free, <4 ps supercontinuum Gbit/s pulse generation over 1535–1560 nm. *Electron. Lett.* **30**, 790–791.

Morioka T., Kawanishi S., Mori K., Saruwatari M. (1994b). Transform-limited, femtosecond WDM pulse generation by spectral filtering of gigahertz supercontinuum. *Electron. Lett.* **30**, 1166–1168.

Morioka T., Uchiyama K., Kawanishi S., Suzuki S., Saruwatari M. (1995). Multiwavelength picosecond pulse source with low jitter and high optical frequency stability based on 200 nm supercontinuum filtering. *Electron. Lett.* **31**, 1064–1066.

Morioka T., Takara H., Kawanishi S. et al. (1996). 1 Tbit/s (100Gbit/s $\times$ 10 channel) OTDM/WDM transmission using a single supercontinuum WDM source. *Electron. Lett.* **32**, 906–907.

Nakazawa M., Tamura K., Kubota H., Yoshida E. (1998). Coherence degradation in the process of supercontinuum generation in an optical fiber. *Opt. Fiber Technol.* **4**, 215–223.

Nguyen H. C., Kuhlmey B. T., Mägi E. C., Steel M. J., Domachuk P., Smith C. L., Eggleton B. J. (2005). Tapered photonic crystal fibres: properties, characterisation and applications. *Appl. Phys. B.* **81**, 377–387.

Nicholson J. W., Abeeluck A. K., Headley C., Yan M. F., Jørgensen C. G. (2003). Pulsed and continuous-wave supercontinuum generation in highly nonlinear, dispersion-shifted fibers. *Appl. Phys. B* **77**, 211–218.

Okuno T., Onishi M., Nishimura M. (1998). Generation of ultra-broad-band supercontinuum by dispersion-flattened and decreasing fiber. *IEEE Photon. Technol. Lett.* **10**, 72–74.

Okuno T., Onishi M., Kashiwada T., Ishikawa S., Nishimura M. (1999). Silica-based functional fibers with enhanced nonlinearity and their applications. *IEEE J. Sel. Top. Quantum Electron.* **5**, 1385–1391.

Pal P., Knox W. H. (2007a). End-sealing short dispersion micromanaged tapered holey fibers by hole-collapsing. *Opt. Express* **15**, 13531–13538.

Pal P., Knox W. H. (2007b). Self referenced Yb-fiber-laser frequency comb using a dispersion micromanaged tapered holey fiber. *Opt. Express* **15**, 12161–12166.

Pelusi M. D., Liu H. F. (1997). Higher order soliton pulse compression in dispersion-decreasing optical fibers. *IEEE J. Quantum Electron.* **33**, 1430–1439.

Reeves W., Skryabin D., Biancalana F. et al. (2003). Transformation and control of ultra-short pulses in dispersion-engineered photonic crystal fibres. *Nature* **424**, 511–515.

Saitoh K., Koshiba M. (2004). Highly nonlinear dispersion-flattened photonic crystal fibers for supercontinuum generation in a telecommunication window. *Opt. Express* **12**, 2027–2032.

Sylvestre T., Vedadi A., Maillotte H., Vanholsbeeck F., Coen S. (2006). Supercontinuum generation using continuous-wave multiwavelength pumping and dispersion management. *Opt. Lett.* **31**, 2036–2038.

Town G. E., Lizier J. T. (2001). Tapered holey fibers for spot-size and numerical aperture conversion. *Opt. Lett.* **26**, 1042–1044.

Travers J. C., Taylor J. R. (2009). Soliton trapping of dispersive waves in tapered optical fibers. *Opt. Lett.* **34**, 115–117.

Travers J. C., Popov S. V., Taylor J. R. (2005). Extended blue supercontinuum generation in cascaded holey fibers. *Opt. Lett.* **30**, 3132–3134.

Travers J. C., Stone J. M., Rulkov A. B. et al. (2007). Optical pulse compression in dispersion decreasing photonic crystal fiber. *Opt. Express* **15**, 13203–13211.

Tse M., Horak P., Poletti F. et al. (2006a). Supercontinuum generation at 1.0 μm in holey fibers with dispersion flattened profiles. *Opt. Express* **14**, 4445–4451.

Tse M., Horak P., Price J. et al. (2006b). Pulse compression at 1.0 μm in dispersion-decreasing holey fibers. *Opt. Lett.* **31**, 3504–3506.

Tse M., Horak P., Poletti F. et al. (2008). Designing tapered holey fibers for soliton compression. *IEEE J. Quantum Electron.* **44**, 192–198.

Wadsworth W., Witkowska A., Leon-Saval S. et al. (2005). Hole inflation and tapering of stock photonic crystal fibres. *Opt. Express.* **13**, 6541–6549.

Xiong C., Witkowska A., Leon-Saval S., Birks T., Wadsworth W. (2006). Enhanced visible continuum generation from a microchip 1064 nm laser. *Opt. Express* **14**, 6188–6193.

Yulin A., Skryabin D., Russell P. (2004). Four-wave mixing of linear waves and solitons in fibers with higher-order dispersion. *Opt. Lett.* **29**, 2411–2413.

13
Supercontinuum generation in chalcogenide glass waveguides

Dong-Il Yeom, Michael Lamont, Barry Luther Davies and
Benjamin J. Eggleton

13.1 Introduction

Supercontinuum (SC) generation, the creation of broadband spectral components from an intense light pulse passing through a nonlinear medium, is of great theoretical interest as well as having numerous applications in optical frequency metrology, bio-imaging and spectroscopy (Dudley et al., 2006). In particular, the demonstration of efficient SC generation in a silica photonic crystal fibre (PCF) and silica fibre tapers using a Ti:Sapphire laser (Birks et al., 2000, Ranka et al., 2000) had a striking impact on this research field. The advent of this new class of waveguide, capable of engineered dispersion and strong confinement of light, facilitated research on the fundamental study of the evolution of ultra-fast pulses in highly nonlinear waveguides, as well as the development of practical broadband light sources using the proper combination of fibres and laser pulses. Although the successful demonstration of ultra-broadband light generation often spanning more than an octave has been made in silica fibre, the small Kerr nonlinear coefficient of silica still limits its practicality. The ideal SC light source would use a compact, low power pulsed laser. This goal has motivated the study of SC generation in waveguides with higher nonlinear coefficients and lower energy thresholds or decreased device length to initiate the nonlinear process. Several approaches based on this idea have been reported utilising highly nonlinear material in a fibre geometry such as lead-silicate, bismuth and chalcogenide fibres (Brambilla et al., 2005, Leong et al., 2006, Mägi et al., 2007), and in a planar waveguide geometry including silicon (Boyraz et al., 2004), AlGaAs (Siviloglou et al., 2006) and chalcogenide waveguides (Psaila et al., 2007, Lamont et al., 2008).

Table 13.1 compares the optical properties for silica and several highly (Kerr-)nonlinear materials which have been previously proposed as several types of waveguides for nonlinear applications. Tapered fibres or holey optical fibres

Supercontinuum Generation in Optical Fibers, ed. by J. M. Dudley and J. R. Taylor. Published by Cambridge University Press.

Table 13.1. *Comparison of properties of several glass materials and silicon at the wavelength of 1.55 μm.*

Material	Silica	Lead silicate	Bismuth oxide	Chalcogenide glass (As_2S_3)	Chalcogenide glass (As_2Se_3)	Silicon
Waveguide type	Holey fibre, Taper	Holey fibre	Holey fibre, Taper	Planar waveguide	Taper	Planar waveguide
Refractive index	1.45*	1.81	2.02	~2.4	~2.8	3.5
n_2($\times 10^{-20}$ m^2/W)	2.7*	41*	32	~300	1100	500~600
Maximum γ reported (/W/m)	0.06	1.9	1.3	10	93.4	~100
Linear propagation loss (dB/m)	0.001	0.3	2~3	20***	60	25
TPA coefficient (m/W)**	–	–	–	6.2×10^{-15}	2.5×10^{-12}	5.7×10^{-12}
Material dispersion (ps/km/nm)	17	−50	−110	−360	−650	−890
IR transmission edge (μm)	~3.2	~2.8	~3.8	~6.5	~10	~12

* measured at 1.06 μm.

** TPA is negligible in several glasses at 1.55 μm.

*** Recently 5 dB/m loss has been reported. References for the values are indicated in main text.

have been fabricated in the glass materials including silica, lead silicate, bismuth and chalcogenide glasses where their core sizes are comparable to the optical wavelengths. Planar waveguides for nonlinear applications have also been reported in chalcogenide glasses and silicon material. Since the light propagating along these waveguides is guided by the refractive-index contrast to the surrounding medium, such as air or a low-index protective polymer, materials possessing a high refractive index enable stronger confinement in the waveguide. As shown in Table 13.1, chalcogenide glasses and silicon exhibit relatively large refractive indices of 2.4~3.5 at a wavelength of 1.55 μm (Klocek, 1991, Saleh and Teich, 2007, Sanghera et al., 2000). The high refractive index is also related to the high Kerr-nonlinear index n_2, which will be discussed later. The Kerr-nonlinear indices n_2 of the materials are also compared in the table. The measured n_2 of silica is 2.7×10^{-20} m^2/W at a wavelength of 1.06 μm (Agrawal, 2001), which is smaller by an order of magnitude than highly nonlinear glasses such as lead silicate and bismuth (Ebendorff-Heidepriem et al., 2004, Leong et al., 2006, Price et al., 2006). Chalcogenide glasses possess an even larger value of n_2; the n_2 of As_2S_3 glass has been reported to be as high as 300×10^{-20} m^2/W (Ruan et al., 2004) and 1100×10^{-20} m^2/W for As_2Se_3 glass (Harbold et al., 2002, Slusher et al., 2004). These values are 2~3 orders larger than that of silica. Large nonlinear indices from 200 to 600×10^{-20} m^2/W have been reported for silicon waveguides from the comparison or direct measurement of nonlinear phase shifts in the spectrum of a pulse at the output of the waveguide or from z-scan measurements (Dulkeith et al., 2006, Tsang et al., 2002, Lin et al., 2007, Bristow et al., 2007). This huge nonlinear index n_2, combined with the enhanced optical intensity due to the tight confinement, can result in an ultra-high nonlinear response, γ, of the waveguide. Silica photonic crystal fibre and tapered wires exhibit a maximum γ $(= 2\pi n_2/(A_{\rm eff}\lambda))$ of 60/W/km at 1.55 μm which is 60 times bigger than that of a conventional silica single-mode fibre (~1.1/W/km). The largest γ reported in chalcogenide glass (Yeom et al., 2008) and silicon (Boyraz et al., 2004, Dulkeith et al., 2006) is up to ~100/W/m, which is 1600 times larger than the maximum value that can be achieved in silica.

The materials compared in Table 13.1 are all transparent at 1.55 μm, so the linear propagation loss due to material absorption is almost negligible in short waveguides. The loss becomes more significant when the waveguide dimensions are reduced to the submicron range. Scattering losses of the optical field at the core-cladding boundary may become more significant; and in the transition regions to the tapered fibre waist, non-adiabatic coupling losses may appear. Typical losses of 20 dB/m and 25 dB/m have been reported in chalcogenide and silicon planar waveguides respectively. As_2Se_3 submicron tapers also exhibit large scattering losses as high as ~60 dB/m (2 dB for 30 mm length with a 0.95 μm diameter), although the original untapered As_2Se_3 fibre had a small transmission loss of only 1 dB/m. In addition to the linear loss, nonlinear losses such as two-photon absorption (TPA) need to be considered. For silica and other glass materials possessing wide bandgaps, TPA can be negligible at 1.55 μm. In some chalcogenide glasses, where the absorption tail from half the bandgap energy extends to infrared (IR) wavelengths, TPA can be significant. The TPA in silicon, not accounting for free carrier absorption (FCA), is strong because 1.55 μm corresponds to a photon energy more than half its bandgap. The measured TPA coefficients for an As_2S_3 waveguide, As_2Se_3 fibre and silicon are reported to be 6.2×10^{-15} m/W, 2.5×10^{-12} m/W (Fu et al., 2005) and 5.7×10^{-12} (Lin et al., 2007, Bristow et al., 2007) respectively. Whilst the value for As_2S_3 is sufficiently low to have negligible effect on beam propagation, in As_2Se_3 and Si waveguides two-photon absorption strongly limits propagation of high intensity light in these waveguides.

Most bulk high nonlinearity glasses as well as silicon have normal group velocity dispersion at 1.55 μm. In general, these high nonlinearity materials are also transparent to longer wavelengths in the IR compared with oxide glasses such as silica, as can be expected from the Kramers–Kronig relations and Miller's rule (Boyd, 2008). For example, As_2Se_3 chalcogenide glass has low multi-phonon energies because of the massive atoms and weak bonding (Yamane and Asahara, 2000), which result in transparency in the mid-IR up to 10 μm and a high linear/nonlinear index with a large normal dispersion of −650 ps/nm/km at 1.55 μm.

Among the nonlinear materials compared in the table, chalcogenide glasses are remarkable for nonlinear optics because they exhibit very high nonlinear coefficients with moderate or low nonlinear absorption. In addition, their high refractive index (2.4∼2.8) allows strong optical field confinement, enabling small form factors in photonic chip design, small waveguide bend radii, as well as enhanced optical intensities for increased nonlinear interaction efficiency. Their transparency from the near to mid-IR wavelengths is another outstanding property of these materials, enabling broad SC generation. They are also attractive in terms of waveguide fabrication because chalcogenide glass can be drawn into high quality fibre (linear loss of less than 1 dB/m for step-indexed fibre at 1550 nm) (Slusher et al., 2004) or fabricated as a planar waveguide on an oxidised silicon wafer (propagation loss of 0.05 dB/cm) (Madden et al., 2007). Although a number of nonlinear optical devices using chalcogenide glass have been demonstrated over the past decade (Asobe et al., 1993, Fu et al., 2005, Ta'eed et al., 2005), efficient SC generation had not been successfully reported until recently due to the problem of the large material dispersion and the difficulty of fabricating submicron size devices.

In this chapter, we review low energy threshold SC generated in a chalcogenide-based fibre and rib waveguides. This was achieved by the proper design of the sub micron chalcogenide tapers and rib waveguides to produce a high nonlinear response with optimally tailored dispersion. In the first section of this chapter, the fundamental material properties of chalcogenide glass for nonlinear optical applications are investigated, which covers the topics from linear optical properties such as refractive index, linear loss and material group velocity dispersion, to nonlinear properties such as the Kerr-nonlinear coefficient, TPA and the Raman response function. In the next section, efficient SC generation in dispersion-engineered submicron As_2Se_3 chalcogenide fibre tapers is demonstrated. The design and fabrication of fibre tapers are introduced, considering the total dispersion and nonlinearity of the taper as well as the selection of optimal pulse duration. We then report SC generation based on optimised experimental conditions and compare this with the numerically modelled result for different experimental conditions. The following section introduces broadband SC generation in a dispersion-managed As_2S_3 planar rib waveguide. First, the design and fabrication of As_2S_3 waveguides are overviewed. The SC experiment and numerical analysis is then discussed. The final section summarises this chapter and looks at future prospects – possible applications for an on-chip nonlinear photonic device with improved performance and extended functionality.

13.2 Material properties of chalcogenide glass for nonlinear optic applications

13.2.1 Linear optical properties

Chalcogenide glasses contain the chalcogen elements S, Se, Te, covalently bonded with elements such as Ge, As, Sb and form stable glasses over a wide range of compositions (Sanghera et al., 2000, Zakery and Elliott, 2003). Their optical properties such as refractive index, nonlinearity, and band-edge can, therefore, be tuned by the appropriate selection of composition. In general, chalcogenide glasses possess relatively high refractive indices, for example As_2Se_3 glass has a refractive index between 2.7 and 2.8 at near-IR wavelengths. For waveguide fabrication a high refractive index is advantageous and allows strong confinement of light, a reduction in the effective mode area and thus an increased nonlinear response. The high index contrast also allows the waveguide bend radius to be reduced, which allows more compact integrated photonic devices. Here we focus on the material properties of two glass compositions, As_2S_3 and As_2Se_3, prevalently used in both fibre and waveguide forms.

Figure 13.1 shows the dispersion of the refractive index of bulk As_2S_3 obtained from the Sellmeier equation (Klocek, 1991). The index of refraction (n) gradually decreases from 2.5 to 2.4 as the wavelength increases into the IR wavelength range. The group velocity dispersion (GVD) of the material can be calculated by the following relationship,

$$D = -\frac{\lambda}{c}\frac{d^2 n}{d\lambda^2}(\text{ps/nm/km}), \tag{13.1}$$

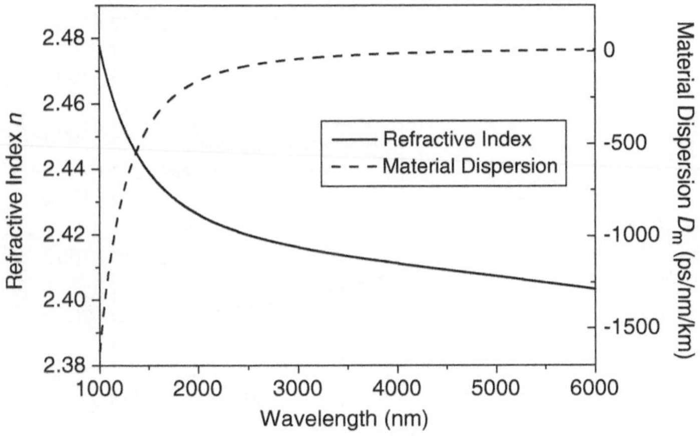

Fig. 13.1. Refractive index and group velocity dispersion of As_2S_3 chalcogenide glass near-IR wavelengths.

where c and λ are the velocity and wavelength of light, respectively. The GVD as a function of wavelength for As_2S_3 is also shown in Fig. 13.1. It should be noted that this material has a zero dispersion point (ZDP) at about 4.9 μm, and exhibits large normal dispersion (GVD < 0) across the near-IR. At 1550 nm the GVD is about -360 ps/nm/km in As_2S_3.

In the case of As_2Se_3 chalcogenide glass, the dispersion of the linear refractive index in the near-IR can be expressed by the following approximation (Slusher et al., 2004):

$$n^2 = 1 + \frac{E_d E_S}{E_s^2 - (h\nu)^2}, \tag{13.2}$$

where E_d and E_s are the electronic oscillator energy ($\sim$26 eV) and the Sellmeier gap energy ($\sim$4.1 eV), respectively. Figure 13.2 shows the calculated refractive index of As_2Se_3 glass for mid-IR wavelengths. The index of refraction varies between 2.7 and 2.8, almost double that of silica. The calculated GVD from Eqns 13.1 and 13.2 also shows large normal dispersion values (-650 ps/nm/km) at 1550 nm. For many nonlinear processes, specifically those related to SC generation, large normal dispersion is detrimental to device performance. This necessitates dispersion control, which is considered in a later section.

Since chalcogenide glasses are transparent in the mid-IR, they have been widely used as a basic material for IR optical components. Previous measurement of relative transmission of several bulk glass samples with a thickness of 2$\sim$3 mm has showed that the chalcogenide glasses (sulphide, selenide, telluride) transmit to wavelengths beyond 12 μm, compared to silica and fluoride glasses exhibiting the transmission band edge of $\sim$4 μm and $\sim$8 μm respectively at the longer wavelength

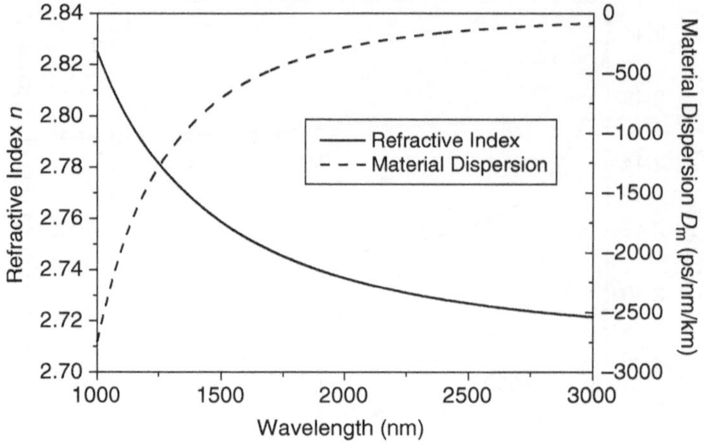

Fig. 13.2. Refractive index and group velocity dispersion of As_2Se_3 chalcogenide glass near-IR wavelengths.

(Sanghera et al., 2000). When these materials are fabricated into fibre or planar waveguide forms, the transmission loss can change depending on the degree of purification, the composition, and the surface roughness of the waveguide. Among these compositions, As-S and As-Se fibres show the best transparency with losses as low as 0.2∼1 dB/m.

13.2.2 Nonlinear optical properties

Recent attention has been paid to chalcogenide glasses for applications in nonlinear optics, owing to their large Kerr-nonlinear coefficients. The Kerr-nonlinear coefficient (n_2), which is responsible for the nonlinear index change for the intensity of the incident light, is directly connected to the real part of the third-order susceptibility, $\chi^{(3)}$. According to Miller's empirical rule (Boyd, 2008), the third-order susceptibility ($\chi^{(3)}$) is proportional to the fourth power of the linear susceptibility ($\chi^{(1)}$). In the case of chalcogenide glasses, numerical work has reported that the nonlinear coefficient n_2 is proportional approximately to n^5 when the refractive index n of the glass lies between 2 and 3 (Slusher et al., 2004). These results imply that the nonlinearity for As_2S_3 and As_2Se_3 chalcogenide glasses will be high because of their high linear refractive indices (2.4∼2.8). Indeed the nonlinear refractive index n_2 of chalcogenide glasses has been measured to be two orders or more greater than silica (refractive index of ∼1.5 at near IR wavelengths). For example, the nonlinear index n_2 of As_2S_3 has been reported to be $3{\sim}5 \times 10^{-18}$ m²/W, whilst for As_2Se_3 glasses values of $1{\sim}2.3 \times 10^{-17}$ m²/W have been reported (Cerqua-Richardson et al., 1998, Harbold et al., 2002, Lenz et al., 2000). These values are 100∼1000

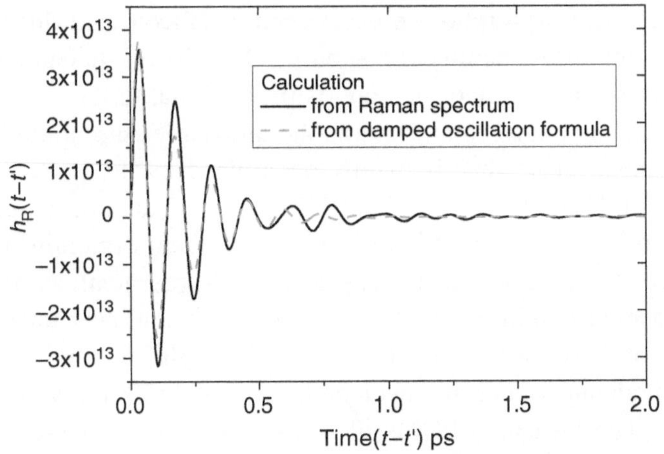

Fig. 13.3. Estimated delayed Raman response function $h(t)$ of As_2Se_3. Estimation from the actual Raman gain spectrum in Slusher et al. (2004) is compared with that obtained from the damped oscillation model.

times greater than silica ($n_2 = 2.7 \times 10^{-20}$ m^2/W), and these large values of n_2 allow a dramatic reduction in the device size or power threshold for nonlinear applications.

The nonlinear third-order susceptibility $\chi^{(3)}$ in the time domain is expressed by (Agrawal, 2001):

$$\chi^{(3)}(t - t_1, t - t_2, t - t_3) = \chi^{(3)} R(t - t_1)\delta(t - t_2)\delta(t_1 - t_3), \qquad (13.3)$$

where the response function, $R(t)$, includes both the electronic and nuclear contribution such that $R(t) = (1 - f_R)\delta(t) + f_R h_R(t)$, and f_R is the fraction of nuclear contribution to $h_R(t)$. The electronic response is usually considered to be almost instantaneous, and described using a delta function, $\delta(t)$. The nuclear contribution shows a relatively long response time due to the nonlinear Raman response, and is about 500 fs in the case of As_2Se_3 glass. The Raman response function can be inferred from the Raman gain spectrum. Figure 13.3 shows the Raman response function deduced from the measured Raman gain spectrum from Slusher et al. (2004). In the figure, two graphs are compared. The first one is directly obtained from the measured Raman gain spectrum and the other is obtained assuming a damped oscillator model; these show good agreement. The Raman contribution to the entire nonlinear process is known to be 12∼14% in the case of chalcogenide glasses. Since the fraction of Raman contribution is similar to that of silica, one can expect chalcogenide glasses to also exhibit large Raman gain.

TPA is a nonlinear absorption process resulting in intensity-dependent loss within a material. The TPA processes also lead to the generation of free carriers and additional free-carrier absorption in cases such as silicon, and this can result in significant slowing of the nonlinear response. The TPA coefficient (β) relates the absorption loss to the light intensity, and is proportional to the imaginary part of $\chi^{(3)}$. Therefore, β is linked to n_2 through the third-order susceptibility $\chi^{(3)}$. The expected nonlinear phase shift is mainly determined by n_2, but can be reduced by the effects of TPA. It is convenient to introduce a nonlinear figure of merit (FOM), defined by FOM $= n_2/\beta\lambda$, to assess the amount of nonlinear phase shift in the presence of TPA. For a device based on third-order nonlinearity, it is desirable to have a high nonlinear FOM, or a large Kerr-nonlinear coefficient and a small amount of TPA. In the case of chalcogenide glasses, the TPA coefficient has small to moderate values at wavelengths around 1550 nm whilst the nonlinearity is large. The measured FOMs of As_2S_3 and As_2Se_3 glasses are >10 and ~ 2 respectively at near-IR wavelengths (Lenz et al., 2000, Ruan et al., 2004). Figure 13.4(a) shows the measured n_2 and β values as a function of wavelength in As_2Se_3 fibre (Nguyen et al., 2006). Both nonlinear coefficients were estimated by fitting measurements to simulations based on the nonlinear Schrödinger equation; the nonlinear power transfer curve gave the value for β, and spectral broadening due to self-phase modulation (SPM) the value for n_2. In the figure, both n_2 and β decrease at wavelengths greater than the half-bandgap ($\sim$1400 nm), but TPA decreases more rapidly which results in a gradual increase of the FOM up to a value of 2 as shown in Fig. 13.4(b). A small FOM can result in significant saturation of the transmitted power before the desired nonlinear phase shift is reached, so large values of FOM are desired to obtain better performances for nonlinear devices. A large FOM can be obtained by adjusting the glass composition in order to shift the electronic bandgap and thus reducing the amount of multi-photon absorption.

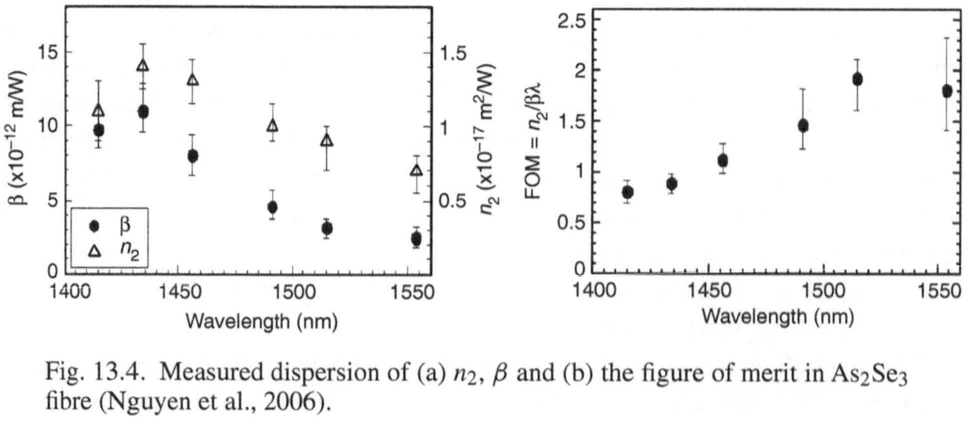

Fig. 13.4. Measured dispersion of (a) n_2, β and (b) the figure of merit in As_2Se_3 fibre (Nguyen et al., 2006).

13.3 Low-energy threshold supercontinuum generation in As_2Se_3 chalcogenide glass fibres

13.3.1 Dispersion and nonlinear properties of As_2Se_3 tapers

Bulk As_2Se_3 chalcogenide glasses possess a high normal dispersion in the near-IR of about −650 ps/nm/km at the wavelength of 1550 nm. Large normal dispersion is not always detrimental to the performance of an optical device, such as when the nonlinear interaction length is very small, and it can even be beneficial for the realisation of an ideal nonlinear transfer function by suppressing spectral oscillations typical of self-phase modulation (Fu et al., 2005). However, normal dispersion is not desirable for many nonlinear optical applications, particularly those related to solitons or parametric processes. For example, soliton-based nonlinear processes are only feasible in the anomalous dispersion regime and parametric processes such as four-wave mixing become much more efficient when the dispersion is anomalous and small. However, normal material dispersion does not mean that the GVD of a mode in a waveguide cannot be anomalous. The modal dispersion is a combination of contributions from the material and the waveguide (the dispersion induced by the waveguide geometry). If the waveguide dispersion is anomalous and comparable in magnitude to the normal material dispersion, it is possible for the two to cancel each other, resulting in net zero dispersion for the waveguide mode. The high refractive index of As_2Se_3 (∼2.8) provides strong mode confinement even at very small transverse dimensions capable of creating large anomalous waveguide dispersion. This enables a shift of the zero-dispersion point to shorter wavelengths, as has been reported in silica fibre tapers (Birks et al., 2000) and air/silica photonic crystal fibres (Ranka et al., 2000). In particular, tapering optical fibres, a process involving the local heating and stretching of the fibre, is a well-known post-processing method for controlling the dispersion of a waveguide while maintaining the very low surface roughness. The reduced size of the waveguide, and thus the effective area of the mode, also causes strong confinement of the optical field and leads to an enhancement of the nonlinear response of the device.

Figure 13.5 shows RSoft FemSIM^{TM} simulations to illustrate how waveguide dispersion affects the mode GVD of a taper as the diameter is reduced, shifting the zero-dispersion point from the mid-IR for bulk As_2Se_3 glass down to near-IR at diameters around 1 μm. This is summarised in Fig. 13.6(a), which shows the dependence of the dispersion at 1550 nm on the taper diameter. These simulations indicate that the dispersion changes from normal to anomalous at a taper diameter of about 1.2 μm, and remains anomalous until the diameter drops below 0.4 μm. The calculated nonlinear parameter (γ) is also shown in Fig. 13.6(b) for decreasing taper diameters, with the dispersion ranges indicated. The confinement of the guided mode becomes stronger as the taper diameter decreases, even when it reaches the

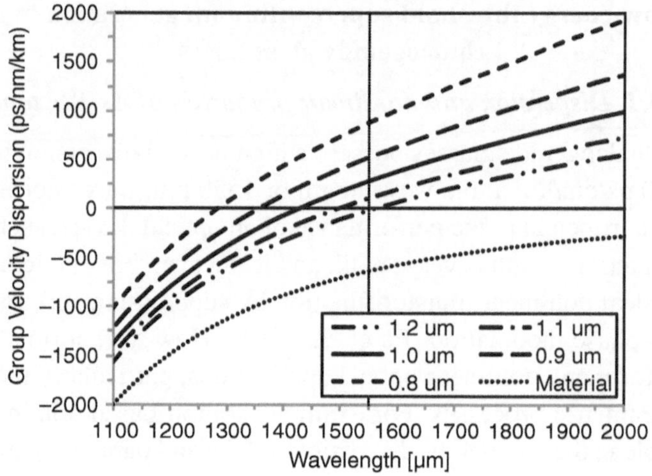

Fig. 13.5. Group velocity dispersion of As_2Se_3 fibre tapers of varying diameters are contrasted with the material dispersion of As_2Se_3. The vertical line indicates a wavelength of 1550 nm.

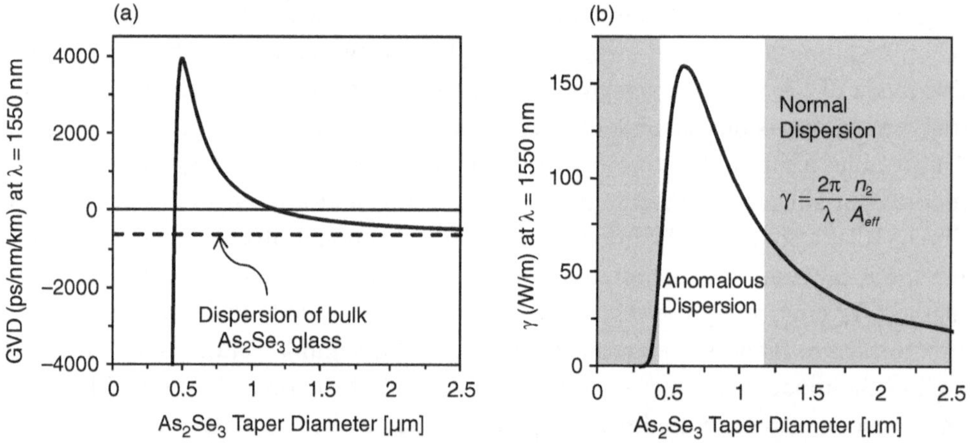

Fig. 13.6. Calculated (a) dispersion and (b) nonlinearity in As_2Se_3 tapers with different diameters at the wavelength of 1550 nm.

submicron range, due to the large index contrast between air and As_2Se_3 glass ($\Delta n \sim 1.7$). The strongly confined optical field, combined with a large n_2 value of 1.1×10^{-17} m^2/W in As_2Se_3 glass, results in an extremely high γ. The maximum value of γ we obtained in this calculation is about ~160/W/m at a taper diameter about 0.6 μm, which is about 150 000 times larger than that in normal silica single-mode fibre (SMF). This is also 2500 times larger than what is possible with silica tapers, quantitatively attributed to the 500 times enhancement in n_2 and a 5 times stronger field confinement (effective area reduction).

13.3.2 Fabrication of chalcogenide tapers

An As_2Se_3 submicron taper was fabricated using a modified version of the standard flame brushing technique, where a resistive heater was employed instead of a flame due to the low melting temperature of As_2Se_3 (below 200°C) (Mägi et al., 2007). Figure 13.7 describes the detailed process of fabrication. Since the As_2Se_3 fibre was initially multi-moded (core/cladding diameter 7.7 μm/170 μm, NA~0.2), we pre-tapered it to a fibre with a 75 μm cladding-diameter waist as depicted in Fig 13.7(a). After cleaving the uniform section of the pre-tapered fibre, UV-cured epoxy was used to butt-couple each end to fibre pigtails consisting of a short length of mode-matched high numerical aperture (NA) silica fibre and standard SMF. The total coupling loss (from input-to-output SMF) was typically −4 dB, and could be further improved using careful adjustment of the pre-tapered diameter to match the effective area of the high NA fibre. We then fabricated the submicron As_2Se_3 wire from this pre-tapered fibre assembly.

The taper profile of the fibre was appropriately designed to obtain an adiabatic transition for the propagating light from the starting diameter to the taper waist. As

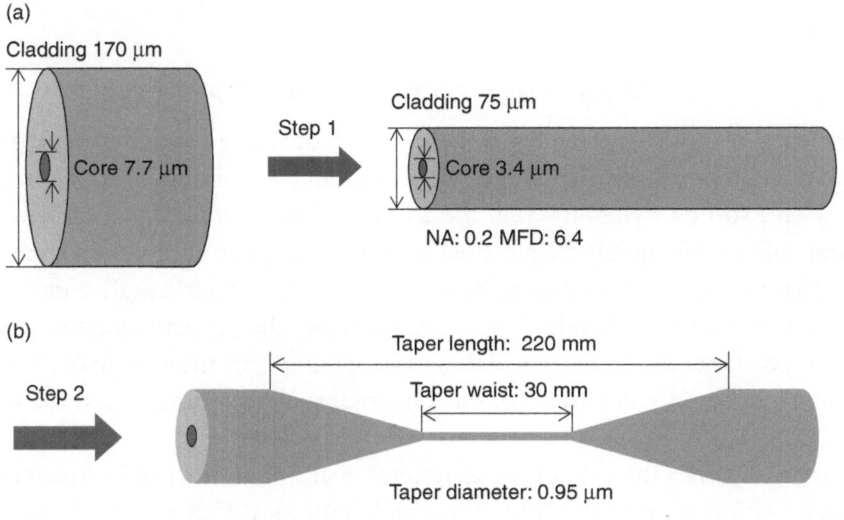

Fig. 13.7. Fabrication process of As_2Se_3 submicron taper. (a) Step 1: initial As_2Se_3 chalcogenide fibre is pre-tapered down to 75 μm to make single-guided mode fibre with matched mode-field diameter. (b) Step 2: pre-tapered fibre connected with silica fibre is then further tapered down to submicron while monitoring transmission power.

Fig. 13.8. SEM scanning images of fabricated As_2Se_3 taper waist show a uniformly tapered waist section with 0.95 μm.

a result, we obtained an As_2Se_3 submicron taper with a total length of ~220 mm, including a 30-mm-long submicron section as described in Fig. 13.7(b). During the tapering process, an additional loss of about −4 dB appeared, which may result from locally non-adiabatic structures in the taper or increased scattering at the air–As_2Se_3 interface. Figure 13.8 shows a scanning electron microscope (SEM) image of the fabricated As_2Se_3 taper waist. The inset image clearly indicates a nearly uniform submicron section with minimum diameter of 0.95 μm.

13.3.3 Design of experimental setup

Low-power threshold SC in an As_2Se_3 taper, with only a few watts of peak power, can be efficiently generated using pulses in the femto second regime (Yeom et al., 2008). Here, soliton dynamics (i.e. the formation and evolution of higher-order temporal solitons) is mainly responsible for the large spectral broadening, and the ultra-high nonlinearity of As_2Se_3 tapers enable the formation of a high-order soliton even at low input power levels. For a given sample, the proper selection of input pulse duration is necessary to control the shape of the spectrum since this determines the soliton number (N) as well as soliton dynamics such as soliton period and fission length (L_{fission}).

Figure 13.9 shows the calculated soliton number and normalised sample length (=30 mm) with respect to the soliton fission length for different input pulse durations. Here, two boundary conditions are imposed to facilitate soliton evolution for a low input peak power of 10 W. First, a soliton number more than 5 is chosen to produce a more apparent pulse break-up from higher-order solitons. Secondly, the sample length is set to be more than double L_{fission} so that the subsequent processes

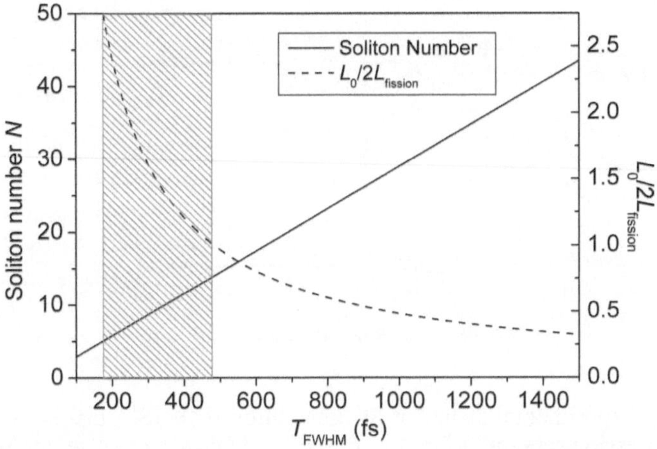

Fig. 13.9. Calculated soliton number N and normalised sample length L_0 with respect to the soliton fission length of $L_{fission}$ for different pulse widths.

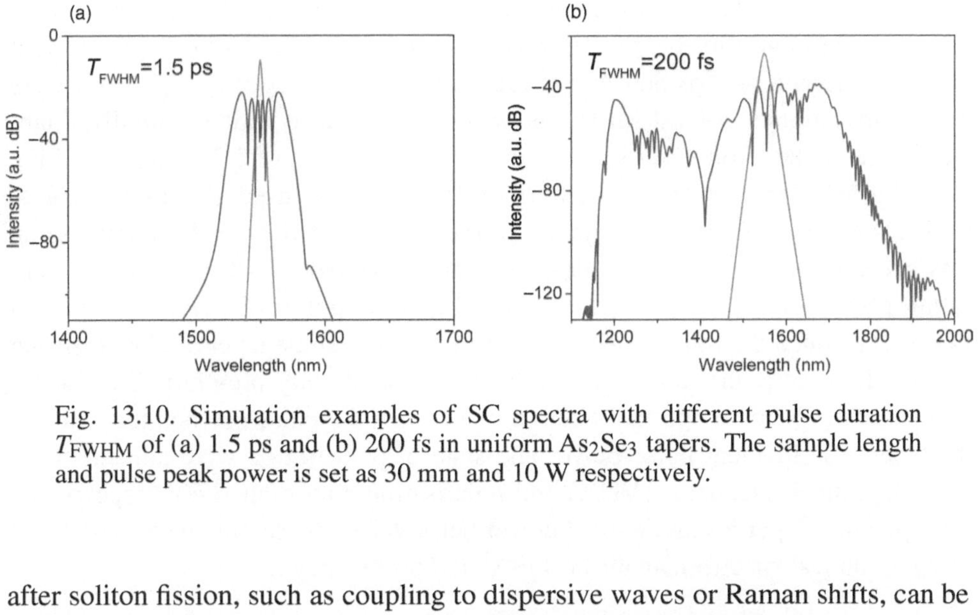

Fig. 13.10. Simulation examples of SC spectra with different pulse duration T_{FWHM} of (a) 1.5 ps and (b) 200 fs in uniform As_2Se_3 tapers. The sample length and pulse peak power is set as 30 mm and 10 W respectively.

after soliton fission, such as coupling to dispersive waves or Raman shifts, can be clearly observed. As shown in the figure, the desired pulse duration ranges from 160 to 480 fs for these two conditions for our sample. Figure 13.10 shows two numerically simulated spectra that support this. Both simulations were carried out for the same conditions (30-mm-long 1 μm-diameter As_2Se_3 taper with 10 W peak power) but with different pulse durations, one outside of the desired range at 1.5 ps and one within the range at 200 fs. Pure SPM-based spectral broadening was observed in the case of the 1.5 ps pulse due to the long dispersion length, as shown in Fig. 13.10(a).

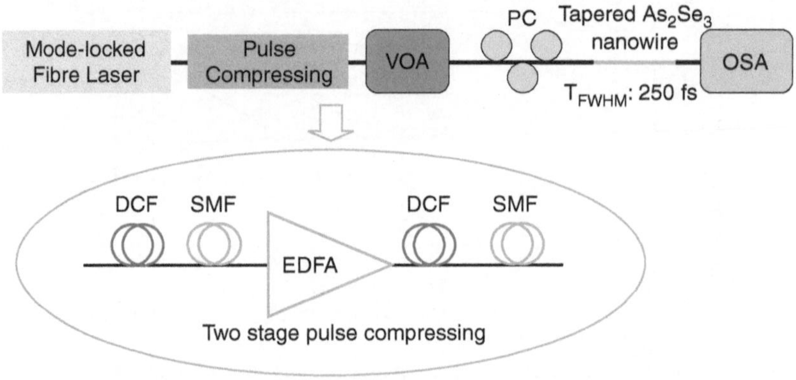

Fig. 13.11. Experimental setup for SC generation in As_2Se_3 tapers. A two-stage pulse compression scheme is employed to provide different pulse durations.

Figure 13.10(b) shows a spectrum typical of SC generated at the same peak power but with an input pulse of 200 fs duration.

A schematic of the experimental setup used to produce a low-power threshold supercontinuum in the As_2Se_3 tapers is shown in Fig. 13.11. Two different pulse durations of 1.2 ps and 250 fs full-width at half maximum (FWHM) were used to investigate spectral variations with different pulse widths. Initially, a lab-built mode-locked fibre laser with measured pulse FWHM of 1.2 ps and repetition rate of 9 MHz at a centre wavelength of 1555 nm was used. A pulse compression scheme was used to generate the shorter duration pulses. Pulse compression was achieved using normal single-mode fibres (SMFs) and dispersion compensating fibres (DCFs) with an erbium-doped fibre amplifier (EDFA), as shown in Fig. 13.11. Since the transition regions of the chalcogenide taper exhibited strong normal dispersion, the pulses were adjusted to be slightly pre-chirped in such a way that they become optimally compressed when the pulse reached the waist of the taper. The incident pulse power and state of polarisation were adjusted by a variable optical attenuator (VOA) and polarisation controller (PC), respectively. The spectral properties and optical power before/after the sample were monitored using an optical spectrum analyser (OSA) and power meter.

13.3.4 Experimental results and analysis

The measured output spectra with different pulse durations are shown in Fig. 13.12 as a function of incident power into the submicron As_2Se_3 taper. The peak power (P_0) at the taper waist was estimated from the measurement of average incident power and taper loss. As in the previous numerical simulations, a pulse duration of

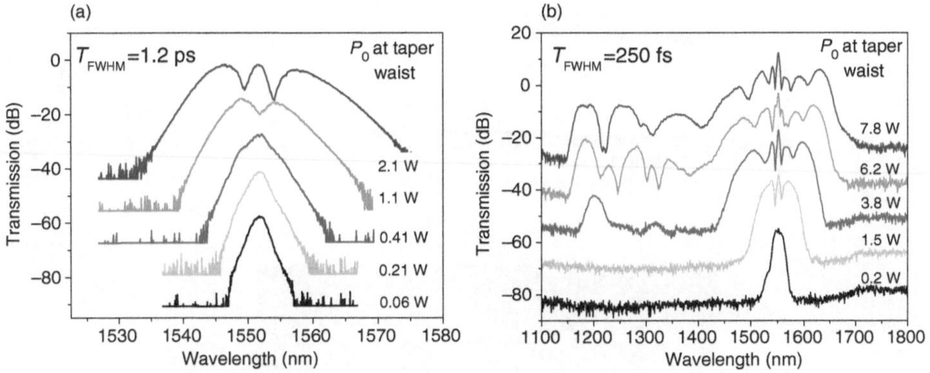

Fig. 13.12. Transmission spectra of the As_2Se_3 taper with different pulse durations of (a) 1.2 ps and (b) 250 fs.

1.2 ps results in spectral broadening predominantly caused by SPM, as the experimental spectra confirm in Fig. 13.12(a). A rough estimation of nonlinearity from the spectrum, without the consideration of TPA effects, gives a γ of 120~140/W/m. Figure 13.12(b) shows the spectra when the incident pulse was compressed down to a FWHM of 250 fs. The initial spectral broadening at peak powers below 1.5 W is similar to that in Fig. 13.12(a). However, as the peak power was increased, a significant boost to the spectral broadening was observed. In particular, the generation and development of non-solitonic radiation at shorter wavelengths was clearly seen from peak powers greater than 3.8 W. This is due to phase matching between soliton pulses, generated by the soliton fission process, with linear dispersive waves (Herrmann et al., 2002). The calculated spectral bandwidth from the intersection of the noise level was more than 500 nm at a peak power of 7.8 W; this corresponds to a pulse energy of ~2.2 pJ.

Figure 13.13 compares the experimental spectra with numerical simulation based on the split-step Fourier method (Agrawal, 2001). The simulation incorporates the whole taper structure, including the 30-mm-long centred taper waist, by gradually varying the nonlinear and dispersion parameters depending on taper diameter. For example, the effective area $A_{\rm eff}$ and dispersion β_2, β_3, β_4 are 0.4773 μm^2, -360 ps^2/km, 3.85 ps^3/km, -0.01 ps^4/km at 1 μm taper diameter. A TPA coefficient (β of 2.5×10^{-12} m/W) (Slusher et al., 2004) and Raman effects in As_2Se_3 glass are also applied to the numerical modelling. In the simulation, shown in Fig. 13.13(b), a nominal value of the axially varying taper waist was fitted to 1 μm instead of the measured value of 0.95 μm to reconcile the simulation with experiment, and provided good agreement with the experimental spectrum shown in Fig. 13.13(a). This slight discrepancy of diameter for the simulation is expected to come from the ambiguity of tapering loss position and dispersion values in the

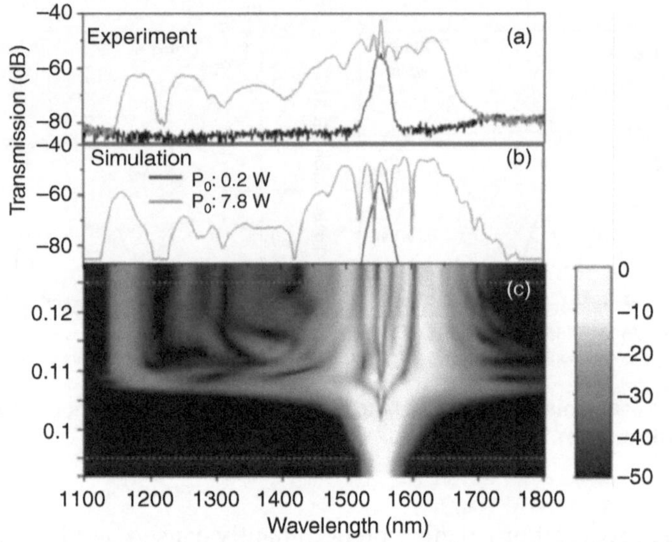

Fig. 13.13. Comparison of (a) experimental results of SC spectra with (b) simulation for the incident pulse with T_{FWHM} of 250 fs. (c) Evolution of the pulse spectrum propagating along the As_2Se_3 taper at $P_0 = 7.8$ W.

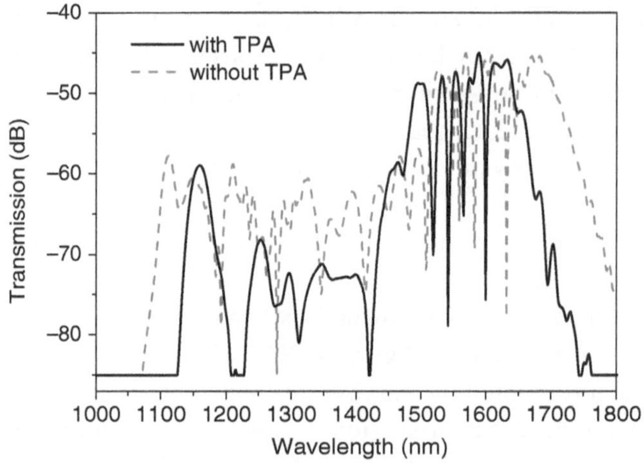

Fig. 13.14. Numerical comparison of the effect of TPA on the SC spectrum at a peak power of 7.8 W.

As_2Se_3 taper. The maximum γ from the simulation is about 93.4/W/m. Although this is slightly smaller than that simply inferred from the spectrum in Fig. 13.12(a), this value is still more than 80 000 times more than the γ of silica SMF, and 1500 times the maximum for silica tapers. The evolution of the SC spectrum is shown in Fig. 13.13(c), focusing on the taper waist section of the sample where a white dotted line indicates the taper waist boundary. It should be noted that there was no appreciable spectral broadening before and after the taper waist section. In addition, the soliton fission process and associated dispersive wave generated at shorter wavelengths occurs in the first 15 mm of the taper waist. Finally, Fig. 13.14 numerically compares the effect of TPA on SC generation in As_2Se_3 tapers. From the simulation, a slight reduction in the spectral bandwidth and increase of the soliton fission length for the SC process are observed in the presence of TPA. However, due to the moderate value of TPA for As_2Se_3 glass, it does not critically affect the fundamental SC process.

13.4 Supercontinuum generation in As_2S_3 chalcogenide planar waveguide

The chalcogenide fibre taper, discussed in Section 13.3, was engineered to have anomalous dispersion despite the normal material dispersion of chalcogenide glass. As well, the nonlinearity of the fibre was enhanced due to a reduced effective area, and this resulted in SC generation at very low powers. In this section, we apply these same principles to a different geometry — chalcogenide planar waveguides — and a different composition of chalcogenide glass — As_2S_3. As_2S_3 has a nonlinear index (n_2) roughly one quarter of As_2Se_3, which is still ≈100 times higher than silica, but has almost negligible TPA (β of 6.2×10^{-15} W/m) and is generally considered a more robust material.

13.4.1 Device design and fabrication

Rib waveguides were fabricated in 0.85 μm thick As_2S_3 films by photolithography and dry etching (Madden et al., 2007). An SEM (scanning electron microscope) image of the resulting As_2S_3 rib waveguide is shown in Fig. 13.15. The films were prepared by thermal evaporation from a Ta baffled box whose temperature was stabilised to 315°C in a chamber pumped to a base pressure of 3×10^{-7} Torr. The films were deposited onto 100 mm diameter ⟨100⟩ orientated silicon wafers with a 1.5 μm thick surface layer of thermal oxide as the bottom cladding for the waveguide. By mounting the wafers in a carousel that underwent planetary motion during evaporation, thickness uniformity of about ±1% was achieved. After deposition, the films were placed in a vacuum oven and annealed at 130°C for 24 hours to relax their bond structure towards the bulk state. During this process, the index rose from 2.29 for as-deposited films to 2.38 at 1550 nm.

Since As_2S_3 is prone to chemical attack by common photo-resist (PR) developers, a protective layer in the form of a bottom anti-reflective coating (BARC, XHiRC-16 from Brewer Science) was coated onto the As_2S_3 film at a spin speed of 5000 rpm and cured at 115°C for 2 minutes on a hot plate. PR patterns were produced on

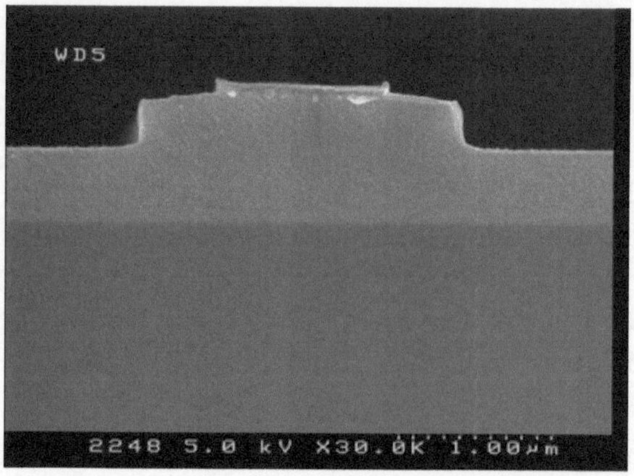

Fig. 13.15. Scanning electron microscope (SEM) image of the end facet of an As_2S_3 waveguide during processing (a small layer of PMMA remains on the As_2S_3), and before IPG™ cladding. Image provided by Duk-Yong Choi of the Laser Physics Centre at the Australian National University.

the BARC using contact photolithography (MA6 from Suss MicroTec) at 365 nm using a high-resolution resist (Clariant AZ MiR701). An inductively coupled plasma reactive ion etching system (ICP-RIE, Oxford Plasmalab 100 ICP) was used to etch the BARC and As_2S_3. The BARC was etched in oxygen plasma with the addition of small amounts of argon and CHF_3, whilst pure CHF_3 was utilised to pattern the As_2S_3 film. The etch rates of the films and PR were measured in situ by a laser interferometer (D-205 Digilem from Jovin Yvon-Sofie).

The use of CHF_3 for etching As_2S_3 offers several advantages over CF_4-O_2 gas chemistries reported previously (Choi et al., 2008). In optimised conditions, CHF_3 etching results in excellent etch selectivity between As_2S_3 and photo-resist. This reduces etch bias whilst sidewall passivation through polymer deposition assists both in obtaining vertical sidewalls and reducing sidewall roughness by limiting chemical etching by F atoms, and results in decoration of nanophase components that exist in the films. The best plasma conditions were 20 SCCM CHF_3, 10 mTorr process pressure, 30 W substrate bias power, and 300 W ICP power (using an Oxford Instruments 100 series ICP etcher). The etched surface roughness obtained by atomic force microscopy was more than halved (from 3.3 nm RMS to 1.5 nm RMS) compared to the CF_4-O_2 process. After resist removal, the waveguides were clad with a 15 μm layer of Inorganic Polymer Glass™ (RPO Pty Ltd) which has an index of 1.53 at 1550 nm. End facets were prepared by hand-cleaving with a diamond scriber.

Although the chalcogenide glass considered here differs from the glass used in the previous section, As_2S_3 still has a large normal dispersion. Refractive index measurements carried out on deposited thin-film As_2S_3 result in the following Sellmeier equation:

$$n^2(\lambda) = 1 + \frac{1.612}{\lambda^2 - 0.0284} + \frac{1.922}{\lambda^2 - 0.0571} + \frac{0.884}{\lambda^2 - 0.110} + \frac{0.127}{\lambda^2 - 0.209}. \quad (13.4)$$

From Eqn. 13.1, the resulting material dispersion at a wavelength of 1550 nm (D_{mat}) is −360 ps/nm/km, about half that of As_2Se_3. To achieve the anomalous dispersion necessary for SC generation, we must reduce the height and width of the waveguide in an analogous manner to that employed with the fibre taper. However, in practice, the width is set by the photomask pattern used in photolithography and is not easily varied. For this reason, the waveguide height and etch depth are the preferred variables used for engineering the dispersion. Unlike the fibre taper, which has circular symmetry, planar waveguides can be polarised horizontally (TE) and vertically (TM). Because a rib waveguide is not fully etched, the TE-mode does not experience as strong waveguide dispersion as the TM-mode. This can be seen in Fig. 13.16, which shows how the GVD of the fundamental TE- and TM-modes varies with decreasing waveguide height, modelled using RSoft FemSIM^{TM}. With an etch depth-to-height ratio of 44%, the TE-mode cannot reach the anomalous waveguide dispersion necessary to compensate the normal material dispersion. However, the GVD of the TM-mode is anomalous for heights between 0.97 and 0.74 μm. Reducing the height of the waveguide also reduces the mode effective

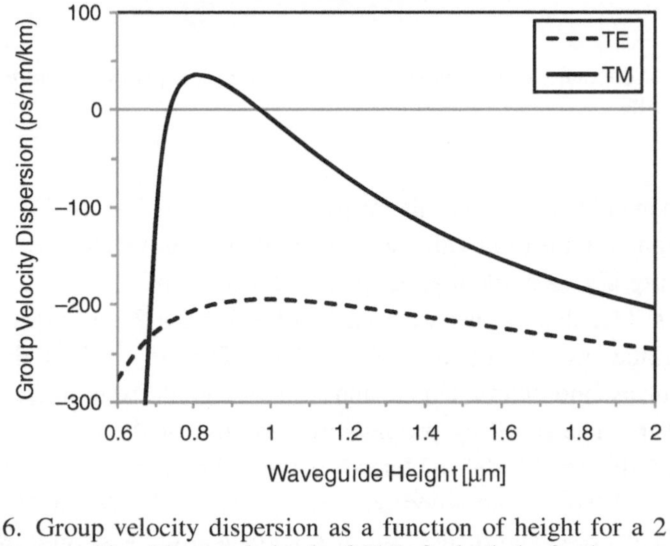

Fig. 13.16. Group velocity dispersion as a function of height for a 2 μm wide As_2S_3 rib waveguide with an etch depth of 44%, for both the fundamental TE- and TM-modes.

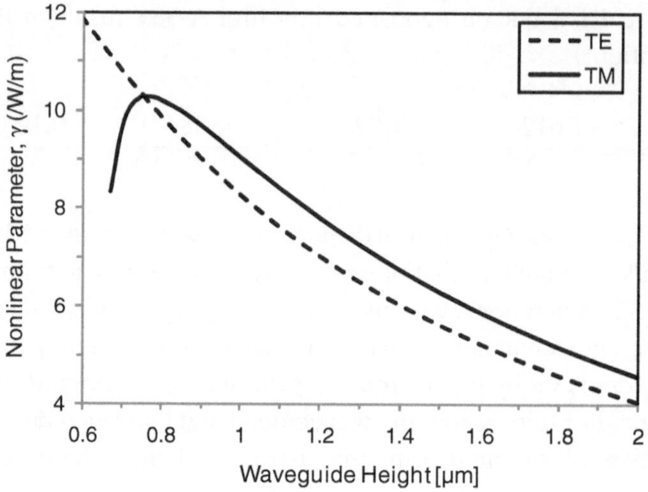

Fig. 13.17. The nonlinear parameter (γ) as a function of height for the same parameters used in Fig. 13.16.

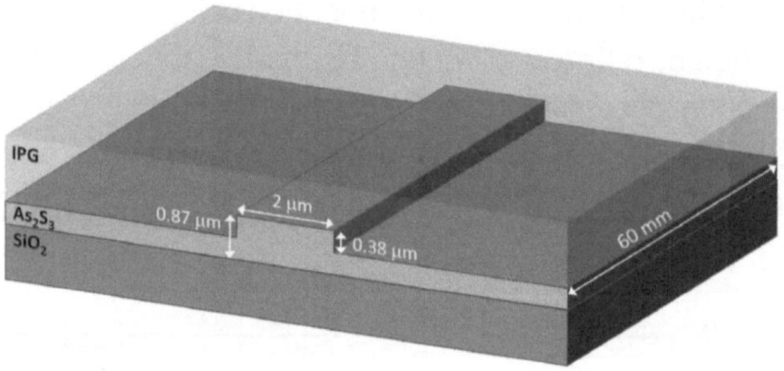

Fig. 13.18. Schematic of the dispersion engineered As_2S_3 rib waveguide used for SC generation.

area, which then enhances the nonlinear parameter (γ), shown in Fig. 13.17. However, if the height is made too small, the mode will lose confinement and the effective area expands rapidly, as evident at heights below 0.8 μm.

The process described above was used to fabricate a 2 μm-wide, 6.0 cm-long As_2S_3 waveguide, as illustrated in Fig. 13.18. The As_2S_3 film thickness was 0.87 μm, ensuring anomalous GVD, and the waveguide was created by etching down by 380 nm. The resulting propagation loss (α) of the waveguide was estimated to be 0.6 dB/cm. Modelling based on the finite element method predicts an effective area of 1.2 μm^2 and a nonlinear parameter of 9.8 $W^{-1}m^{-1}$ or 9100 $\times\ \gamma$ of standard SMF fibre. The dispersion of this waveguide is shown in Fig. 13.19. The TE-mode GVD remains normal, −210 ps/nm/km at a wavelength of 1550 nm;

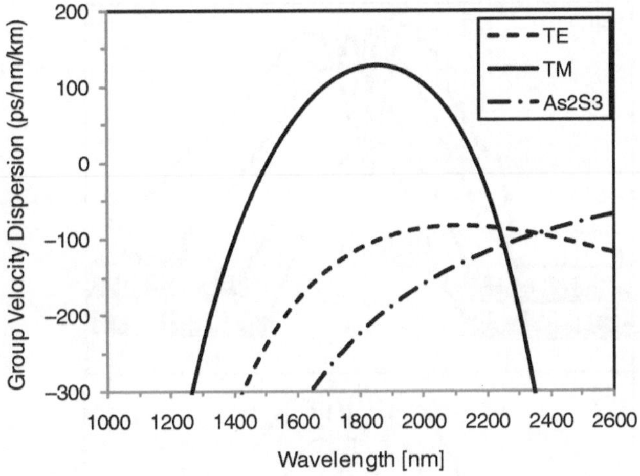

Fig. 13.19. GVD of the dispersion engineered As_2S_3 waveguide as a function of wavelength.

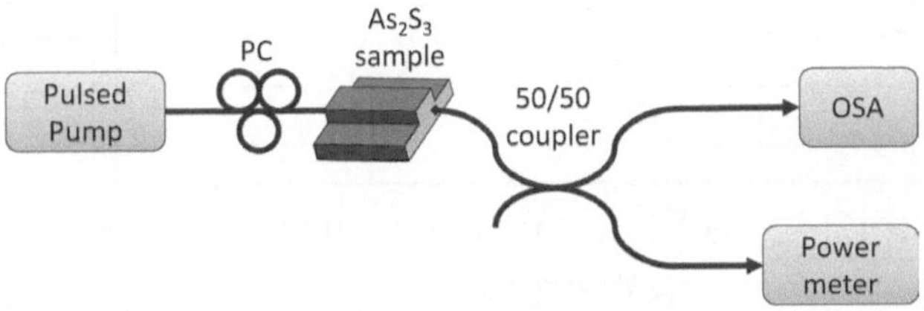

Fig. 13.20. Experimental setup of SC generation in the As_2S_3 waveguide. PC is the polarisation controller, and OSA is the optical spectrum analyser.

while the TM-mode has two zero-dispersion wavelengths at 1510 nm and 2170 nm, and anomalous dispersion of +29 ps/nm/km at 1550 nm.

13.4.2 Experiment and analysis

The SC generation experiment was performed using a 10 GHz mode-locked fibre laser with a FWHM of 610 fs after pulse compression through SMF fibre. Light from the laser was sent through a polarisation controller to select either the TE- or TM-mode and then coupled into the waveguide using a lensed fibre with coupling losses estimated at 3.7 dB per facet. The maximum coupled input power was ~0.6 mW average or 68 W peak, and was varied by moving the lensed fibre further away from the waveguide. The output light was collected by another lensed fibre and sent to both a power meter and an optical spectrum analyser using a 50/50 fibre coupler, as seen in the experimental setup Fig. 13.20.

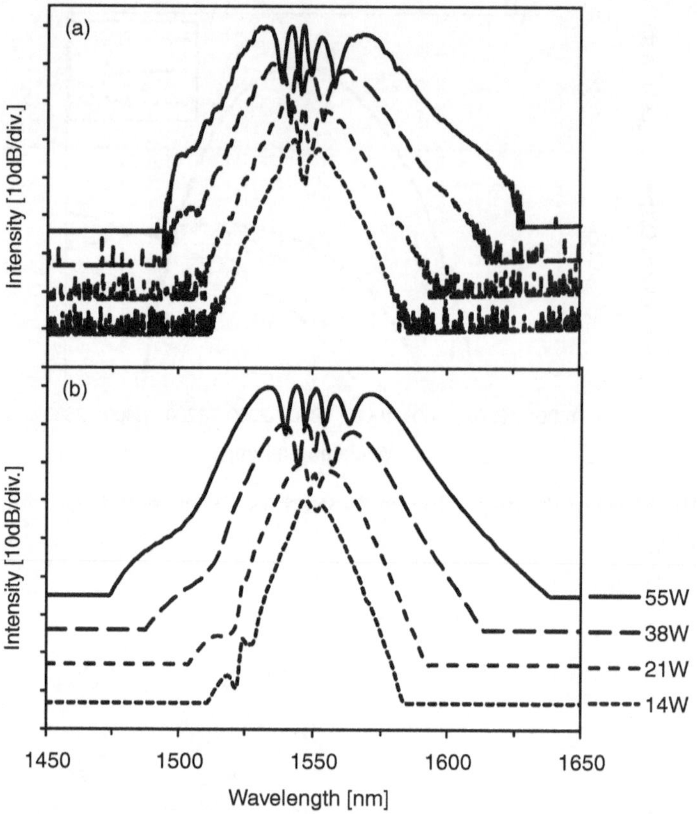

Fig. 13.21. (a) Experimentally measured spectra at increasing input power levels after propagation through the TE-mode of the As_2S_3 waveguide, which has normal GVD at a wavelength of 1550 nm. (b) SSFM simulated spectra for the same polarisation and input powers.

Figure 13.21(a) shows the experimentally measured output spectrum when the input light was TE-polarised. Because the GVD remained normal, no solitonic behaviour or four-wave mixing (FWM) occurs and, as the input power is increased, the spectral broadening is primarily due to SPM. This is in good agreement with simulations, shown in Fig. 13.21(b), by solving the nonlinear Schrödinger equation using the split-step Fourier method with an adaptive step-size (Sinkin et al., 2003). The simulation includes Raman scattering and self-steepening effects, and both linear and nonlinear losses due to TPA. The Raman gain spectrum was modelled following the results from Li et al. (2005) with a fractional Raman contribution of $f_R = 0.11$.

Although the TE-mode resulted in only SPM due to its normal GVD, the TM-mode is engineered to meet the required dispersion for SC generation. This is

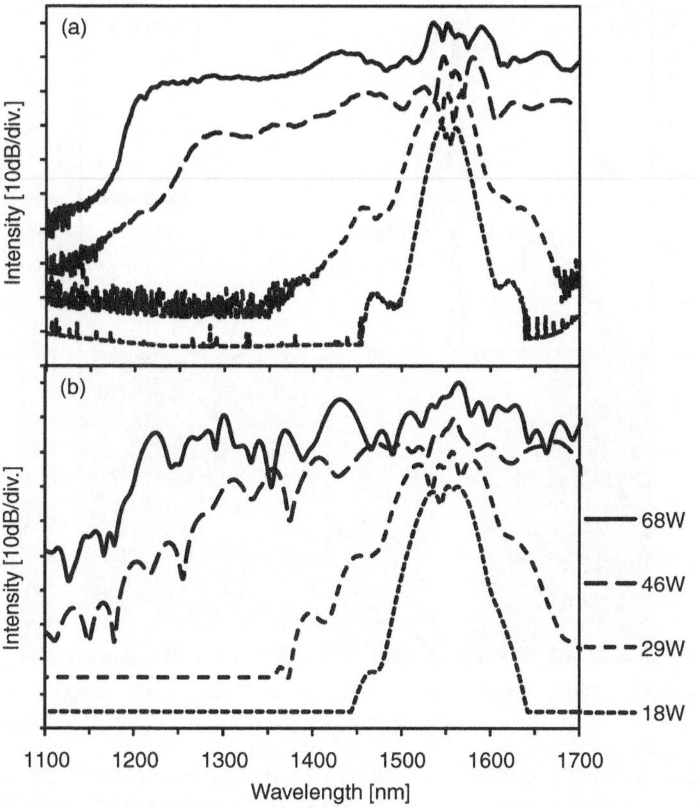

Fig. 13.22. (a) Experimentally measured spectra at increasing input power levels after propagation through the TM-mode of the As_2S_3 waveguide, which is engineered to have anomalous GVD at a wavelength of 1550 nm. (b) SSFM simulated spectra for the same polarisation and input powers.

confirmed by the experimentally measured spectra in Fig. 13.22(a). As the input power increases, the output spectrum expands rapidly beyond the maximum wavelength detectable on the OSA. The same SSFM simulations agree closely to the experimental spectra at all the measured power levels, as shown in Fig. 13.22(b). From this, the broadest SC spectrum spans 1400 nm (1.4 octaves) at −60 dB, just above the experimental noise floor, and the −30 dB spectral width spans 750 nm.

13.4.3 Discussion

The SC spectrum shown in Fig. 13.22(a) has a greater symmetry than is generally observed (Dudley et al., 2006). The reason for this becomes clear by considering the various linear and nonlinear length scales involved. The dispersion length, $L_D = T_0^2/|\beta_2|$, for the TM-mode is 2.97 m, orders of magnitude longer than the device.

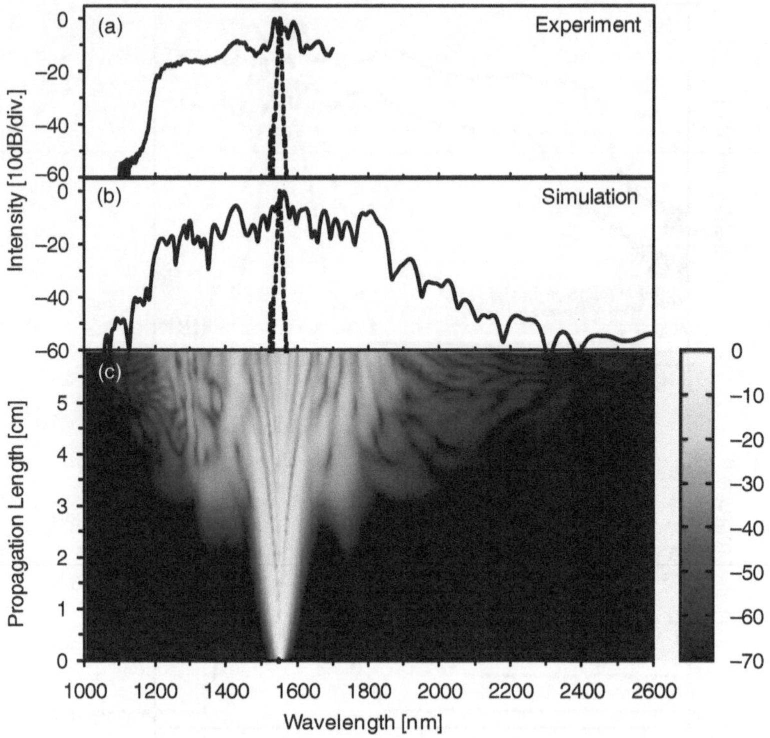

Fig. 13.23. (a) Input and output experimentally measured spectra for an input peak power of 68 W propagating in the TM-mode. (b) Simulated spectra for the same parameters. (c) Simulated spectral evolution for the same input pulse propagating through 6 cm of dispersion engineered As_2S_3 rib waveguide.

However, the nonlinear length, $L_{NL} = 1/\gamma P$, is 1.5 mm at the highest peak power. This results in a very high soliton number, $N = (L_D/L_{NL})^{1/2}$ (greater than 40) and a soliton fission length, $N_{fission} = L_D/N$, of 67 mm which is comparable to the device length. Because of this, the primary nonlinear process driving the SC generation is FWM rather than soliton fission. This is supported by the spectral evolution shown in Fig. 13.23(c), in which FWM idler terms appear after 2 cm of propagation. The apparent asymmetry occurs because FWM is balanced in frequency, and not wavelength; however, Raman scattering does introduce a slight shift of energy toward longer wavelengths.

13.5 Summary and prospective

Highly nonlinear materials such as silicon and chalcogenide glasses are a promising platform for the realisation of compact, on-chip SC generation. However, the strong normal dispersion of these materials makes dispersion engineering necessary to achieve the anomalous dispersion needed for efficient four-wave mixing and solitonic effects, key processes for efficient SC generation. In this chapter, we have demonstrated the fabrication of anomalous dispersion chalcogenide devices, in both planar waveguide and tapered fibre geometries, and both with enhanced nonlinear parameters. In the fibre taper, low-threshold SC generation was shown, enabled by the very high nonlinear parameter of 93.4/W/m. The pulse energy used was $\sim$2.2 pJ, orders of magnitude lower than more typical values. In the planar waveguide geometry, we used slightly higher pulse energies, $\sim$60 pJ, to obtain over an octave of SC bandwidth. This can be compared to a maximum of 0.3 octaves, or >300 nm at a similar wavelength, achieved using 100 fs pulses in a dispersion engineered silicon waveguide (Hsieh et al., 2007). Because of the low TPA and lack of free-carriers in As_2S_3, much high peak powers can be utilised to generate a full SC spectrum, and with further reduction in the effective area of the waveguide this peak power may be reduced while maintaining a comparable SC bandwidth.

Acknowledgements

This work was produced with the assistance of the Australian Research Council (ARC). The Centre for Ultrahigh-bandwidth Devices for Optical Systems (CUDOS) is an ARC Centre of Excellence.

References

Agrawal, G. P. (2001) Chapter 2. *Nonlinear Fiber Optics*, 3rd edn, San Diego, CA, Academic Press.

Asobe, M., Itoh, H., Miyazawa, T. & Kanamori, T. (1993) Efficient and ultrafast all-optical switching using high delta-N, small core chalcogenide glass-fiber. *Electronics Letters*, **29**, 1966–1968.

Birks, T. A., Wadsworth, W. J. & Russell, P. S. (2000) Supercontinuum generation in tapered fibers. *Optics Letters*, **25**, 1415–1417.

Boyd, R. W. (2008) *Nonlinear Optics*, Academic Press.

Boyraz, O., Indukuri, T. & Jalali, B. (2004) Self-phase-modulation induced spectral broadening in silicon waveguides. *Optics Express*, **12**, 829–834.

Brambilla, G., Koizumi, F., Finazzi, V. & Richardson, D. J. (2005) Supercontinuum generation in tapered bismuth silicate fibres. *Electronics Letters*, **41**, 795–797.

Bristow, A. D., Rotenberg, N. & Van Driel, H. M. (2007) Two-photon absorption and Kerr coefficients of silicon for 850–2200 nm. *Applied Physics Letters*, **90**, 191104.

Cerqua-Richardson, K. A., Mckinley, J. M., Lawrence, B., Joshi, S. & Villeneuve, A. (1998) Comparison of nonlinear-optical properties of sulfide glasses in bulk and thin film form. *Optical Materials*, **10**, 155–159.

Choi, D. Y., Madden, S., Rode, A., Wang, R. P. & Luther-Davies, B. (2008) Plasma etching of As_2S_3 films for optical waveguides. *Journal of Non-Crystalline Solids*, **354**, 3179–3183.

Dudley, J. M., Genty, G. & Coen, S. (2006) Supercontinuum generation in photonic crystal fiber. *Reviews of Modern Physics*, **78**, 1135–1184.

Dulkeith, E., Vlasov, Y. A., Chen, X. G., Panoiu, N. C. & Osgood, R. M. (2006) Self-phase-modulation in submicron silicon-on-insulator photonic wires. *Optics Express*, **14**, 5524–5534.

Ebendorff-Heidepriem, H., Petropoulos, P., Asimakis, S., Finazzi, V., Moore, R. C., Frampton, K., Koizumi, F., Richardson, D. J. & Monro, T. M. (2004) Bismuth glass holey fibers with high nonlinearity. *Optics Express*, **12**, 5082–5087.

Fu, L. B., Rochette, M., Ta'eed, V. G., Moss, D. J. & Eggleton, B. J. (2005) Investigation of self-phase modulation based optical regeneration in single mode As_2Se_3 chalcogenide glass fiber. *Optics Express*, **13**, 7637–7644.

Harbold, J. M., Ilday, F. O., Wise, F. W., Sanghera, J. S., Nguyen, V. Q., Shaw, L. B. & Aggarwal, I. D. (2002) Highly nonlinear As-S-Se glasses for all-optical switching. *Optics Letters*, **27**, 119–121.

Herrmann, J., Griebner, U., Zhavoronkov, N., Husakou, A., Nickel, D., Knight, J. C., Wadsworth, W. J., Russell, P. S. J. & Korn, G. (2002) Experimental evidence for supercontinuum generation by fission of higher-order solitons in photonic fibers. *Physical Review Letters*, **88**, 173901 1–4.

Hsieh, I. W., Chen, X. G., Liu, X. P., Dadap, J. I., Panoiu, N. C., Chou, C. Y., Xia, F. N., Green, W. M., Vlasov, Y. A. & Osgood, R. M. (2007) Supercontinuum generation in silicon photonic wires. *Optics Express*, **15**, 15242–15249.

Klocek, P. (Ed.) (1991) *Handbook of Infrared Optical Materials*, New York, Marcel Dekker.

Lamont, M. R. E., Luther-Davies, B., Choi, D.-Y., Madden, S. J. & Eggleton, B. J. (2008) Supercontinuum generation in dispersion engineered highly nonlinear ($\gamma = 10/\mathrm{W/m}$) As_2S_3 chalcogenide planar waveguides. *Optics Express*, **16**, 14938–14944.

Lenz, G., Zimmermann, J., Katsufuji, T., Lines, M. E., Hwang, H. Y., Spalter, S., Slusher, R. E., Cheong, S. W., Sanghera, J. S. & Aggarwal, I. D. (2000) Large Kerr effect in bulk Se-based chalcogenide glasses. *Optics Letters*, **25**, 254–256.

Leong, J. Y. Y., Petropoulos, P., Price, J. H. V., Ebendorff-Heidepriem, H., Asimakis, S., Moore, R. C., Frampton, K. E., Finazzi, V., Feng, X., Monro, T. M. & Richardson, D. J. (2006) High-nonlinearity dispersion-shifted lead-silicate holey fibers for efficient 1-mu m pumped supercontinuum generation. *Journal of Lightwave Technology*, **24**, 183–190.

Li, W., Seal, S., Rivero, C., Lopez, C., Richardson, K., Pope, A., Schulte, A., Myneni, S., Jain, H., Antoine, K. & Miller, A. C. (2005) Role of S/Se ratio in chemical bonding of As-S-Se glasses investigated by Raman, x-ray photoelectron, and extended x-ray absorption fine structure spectroscopies. *Journal of Applied Physics*, **98**, 053503–11.

Lin, Q., Zhang, J., Piredda, G., Boyd, R. W., Fauchet, P. M. & Agrawal, G. P. (2007) Dispersion of silicon nonlinearities in the near infrared region. *Applied Physics Letters*, **91**, 021111.

Madden, S. J., Choi, D. Y., Bulla, D. A., Rode, A. V., Luther-Davies, B., Ta'eed, V. G., Pelusi, M. D. & Eggleton, B. J. (2007) Long, low loss etched As_2S_3 chalcogenide waveguides for all-optical signal regeneration. *Optics Express*, **15**, 14414–14421.

Mägi, E. C., Fu, L. B., Nguyen, H. C., Lamont, M. R. E., Yeom, D. I. & Eggleton, B. J. (2007) Enhanced Kerr nonlinearity in sub-wavelength diameter As_2Se_3 chalcogenide fiber tapers. *Optics Express*, **15**, 10324–10329.

Nguyen, H. C., Finsterbusch, K., Moss, D. J. & Eggleton, B. J. (2006) Dispersion in nonlinear figure of merit of As_2Se_3 chalcogenide fibre. *Electronics Letters*, **42**, 571–572.

Price, J. H. V., Monro, T. M., Ebendorff-Heidepriem, H., Poletti, F., Finazzi, V., Leong, J. Y. Y., Petropoulos, P., Flanagan, J. C., Brambilla, G., Feng, X. & Richardson, D. J. (2006) Non-silica microstructured optical fibers for mid-IR supercontinuum generation from 2 μm–5 μm. *Proceedings of the SPIE*, **6102**, 54–68.

Psaila, N. D., Thomson, R. R., Bookey, H. T., Shen, S. X., Chiodo, N., Osellame, R., Cerullo, G., Jha, A. & Kar, A. K. (2007) Supercontinuum generation in an ultrafast laser inscribed chalcogenide glass waveguide. *Optics Express*, **15**, 15776–15781.

Ranka, J. K., Windeler, R. S. & Stentz, A. J. (2000) Visible continuum generation in air–silica microstructure optical fibers with anomalous dispersion at 800 nm. *Optics Letters*, **25**, 25–27.

Ruan, Y. L., Li, W. T., Jarvis, R., Madsen, N., Rode, A. & Luther-Davies, B. (2004) Fabrication and characterization of low loss rib chalcogenide waveguides made by dry etching. *Optics Express*, **12**, 5140–5145.

Saleh, B. E. A. & Teich, M. C. (2007) *Fundamentals of Photonics*, John Wiley and Sons.

Sanghera, J. S., Shaw, L. B., Busse, L. E., Nguyen, V. Q., Pureza, P. C., Cole, B. C., Harbison, B. B., Aggarwal, I. D., Mossadegh, R., Kung, F., Talley, D., Roselle, D. & Miklos, R. (2000) Development and infrared applications of chalcogenide glass optical fibers. *Fiber and Integrated Optics*, **19**, 251–274.

Sinkin, O. V., Holzlohner, R., Zweck, J. & Menyuk, C. R. (2003) Optimization of the split-step Fourier method in modeling optical-fiber communications systems. *Journal of Lightwave Technology*, **21**, 61–68.

Siviloglou, G. A., Suntsov, S., El-Ganainy, R., Iwanow, R., Stegeman, G. I., Christodoulides, D. N., Morandotti, R., Modotto, D., Locatelli, A., De Angelis, C., Pozzi, F., Stanley, C. R. & Sorel, M. (2006) Enhanced third-order nonlinear effects in optical AlGaAs nanowires. *Optics Express*, **14**, 9377–9384.

Slusher, R. E., Lenz, G., Hodelin, J., Sanghera, J., Shaw, L. B. & Aggarwal, I. D. (2004) Large Raman gain and nonlinear phase shifts in high-purity As_2Se_3 chalcogenide fibers. *Journal of the Optical Society of America B: Optical Physics*, **21**, 1146–1155.

Ta'eed, V. G., Shokooh-Saremi, M., Fu, L. B., Moss, D. J., Rochette, M., Littler, I. C. M., Eggleton, B. J., Ruan, Y. L. & Luther-Davies, B. (2005) Integrated all-optical pulse regenerator in chalcogenide waveguides. *Optics Letters*, **30**, 2900–2902.

Tsang, H. K., Wong, C. S., Liang, T. K., Day, I. E., Roberts, S. W., Harpin, A., Drake, J. & Asghari, M. (2002) Optical dispersion, two-photon absorption and self-phase modulation in silicon waveguides at 1.5 μm wavelength. *Applied Physics Letters*, **80**, 416–418.

Yamane, M. & Asahara, Y. (2000) *Glasses for Photonics*, Cambridge, Cambridge University Press.

Yeom, D.-I., Mägi, E. C., Lamont, M. R. E., Roelens, M. A. F., Fu, L. & Eggleton, B. J. (2008) Low-threshold supercontinuum generation in highly nonlinear chalcogenide nanowires. *Optics Letters*, **33**, 660–662.

Zakery, A. & Elliott, S. R. (2003) Optical properties and applications of chalcogenide glasses: a review. *Journal of Non-Crystalline Solids*, **330**, 1–12.

14

Supercontinuum generation for carrier-envelope phase stabilization of mode-locked lasers

S. T. Cundiff

14.1 Introduction

The demonstration of continuum generation in microstructure fiber by Ranka et al. (2000) coincided with efforts to measure and control the carrier-envelope phase evolution of mode-locked lasers (Xu et al., 1996; Reichert et al., 1999). It was recognized by Telle et al. (1999) that a broad spectrum, preferably spanning an octave of bandwidth, would enable new methods for stabilizing mode-locked lasers. Shortly thereafter, Jones et al. (2000) demonstrated the use of continuum generation in microstructure fiber to stabilize the carrier-envelope phase of a mode-locked laser in both the time and frequency domains. Stabilization of the carrier-envelope phase evolution means that the frequency spectrum of the laser corresponds to a comb of well-defined frequencies, which can be used to perform optical frequency metrology or build optical atomic clocks (Cundiff and Ye, 2003). It has also enabled the production of single attosecond pulses (Corkum and Krausz, 2007).

The key concept for this body of work is the carrier-envelope phase of an ultrashort pulse and its evolution. This concept is based on the decomposition of the pulse into an envelope function, $\hat{E}(t)$, that is superimposed on a continuous carrier wave with angular frequency ω_c, so that the electric field of the pulse is written $E(t) = \hat{E}(t)e^{i\omega_c t}$. The carrier-envelope phase, ϕ_{ce}, is the phase shift between the peak of the envelope and the closest peak of the carrier wave, as illustrated in Fig. 14.1. In any dispersive material, the difference between group and phase velocities will cause ϕ_{ce} to evolve as the pulse propagates.

This chapter will begin with a brief review of the optical spectrum of a mode-locked laser, how it is affected by evolution of ϕ_{ce} and how the evolution can be measured using a broadband continuum. The effects of the continuum generation on the measurements will then be discussed. The last section will briefly review the

Supercontinuum Generation in Optical Fibers, ed. by J. M. Dudley and J. R. Taylor. Published by Cambridge University Press.

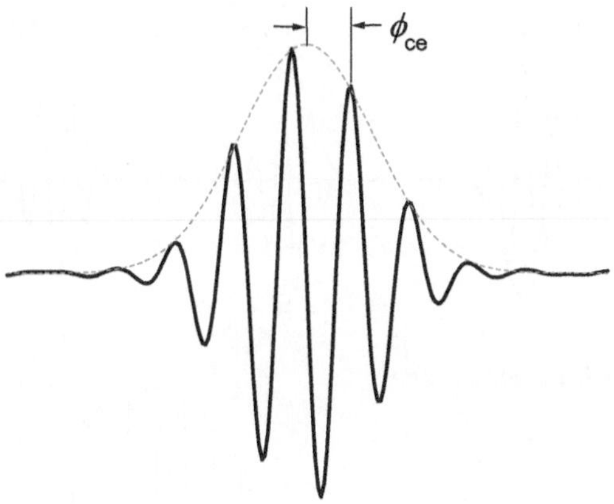

Fig. 14.1. An ultrashort optical pulse showing electric field (solid), envelope (dashed) and the carrier-envelope phase, ϕ_{ce}.

applications enabled by carrier-envelope phase stabilization. The book by Ye and Cundiff (2005) provides a more in-depth discussion of frequency comb technology.

14.2 Optical spectrum of a mode-locked laser

Mode-locked lasers generate trains of short optical pulses by establishing a fixed phase relationship between multiple longitudinal lasing modes. The pulses are separated in time by the cavity round-trip time, which is the same as the frequency spacing between the longitudinal modes. The duration of the individual pulses is inversely proportional to the number of modes, i.e., the optical bandwidth.

If the pulses emitted by the laser were identical, then the optical spectrum would be easily written as a Fourier series, giving a sequence of sharp lines at integer multiples of the repetition rate f_{rep} with an overall envelope that is the spectrum of a single pulse. However, the output pulses are not identical because ϕ_{ce} of the pulse circulating inside the cavity evolves. Thus each output pulse has a phase that is different from the preceding one by an amount $\Delta\phi_{ce}$. Fourier analysis of the resulting pulse train (see for example, Cundiff 2002), shows that the output is a comb of lines at frequencies $\nu_n = nf_{rep} + f_0$. The offset frequency, $f_0 = \frac{1}{2\pi} f_{rep} \Delta\phi_{ce}$, is due to the evolution of the carrier-envelope phase. The relationship between a pulse train with evolving carrier-envelope phase and optical frequency comb spectrum is shown in Fig. 14.2.

The pulse-to-pulse change in the phase for a train of pulses emitted by a mode-locked laser occurs because the phase and group velocities are different. The difference in velocities inside the cavity arises because of dispersion of the optical elements. Since the pulse is sampled once per round-trip when it hits the output

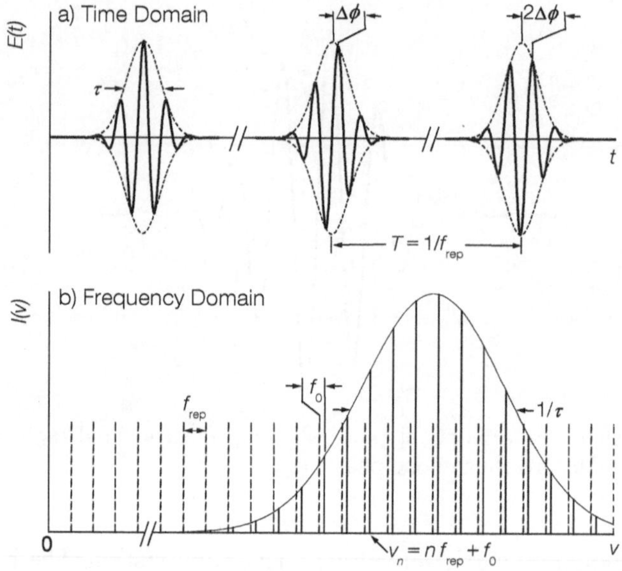

Fig. 14.2. The correspondence between the (a) time and (b) frequency domains for a pulse train with evolving phase.

coupler, it is only the phase change modulo 2π that matters. Specifically

$$\Delta\phi_{\mathrm{ce}} = \left(\frac{1}{v_{\mathrm{g}}} - \frac{1}{v_{\mathrm{p}}}\right) l_{\mathrm{c}}\omega_{\mathrm{c}} \bmod 2\pi \tag{14.1}$$

where v_{g} (v_{p}) is the mean group (phase) velocity in the laser cavity, l_{c} is the cavity length and ω_{c} is the carrier frequency.

It is relatively easy to measure f_{rep}, thus if f_0 could be measured, the pulse train would be fully characterized in time and frequency. Comparison of a single comb line to a continuous-wave (cw) laser of known optical frequency would do the trick. However, it turns out to be very difficult to determine the absolute frequency of a cw laser with sufficient accuracy. Indeed turning that problem on its head, using the frequency comb produced by a mode-locked laser to measure an absolute optical frequency, has produced much of the excitement in the field, and was mentioned in the citation for the 2005 Nobel Prize in Physics (Hall, 2006; Hänsch, 2006). Thus the question is how to measure f_0 without using an external reference.

14.2.1 Self-referencing measurement of f_0

Measurement of f_0 without an external reference is possible by comparing the comb to itself. The most common method, known as ν-to-2ν self-referencing, compares the comb to its second harmonic (Jones et al., 2000). However, it is also possible to

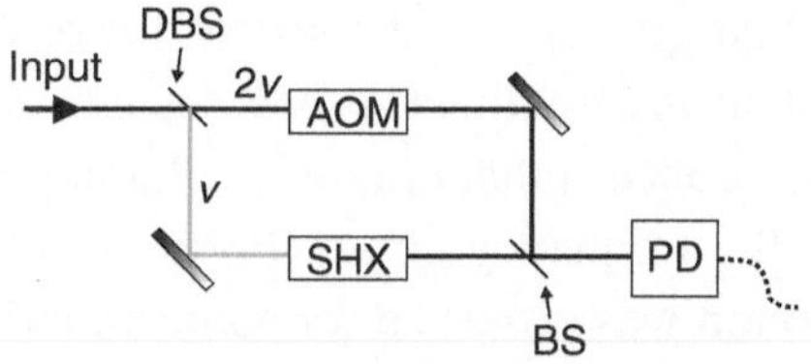

Fig. 14.3. Schematic of a ν-to-2ν interferometer. The input light is divided into two arms by a dichroic beam splitter (DBS). The low frequency (ν) part of the spectrum is frequency doubled by a second harmonic crystal (SHX). The high frequency part (2ν) can be frequency shifted by an acousto-optic modulator (AOM), which is not necessary but allows f_0 to be locked to arbitrary values including $f_0 = 0$. The two arms are then recombined on an ordinary beam splitter (BS) and detected with a photodiode (PD).

use higher order harmonics (Telle et al., 1999; Ramond et al., 2002) or to compare the comb to the harmonics of a cw laser (Diddams et al., 2000a,b).

The prerequisite for ν-to-2ν self-referencing is an optical spectrum that "spans" an octave. If such a broad spectrum is frequency doubled, then the high frequency wing of the fundamental spectrum will overlap with the low frequency spectrum of the second harmonic. In the overlap region, the frequency difference between near by comb lines can easily be determined by measuring the frequency of the heterodyne beat they produce. The difference frequency will be

$$2f_n - f_{2n} = 2(nf_{\mathrm{rep}} + f_0) - (2nf_{\mathrm{rep}} + f_0) = f_0. \tag{14.2}$$

Thus an octave spanning spectrum enables simple measurement of f_0. Note that the octave does not need to be achieved at the 3 dB points. Typically self-referencing can be implemented even with spectra that are down by 40 dB at the octave points from the peak. Thus the definition of "octave spanning" is usually taken to be an operational one that it is possible to implement ν-to-2ν self-referencing. Self-referencing is typically implemented as a "ν-to-2ν interferometer," which is shown schematically in Fig. 14.3.

Although octave-spanning lasers have been demonstrated (Fortier et al., 2003, 2006; Matos et al., 2004), they were not available when self-referencing was first developed. Rather, the early demonstrations used continuum generation in microstructure fiber to broaden the laser spectrum. Continuum generation is still used today in many cases; the relative merits of it versus an octave spanning laser are still being debated.

14.3 Continuum generation for self-referencing

The successful demonstration of self-referencing using continuum generation in microstructure fiber by Jones et al. (2000) led to significant effort to better understand the nonlinear processes that give rise to the spectral broadening. The broadening clearly demonstrates that strong nonlinear processes are at work. The presence of strong nonlinearity causes concern that amplitude fluctuations will produce phase fluctuations that corrupt the frequency comb (Fortier et al., 2002). In addition, a broadband noise background was observed for some situations (Ames et al., 2003; Corwin et al., 2003a,b). Understanding both of these issues was an important goal.

14.3.1 Amplitude-to-phase noise conversion

Amplitude noise on the input pulses to the microstructure fiber used for spectral broadening will be converted to phase noise that might be sufficiently strong to overwhelm the actual evolution of the carrier-envelope phase. The phase noise occurs because the index of refraction in the fiber depends on intensity due to the Kerr effect, $n(I) = n_0 + n_2 I$, where n_0 is the linear index of refraction, n_2 is the nonlinear index of refraction and I is the light intensity in the core of the microstructure fiber. A carrier-envelope phase shift of $\Delta\phi_{\mathrm{ce}} = \omega l n_{\mathrm{g}}/c - \omega l n_{\mathrm{p}}/c = \omega^2 l/c(dn_0/d\omega + I dn_2/d\omega)$ occurs during propagation over a distance l for light of frequency ω. The group index is $n_{\mathrm{g}} = n + \omega dn/d\omega$, $n_{\mathrm{p}} = n$ is the phase index and c is the speed of light. The change in ϕ_{ce} due to nonlinearity is proportional to the dispersion of n_2 and ΔI, the change in intensity. The coupling coefficient, C_{AP} between changes in phase and changes in power is defined by the relation $\Delta\phi_{\mathrm{ce}}^{\mathrm{NL}} = \omega^2 l/c(dn_2/d\omega)\Delta I = C_{\mathrm{AP}}\Delta P$.

Typically, the output of a ν-to-2ν interferometer is used to actively stabilize the offset frequency of a laser using a feedback loop. Thus phase noise generated in the fiber by amplitude fluctuations will be written onto the laser output by the feedback loop. The effects of amplitude-to-phase noise conversion can be seen by running two ν-to-2ν interferometers side-by-side, each using a separate piece of microstructure fiber (see Fig. 14.4). One interferometer locks f_0 of the laser, while the second, out-of-loop, interferometer measures the fluctuations in f_0.

To measure the efficiency of converting amplitude changes to phase changes, a sinusoidal modulation at frequency ω_{m} is imposed on the laser power prior to the microstructure fiber used to produce the octave spanning spectrum for locking the laser. The time varying amplitude results in a time varying phase $\phi_{\mathrm{ce}}(t) = C_{\mathrm{AP}}P(t)$, which corresponds to an offset frequency $f_0(t) = \frac{1}{2\pi}d\phi_{\mathrm{ce}}/dt = \frac{1}{2\pi}\omega_{\mathrm{m}}C_{\mathrm{AP}}\Delta P\cos(\omega_{\mathrm{m}}t)$ where ΔP is the modulation depth. The feedback loop adjusts f_0 of the laser to compensate for the frequency shift induced by the amplitude-to-phase conversion in the fiber. The measurement interferometer then records the varying f_0. In addition to the imposed power variation, there will be intrinsic phase noise that adds in quadrature, thus the total fluctuations will be $\Delta f_0 = [\eta^2 + (C_{\mathrm{AP}}\Delta P)^2]^2$. Figure 14.5(a) shows the measured fluctuations as a function of the strength of the modulation. Each data point was produced using a

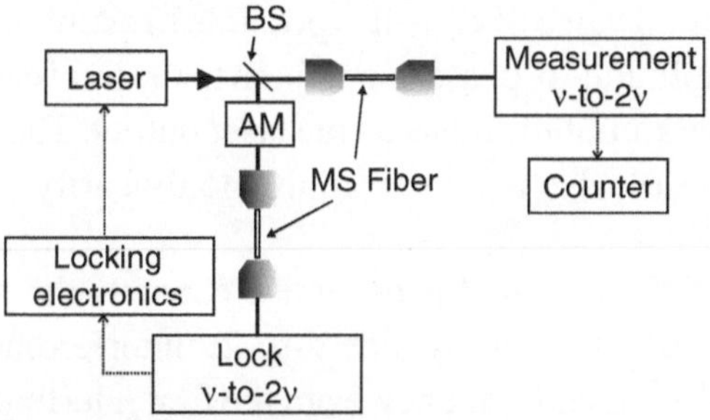

Fig. 14.4. Schematic of the dual ν-to-2ν interferometers used to measure the conversion of amplitude to phase fluctuations in microstructure fiber. The amplitude modulator (AM) can be used to systematically vary the input power to the lock interferometer.

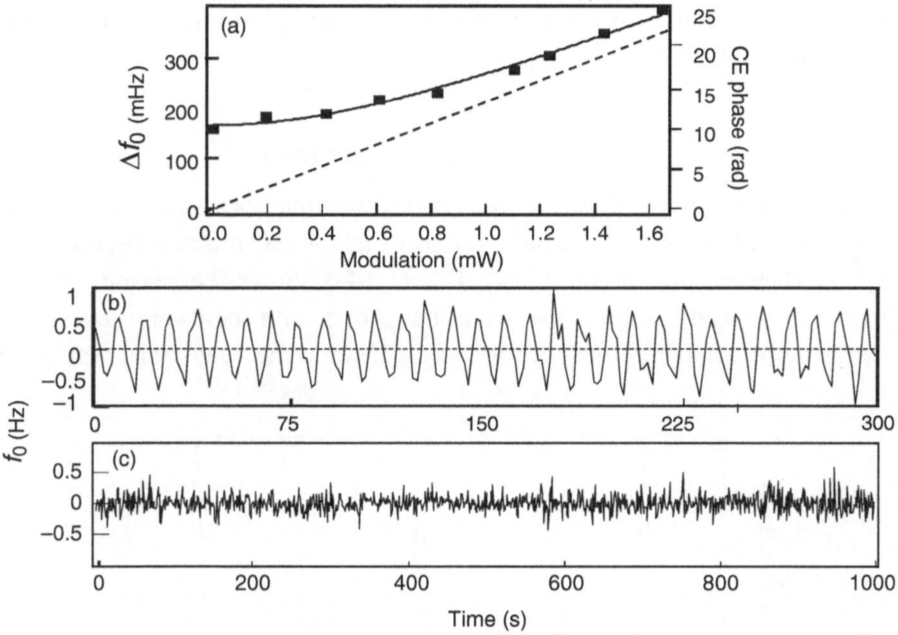

Fig. 14.5. (a) rms deviation Δf_0 (left axis) and accumulated carrier-envelope phase (right axis) as function of modulation power (squares). Solid curve is a fit to $[\eta^2 + (C_{AP}\Delta P)^2]^2$; dashed curve is $C_{AP}\Delta P$. (b) Time record of f_0 measured by the second ν-to-2ν interferometer relative to the average value. The sinusoidal variation of f_0 results from an applied power modulation depth of 5% at 0.1 Hz. (c) Time record of f_0 recorded with 1.0 s gate time. Reproduced from Fortier et al. (2002).

1s gate time for the counter and a 200 s record, such as shown in Fig. 14.5(b). The fit yields a conversion coefficient of 37 rad/mW for a repetition rate of 100 MHz and 43 mW of average power coupled into the fiber.

Based on the measured value of C_{AP}, it is possible to estimate the contributions of phase noise due to amplitude-to-phase conversion to the total carrier-envelope phase noise by measuring the amplitude noise on the laser output. The result is that approximately 0.5 rad of phase noise results from the nonlinearity of the microstructure fiber.

To determine if the accumulated phase noise corrupts the long-term phase stability of the stabilization, the side-by-side ν-to-2ν interferometers were used in a further experiment. The offset frequency from the first interferometer, f_0^L, was used to lock the laser while the offset frequency from the second interferometer, f_0^M, was recorded. All sources of noise, including amplitude-to-phase noise conversion in both fibers and residual laser noise not corrected by the feedback loop contribute to the fluctuations in f_0^M. Using a 1 second gate time, f_0^M was recorded for 1000 s, as shown in Fig. 14.5. The root-mean-square fluctuations were 134 mHz. Analysis of the Allan deviation shows that the carrier-envelope phase jitter is due to white noise (Fortier et al., 2002).

14.3.2 Excess noise generation

The first observation by Ranka et al. (2000) of supercontinuum generation in microstructure fiber used 100 fs pulses. However, for self-referencing one always uses pulse durations of 30 fs or less. Figure 14.6 shows the reason, namely a prominent noise floor in the rf spectrum that grows with increasing pulse energy for 100 fs pulses. Self-referencing produces a heterodyne beat signal that is weak

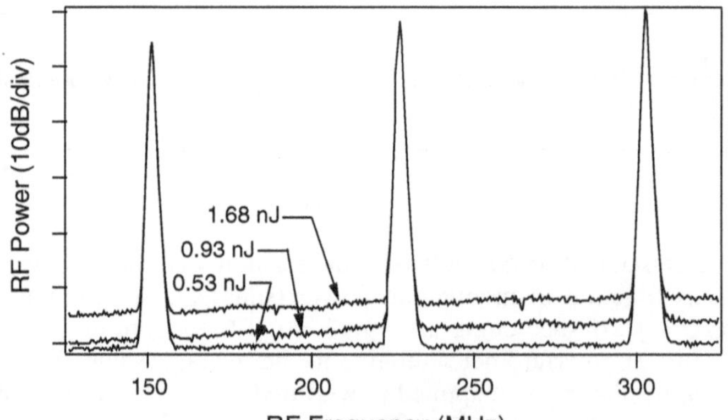

Fig. 14.6. rf spectra of the output of a microstructured fiber for 100 fs input pulses for several pulse energies. The peaks correspond to multiples of f_{rep} and are normalized in amplitude. Note the relative increase in the noise between the peaks. For the lowest pulse energy spectrum, the noise floor is due to the electronics. Reproduced from Ames et al. (2003).

compared to the signals at multiples of the repetition rate and thus can be masked by the noise floor. For shorter duration pulses the generated noise is weaker, reaching the point of being undetectable for 10 fs pulses.

The experimental apparatus for characterizing the noise generated by a 100 fs pulse in microstructure fiber is fairly simple. The output of a mode-locked Ti:sapphire laser was coupled into a 9.2 cm length of microstructure fiber. The laser had a repetition rate of 75 MHz. The polarization was adjusted to maximize the spectral width of the generated continuum. The launched power was adjusted by inserting neutral density (ND) filters prior to the objective that coupled the laser light into the fiber. No compensation for dispersion in optical elements was included. The microstructure fiber was the same as that used by Ranka et al. (2000) for the first demonstration of continuum generation in microstructure fiber. It had zero group velocity dispersion at a wavelength of approximately 770 nm. The output of the fiber was collimated and then focused onto a photodiode. The current output of the photodiode was converted to a voltage by a transimpedance amplifier optimized for 300 MHz. The amplifier also provided a voltage proportional to the dc output of the diode, which was used to monitor the average optical fiber coupled into the fiber. Saturation of the photodiode was prevented by using reflective ND filters to attenuate the output of the microstructure fiber. The attenuation was adjusted over the course of the measurement. A second stage of amplification was used to further boost the signal.

To isolate the noise, the signal from the second amplifier was passed through a bandpass filter with a bandwidth of 8 MHz and tuned to a center frequency of 256 MHz, which was roughly 3.5 times the repetition rate. A 300 MHz low-pass filter followed the bandpass filter. The output of the bandpass filter was detected by a rf spectrum analyzer. The rf filtering was necessary to prevent the generation of spurious mixing products in the spectrum analyzer. The rf spectrum was integrated over a 7.0 MHz band that corresponded to the pass band of the rf filter. Typical results for the integrated noise power as a function of launched optical power are shown in Fig. 14.7(a).

By removing the fiber from the setup, the background noise could be measured and was consistent with the level expected from shot noise. The shot noise was estimated from the average optical power when the power was high enough that electronic noise was negligible. This result showed that the laser intensity was dominated by shot noise, not technical noise. The combined electronic noise due to the detector, amplifiers and spectrum analyzer was also characterized and was subtracted from the measurements. The lowest usable optical power was limited by needing to have the light noise exceed the electronic noise.

Comparing the results for different levels of attenuation prior to detection requires care because the shot noise contribution scales as the square root of the detected

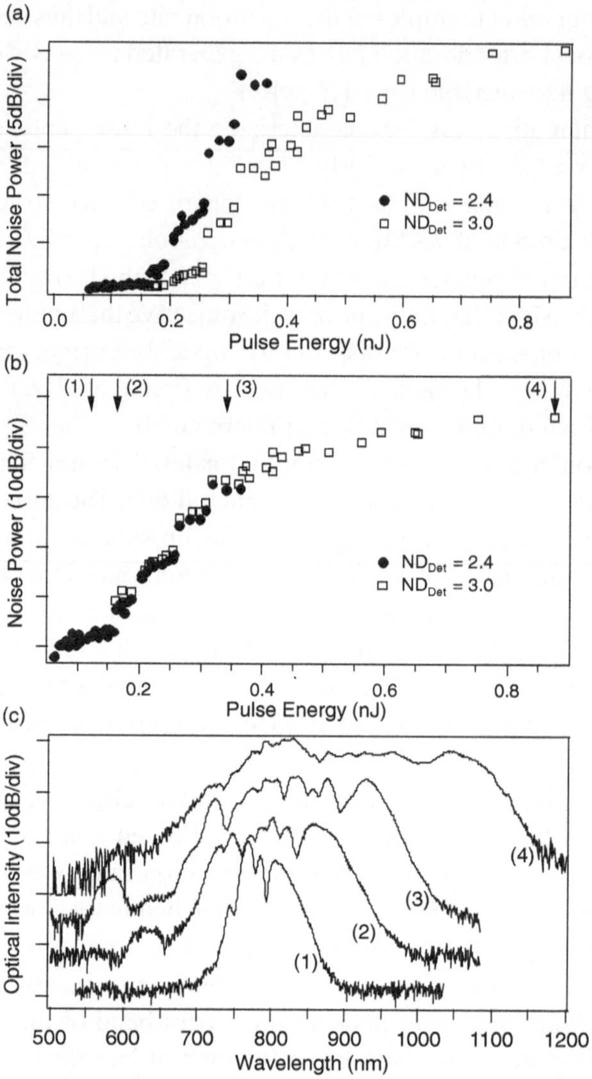

Fig. 14.7. (a) Total noise data without compensating for the change in detector attenuation. (b) Data after combining results from two different detector attenuations and removal of the electronic noise. (c) Output spectra at four coupled powers. The corresponding pulse energies are marked by arrows in (b). All data in this figure are for a laser wavelength of 785 nm. Reproduced from Ames et al. (2003).

power, while the excess noise is attenuated by the ND filters in proportion to the optical power. The voltage produced by the detector is proportional to the optical power, thus the shot noise contribution to the rf power spectrum is also proportional to the optical power, whereas the excess noise scales as the square of the optical power. To correct for the ND filters, the noise components must first be separated. Figure 14.7(b) shows the combination of the two curves in Fig. 14.7(a) after correction and subtracting the electronic noise floor.

The optical spectrum of the generated continuum is shown in Fig. 14.7(c). The significant structure must be taken into account when defining a quantitative measure of its width, which was chosen to be the width at which the intensity is 12 dB down from the peak value.

The generation of excess noise in fiber is based on the nonlinear Schrödinger equation (NLSE). Quantum effects on pulse propagation in a nonlinear medium are based on the development of quantized versions of the NLSE (Potasek and Yurke, 1988; Lai and Yu, 1995; Drummond and Corney, 2001). An alternative approach is to insert the quantized field in the NLSE (Ames et al., 2003). The noise properties of the propagating pulse are modeled by linearizing the NLSE and solving in Fourier space. The resulting description can be interpreted as describing the evolution of a single spectral mode coupled to a reservoir consisting of all other spectral modes. The results of the simulations for 100 fs pulses, and using dispersion and nonlinearity values for microstructure fiber, show that the excess noise grows with increasing power. The results agree qualitatively with the experimental results, including the wavelength dependence, however, greater excess noise is present in the experiments than in the simulations. This discrepancy is attributed to the neglect of Raman effects in the simulations (Coen et al., 2002). Measurements with 10 fs pulses showed much lower excess noise, indeed it was unmeasurable for chirp-optimized pulses. The much lower excess noise for broad band pulses is one reason for their prevalence in optical frequency metrology applications.

A careful comparison between experiment and theory by Corwin et al. (2003a) showed that good agreement was obtained when the Raman terms are included. The calculations are based on a generalized stochastic nonlinear Schrödinger equation and include both shot noise on the input and spontaneous Raman scattering along the fiber. The noise was measured and simulated as a function of input pulse energy and pulse chirp (Corwin et al., 2003a,b).

In addition to excess noise produced by quantum processes in the microstructure fiber, amplitude noise on the input pulse can be amplified during continuum generation. Newbury et al. (2003) quantified this effect through experimental measurements and numerical simulations, including its dependence on wavelength in the continuum and pulse energy and duration. It occurs primarily at low frequencies and differs significantly from the effects of shot noise on the input pulse.

14.4 Applications of mode-locked lasers with locked carrier-envelope phase evolution

The primary applications of mode-locked lasers with locked carrier-envelope phase evolution can be divided into two categories: those applications that use the resulting stable frequency comb and those that need pulses with a well-defined electric field. The most developed applications in the first class are optical frequency metrology and optical atomic clocks, where femtosecond frequency combs have been revolutionary. Another nascent application is the production of arbitrary optical waveforms, where pulse shaping techniques are pushed to the limit of having resolution that matches the comb spacing (Jiang et al., 2007) and ultimately modulation rates that match the repetition rate. Applications that rely on the well-defined phase fall in the area of extreme nonlinear optics, where the electric field of the pulse is strong enough to remove electrons from atoms or molecules.

14.4.1 Optical frequency metrology and optical atomic clocks

The absolute measurement of an optical frequency must be referenced to the definition of the second, which is the cesium clock. Bridging the gap between the frequency of the cesium clock, approximately 9 GHz, and optical frequencies has been a technological challenge. Until the late 1990s this feat was achieved using frequency multiplication chains (Schnatz et al., 1996). Frequency chains suffer from complexity and the fact that phase noise is multiplied up in parallel with the frequency.

It was recognized early on that mode-locked lasers presented an alternate approach based on frequency division. Early work by Eckstein et al. (1978) demonstrated the use of a mode-locked laser to measure the hyperfine splitting of sodium. The limited bandwidth of early mode-locked lasers restricted their usefulness for this application. The development of mode-locked Ti:sapphire lasers in the 1990s renewed interest in using mode-locked lasers for frequency metrology (Udem et al., 1999; Telle et al., 1999; Diddams et al., 2000a,b). The experimental efforts focused on showing that there was a well-defined frequency comb (Udem et al., 1999) and measuring a 104 THz frequency difference (Diddams et al., 2000a).

The demonstration of optical frequency metrology based on self-referencing techniques was a breakthrough (Jones et al., 2000; Diddams et al., 2000a; Reichert et al., 2000) in that it allowed referencing of an optical frequency measurement to a cesium clock in a single step that was greatly simplified compared to using a frequency chain. As a result, the number of measurements increased and their accuracy rapidly improved (see Fig. 14.8). Comparisons between combs by Ma et al. (2004) showed that their intrinsic uncertainty did not limit the measurements, but rather it was limited by the intrinsic fluctuations in the quantum transitions being measured and the techniques used to probe it.

In Fig. 14.8, the progress in cesium clocks is also plotted, showing that uncertainty of optical frequency measurements is surpassing that of cesium clocks. This transition clearly motivates the development of new atomic clocks based on optical, rather than microwave, transitions. Frequency combs would serve as the "gear work" that

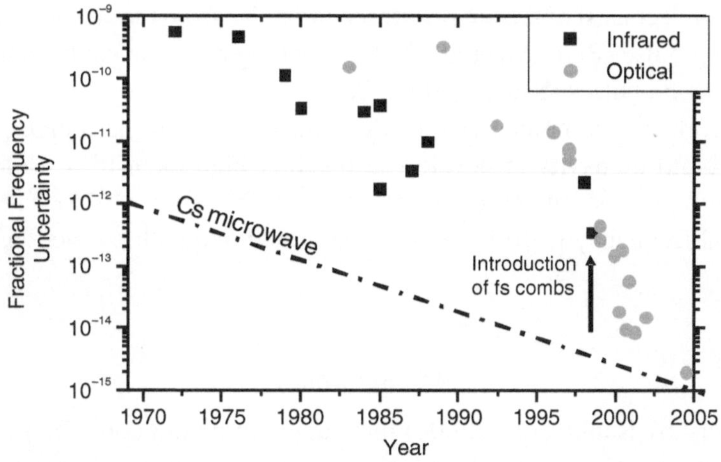

Fig. 14.8. Historical progress in the uncertainty of optical frequency metrology. The uncertainty of the best cesium clocks is also shown. Courtesy of Leo Hollberg.

translates the optical frequencies down to radio frequencies. Early demonstrations of optical atomic clocks were based on single trapped mercury ions (Diddams et al., 2001) and iodine molecules (Ye et al., 2001). Work continues on trapped ions (Rosenband et al., 2008), however, neutral atoms in an optical lattice are producing just as low uncertainties (Ludlow et al., 2008).

14.4.2 Extreme nonlinear optics and attosecond pulse generation

Extreme nonlinear optics refers to the situation where the optical pulse is so intense that the electric field of the pulse is relevant, not the intensity profile, but in the perturbative regime (Brabec and Krausz, 2000). In this regime, the electric field of the pulse can be strong enough that it overcomes the Coulomb attraction binding an electron to an atom and ionizes it. The ionization typically displays a threshold with respect to the electric field of the pulse. This threshold causes the ionization to depend on the phase of the pulse for sufficiently short pulses if the threshold is close to the maximum field in the pulse. Measurements of the photoelectron yield in opposite directions have provided direct measurements of the carrier-envelope phase of amplified pulses (Paulus et al., 2003).

If the ionized electron slams back into the ion, a pulse of extreme ultraviolet to soft x-ray light can be generated (Brabec and Krausz, 2000). This process is known as high harmonic generation as an x-ray/UV pulse is emitted for every half-cycle of the field, resulting in an emission spectrum that consists of lines at harmonics of the incident laser light. Each x-ray pulse has a duration of approximately 100 attoseconds (as). The highest frequencies of the spectrum, known at the "cutoff-region", consist of the single attosecond pulse from the strongest half-cycle of the driving pulse (Baltuska et al., 2003). Locking $\phi_{\rm ce}$ enables the generation of single attosecond pulses (Sansone et al., 2006).

Recent work has combined frequency combs with high harmonic generation by using a build-up cavity to develop sufficiently high intensities while preserving the high repetition rate (Jones et al., 2005; Göhle et al., 2005). These results allow optical frequency metrology to be performed in the ultraviolet region of the spectrum.

14.5 Summary

The discovery by Ranka et al. (2000) that supercontinuum could be generated in microstructure fiber by nanojoule pulses enabled techniques to measure and stabilize the carrier-envelope phase evolution in mode-locked lasers. In turn, these techniques resulted in revolutionary advances in optical frequency metrology, enabled the development of optical atomic clocks and produced breakthroughs in the generation of attosecond x-ray pulses. Further developments are expected in these fields, as well as the opening of new fields such as arbitrary optical waveform generation.

Acknowledgments

I would like to acknowledge all who helped with the work on understanding noise produced during continuum generation and its effect on carrier-envelope phase stabilization in my lab at JILA, especially T.M. Fortier and J.N. Ames, who collected the data presented in Sections 14.3.1 and 14.3.2, respectively. I would also like to acknowledge J. Ye, J.L. Hall and D.J. Jones for their contributions and R.S. Windeler for providing the microstructure fiber. The work was funded by NIST, ONR and NSF.

References

Ames J., Ghosh S., Windeler R., Gaeta A., Cundiff S. (2003). Excess noise generation during spectral broadening in a microstructured fiber. *Appl. Phys. B* **77**, 279–284.

Baltuska A., Udem T., Uiberacker M. et al. (2003). Attosecond control of electronic processes by intense light fields. *Nature* **421**, 611–615.

Brabec T., Krausz F. (2000). Intense few-cycle laser fields: Frontiers of nonlinear optics. *Rev. Mod. Phys.* **72**, 545–591.

Coen S., Chau A., Leonhardt R. et al. (2002). Supercontinuum generation by stimulated Raman scattering and parametric four-wave mixing in photonic crystal fibers. *J. Opt. Soc. Am. B* **19**, 753–764.

Corkum P. B., Krausz F. (2007). Attosecond science. *Nat. Phys.* **3**, 381–387.

Corwin K., Newbury N., Dudley J. M. et al. (2003a). Fundamental noise limitations to supercontinuum generation in microstructure fiber. *Phys. Rev. Lett.* **90**, 113904.

Corwin K., Newbury N., Dudley J. M. et al. (2003b). Fundamental amplitude noise limitations to supercontinuum spectra generated in a microstructured fiber. *App. Phys. B* **77**, 269–277.

Cundiff S. (2002). Phase stabilization of ultrashort optical pulses. *J. Phys. D* **35**, R43–R59.

Cundiff S., Ye J. (2003). Colloquium: Femtosecond optical frequency combs. *Rev. Mod. Phys.* **75**, 325–342.

Diddams S., Jones D., Ma L., Cundiff S., Hall J. (2000a). Optical frequency measurement across a 104-THz gap with a femtosecond laser frequency comb. *Opt. Lett.* **25**, 186–188.

Diddams S., Jones D., Ye J. et al. (2000b). Direct link between microwave and optical frequencies with a 300 THz femtosecond laser comb. *Phys. Rev. Lett* **84**, 5102–5105.

Diddams S., Udem T., Bergquist J. et al. (2001). An optical clock based on a single trapped $^{199}Hg^{+}$ ion. *Science* **293**, 825–828.

Drummond P., Corney J. (2001). Quantum noise in optical fibers. I. Stochastic equations. *J. Opt. Soc. Am. B* **18**, 139–152.

Eckstein J., Ferguson A., Hänsch T. (1978). High-resolution 2-photon spectroscopy with picosecond light pulses. *Phys. Rev. Lett.* **40**, 847–850.

Fortier T., Bartels A., Diddams S. (2006). Octave-spanning Ti:sapphire laser with a repetition rate <1 GHz for optical frequency measurements and comparisons. *Opt. Lett.* **31**, 1011–1013.

Fortier T., Jones D., Cundiff S. (2003). Phase stabilization of an octave-spanning Ti:sapphire laser. *Opt. Lett.* **28**, 2198–2200.

Fortier T., Ye J., Cundiff S., Windeler R. (2002). Nonlinear phase noise generated in air–silica microstructure fiber and its effect on carrier-envelope phase. *Opt. Lett.* **27**, 445–447.

Göhle C., Udem T., Herrmann M. et al. (2005). A frequency comb in the extreme ultraviolet. *Nature 436*, 7048, 234–237.

Hall J. L. (2006). Nobel lecture: Defining and measuring optical frequencies. *Rev. Mod. Phys.* **78**, 1279.

Hänsch T. W. (2006). Nobel lecture: Passion for precision. *Rev. Mod. Phys.* **78**, 1297.

Jiang Z., Huang C.-B., Leaird D. E., Weiner A. M. (2007). Optical arbitrary waveform processing of more than 100 spectral comb lines. *Nat. Photonics* **1**, 463–467.

Jones D., Diddams S., Ranka J. et al. (2000). Carrier-envelope phase control of femtosecond mode-locked lasers and direct optical frequency synthesis. *Science* **288**, 635–639.

Jones R., Moll K., Thorpe M., Ye J. (2005). Phase-coherent frequency combs in the vacuum ultraviolet via high-harmonic generation inside a femtosecond enhancement cavity. *Phys. Rev. Lett.* **94**, 193201.

Lai Y., Yu S. (1995). General quantum-theory of nonlinear optical pulse propagation. *Phys. Rev. A* **51**, 817–829.

Ludlow A. D., Zelevinsky T., Campbell G. K. et al. (2008). Sr lattice clock at $1 \times 10(-16)$ fractional uncertainty by remote optical evaluation with a Ca clock. *Science* **319**, 1805–1808.

Ma L., Bi Z., Bartels A. et al. (2004). Optical frequency synthesis and comparison with uncertainty at the $10(-19)$ level. *Science* **303**, 1843–1845.

Matos L., Kleppner D., Kuzucu O. et al. (2004). Direct frequency comb generation from an octave-spanning, prismless Ti:sapphire laser. *Opt. Lett.* **29**, 1683–1685.

Newbury N., Washburn B., Corwin K., Windeler R. (2003). Noise amplification during supercontinuum generation in microstructure fiber. *Opt. Lett.* **28**, 944–946.

Paulus G., Lindner F., Walther H. et al. (2003). Measurement of the phase of few-cycle laser pulses. *Phys. Rev. Lett.* **91**, 253004.

Potasek M., Yurke B. (1988). Dissipative effects on squeezed light generated in systems governed by the nonlinear Schrödinger equation. *Phys. Rev. A* **38**, 1335–1348.

Ramond T., Diddams S., Hollberg L., Bartels A. (2002). Phase-coherent link from optical to microwave frequencies by means of the broadband continuum from a 1-GHz Ti:sapphire femtosecond oscillator. *Opt. Lett.* **27**, 1842–1844.

Ranka J., Windeler R., Stentz A. (2000). Visible continuum generation in air–silica microstructure optical fibers with anomalous dispersion at 800 nm. *Opt. Lett.* **25**, 25–27.

Reichert J., Holzwarth R., Udem T., Hänsch T. (1999). Measuring the frequency of light with mode-locked lasers. *Opt. Commun.* **172**, 59–68.

Reichert J., Niering M., Holzwarth R., Weitz M., Udem T., Hänsch T. (2000). Phase coherent vacuum-ultraviolet to radio frequency comparison with a mode-locked laser. *Phys. Rev. Lett.* **84**, 3232–3235.

Rosenband T., Hume D. B., Schmidt P. O. et al. (2008). Frequency ratio of Al+ and Hg+ single-ion optical clocks; Metrology at the 17th decimal place. *Science* **319**, 1808–1812.

Sansone G., Benedetti E., Calegari F. et al. (2006). Isolated single-cycle attosecond pulses. *Science* **314**, 443–446.

Schnatz H., Lipphardt B., Helmcke J., Riehle F., Zinner G. (1996). First phase-coherent frequency measurement of visible radiation. *Phys. Rev. Lett.* **76**, 18–21.

Telle H., Steinmeyer G., Dunlop A., Stenger J., Sutter D., Keller U. (1999). Carrier-envelope offset phase control: A novel concept for absolute optical frequency measurement and ultrashort pulse generation. *Appl. Phys. B* **69**, 327–332.

Udem T., Reichert J., Holzwarth R., Hänsch T. (1999). Accurate measurement of large optical frequency differences with a mode-locked laser. *Opt. Lett.* **24**, 881–883.

Xu L., Spielmann C., Poppe A., Brabec T., Krausz F., Hansch T. (1996). Route to phase control of ultrashort light pulses. *Opt. Lett.* **21**, 2008–2010.

Ye J., Cundiff S. T. (2005). *Femtosecond Optical Frequency Comb Technology*. New York: Springer-Verlag.

Ye J., Ma L., Hall J. (2001). Molecular iodine clock. *Phys. Rev. Lett.* **87**, 270801.

15

Biophotonics applications of supercontinuum generation

C. Dunsby and P. M. W. French

15.1 Introduction

Continua generated using high pulse energy laser systems to create broad spectra (Alfano and Shapiro, 1970) have been used for spectroscopy for many years (e.g. Busch et al., 1973). The application of fibre-generated continua to spectroscopy was suggested as early as 1976 in work by Lin and Stolen (1976) where a continuum spanning ~450–600 nm was generated in a step-index fibre pumped by a nitrogen pumped dye laser. Since the demonstration of supercontinuum generation in microstructured optical fibres (MOF), however, the range of spectroscopic and imaging applications has increased enormously, owing to the high average powers, unprecedented spectral width and relatively low cost and low complexity of such sources. This chapter specifically focuses on the applications of supercontinua generated in MOFs and, in particular, on applications in biophotonics.

MOF supercontinuum sources can be broadly grouped into three categories according to whether the laser pump source emits femtosecond pulses, picosecond–nanosecond pulses or cw radiation. In general terms, sub-ps pulses can produce broad supercontinua spanning from the UV to the NIR but the peak intensity damage threshold at the input end of the microstructured optical fibre limits the maximum average power that can be obtained in the supercontinuum to ~<0.5 W with typically ~<0.5 mW/nm available in the visible spectrum. The use of pump lasers with longer ps–ns pulses can significantly increase the maximum available average power before the onset of damage in the MOF such that high power supercontinua with several mW/nm in the visible can be achieved (e.g. Rulkov et al., 2005). Supercontinuum sources pumped with cw lasers can achieve spectral power densities of 10's mW/nm, although the spectral coverage does not extend so far into the visible (e.g. Travers et al., 2008). The use of pulsed lasers to pump supercontinua provides

Supercontinuum Generation in Optical Fibers, ed. by J. M. Dudley and J. R. Taylor. Published by Cambridge University Press.

broadly tunable ultrashort pulses that are useful for time-resolved measurements or nonlinear optical applications requiring high peak power, such as multiphoton microscopy. The capability to implement rapid tuning and/or simultaneous emission at multiple wavelengths via spectral selection of supercontinua is particularly interesting when considering the cost and complexity of alternative approaches based on synchronous mode-locked lasers or optical parametric sources.

MOF supercontinuum sources have already had a wide impact in the field of biophotonics, having been applied to spectroscopic measurements and implemented in imaging instrumentation. Typically, biophotonics involves the analysis of biological tissues, cells and liquid solutions, for which contrast can be obtained from reflected light, absorption, elastic scattering, polarization, Raman scattering and fluorescence. The interaction of optical radiation with biological samples can provide molecular contrast while structural information can be obtained from a range of microscopy techniques, many of which exploit the characteristic properties of laser radiation. Supercontinuum sources therefore have the potential to provide spectral versatility and spatially coherent, high brightness "laser" radiation for a wide range of biophotonics applications – often at a significantly lower cost and smaller footprint than established laser technology.

15.2 Spectroscopy

By definition, supercontinuum (SC) sources provide broad spectral coverage; they also provide spatially coherent radiation and can provide ultrashort pulses with durations of a few picoseconds. For spectroscopic metrology and imaging applications, they offer the spectral versatility of lamp based sources with the advantages of laser radiation, including higher brightness for improved signal-to-noise (S/N) measurements and ultrashort pulse operation, while avoiding the complexity and costs associated with tunable visible laser systems that typically involve mode-locked lasers with second harmonic generation (SHG) or optical parametric generation. These advantages have led to their incorporation in a range of instrumentation for spectroscopy.

One of the first applications of a supercontinuum source to spectroscopy exploited a near infrared (NIR) supercontinuum that was temporally dispersed in a length of optical fibre to provide a rapidly spectrally swept source for absorption spectroscopy of gases (Kelkar et al., 1999). Spectral scan widths of 200 nm could be swept in ~20 ns using this technique at a wavelength of 1.5 μm (Sanders, 2002) and spectral resolutions of down to 40 pm have been obtained using this method (Hult et al., 2007). More recently, a supercontinuum source has also been applied to cavity enhanced gas absorption measurements (Langridge et al., 2008). Another potential application is for performing absorption measurements in small microfluidic channels for lab-on-a-chip applications (Domachuk et al., 2007). For a more detailed review of supercontinuum sources applied to gas sensing and spectroscopy, see Kaminski et al. (2008).

Supercontinuum sources have also been applied to scattering spectroscopy of small particles. Lindfors et al. (2004) used a confocal detection system to image and study the interferometric scattering spectrum of gold nanospheres and to determine the plasmon resonance for individual nanoparticles. Supercontinuum sources have also been applied to studying larger dielectric particles causing Mie scattering, for example the work of Li et al. (2005b) investigated the Mie scattering spectrum from single microspheres. These methods provide accurate information on the size and shape of individual microscopic particles and, as information can be obtained from individual particles, it is also then possible to determine particle size distributions. If the average power of the supercontinuum is increased and the beam is tightly focused, then supercontinuum sources can also be used for simultaneous optical trapping and spectroscopy of individual particles in the liquid or gas phase (Li et al., 2005a, Guillon et al., 2008).

Supercontinuum generation can also be applied to fluorescence spectroscopy for which the pulsed nature of SC radiation can be exploited. Traditionally, ultrafast laser sources made time-resolved fluorometers relatively large, complex and expensive. An important alternative approach is based on the use of high-speed (nanosecond) pulsed flashlamps that can provide lifetime measurements with picosecond resolution over a broad range of excitation wavelengths via spectral selection (Birch and Imhof, 1977). The performance of such flashlamps in terms of intensity, stability and repetition rate, however, limits the achievable measurement precision compared to instruments using high repetition rate ultrafast lasers. Today computer-controlled mode-locked Ti:Sapphire lasers, together with harmonic generation and optical parametric oscillator technology (Maus et al., 2001), can provide superior precision and experimental convenience but such systems add very significant cost, size and complexity to a fluorometer.

Recently, a compact multidimensional spectrofluorometer using a fibre-laser-pumped supercontinuum source has been developed that combines the spectral tunability and temporal resolution of the SC for multidimensional fluorescence spectroscopy (Manning et al., 2008). This compact system, which is shown in Fig. 15.1, has a footprint of only 60×90 cm and is capable of resolving fluorescence excitation spectra, emission spectra (using either 1- or 16-channel detection), fluorescence decay profiles and polarization for time-resolved fluorescence anisotropy decays. This instrument has been applied to characterize fluorescence lifetime probes, e.g. for sensing membrane lipid order and for Forster resonant energy transfer (FRET) in cell biology, and to the study of biological tissue autofluorescence.

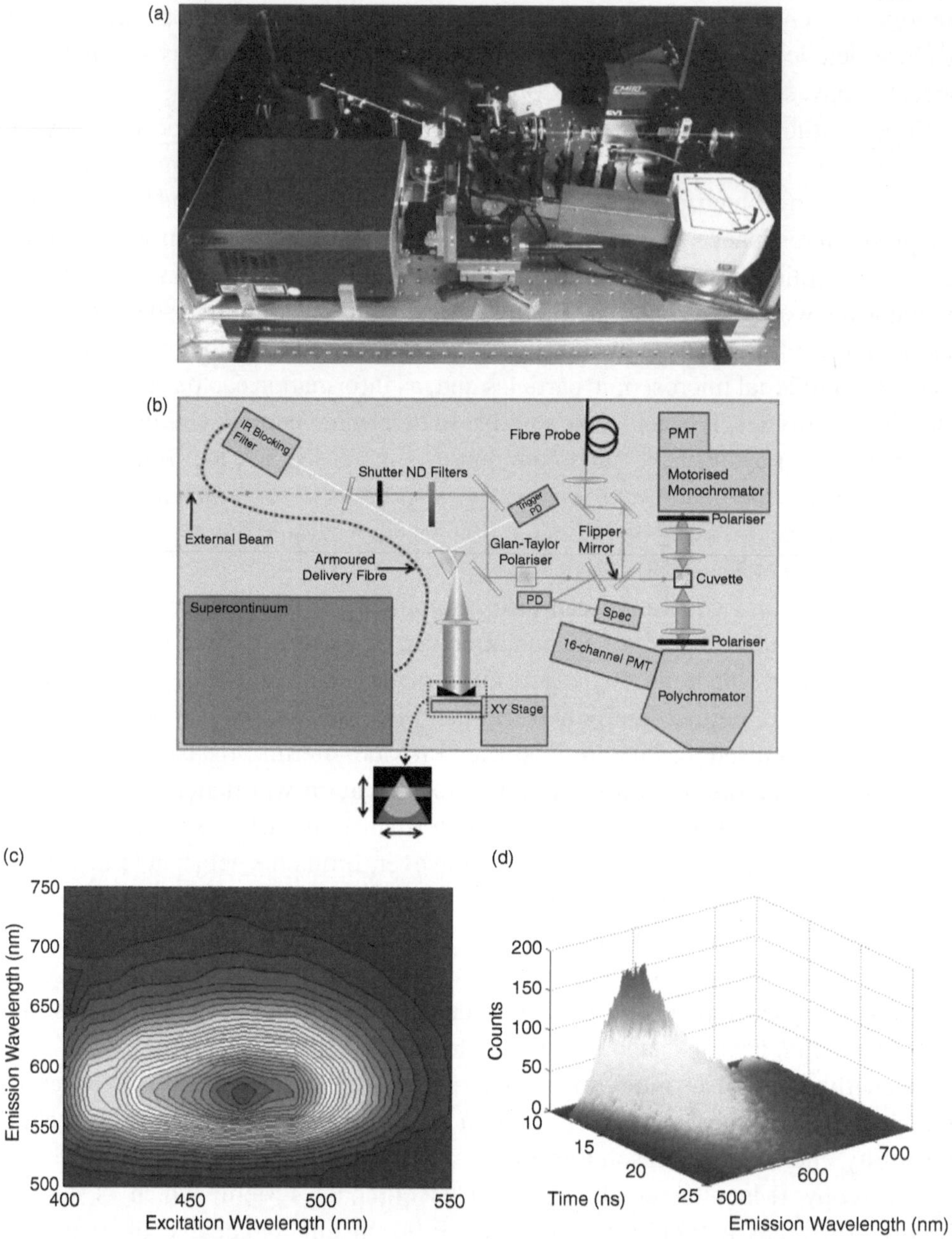

Fig. 15.1. (a) Photograph and (b) experimental layout of a compact multidimensional spectrofluorometer based around a commercially available all-fibre supercontinuum source. (c) Excitation-emission matrix and (d) spectrally resolved fluorescence decay profile for lipid probe di-4-ANEPPDHQ stained unilamellar vesicles in the ordered phase. From Manning et al. (2008). Copyright Wiley-VCH Verlag GmbH & Co. KGaA. Reproduced with permission.

Fluorescence spectroscopy commonly uses a single wavelength to excite fluorescence that is then detected at longer wavelengths by an appropriate detector. Typically, spectral discrimination is employed, which may be implemented using filters, spectrographs or scanning spectrometers, to separate the fluorescence signal from the excitation beam and to prevent excitation light from reaching the detector. This approach, which suits the capabilities of a tunable laser, requires that multiple fluorophores are excited sequentially but it is often desirable to measure multiple fluorophores simultaneously. Recent work by Ye et al. (2007) showed that fluorescence from multiple fluorophores with different excitation spectra can be simultaneously excited using broadband SC excitation and that the resulting fluorescence can then be detected using the temporal discrimination of a streak camera to separate the excitation pulse from the fluorescence.

As well as fluorescence lifetime measurements, MOF SC sources can also be applied to time-resolved pump-probe spectroscopy. The first demonstration of a supercontinuum for femtosecond pump-probe experiments (Nagarajan et al., 2002) used a probe beam generated by pumping a microstructured optical fibre with a cavity dumped Ti:Sapphire laser and was applied to studies of a number of samples including the transient absorption spectrum of green fluorescent protein (GFP).

The inherent ultrafast nature of supercontinuum radiation can also be exploited for nonlinear optical spectroscopy including two-photon excitation of fluorescence and nonlinear Raman spectroscopy such as coherent anti-Stokes Raman spectroscopy (CARS) (Volkmer, 2005). Two-photon excitation of fluorescence was first demonstrated by Jureller et al. (2003) and Shi et al. demonstrated an elegant utilization of a supercontinuum source to measure an entire two-photon absorption cross-section spectrum in a single shot (Shi et al., 2006a). This was achieved through a carefully designed spatial dispersion of the supercontinuum in the sample with the resulting spatially varying fluorescence signal being imaged back to a CCD camera.

MOF supercontinua have been applied to coherent Raman spectroscopy (Kano and Hamaguchi, 2003) and frequency- and time-resolved CARS (Kano and Hamaguchi, 2004, Konorov et al., 2004, Andresen et al., 2005, Kee et al., 2006, Zheltikov, 2007) for which synchronous trains of ultrashort pulses at different wavelengths are required. The use of a supercontinuum source removes the need for relatively complex laser systems, usually based on two synchronized laser oscillators operating at different centre wavelengths or on a synchronously pumped optical parametric oscillator. SC sources have been developed that are specifically optimized for CARS applications. For example, by optimizing the length of the microstructured optical fibre, it is possible to minimize the pulse length of the SC source and increase the temporal accuracy of such systems for time-resolved CARS (Kano and Hamaguchi, 2006a). A compact and experimentally robust approach is to use adaptive pulse shaping to split the output from a single SC source into two separate optimized pulse trains at different wavelengths (von Vacano and Motzkus, 2006, von Vacano et al., 2006). The use of fibre-laser pumped SC or sub-nanosecond microchip laser pumped SC (Okuno et al., 2007) sources can also further reduce the overall complexity of CARS spectroscopy systems.

15.3 Microscopy

Optical microscopy can be broadly divided into "conventional" wide-field microscopes and laser scanning confocal or multiphoton microscopes. Wide-field microscopy is well served by thermal and arc lamp based sources that provide broad spectral coverage, although they are limited with respect to time-resolved imaging techniques. Confocal microscopy is a method for obtaining optically sectioned images. It provides high resolution 3D imaging of cellular and tissue structures (Pawley, 1995) and confocal fluorescence microscopy has become the gold standard for optical biological imaging. Confocal imaging can be applied in reflection or transmission imaging modes and can achieve contrast using scattered light, polarization, fluorescence or Raman based contrast.

Optical sectioning in a confocal microscope is achieved by scanning a point of illumination over the specimen and collecting light back through a confocal pinhole. Light from outside the excitation volume is not focused onto the confocal detector pinhole, is not efficiently collected by the confocal detector and so is rejected from the final image. As the resolution of confocal microscopy relies on forming a small diffraction limited spot in the specimen, then diffraction limited spatially coherent illumination sources are required and this need is usually met by one or more lasers. Unfortunately the spectral coverage in the visible is limited – traditionally, confocal microscopy has employed a range of gas and ion lasers to provide a range of discrete visible laser lines and a commercial system might contain three or more laser sources of this type, which leads to significant running costs and system complexity. When using confocal fluorescence microscopes, one of the main selection criteria for choosing fluorophores has been the overlap of the fluorophore excitation spectrum with the laser lines available on the instrument, but in many cases this match is a compromise. The use of supercontinuum sources extends the spectral coverage for all forms of confocal microscopy and for fluorescence imaging allows fluorophores to be excited at their optimum wavelength. Pulsed SC sources can provide the excitation for fluorescence lifetime imaging. Ultimately one might expect a single high power supercontinuum source to replace all the lasers in a confocal microscope, providing enhanced versatility while reducing complexity and costs.

15.3.1 Confocal reflectance microscopy

A number of methods have been proposed to increase the speed with which depth-resolved confocal reflectance images can be obtained, and such methods often aim

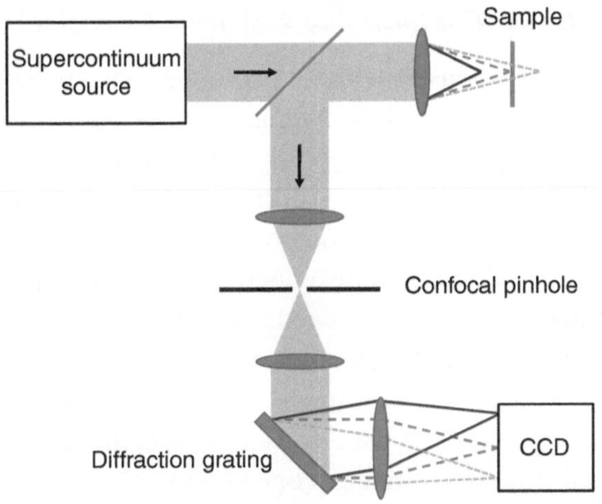

Fig. 15.2. Schematic for chromatic confocal microscopy. Light from a continuum source is focused onto the sample by a lens exhibiting longitudinal chromatic aberration. Light reflected from the sample is reflected via a beamsplitter, focused through a confocal pinhole and input to a spectrometer. Blue, green and red rays are indicated by solid, long-dashed and short-dashed lines respectively.

to reduce the requirement to scan the confocal point in all three spatial dimensions. One way of achieving this is to make use of longitudinal chromatic aberration in an objective lens, in combination with a spectrally broad source, to simultaneously probe multiple depths in the specimen. The light reflected by the specimen is recorded by a spectrometer, and the wavelength of the reflected light encodes the height of the specimen. Figure 15.2 shows a schematic of this technique, which requires a high brightness, spectrally broad source. The first application of a supercontinuum to this technique was performed by Garzon et al. (2004) who used a spatially coherent NIR supercontinuum generated in a single mode step index fibre pumped by a Q-switched 532 nm laser. The spectral range of 700–950 nm was focused to a range of axial positions over 200 μm using the axial chromatic aberration produced by a diffractive Fresnel lens, and an axial resolution of ~1 μm was achieved. Following this work, Shi et al. (2004) used a broader supercontinuum generated in a microstructured optical fibre to increase the resolution of this technique. An axial range of 7 μm was achieved with an axial resolution of ~0.7 μm. A similar method has also been used to image many laterally separated points simultaneously through lateral, rather than axial, dispersion (Shi et al., 2006b).

More recently, a supercontinuum source has been applied to spectrally resolved confocal reflectance imaging (Booth et al., 2008). This system acquires a high resolution reflectance spectrum for each pixel in the image and was employed to image thin-film interference effects in integrated circuits and to imaging interference effects in natural specimens, including scales on butterfly wings and iridescence of shells.

15.3.2 Confocal fluorescence microscopy

Fluorescence microscopy is used widely in biology and medicine, as biological molecules can be tagged using a broad range of fluorescent labelling techniques and with high molecular specificity. Typically, biological specimens can be labelled with two or more different fluorophores corresponding to two or more molecules of interest. The main advantage of a broad supercontinuum source for confocal microscopy is that it provides a spatially coherent source that, when combined with a method for spectral selection, can be used to excite a wide range of fluorophores. Figure 15.3 shows a schematic of a general configuration for the use of a supercontinuum source with a confocal microscope. Supercontinuum sources pumped by amplified mode-locked fibre laser oscillators have recently been incorporated into commercially available confocal microscope systems.

To the best of our knowledge, the application of a supercontinuum source to confocal microscopy was first proposed in a patent filed by Birk et al. (Birk and Storz, 2001). However, the first demonstration of a supercontinuum source applied to laser scanning microscopy was for multiphoton microscopy (Jureller et al., 2003, Deguil et al., 2004), which is described in more detail in Section 15.3.4. This was followed by the first report of a confocal microscope using a supercontinuum source (mode-locked Ti:Sapphire oscillator pumped MOF), which operated at a single excitation wavelength (McConnell, 2004), and the first demonstration of confocal microscopy and fluorescence lifetime imaging (FLIM) with a tunable supercontinuum source (Dunsby et al., 2004).

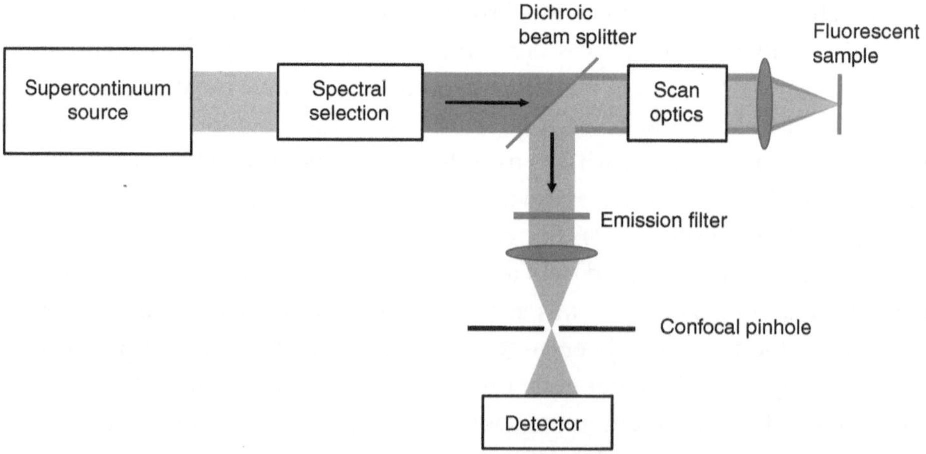

Fig. 15.3. Schematic for fluorescence confocal microscopy with a supercontinuum source. Following spectral selection, the SC is focused onto the sample and the resulting fluorescence is collected via a dichroic beam splitter, emission filter and confocal pinhole onto the detector.

Spectral selection from the supercontinuum can be performed using a range of techniques including an appropriate band-pass filter (McConnell, 2004), tunable prism-based spectrometer (Dunsby et al., 2004), acousto-optic beam splitter (AOBS) (Betz et al., 2005), digital micro-mirror device (DMD) (McConnell et al., 2006) or acousto-optic tunable filter (AOTF) (Frank et al., 2007). If the excitation source is swept through multiple wavelengths, then it is possible to acquire excitation resolved spectra of the specimen (Dunsby et al., 2004) or even excitation-emission resolved spectra for each pixel of the image, as implemented in line-scanning (Owen et al., 2007) and confocal (Kaminski et al., 2008) microscopes. Figure 15.4 shows a false-colour confocal fluorescence image and an excitation-emission plot corresponding to a small region of the image, which was obtained using a confocal microscope with a tunable continuum source and a spectrally resolved detector.

Supercontinua generated in microstructured fibres exhibit broadband noise as a result of nonlinear amplification of photon shot noise and spontaneous Raman scattering (Corwin et al., 2003, Newbury et al., 2003). While this may be a problem for some applications, it is not a significant issue for cw fluorescence intensity measurements or fluorescence lifetime measurements due to the small number of fluorescent photons that are detected per pixel per excitation pulse in a typical fluorescence measurement. For example, a 20% standard deviation in the source pulse energy in a fluorescence imaging system would only be significant compared to the shot noise on the fluorescence signal if more than 25 photons are detected

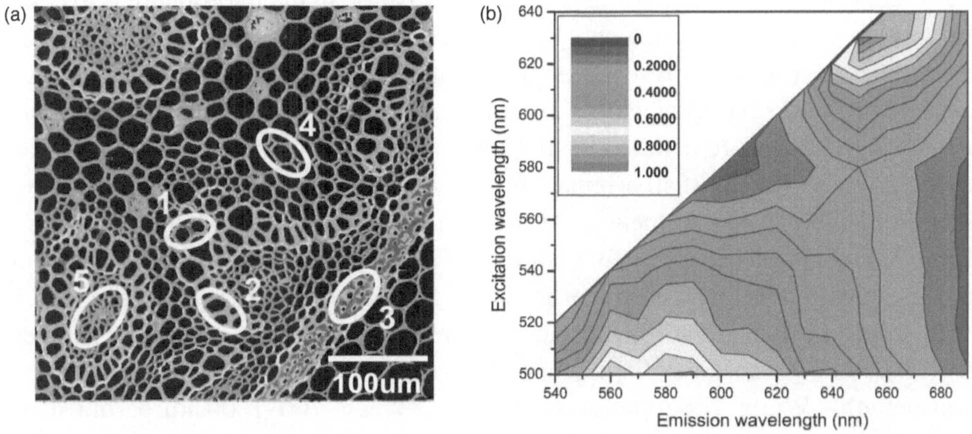

Fig. 15.4. (a) False-colour confocal fluorescence image where the false colour scale encodes the mean emission wavelength. (b) Excitation-emission plot for region 1 in (a). Reproduced with kind permission from Kaminski et al. (2008), © Springer-Verlag.

per excitation pulse per pixel, which is much higher than generally achieved in fluorescence imaging (Dunsby et al., 2004).

15.3.3 Fluorescence lifetime imaging microscopy

Fluorescence lifetime imaging (FLIM) makes use of fast detection schemes to record the fluorescence decay profile at each pixel in an image on a sub-nanosecond time-scale. The recorded decay profiles are then analyzed to calculate the fluorescence lifetime using an appropriate decay model. FLIM can be used to map the localization of fluorophores with different lifetimes or to report spatial variations in fluorophore environment, e.g. temperature, viscosity, [Ca^{2+}] or proximity to other fluorophores via fluorescence resonance energy transfer (FRET) (Lakowicz, 1999, Cubeddu et al., 2002).

Because of the pulsed nature of SC sources, it is straightforward to perform fluorescence lifetime imaging (FLIM) in confocal (Dunsby et al., 2004), wide-field time-gated (Dunsby et al., 2004), spinning-disc (Grant et al., 2005) and slit-scanning (Owen et al., 2007) microscopes. By utilizing a tapered MOF (Kudlinski et al., 2006), excitation-emission-lifetime-resolved imaging is possible with pulsed excitation tuning from the near UV and throughout the visible, see Fig. 15.5 (Owen et al., 2007).

15.3.4 Multiphoton fluorescence microscopy

Multiphoton microscopy (Denk et al., 1990, Zipfel et al., 2003) is similar to confocal microscopy, except that the fluorescence transition in the fluorophore is excited by a two-photon rather than a one-photon absorption. If ultrafast excitation pulses are employed, useful fluorescence signals can be excited using only milliwatts of average power. As the efficiency of a multiphoton fluorescence excitation process is nonlinear, it is confined to the focal spot and therefore all fluorescent photons can be collected since they must all originate from the focal volume. This provides optical sectioning with no requirement to “de-scan” the fluorescence back through a confocal aperture, as is necessary for confocal microscopy. As the nonlinear excitation process only reaches reasonable efficiencies for the high peak powers associated with picosecond or femtosecond pulses therefore tunable ultrafast laser sources are desirable in order to excite a range of fluorescent labels at different excitation wavelengths. While Ti:Sapphire lasers tuning over ~700–1000 nm permit dyes with single photon absorption in the range ~350–500 nm to be excited, they are somewhat expensive and the availability of more compact and lower cost alternatives could make multiphoton microscopy more widely available. SC sources can also extend the spectral coverage of multiphoton microscopy.

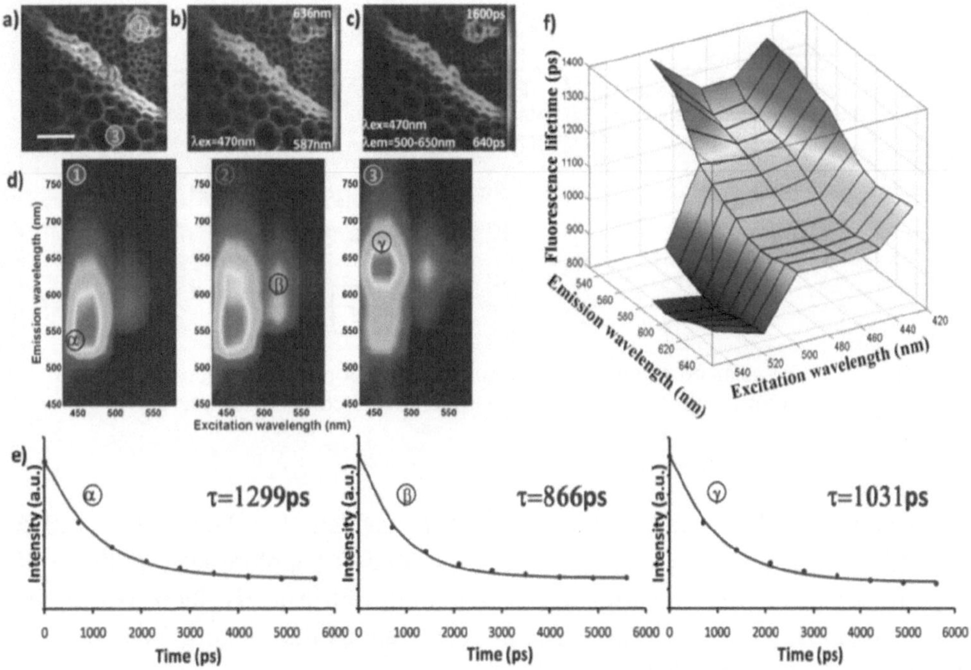

Fig. 15.5. Excitation-emission-lifetime (EEL)-resolved fluorescence imaging of stained *Convallaria*: (a) fluorescence intensity image integrated over EEL; (b) mean emission wavelength image and (c) FLIM image for excitation at 470 nm; (d) excitation-emission matrices corresponding to regions indicated in (a); (e) selected fluorescence decay profiles corresponding to regions α, β and γ respectively indicated in (d); (f) plot of mean fluorescence lifetime as a function of emission and excitation wavelength. Reproduced from Owen et al. (2007), with permission.

Multiphoton microscopy with a spectrally selected SC was first performed by Jureller et al. who used a cavity dumped Ti:Sapphire laser to pump a 17 cm length of tapered fibre with an external diameter of 2.5 μm (Jureller et al., 2003). The continuum produced in the taper was temporally compressed to <50 fs using a prism-pair pulse compressor and a portion of the SC was selected using a slit placed in the spectrally dispersed end of the compressor. Figure 15.6 shows a generalized schematic of a multiphoton microscope employing a SC excitation source. Multiphoton microscopy without a Ti:Sapphire laser was demonstrated by Deguil et al. who used a mode-locked femtosecond Yb:KGW oscillator to pump a MOF and selected from the SC using an interference filter to give tunable pulses of 950–1050 nm with pulse widths of <300 fs (Deguil et al., 2004). Similarly, Yokoyama et al. demonstrated that frequency doubling of a high power fibre laser source at 1.5 μm to 774 nm could be used to produce a SC spanning from 600 to 1200 nm, which was then used for two-photon microscopy (Yokoyama et al., 2007).

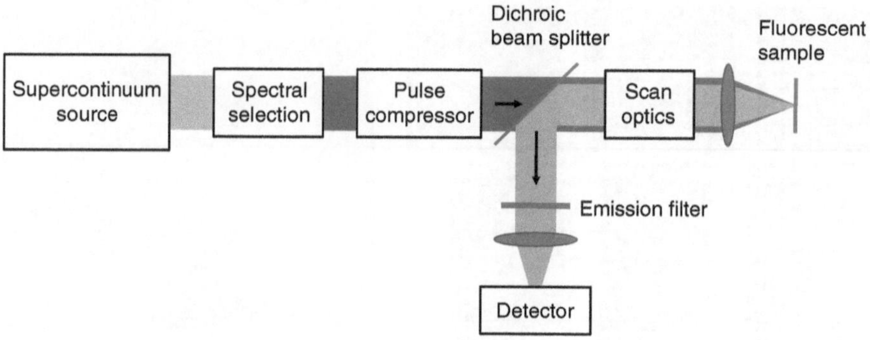

Fig. 15.6. Generalized schematic for multiphoton fluorescence microscopy with a supercontinuum source, including an optional pulse compressor stage.

To extend multiphoton microscopy to shorter wavelengths, Palero et al. (2005) utilized spectral selection at 550 nm from a visible Ti:Sapphire laser-pumped SC to perform two-photon excitation microscopy of the intrinsically fluorescent amino acid tryptophan in the UV – an important endogenous fluorophore that cannot be directly excited using a Ti:Sapphire laser.

15.3.5 Stimulated emission depletion microscopy with supercontinuum based sources

Stimulated emission depletion (STED) microscopy (Hell, 2007) allows the conventional diffraction limited resolution of a confocal microscope to be surpassed. Fluorophores are initially excited by a diffraction limited excitation beam and then selectively depleted by a second beam via the process of stimulated emission. By using a ring-shaped depletion beam to preferentially "switch-off" fluorophores located at the periphery of the excitation point spread function, it is possible to decrease the spatial extent of excited fluorophores (i.e. to decrease the width of the point spread function) and thus achieve imaging at below the diffraction limit. To date, this technique has achieved resolutions of below 50 nm for live-cell imaging (Hein et al., 2008).

STED microscopy is normally implemented using pulsed excitation and depletion beams, which requires the use of temporally synchronized pulsed laser sources at different wavelengths. Supercontinuum based sources therefore provide an opportunity to reduce the complexity of the laser source and to provide flexibility in terms of both the excitation and depletion wavelengths, as shown in the schematic in Fig. 15.7. STED microscopy with a supercontinuum source was first demonstrated with an excitation beam provided by spectral selection from

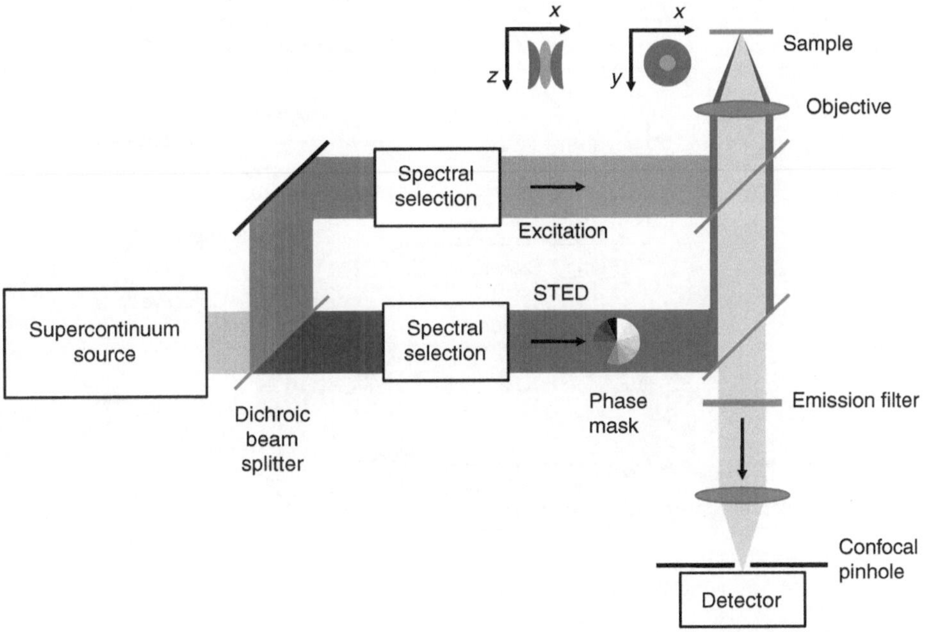

Fig. 15.7. Generalized schematic for STED microscopy with a supercontinuum source. The SC source is split into the excitation and STED beams, which both then undergo further spectral selection. The "donut shaped" depletion beam is generated through phase shaping of the STED beam, either using a phase mask (as shown), a diffractive optic or a spatial light modulator. The two beams are then recombined and focused onto the specimen. Fluorescence is collected back through an emission filter and confocal pinhole onto the detector.

a supercontinuum generated in a Ti:Sapphire laser-pumped MOF and the same mode-locked Ti:Sapphire laser was used to provide the depletion beam (Auksorius et al., 2008). Spectral selection of two different portions of the same supercontinuum source to provide widely tunable excitation and depletion beams has also been demonstrated (Wildanger et al., 2008) and shown to achieve resolutions down to ~30 nm. It seems likely that high power supercontinuum sources will play an important role in future STED microscopes, increasing the spectral versatility and perhaps lowering the cost and complexity.

15.3.6 Coherent anti-stokes Raman scattering microscopy

The use of supercontinuum sources for coherent Raman and CARS spectroscopy was introduced in Section 15.2 and SC sources are also applicable to CARS microscopy, as depicted in Fig. 15.8. The first demonstration of a supercontinuum source applied to CARS microscopy was used to image polystyrene spheres

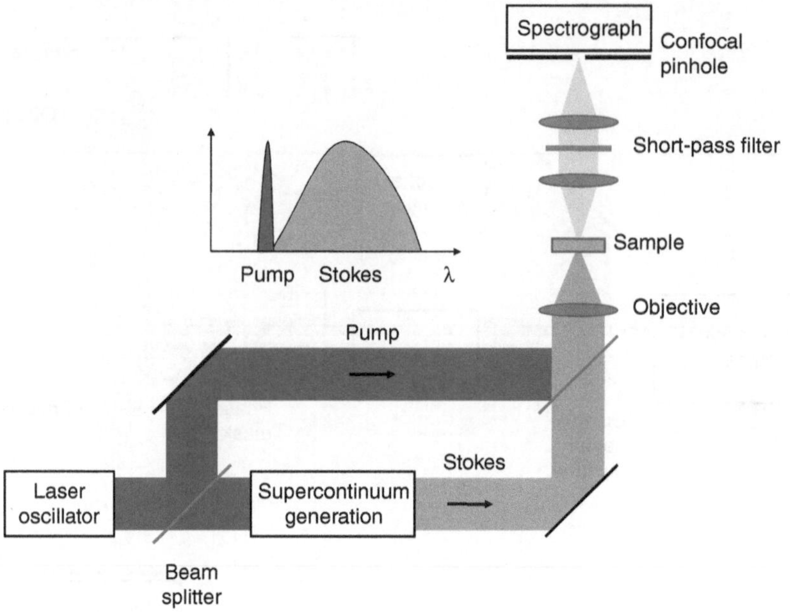

Fig. 15.8. Generalized schematic for multiplexed CARS microscopy with a supercontinuum source.

at a single Raman shift using a pump beam at 643 nm and a Stokes wavelength of 795 nm. The Raman pump at 643 nm was selected from a MOF supercontinuum pumped by a mode-locked laser at 795 nm (Paulsen et al., 2003). Later multiplexed CARS was demonstrated using a mode-locked laser as the Raman pump and a broadband SC source to provide the Stokes beam to produce a full Raman spectrum spanning $\sim 2500\,\text{cm}^{-1}$ for every pixel of the image (Kee and Cicerone, 2004, Kano and Hamaguchi, 2005a).

The use of a SC source for CARS microscopy of cells has also been demonstrated (Kano and Hamaguchi, 2005b). This can be combined with other imaging modalities such as multiphoton microscopy (Kano and Hamaguchi, 2006b, Kano and Hamaguchi, 2007) such that intrinsic two-photon excited fluorescence and CARS contrast permit different cellular structures to be monitored simultaneously. As for single-point CARS spectroscopy, it is also possible to use microchip laser pumped SC sources for CARS microscopy (Okuno et al., 2008).

15.4 Optical coherence tomography

One of the most important and active areas within biophotonics is the development of optical coherence tomography (OCT) (Fujimoto, 2003, Drexler, 2004). OCT is the optical analogue of ultrasound imaging, and provides depth resolved images of

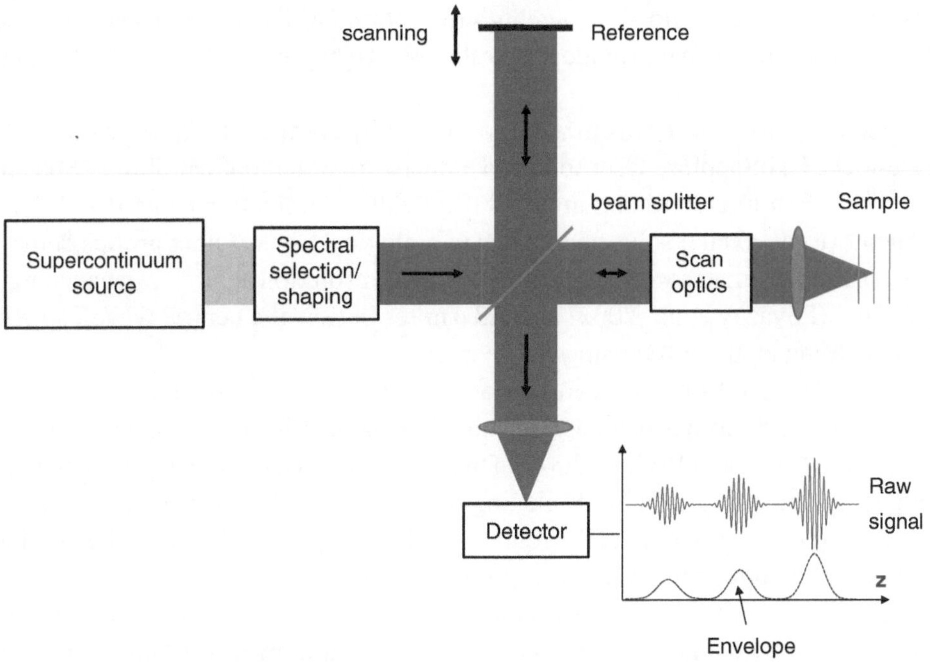

Fig. 15.9. Generalized diagram for OCT with an ultrashort coherence length supercontinuum source implemented using bulk optics. Broadband OCT systems can also be implemented using fibre optics but care must be taken to balance the dispersion in the two arms of the interferometer.

biological tissue using the photon time of flight to determine the depth in a specimen from which incident light is reflected or scattered. The path length of the reflected photons is measured using an interferometric detection technique, as depicted in Fig. 15.9, which requires a spatially coherent source with a short coherence length. The coherence length defines the axial resolution that can be achieved, which in turn is determined by the spectral width of the source. Thus OCT requires spatially coherent sources with broad spectral widths to obtain the highest axial resolutions. A further requirement is high average power since the coherent signal (conveyed by "ballistic" or unscattered photons) is strongly attenuated in scattering media such as biological tissue. These requirements have traditionally been provided by mode-locked laser sources with short pulse durations and correspondingly large bandwidths. Although axial resolutions down to $\sim 1\ \mu\text{m}$ have been achieved using highly specialized broad bandwidth Ti:Sapphire sources (Drexler et al., 1999), there is a need for more compact and lower cost sources for OCT. Ultimately, the spectral width of laser radiation – and therefore the OCT resolution – is limited by the bandwidth of the laser gain medium. The extreme nonlinear spectral broadening provided by SC generation permits high resolution OCT from relatively compact and low cost sources that provide spectral coverage beyond that of the Ti:Sapphire laser.

SC source based OCT was first demonstrated by Hartl et al. (2001) who used a mode-locked Ti:Sapphire laser to pump a microstructured optical fibre at 800 nm, which was then filtered to obtain broad bandwidth illumination centred at 1.3 μm that achieved an axial resolution of 2 μm in biological tissue. Other groups demonstrated sub-micrometer (~0.5 μm in tissue) axial resolution at a central wavelength of 730 nm (Povazay et al., 2002) and 1.3 μm resolution at a central wavelength of 1.1 μm (Wang et al., 2003) using SC sources.

Spectroscopic information can be obtained from OCT by imaging at different wavelengths and this has been demonstrated using dual-band SC sources (Aguirre et al., 2006, Wang and Rollins, 2007). Dual-band OCT has also been implemented using temporal and spectral multiplexing of two separate bands spectrally selected from a commercially available SC source (Spoler et al., 2007). For a detailed review of SC sources applied to OCT see Unterhuber et al. (2004).

While SC sources can achieve comparable, or even superior, resolution and broader spectral coverage in OCT systems compared to mode-locked lasers, they do present some less welcome issues that need to be addressed. To achieve high performance OCT, it is important that the OCT source has a smooth spectral profile in order to avoid the appearance of side-lobes on the axial point spread function. This requires appropriate shaping of the spectrally filtered supercontinuum (Apolonski et al., 2002). Another important issue is the significant pulse-to-pulse variation in the SC spectral profile and energy (Gu et al., 2002), which can result in significant excess noise (Corwin et al., 2003, Newbury et al., 2003). To address this, balanced detection schemes can be implemented in OCT systems that employ a SC source. This is relatively straightforward to implement for time-domain OCT, and can also be achieved for frequency-domain OCT (Moon and Kim, 2007). An alternative approach is to generate lower noise SC using normally dispersive fibre (Nishizawa et al., 2004) or fibres with a parabolic dispersion profile pumped near the minimum dispersion wavelength (Wang et al., 2007).

15.5 Diffuse optical spectroscopy and tomography

Diffuse optical spectroscopy (DOS) involves making quantitative measurements of the scattering, absorption or fluorescence properties of highly scattering samples for which the propagation of light is isotropic and can be modelled using the diffusion approximation (Gibson et al., 2005). DOS has several medical applications including breast cancer detection (Dehghani et al., 2003) and monitoring haemoglobin saturation in the brain (Benaron et al., 2000). DOS measurements are often time-resolved because the determination of the mean photon path length can improve the precision of measurements of scattering and absorption coefficients. It is also often desirable to make spectrally resolved DOS measurements to detect and quantify specific analytes with known spectral properties. This is typically undertaken using a number of discrete illumination wavelengths with a detector for each spectral channel or using spectrally broad illumination coupled with spectrally resolved detection using, e.g. a polychromator. DOS can be extended to diffuse optical tomography through the use of multiple sources and detectors. By considering the propagation of photons between different source–detector pairs, it is possible to use inverse scattering calculations to determine the spatial localization of scattering and absorbing regions within a specimen. The precision of diffuse optical tomography can be improved using spectrally or temporally resolved measurements.

The first applications of an SC source to time resolved diffuse optical spectroscopy (TR-DOS) were performed by Bassi et al. (2004) and Abrahamsson et al. (2004). Figure 15.10 shows a schematic of time-resolved TR-DOS using an SC source. This work was extended to parallel multispectral detection in 16 channels (Swartling et al., 2005) and a 32-channel detector system was applied to mammography applications (Bassi et al., 2006). Other developments using SC

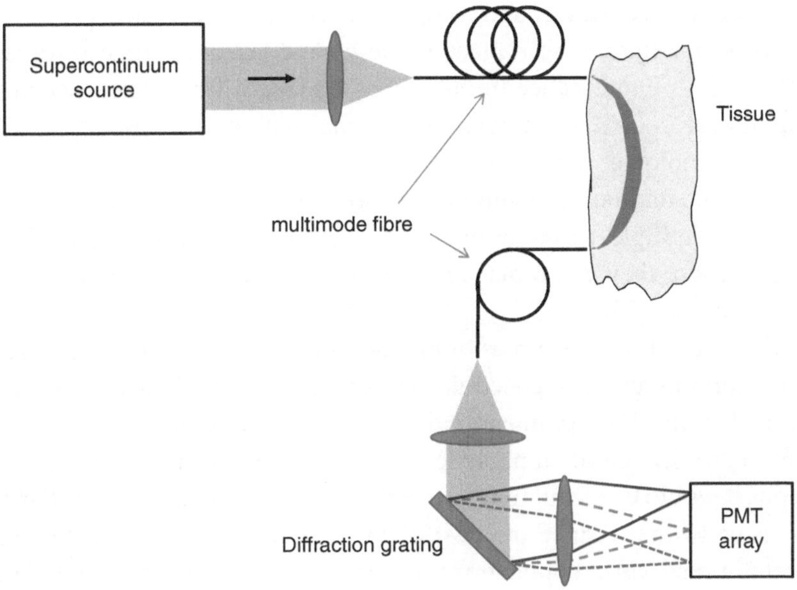

Fig. 15.10. Generalized diagram for diffuse optical spectroscopy with a supercontinuum source. Blue, green and red rays are indicated by solid, long-dashed and short-dashed lines respectively.

sources include measurements of anisotropic light propagation in human skeletal muscle (Binzoni et al., 2006) and the development of a compact TR-DOS instrument using a commercially available SC source (Bassi et al., 2007). SC sources have also been applied to 3D fluorescence lifetime imaging via diffuse fluorescence tomography (Soloviev et al., 2007). In addition fluorescence lifetime tomography has been realized using an SC excitation source applied to optically cleared samples including mouse embryos (McGinty et al., 2008) and plants using the technique of optical projection tomography (OPT) (Sharpe et al., 2002).

15.6 Conclusions and outlook

Supercontinuum sources have many applications in the field of biophotonics due partly to their spectral coverage, high spatial coherence, high spectral power density and the availability of ultrashort pulses, and also due to their relatively low cost, small footprint and low energy consumption. One can imagine that a turnkey compact, low-cost SC source could become almost as ubiquitous as was the HeNe laser in many branches of science. For biophotonics, SC sources can often directly replace lamp-based sources to offer higher S/N measurements over the visible and NIR spectrum. Suitably robust SC sources could also directly replace gas and solid-state lasers in many standard instruments such as fluorometers, cytometers, microscopes and multiwell plate readers, offering enhanced spectral versatility at significantly reduced cost, size and complexity. The ultrashort pulse potential of SC sources has been exploited to replace mode-locked lasers for time-resolved measurements, such as fluorescence lifetime imaging and diffuse optical spectroscopy, and for nonlinear optical techniques such as multiphoton microscopy and CARS. Increasingly techniques are being designed to simultaneously exploit the broad spectral coverage and pulsed nature of SC sources, e.g. to generate synchronized pulse trains at multiple wavelengths for pump probe spectroscopy or for STED microscopy, where they can replace much more complicated laser systems based on optical parametric generation.

While the future for SC sources in biophotonics looks very bright, there is still work to be done in terms of spectral coverage and power density, particularly at shorter wavelengths. For nonlinear optical applications, robust compression of the SC output to provide readily tunable femtosecond pulses is desirable and for many applications it would be useful to improve the SC noise properties. Most important, however, is to further reduce the cost of SC sources to permit their widespread deployment in real-world applications. Progress in terms of yet smaller footprints and robust operation reliability is also desirable. Nevertheless, it is already apparent that supercontinuum sources represent a "step-change" technology for biophotonics and many other applications.

References

Abrahamsson, C., Svensson, T., Svanberg, S., Andersson-Engels, S., Johansson, J. & Folestad, S. (2004) Time and wavelength resolved spectroscopy of turbid media using light continuum generated in a crystal fiber. *Optics Express*, **12**, 4103–4112.

Aguirre, A. D., Nishizawa, N., Fujimoto, J. G., Seitz, W., Lederer, M. & Kopf, D. (2006) Continuum generation in a novel photonic crystal fiber for ultrahigh resolution optical coherence tomography at 800 nm and 1300 nm. *Optics Express*, **14**, 1145–1160.

Alfano, R. R. & Shapiro, S. L. (1970) Observation of self-phase modulation and small-scale filaments in crystals and glasses. *Physical Review Letters*, **24**, 592–594.

Andresen, E. R., Paulsen, H. N., Birkedal, V., Thogersen, J. & Keiding, S. R. (2005) Broadband multiplex coherent anti-Stokes Raman scattering microscopy employing photonic-crystal fibers. *Journal of the Optical Society of America B-Optical Physics*, **22**, 1934–1938.

Apolonski, A., Povazay, B., Unterhuber, A., Drexler, W., Wadsworth, W. J., Knight, J. C. & Russell, P. S. (2002) Spectral shaping of supercontinuum in a cobweb photonic-crystal fiber with sub-20-fs pulses. *Journal of the Optical Society of America B-Optical Physics*, **19**, 2165–2170.

Auksorius, E., Boruah, B. R., Dunsby, C., Lanigan, P. M. P., Kennedy, G., Neil, M. A. A. & French, P. M. W. (2008) Stimulated emission depletion microscopy with a supercontinuum source and fluorescence lifetime imaging. *Optics Letters*, **33**, 113–115.

Bassi, A., Farina, A., D'Andrea, C., Pifferi, A., Valentini, G. & Cubeddu, R. (2007) Portable, large-bandwidth time-resolved system for diffuse optical spectroscopy. *Optics Express*, **15**, 14482–14487.

Bassi, A., Spinelli, L., D'Andrea, C., Giusto, A., Swartling, J., Pifferi, A., Torricelli, A. & Cubeddu, R. (2006) Feasibility of white-light time-resolved optical mammography. *Journal of Biomedical Optics*, **11**, 054035.

Bassi, A., Swartling, J., D'Andrea, C., Pifferi, A., Torricelli, A. & Cubeddu, R. (2004) Time-resolved spectrophotometer for turbid media based on supercontinuum generation in a photonic crystal fiber. *Optics Letters*, **29**, 2405–2407.

Benaron, D. A., Hintz, S. R., Villringer, A., Boas, D., Kleinschmidt, A., Frahm, J., Hirth, C., Obrig, H., Van Houten, J. C., Kermit, E. L., Cheong, W. F. & Stevenson, D. K. (2000) Noninvasive functional imaging of human brain using light. *Journal of Cerebral Blood Flow and Metabolism*, **20**, 469–477.

Betz, T., Teipel, J., Koch, D., Hartig, W., Guck, J., Kas, J. & Giessen, H. (2005) Excitation beyond the monochromatic laser limit: simultaneous 3-D confocal and multiphoton microscopy with a tapered fiber as white-light laser source. *Journal of Biomedical Optics*, **10**, 054009.

Binzoni, T., Courvoisier, C., Giust, R., Tribillon, G., Gharbi, T., Hebden, J. C., Leung, T. S., Roux, J. & Delpy, D. T. (2006) Anisotropic photon migration in human skeletal muscle. *Physics in Medicine and Biology*, **51**, N79–N90.

Birch, D. J. S. & Imhof, R. E. (1977) A single-photon counting fluorescence decay-time spectrometer. *Journal of Physics E: Scientific Instruments*, **10**, 1044–1049.

Birk, H. & Storz, R. (2001) Illuminating device and microscope. United States, Leica Microsystems; Heidelberg GmbH, Germany.

Booth, M. J., Juskaitis, R. & Wilson, T. (2008) Spectral confocal reflection microscopy using a white light source. *Journal of the European Optical Society-Rapid Publications*, **3**, 08026.

Busch, G. E., Jones, R. P. & Rentzepi, P. M. (1973) Picosecond spectroscopy using a picosecond continuum. *Chemical Physics Letters*, **18**, 178–185.

Corwin, K. L., Newbury, N. R., Dudley, J. M., Coen, S., Diddams, S. A., Weber, K. & Windeler, R. S. (2003) Fundamental noise limitations to supercontinuum generation in microstructure fiber. *Physical Review Letters*, **90**, 113904.

Cubeddu, R., Comelli, D., D'Andrea, C., Taroni, P. & Valentini, G. (2002) Time-resolved fluorescence imaging in biology and medicine. *Journal of Physics D-Applied Physics*, **35**, R61–R76.

Deguil, N., Mottay, E., Salin, F., Legros, P. & Choquet, D. (2004) Novel diode-pumped infrared tunable laser system for multi-photon microscopy. *Microscopy Research and Technique*, **63**, 23–26.

Dehghani, H., Pogue, B. W., Poplack, S. P. & Paulsen, K. D. (2003) Multiwavelength three-dimensional near-infrared tomography of the breast: initial simulation, phantom, and clinical results. *Applied Optics*, **42**, 135–145.

Denk, W., Strickler, J. H. & Webb, W. W. (1990) 2-Photon laser scanning fluorescence microscopy. *Science*, **248**, 73–76.

Domachuk, P., Perry, H., Cronin-Golomb, M. & Omenetto, F. G. (2007) Towards an integrated optofluidic diffractive spectrometer. *IEEE Photonics Technology Letters*, **19**, 1976–1978.

Drexler, W. (2004) Ultrahigh-resolution optical coherence tomography. *Journal of Biomedical Optics*, **9**, 47–74.

Drexler, W., Morgner, U., Kärtner, F. X., Pitris, C., Boppart, S. A., Li, X. D., Ippen, E. P. & Fujimoto, J. G. (1999) In vivo ultrahigh-resolution optical coherence tomography. *Opt. Lett.*, **24**, 1221–1223.

Dunsby, C., Lanigan, P. M. P., McGinty, J., Elson, D. S., Requejo-Isidro, J., Munro, I., Galletly, N., McCann, F., Treanor, B., Onfelt, B., Davis, D. M., Neil, M. A. A. & French, P. M. W. (2004) An electronically tunable ultrafast laser source applied to fluorescence imaging and fluorescence lifetime imaging microscopy. *Journal of Physics D-Applied Physics*, **37**, 3296–3303.

Frank, J. H., Elder, A. D., Swartling, J., Venkitaraman, A. R., Jeyasekharan, A. D. & Kaminski, C. F. (2007) A white light confocal microscope for spectrally resolved multidimensional imaging. *Journal of Microscopy–Oxford*, **227**, 203–215.

Fujimoto, J. G. (2003) Optical coherence tomography for ultrahigh resolution in vivo imaging. *Nat Biotech*, **21**, 1361–1367.

Garzon, J., Meneses, J., Tribillon, G., Gharbi, T. & Plata, A. (2004) Chromatic confocal microscopy by means of continuum light generated through a standard single-mode fibre. *Journal of Optics A-Pure and Applied Optics*, **6**, 544–548.

Gibson, A. P., Hebden, J. C. & Arridge, S. R. (2005) Recent advances in diffuse optical imaging. *Physics in Medicine and Biology*, **50**, R1–R43.

Grant, D. M., Elson, D. S., Schimpf, D., Dunsby, C., Requejo-Isidro, J., Auksorius, E., Munro, I., Neil, M. A. A., French, P. M. W., Nye, E., Stamp, G. & Courtney, P. (2005) Optically sectioned fluorescence lifetime imaging using a Nipkow disk microscope and a tunable ultrafast continuum excitation source. *Optics Letters*, **30**, 3353–3355.

Gu, X., Xu, L., Kimmel, M., Zeek, E., O'Shea, P., Shreenath, A. P., Trebino, R. & Windeler, R. S. (2002) Frequency-resolved optical gating and single-shot spectral measurements reveal fine structure in microstructure-fiber continuum. *Optics Letters*, **27**, 1174–1176.

Guillon, M., Dholakia, K. & Mcgloin, D. (2008) Optical trapping and spectral analysis of aerosols with a supercontinuum laser source. *Optics Express*, **16**, 7655–7664.

Hartl, I., Li, X. D., Chudoba, C., Ghanta, R. K., Ko, T. H., Fujimoto, J. G., Ranka, J. K. & Windeler, R. S. (2001) Ultrahigh-resolution optical coherence tomography using continuum generation in an air–silica microstructure optical fiber. *Optics Letters*, **26**, 608–610.

Hein, B., Willig, K. I. & Hell, S. W. (2008) Stimulated emission depletion (STED) nanoscopy of a fluorescent protein-labeled organelle inside a living cell. *Proceedings of the National Academy of Sciences of the United States of America*, **105**, 14271–14276.

Hell, S. W. (2007) Far-field optical nanoscopy. *Science*, **316**, 1153–1158.

Hult, J., Watt, R. S. & Kaminski, C. F. (2007) High bandwidth absorption spectroscopy with a dispersed supercontinuum source. *Optics Express*, **15**, 11385–11395.

Jureller, J. E., Scherer, N. F., Birks, T. A., Wadsworth, W. J. & Russell, P. S. J. (2003) Widely tunable femtosecond pulses from a tapered fiber for ultrafast microscopy and multiphoton applications. In Miller, R. J. D., Murnane, M. M., Scherer, N. F. & Weiner, A. M. (Eds.) *Ultrafast Phenomena XIII,* Berlin, Springer-Verlag.

Kaminski, C. F., Watt, R. S., Elder, A. D., Frank, J. H. & Hult, J. (2008) Supercontinuum radiation for applications in chemical sensing and microscopy. *Applied Physics B-Lasers and Optics*, **92**, 367–378.

Kano, H. & Hamaguchi, H. (2003) Characterization of a supercontinuum generated from a photonic crystal fiber and its application to coherent Raman spectroscopy. *Optics Letters*, **28**, 2360–2362.

Kano, H. & Hamaguchi, H. (2004) Femtosecond coherent anti-Stokes Raman scattering spectroscopy using supercontinuum generated from a photonic crystal fiber. *Applied Physics Letters*, **85**, 4298–4300.

Kano, H. & Hamaguchi, H. (2005a) Ultrabroadband ($>2500\,cm^{-1}$) multiplex coherent anti-Stokes Raman scattering microspectroscopy using a supercontinuum generated from a photonic crystal fiber. *Applied Physics Letters*, **86**, 121113.

Kano, H. & Hamaguchi, H. (2005b) Vibrationally resonant imaging of a single living cell by supercontinuum-based multiplex coherent anti-Stokes Raman scattering microspectroscopy. *Optics Express*, **13**, 1322–1327.

Kano, H. & Hamaguchi, H. (2006a) Dispersion-compensated supercontinuum generation for ultrabroadband multiplex coherent anti-Stokes Raman scattering spectroscopy. *Journal of Raman Spectroscopy*, **37**, 411–415.

Kano, H. & Hamaguchi, H. (2006b) In-vivo multi-nonlinear optical imaging of a living cell using a supercontinuum light source generated from a photonic crystal fiber. *Optics Express*, **14**, 2798–2804.

Kano, H. & Hamaguchi, H. O. (2007) Supercontinuum dynamically visualizes a dividing single cell. *Analytical Chemistry*, **79**, 8967–8973.

Kee, T. W. & Cicerone, M. T. (2004) Simple approach to one-laser, broadband coherent anti-Stokes Raman scattering microscopy. *Optics Letters*, **29**, 2701–2703.

Kee, T. W., Zhao, H. X. & Cicerone, M. T. (2006) One-laser interferometric broadband coherent anti-Stokes Raman scattering. *Optics Express*, **14**, 3631–3640.

Kelkar, P. V., Coppinger, F., Bhushan, A. S. & Jalali, B. (1999) Time-domain optical sensing. *Electronics Letters*, **35**, 1661–1662.

Konorov, S. O., Akimov, D. A., Ivanov, A. A., Alfimov, M. V. & Zheltikov, A. M. (2004) Microstructure fibers as frequency-tunable sources of ultrashort chirped pulses for coherent nonlinear spectroscopy. *Applied Physics B-Lasers and Optics*, **78**, 565–567.

Kudlinski, A., George, A. K., Knight, J. C., Travers, J. C., Rulkov, A. B., Popov, S. V. & Taylor, J. R. (2006) Zero-dispersion wavelength decreasing photonic crystal fibers for ultraviolet-extended supercontinuum generation. *Optics Express*, **14**, 5715–5722.

Lakowicz, J. R. (1999) *Principles of Fluorescence Spectroscopy*, New York, Kluwer Academic/Plenum Publishers.

Langridge, J. M., Laurila, T., Watt, R. S., Jones, R. L., Kaminski, C. F. & Hult, J. (2008) Cavity enhanced absorption spectroscopy of multiple trace gas species using a supercontinuum radiation source. *Optics Express*, **16**, 10178–10188.

Li, P., Shi, K. B. & Liu, Z. W. (2005a) Manipulation and spectroscopy of a single particle by use of white-light optical tweezers. *Optics Letters*, **30**, 156–158.

Li, P., Shi, K. B. & Liu, Z. W. (2005b) Optical scattering spectroscopy by using tightly focused supercontinuum. *Optics Express*, **13**, 9039–9044.

Lin, C. & Stolen, R. H. (1976) New nanosecond continuum for excited-state spectroscopy. *Applied Physics Letters*, **28**, 216–218.

Lindfors, K., Kalkbrenner, T., Stoller, P. & Sandoghdar, V. (2004) Detection and spectroscopy of gold nanoparticle using supercontinuum white light confocal microscopy. *Physical Review Letters*, **93**, 037401.

Manning, H. B., Kennedy, G. T., Owen, D. M., Grant, D. M., Magee, A. I., Neil, M. A. A., Itoh, Y., Dunsby, C. & French, P. M. W. (2008) A compact, multidimensional spectrofluorometer exploiting supercontinuum generation. *Journal of Biophotonics*, **1**, 494–505.

Maus, M., Rousseau, E., Cotlet, M., Schweitzer, G., Hofkens, J., Auweraer, M. V. D., Schryver, F. C. D. & Krueger, A. (2001) New picosecond laser system for easy tunability over the whole ultraviolet/visible/near infrared wavelength range based on flexible harmonic generation and optical parametric oscillation. *Review of Scientific Instruments*, **72**, 36–40.

McConnell, G. (2004) Confocal laser scanning fluorescence microscopy with a visible continuum source. *Optics Express*, **12**, 2844–2850.

McConnell, G., Poland, S. & Girkin, J. M. (2006) Fast wavelength multiplexing of a white-light supercontinuum using a digital micromirror device for improved three-dimensional fluorescence microscopy. *Review of Scientific Instruments*, **77**, 013702.

McGinty, J., Tahir, K. R., Laine, R., Talbot, C. B., Dunsby, C., Neil, M. A. A., Quintana, L., Swoger, J., Sharpe, J. & French, P. M. W. (2008) Fluorescence lifetime optical projection tomography. *Journal of Biophotonics*, **1**, 390–394.

Moon, S. & Kim, D. Y. (2007) Normalization detection scheme for high-speed optical frequency-domain imaging and reflectometry. *Optics Express*, **15**, 15129–15146.

Nagarajan, V., Johnson, E., Schellenberg, P., Parson, W. & Windeler, R. (2002) A compact versatile femtosecond spectrometer. *Review of Scientific Instruments*, **73**, 4145–4149.

Newbury, N. R., Washburn, B. R., Corwin, K. L. & Windeler, R. S. (2003) Noise amplification during supercontinuum generation in microstructure fiber. *Optics Letters*, **28**, 944–946.

Nishizawa, N., Chen, Y., Hsiung, P., Ippen, E. P. & Fujimoto, J. G. (2004) Real-time, ultrahigh-resolution, optical coherence tomography with an all-fiber, femtosecond fiber laser continuum at 1.5 μm. *Optics Letters*, **29**, 2846–2848.

Okuno, M., Kano, H., Leproux, P., Couderc, V. & Hamaguchi, H. (2007) Ultrabroadband ($>2000\,\mathrm{cm}^{-1}$) multiplex coherent anti-Stokes Raman scattering spectroscopy using a subnanosecond supercontinuum light source. *Optics Letters*, **32**, 3050–3052.

Okuno, M., Kano, H., Leproux, P., Couderc, V. & Hamaguchi, H. (2008) Ultrabroadband multiplex CARS microspectroscopy and imaging using a subnanosecond supercontinuum light source in the deep near infrared. *Optics Letters*, **33**, 923–925.

Owen, D. M., Auksorius, E., Manning, H. B., Talbot, C. B., De Beule, P. A. A., Dunsby, C., Neil, M. A. A. & French, P. M. W. (2007) Excitation-resolved hyperspectral fluorescence lifetime imaging using a UV-extended supercontinuum source. *Optics Letters*, **32**, 3408–3410.

Palero, J. A., Boer, V. O., Vijverberg, J. C., Gerritsen, H. C. & Sterenborg, H. (2005) Short-wavelength two-photon excitation fluorescence microscopy of tryptophan with a photonic crystal fiber based light source. *Optics Express*, **13**, 5363–5368.

Paulsen, H. N., Hilligsoe, K. M., Thogersen, J., Keiding, S. R. & Larsen, J. J. (2003) Coherent anti-Stokes Raman scattering microscopy with a photonic crystal fiber based light source. *Optics Letters*, **28**, 1123–1125.

Pawley, J. B. (1995) *Handbook of Biological Confocal Microscopy*, New York, Plenum Press.

Povazay, B., Bizheva, K., Unterhuber, A., Hermann, B., Sattmann, H., Fercher, A. F., Drexler, W., Apolonski, A., Wadsworth, W. J., Knight, J. C., Russell, P. S. J., Vetterlein, M. & Scherzer, E. (2002) Submicrometer axial resolution optical coherence tomography. *Optics Letters*, **27**, 1800–1802.

Rulkov, A. B., Vyatkin, M. Y., Popov, S. V., Taylor, J. R. & Gapontsev, V. P. (2005) High brightness picosecond all-fiber generation in 525–1800 nm range with picosecond Yb pumping. *Optics Express*, **13**, 377–381.

Sanders, S. T. (2002) Wavelength-agile fiber laser using group-velocity dispersion of pulsed super-continua and application to broadband absorption spectroscopy. *Applied Physics B-Lasers and Optics*, **75**, 799–802.

Sharpe, J., Ahlgren, U., Perry, P., Hill, B., Ross, A., Hecksher-Sorensen, J., Baldock, R. & Davidson, D. (2002) Optical projection tomography as a tool for 3D microscopy and gene expression studies. *Science*, **296**, 541–545.

Shi, K. B., Heikal, A. A. & Liu, Z. W. (2006a) A single-shot approach for measuring two-photon action cross-section of fluorescent markers. *Optics Express*, **14**, 8722–8727.

Shi, K. B., Li, P., Yin, S. Z. & Liu, Z. W. (2004) Chromatic Confocal microscopy using supercontinuum light. *Optics Express*, **12**, 2096–2101.

Shi, K. B., Nam, S. H., Li, P., Yin, S. Z. & Liu, Z. W. (2006b) Wavelength division multiplexed confocal microscopy using supercontinuum. *Optics Communications*, **263**, 156–162.

Soloviev, V. Y., McGinty, J., Tahir, K. B., Neil, M. A. A., Sardini, A., Hajnal, J. V., Arridge, S. R. & French, P. M. W. (2007) Fluorescence lifetime tomography of live cells expressing enhanced green fluorescent protein embedded in a scattering medium exhibiting background autofluorescence. *Optics Letters*, **32**, 2034–2036.

Spoler, F., Kray, S., Grychtol, P., Hermes, B., Bornemann, J., Forst, M. & Kurz, H. (2007) Simultaneous dual-band ultra-high resolution optical coherence tomography. *Optics Express*, **15**, 10832–10841.

Swartling, J., Bassi, A., D'Andrea, C., Pifferi, A., Torricelli, A. & Cubeddu, R. (2005) Dynamic time-resolved diffuse spectroscopy based on supercontinuum light pulses. *Applied Optics*, **44**, 4684–4692.

Travers, J. C., Rulkov, A. B., Cumberland, B. A., Popov, S. V. & Taylor, J. R. (2008) Visible supercontinuum generation in photonic crystal fibers with a 400 W continuous wave fiber laser. *Optics Express*, **16**, 14435–14447.

Unterhuber, A., Povazay, B., Bizheva, K., Hermann, B., Sattmann, H., Stingl, A., Le, T., Seefeld, M., Menzel, R., Preusser, M., Budka, H., Schubert, C., Reitsamer, H., Ahnelt, P. K., Morgan, J. E., Cowey, A. & Drexler, W. (2004) Advances in broad bandwidth light sources for ultrahigh resolution optical coherence tomography. *Physics in Medicine and Biology*, **49**, 1235–1246.

Volkmer, A. (2005) Vibrational imaging and microspectroscopies based on coherent anti-Stokes Raman scattering microscopy. *Journal of Physics D-Applied Physics*, **38**, R59–R81.

Von Vacano, B. & Motzkus, M. (2006) Time-resolved two color single-beam CARS employing supercontinuum and femtosecond pulse shaping. *Optics Communications*, **264**, 488–493.

Von Vacano, B., Wohlleben, W. & Motzkus, M. (2006) Actively shaped supercontinuum from a photonic crystal fiber for nonlinear coherent microspectroscopy. *Optics Letters*, **31**, 413–415.

Wang, H. & Rollins, A. M. (2007) Optimization of dual-band continuum light source for ultrahigh-resolution optical coherence tomography. *Applied Optics*, **46**, 1787–1794.

Wang, H., Fleming, C. P. & Rollins, A. M. (2007) Ultrahigh-resolution optical coherence tomography at 1.15 μm using photonic crystal fiber with no zero-dispersion wavelengths. *Optics Express*, **15**, 3085–3092.

Wang, Y. M., Zhao, Y. H., Nelson, J. S., Chen, Z. P. & Windeler, R. S. (2003) Ultrahigh-resolution optical coherence tomography by broadband continuum generation from a photonic crystal fiber. *Optics Letters*, **28**, 182–184.

Wildanger, D., Rittweger, E., Kastrup, L. & Hell, S. W. (2008) STED microscopy with a supercontinuum laser source. *Optics Express*, **16**, 9614–9621.

Ye, J. Y., Divin, C. J., Baker, J. R. & Norris, T. B. (2007) Whole spectrum fluorescence detection with ultrafast white light excitation. *Optics Express*, **15**, 10439–10445.

Yokoyama, H., Tsubokawa, H., Guo, H. C., Shikata, J., Sato, K., Takashima, K., Kashiwagi, K., Saito, N., Taniguchi, H. & Ito, H. (2007) Two-photon bioimaging utilizing supercontinuum light generated by a high-peak-power picosecond semiconductor laser source. *Journal of Biomedical Optics*, **12**, 054019.

Zheltikov, A. M. (2007) Time-resolved coherent Raman and sum-frequency generation spectroscopy with wavelength-tunable, short-pulse, photonic-crystal fiber light sources. *Journal of Raman Spectroscopy*, **38**, 1052–1063.

Zipfel, W. R., Williams, R. M. & Webb, W. W. (2003) Nonlinear magic: multiphoton microscopy in the biosciences. *Nat. Biotech.*, **21**, 1369–1377.

16

Fiber sources of tailored supercontinuum in nonlinear microspectroscopy and imaging

A. M. Zheltikov

16.1 Introduction

Over the past few years, new optical fibers with enhanced nonlinearity and tailored dispersion (Russell 2003; Knight 2003) have been providing a constantly growing platform for the development of advanced fiber-format devices and components for optical metrology (Jones et al. 2000; Udem et al. 2002), ultrashort-pulse laser technologies (Zheltikov 2007a), biomedicine (Hartl et al. 2001), quantum optics (Rarity et al. 2005), spectroscopy (Sidorov-Biryukov et al. 2006), and microscopy (Paulsen et al. 2003). Unique options offered by photonic-crystal fiber (PCF) technology (Russell 2006), such as dispersion management through fiber structure engineering (Reeves et al. 2003) and enhancement of optical nonlinearity due to a strong field confinement in a small-size fiber core (Fedotov et al. 2001), have been pushing the frontiers of fiber optics, allowing the creation of efficient sources of supercontinuum radiation (Ranka et al. 2000; Dudley et al. 2006; Zheltikov 2006), novel compact fiber lasers (Lim et al. 2002; Limpert et al. 2006), as well as frequency converters (Akimov et al. 2003), pulse compressors (Südmeyer et al. 2003), fiber components for biomedical optics (Flusberg et al. 2005a, 2005b), and optical sensors (Monro et al. 2001).

In the rapidly expanding field of nonlinear microscopy and spectroscopy, PCFs have been shown to possess a tremendous potential for making laser microscopes and spectrometers simpler and much more compact through the replacement of wavelength-tunable laser sources, such as optical parametric amplifiers and dye lasers, by a specifically designed segment of fiber. Paulsen et al. (2003) have simplified microscopy based on coherent anti-Stokes Raman scattering (CARS) by using a blue-shifted dispersive-wave emission from a PCF as a pump field and employing Ti:sapphire laser pulses as a Stokes field. Efficient frequency conversion

Supercontinuum Generation in Optical Fibers, ed. by J. M. Dudley and J. R. Taylor. Published by Cambridge University Press.

and supercontinuum generation in PCFs have been shown to enhance the capabilities of chirped-pulse CARS (Konorov et al. 2004a) and coherent inverse Raman spectroscopy (Kano & Hamaguchi 2003). Efficient spectral broadening of ultrashort pulses in PCFs with carefully engineered dispersion profiles makes these fibers ideal light sources for pump–supercontinuum probe time- and frequency-resolved nonlinear-optical measurements (Zheltikov 2006) and offer interesting new options for multiplex CARS microscopy (Kano & Hamaguchi 2005; Andresen et al. 2005). In recent experiments (von Vacano et al. 2006), a single-beam CARS approach (Dudovich et al. 2002) has been implemented in a new and elegant way through the generation of supercontinuum in a PCF followed by a compression and pulse shaping of this supercontinuum with the help of a pulse shaper based on a spatial light modulator. In this chapter, we provide an overview of novel fiber supercontinuum-generating and wavelength-shifting components and their applications in the rapidly growing field of nonlinear microspectroscopy and imaging.

16.2 Photonic-crystal fiber light sources for nonlinear Raman microspectroscopy

Coherent anti-Stokes Raman scattering (Eesley 1981) has long proved to be a powerful technique for temperature and concentration measurements in excited gases, combustion, and flames (Eckbreth 1988), gas-phase analysis (Zheltikov & Koroteev 1999), and high-resolution molecular spectroscopy. In the femtosecond era, short-pulse CARS has been shown to allow time-resolved analysis and quantum control of ultrafast processes and wave-packet dynamics in molecular systems. The nonlinear nature and the spectral selectivity of CARS make this method an ideal tool for nonlinear microscopy (Zumbusch et al. 1999), opening new horizons in bioimaging (Evans et al. 2005). The recently demonstrated coherence-controlled CARS (Dudovich et al. 2002) improves the sensitivity and the selectivity of nonlinear Raman spectroscopy and enhances the potential of CARS microscopy.

Light sources based on photonic-crystal fibers have been recently shown to offer attractive fiber-optic solutions for simplification and miniaturization of CARS microscopes (Paulsen et al. 2003) and spectrometers (Sidorov-Biryukov et al. 2006). Fibers of this type can provide high efficiencies of nonlinear-optical spectral transformations of ultrashort laser pulses due to the strong confinement of the light field in a small-area fiber core and tailored dispersion of guided modes, which can be optimized for a given type of nonlinear-optical interaction by modifying the fiber structure. Specially designed PCF frequency converters for ultrashort laser pulses have been used to demonstrate frequency-resolved optically gated CARS (Konorov et al. 2004a), where a blue-shifted output of PCFs pumped with femtosecond Cr:forsterite laser pulses was used as a Stokes field and the second harmonic of

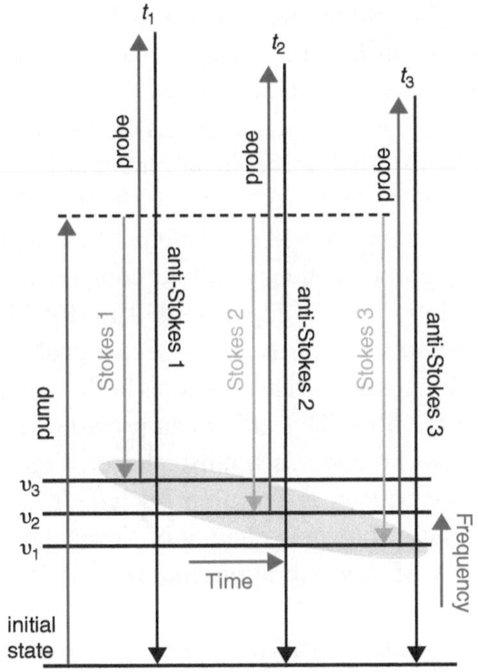

Fig. 16.1. Multiplex CARS microspectroscopy with a chirped short-pulse output of a photonic-crystal fiber (represented by an ellipse on the time–frequency plane). The PCF output is chirped in such a way as to provide three different Stokes frequencies for a selective Raman excitation of vibrational states v_1, v_2, and v_3 at the instants of time t_1, t_2, and t_3, respectively. A fixed-frequency field is scattered off these excitations, giving rise to a multiplex CARS signal, with the information on vibrational states v_1, v_2, and v_3 encoded in three different anti-Stokes frequencies.

the Cr:forsterite laser was employed as a pump field. Supercontinuum-generating PCFs have been shown to provide a broad range of wavelength tunability, offering new interesting options for multiplex CARS microscopy (Kano & Hamaguchi 2005). One of the approaches in multiplex CARS microspectroscopy is illustrated in Fig. 16.1. Here, a chirped short-pulse output of a PCF (represented by an ellipse on the time–frequency plane) provides three different Stokes frequencies for a selective Raman excitation of vibrational states v_1, v_2, and v_3 at the instants of time t_1, t_2, and t_3, respectively. A fixed-frequency field is scattered off these excitations, giving rise to a multiplex CARS signal, with the information on vibrational states v_1, v_2, and v_3 encoded in three different anti-Stokes frequencies.

Four-wave-mixing processes in highly nonlinear PCFs, on the other hand, enable a fiber-format integration of efficient frequency conversion and pulse shaping of laser fields (Zheltikov 2007b). Such fiber pulse shapers are especially promising for a coherence control of nonlinear Raman processes and single-beam CARS microspectroscopy. A light field propagating through an optical fiber in the region of anomalous dispersion can become unstable with respect to a weak periodic modulation perturbing the field waveform (Agrawal 2001). In this regime, called modulation instability, such periodic modulations can grow exponentially from the level of noise or a weak external probe signal, transforming the input laser field into a sequence of light pulses. In the spectral domain, this effect can be understood as parametric four-wave mixing (FWM) of the input field with a central frequency ω_0, serving as a pump, and a frequency-shifted component $\omega_0 \pm \Omega$ corresponding to a periodic modulation of the input field. This FWM process generates a Stokes or anti-Stokes counterpart of the weak seed signal at $\omega_0 \pm \Omega$ through the $\omega_0 \mp \Omega = 2\omega_0 - (\omega_0 \pm \Omega)$ interaction and amplifies both sidebands as the field propagates further on down the fiber. This effect can provide a mechanism whereby a highly nonlinear PCF can function as a pulse shaper. Figure 16.2 shows how a sequence of light pulses produced as a result of such FWM in a PCF (upper panel) leads to a selective in-phase Raman excitation of molecular vibrations (Zheltikov 2007b). The amplitude Q_0 of these vibrations increases as Ω approaches the Raman transition frequency Ω_0.

In a recent experiment, Sidorov-Biryukov et al. (2006) have demonstrated time-resolved CARS spectroscopy using a frequency-tunable soliton output of a silica PCF as a Stokes field for two-color time-resolved CARS. In these experiments, a Ti:sapphire laser oscillator pumped with a second-harmonic output of a diode-pumped Nd:YVO_4 laser generated light pulses with a typical temporal width of

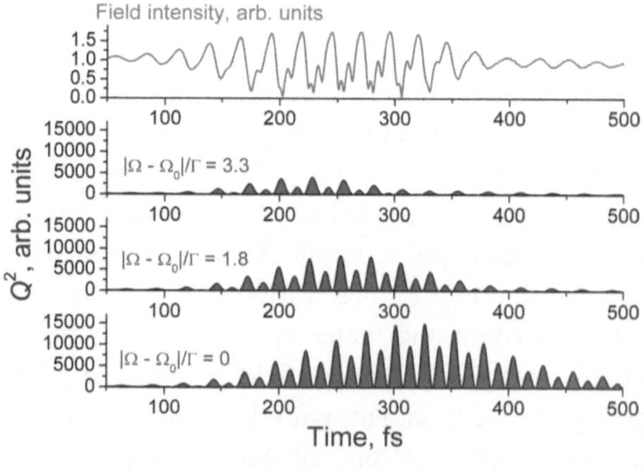

Fig. 16.2. Selective in-phase Raman excitation of molecular vibrations with a resonance sequence of ultrashort light pulses produced through FWM in a PCF (upper panel). The amplitude Q_0 of these vibrations is shown as a function of time for $|\Omega - \Omega_0|/\Gamma$ taking the values of 0, 1.8, and 3.3 (from top to bottom), Γ being the Raman transition linewidth, $\Omega_0/\Gamma \approx 24.2$.

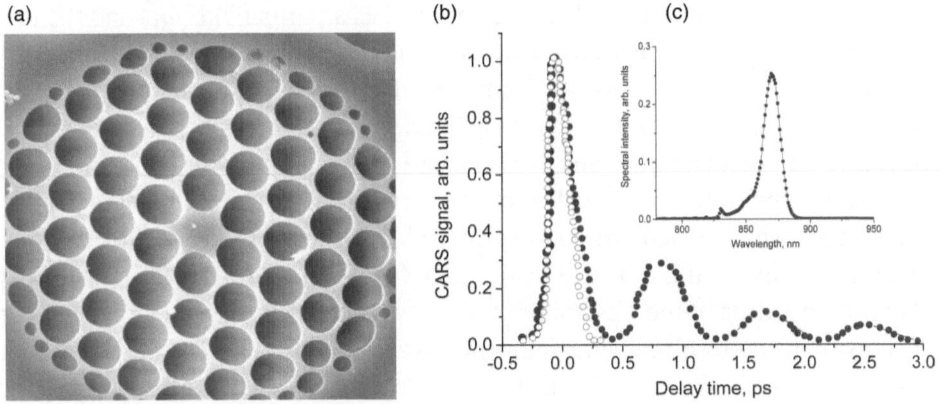

Fig. 16.3. (a) An SEM image of PCF. (b) CARS signal measured as a function of the delay time between the probe pulse and the pump–Stokes pulse dyad is shown by the filled circles. Open circles display the cross-correlation of the red-shifted PCF output and a 50-fs Ti:sapphire laser pulse measured through sum-frequency generation in a BBO crystal. (c) The spectrum of the red-shifted PCF output reflected off a diffraction grating.

about 30 fs, an energy up to 3 nJ, and a central wavelength of 800 nm at a pulse repetition rate of 100 MHz. The Ti:sapphire laser output was divided into two beams with a 70 : 30 beam splitter. The laser pulses of the first beam had an energy of about 2 nJ, providing pump and probe photons for CARS measurements. The Ti:sapphire laser pulse width increased up to 50 fs as the beam was transmitted through the delay line and focusing optics.

The laser pulses of the second beam from the beam splitter were coupled into the fundamental mode of a fused silica PCF with a core diameter of 1.8 μm (Fig. 16.3(b)). For the fundamental mode of the fiber, the group-velocity dispersion (GVD) vanishes at 754 nm, providing anomalous dispersion for 800-nm Ti:sapphire laser pulses. The effective area of the fundamental mode at 800 nm is 2.1 μm^2, corresponding to the nonlinearity coefficient $\gamma \approx 110\,W^{-1}\,km^{-1}$. Ti:sapphire laser pulses tend to form optical solitons as they propagate through the PCF. These solitons undergo a continuous frequency down-shift due to the soliton self-frequency shift.

The central wavelength of the red-shifted soliton at the output of the fiber is controlled by the input peak power and the fiber length. With the peak power of input laser pulses increased from 1 to 6 kW, the central wavelength of the red-shifted soliton at the output of a 12-cm PCF is tuned from 840 to 930 nm. As the soliton is shifted toward longer wavelengths, its dispersion changes and effective mode area increases. With the input peak power ranging from 3 to 11 kW, the SSFS in the PCF with a length of 5–20 cm can provide the wavelength tunability range from 870 to 1180 nm for the red-shifted soliton at the output of the fiber. For the $\omega_a = 2\omega_1 - \omega_2$

CARS process with Ti:sapphire laser pulses used as a pump field (ω_1) and the red-shifted soliton from the PCF used as a Stokes field (ω_2), the demonstrated tunability range gives access to the 1000–4025-cm^{-1} region of wavenumbers characteristic of fingerprint Raman transitions in a broad variety of molecular systems.

In contrast to dispersive waves, which sense the normal dispersion and tend to spread out in time, solitons are well-localized in the time domain, suggesting a way of achieving a femtosecond time resolution in CARS spectroscopy and microscopy. To assess the pulse width of the red-shifted PCF output, we mixed this signal with 50-fs 800-nm pulses of the Ti:sapphire laser in a 4-mm BBO crystal and measured the intensity of the resulting sum-frequency signal as a function of the delay time between the pulses. Open circles in Fig. 16.3(b) display such a cross-correlation trace for the red-shifted PCF output centered at $\lambda_s \approx 870$ nm (see Fig. 16.3(c)). The pulse width of this soliton at the output of the fiber is estimated as 160 fs. The uncertainty in the time-domain localization of the red-shifted soliton PCF output, originating from the timing jitter of the soliton induced by fluctuations in the input peak power, was estimated as 30 fs.

To demonstrate time-resolved CARS with a Stokes field provided by the red-shifted solitonic output of a PCF, time-delay CARS measurements were performed on a doublet of Raman transitions in liquid-phase pyridine with Raman wave numbers $\Omega/2\pi c$ (c being the speed of light) of 988 and 1028 cm^{-1}. These transitions have close Raman cross-sections, giving rise to well-pronounced quantum beats in the nonlinear response. Experiments were performed with 50-fs pulses of 800-nm Ti:sapphire laser radiation as a pump (ω_1) and the 870-nm solitonic PCF output as a Stokes field (ω_2). The spectrum of the Stokes pulse reflected off a diffraction grating is presented in Fig. 16.3(c). The 870-nm solitonic PCF output gives access to both 988- and 1028-cm^{-1} Raman modes, as the Stokes wavelengths corresponding to the exact $\Omega = \omega_1 - \omega_2$ Raman resonance with these Raman modes are equal to 868.7 and 871.7 nm, respectively. Time-delayed 50-fs Ti:sapphire laser pulses were employed to probe the Raman vibrations excited by the pump and Stokes fields in the folded CARS geometry, allowing the CARS signal to be spatially separated from the pump, Stokes, and probe fields.

Figure 16.3 presents the intensity of the $\omega_a = 2\omega_1 - \omega_2$ CARS signal measured as a function of the delay time τ between the probe pulse and the pump–Stokes pulse pair. The sharp peak observed in the region of small τ represents the nonresonant CARS signal related to electronic transitions in the medium. Since both the response time of the nonresonant part of cubic nonlinearity and the pulse width of the pump are much shorter than the duration of the Stokes pulse, the time response in the region of small τ (filled circles in Fig. 16.3(b)) closely follows the cross-correlation of the Stokes and probe pulses (open circles). For delay times τ exceeding the duration of this cross-correlation, nonresonant CARS is suppressed, and the time-domain

CARS response displays a characteristic oscillatory behavior, visualizing the quantum beats of the 988- and 1028-cm^{-1} Raman modes. Red-shifted soliton output provided by the PCF has been also applied to probe C–H vibrations in ethanol ($\Omega/2\pi c \approx 2930\,\text{cm}^{-1}$, $\lambda_s \approx 1.045\,\mu\text{m}$) and the 1210-cm^{-1} Raman resonance in toluene ($\lambda_s \approx 886\,\text{nm}$). In both cases, the nonresonant background was efficiently suppressed in time-domain CARS, similar to the result of Fig. 16.3(b), by introducing a time delay between the probe pulse and the pump–Stokes pulse dyad.

16.3 Frequency-shifted megawatt soliton output of a hollow photonic-crystal fiber for time-resolved CARS microspectroscopy

Hollow-core photonic-crystal fibers (Cregan et al. 1999) enable the generation of wavelength-tunable ultrashort pulses in the megawatt range of peak powers, offering a promising solution for a fiber-format source of the Stokes field for CARS microspectroscopy and bioimaging. Electromagnetic radiation can be guided in the hollow core of such fibers by photonic band gaps or a lower density of modes of a microstructured cladding, providing a low-loss waveguiding of high-peak-power laser pulses (Smith et al. 2003). Hollow PCFs have been recently shown to support megawatt optical solitons (Ouzounov et al. 2003), enhance a variety of nonlinear-optical processes (Benabid et al. 2002; Konorov et al. 2003), and allow the delivery of high-power laser pulses for technological and biomedical applications (Konorov et al. 2004b).

In the experiments reported by Ivanov et al. (2006a), hollow PCFs were used for the delivery and soliton frequency shifting of 2.8-MW femtosecond laser pulses, providing a convenient source of a high-peak-power Stokes field for CARS spectroscopy, imaging, and time-resolved measurements. The laser system used in these experiments consisted of a Cr^{4+}:forsterite master oscillator, a stretcher, an optical isolator, a regenerative amplifier, and a compressor. The master oscillator, pumped with a fiber ytterbium laser, generated 30–60-fs light pulses of radiation with a wavelength of 1.23–1.25 μm at a repetition rate of 120 MHz. These pulses were amplified in a Nd:YLF-laser-pumped amplifier and recompressed to the 170-fs pulse duration with the maximum laser pulse energy up to 30 μJ at 1 kHz. A 1-mm-thick BBO crystal was used for second-harmonic generation. Second-harmonic pulses with an energy ranging from 1 to 1000 nJ and a central wavelength of 618 nm (their spectrum is shown by the dotted line in Fig. 16.4(a)) were coupled into a hollow PCF with a core diameter of 12 μm and a cross-section structure shown in Fig. 16.4(b). The fiber was fabricated from S-93 soft glass and provided a transmission band from 0.60 to 0.66 μm. With no true closed PBG achieved in the considered wavelength range, the leakage loss of the air-guided modes was quite high (3.1 dB/m in the peak of transmission).

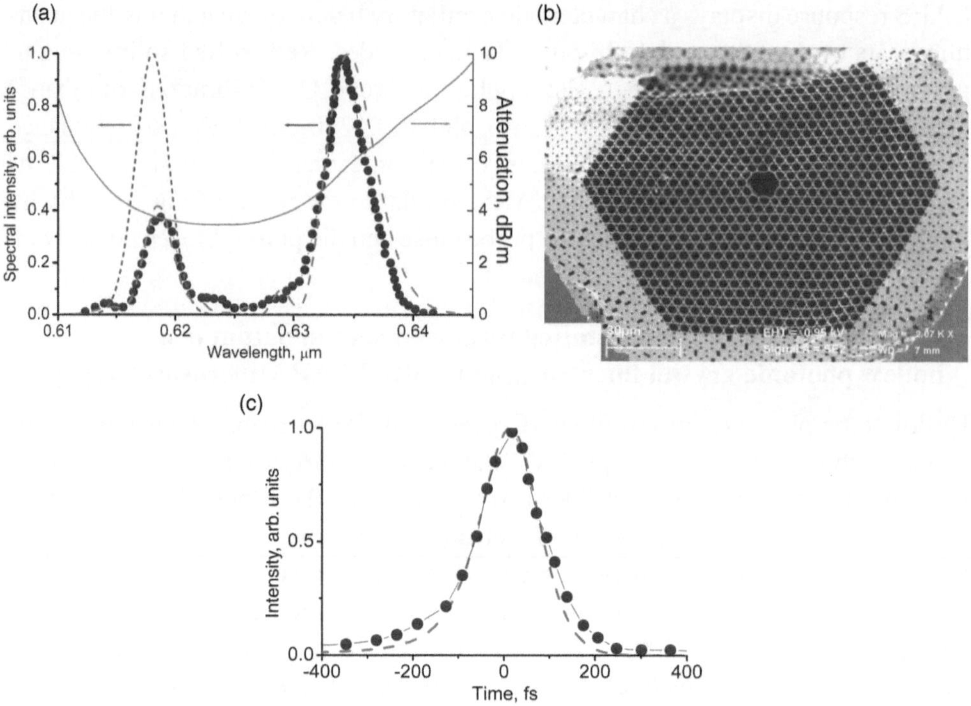

Fig. 16.4. (a) Filled circles show the spectrum of the second-harmonic Cr:forsterite laser pulse transmitted through a hollow PCF (shown in b) with a length of 50 cm filled with atmospheric-pressure air. The input pulse (its spectrum shown by the dotted line) has a central wavelength of 618 nm and an initial energy of 300 nJ. The dashed line represents the numerical solution of the generalized nonlinear Schrödinger equation. Also shown is the wavelength dependence of PCF attenuation. (c) The temporal envelope of the blue-shifted soliton PCF output reconstructed from XFROG (filled circles) and calculated by solving the GNSE (dashed line).

Since the central wavelength of the input light pulses falls in the range of anomalous dispersion (the wavelength of zero group-velocity dispersion (GVD) for our PCF is $\lambda_z \approx 614$ nm), the light pulses tend to generate optical solitons as they propagate through the fiber. For the solitons defined as solutions to the nonlinear Schrödinger equation (NSE), the peak power P_s of the fundamental soliton with a pulse width τ_s is given by $P_s = 0.079\lambda^3 D S_{\text{eff}} \tau_s^{-2}(cn_2)^{-1}$, where D is the GVD, λ is the radiation wavelength, S_{eff} is the effective mode area, c is the speed of light in vacuum, and n_2 is the nonlinear refractive index of the gas filling the core of the PCF. With $\lambda = 618$ nm, $D \approx 70$ ps/(nm · km), $S_{\text{eff}} \approx 100\,\mu\text{m}^2$, $n_2 \approx 3 \cdot 10^{-19}$ cm^2/W (for atmospheric-pressure air), we arrive at an estimate $P_s \approx 1.4$ MW for the peak power of the fundamental NSE soliton with a pulse width $\tau_s \approx 100$ fs.

The spectral and temporal parameters of the PCF output were measured in our experiments by using cross-correlation frequency-resolved optical gating (XFROG). To this end, the pulse transmitted through the PCF was mixed through sum-frequency generation with a 618-nm 90-fs second-harmonic output of the Cr:forsterite laser in a BBO crystal. The spectrum of the PCF output was additionally measured using an Ocean Optics spectrometer. In agreement with the above estimate for the peak power of the fundamental NSE soliton, we observed solitonic phenomena for 90-fs 618-nm input pulses with energies exceeding 180 nJ. Figure 16.4(a) displays a typical spectrum and temporal envelope of the PCF output, measured for an input pulse width of 90 fs and an input pulse energy of 250 nJ. In contrast to low-peak-power input pulses, which tend to spread out in time as they propagate through the fiber due to the fiber dispersion, the soliton PCF output remains localized in the time domain (Fig. 16.4(c)). The pulse width of the PCF output is estimated as 120 fs, and its energy is 130 nJ, corresponding to a peak power of 1.1 MW. The retarded (Raman) part of optical nonlinearity of the gas filling the PCF core down-shifts the soliton frequency – a phenomenon known as the soliton self-frequency shift (SSFS). Similar to solitons in standard, solid-core silica fibers, the SSFS of megawatt solitons is controlled by the peak power of input pulses, as well as by the fiber length and dispersion. In hollow PCFs, the SSFS can be additionally controlled by varying the pressure and the content of the gas filling the fiber core – an option that is unavailable with conventional fiber technologies. We verified the above arguments by numerically solving the generalized nonlinear Schrödinger equation (GNSE) for high-peak-power pulses propagating in a gas-filled hollow PCF. These computations provide a nearly perfect fit for the spectrum (cf. the filled circles and the dashed line in Fig. 16.4(a)) and the temporal envelope (Fig. 16.4(c)) of the hollow-PCF output in our experiments. The high leakage loss of our PCF results in a dissipation of soliton energy, eventually inducing a decay of solitons, thus limiting the attainable soliton frequency shift. We therefore expect that the tunability range of PCF soliton frequency shifters of high-power laser pulses can be further expanded by optimizing the PCF structure for the low-loss waveguiding.

The frequency-shifted megawatt soliton output of the hollow PCF was employed as a Stokes pulse in time-resolved two-color CARS measurements (Fig. 16.5(a)). In two-color CARS, the pump and Stokes fields with frequencies ω_1 and ω_2 coherently excite Raman-active molecular vibrations with a frequency $\Omega = \omega_1 - \omega_2$. The probe field with the frequency ω_1 is then scattered off the coherence induced in the medium by the pump and Stokes fields, giving rise to an anti-Stokes signal at the frequency $\omega_a = 2\omega_1 - \omega_2$. With the 90-fs 618-nm second-harmonic output of the Cr:forsterite laser used as a pump field (ω_1), the frequency shift of megawatt solitons in the hollow PCF was adjusted in such a way as to tune the frequency difference $\omega_1 - \omega_2$ to

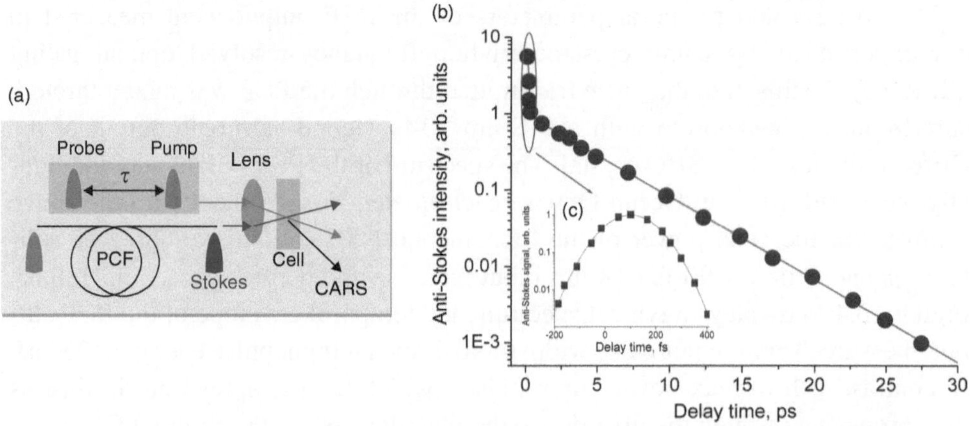

Fig. 16.5. (a) Diagram of the CARS setup. (b) Filled circles represent the intensity of the anti-Stokes signal from a CCl_4 cell measured as a function of the delay time τ between the probe pulse and the pump–Stokes pulse dyad. The solid line shows the $\exp(-2\tau/T_2)$ dependence. (c) The filled squares show a close-up of the $P_a(\tau)$ dependence measured for small τ. The solid line in this inset represents the cross-correlation function for the pump, Stokes, and probe pulses.

a resonance with the Raman-active tetrahedron A_{1g} vibration of carbon tetrachloride molecules ($\Omega/(2\pi c) \approx 459\,\text{cm}^{-1}$). Another 90-fs second-harmonic pulse of the Cr:forsterite laser, applied with a delay time τ with respect to the pump and Stokes pulses, was employed as a probe field in our experiments. In the noncollinear CARS geometry used in our experiments, the anti-Stokes signal can be conveniently separated in space from the pump, Stokes, and probe fields (Fig. 16.5(a)).

The filled circles in Fig. 16.5(b) present the power of the anti-Stokes signal P_a from the liquid-phase CCl_4 sample measured as a function of the delay time τ. For small τ, the $P_a(\tau)$ dependence is dominated by the nonresonant part of the CARS signal. This section of the $P_a(\tau)$ dependence (Fig. 16.5(c)) recovers the cross-correlation trace of the pump, Stokes, and probe pulses. For delay times τ exceeding the pulse widths of the input light fields, the $P_a(\tau)$ dependence visualizes the exponential decay of the Raman-resonant response, originating from the dephasing of coherent vibrations of CCl_4. The high peak power of the Stokes pulse provided by the soliton output of the hollow PCF provides a broad dynamic range of CARS signal detection (about four decades in our experiments), allowing time-resolved CARS measurements within the range of delay times from tens of femtoseconds up to tens of picoseconds. The decay of the coherent response of the 459-cm^{-1} CCl_4 vibration (filled circles in Fig. 16.5(b)) is well approximated by an exponential $\exp(-2\tau/T_2)$ with the phase relaxation time $T_2/2 \approx 4$ ps (the solid line in Fig. 16.5(b)).

Experimental studies thus show that, with hollow-core PCFs, the fiber format of beam delivery becomes fully compatible with the requirements of CARS microspectroscopy. In the experiments presented in this work, hollow PCFs have been used for the delivery and soliton frequency shifting of 2.8-MW femtosecond pulses with an input central wavelength of 618 nm. The frequency-shifted megawatt soliton output of hollow PCFs is shown to be ideally suited as a Stokes field for coherent Raman spectroscopy and imaging, as well as for time-resolved CARS measurements. The high peak power of the Stokes pulse provided by the soliton output of the hollow PCF provides a dynamic range of CARS signal detection of about four decades, allowing time-resolved CARS studies of ultrafast relaxation processes on time scales from tens of femtoseconds up to tens of picoseconds.

As recently demonstrated by Serebryannikov & Zheltikov (2007), ionization phenomena can substantially modify the soliton propagation dynamics of high-peak-power laser pulses in hollow-core PCFs. Numerical solution of the pulse-evolution equation for a high-peak-power laser field in an ionizing gas medium reveals two qualitatively different scenarios of soliton evolution in a hollow PCF controlled by the ionization potential I_p of the gas filling the fiber core. Hollow PCFs filled with high-I_p gases are shown to allow formation of gigawatt soliton features, which remain stable over large propagation distances, with their spectrum undergoing a continuous red shift due to the retarded nonlinearity of the fiber cladding. In hollow PCFs filled with low-I_p gases, the ionization-induced change in the refractive index of the gas leads to a blue shifting of soliton transients, pushing their spectrum beyond the point of zero group-velocity dispersion, thus preventing formation of stable high-peak-power solitons. Hollow waveguides capable of providing soliton transmission regimes for gigawatt laser pulses suggest attractive solutions for long-distance transmission of high-power optical signals, creation of wavelength-tunable sources of high-peak-power ultrashort light pulses, fiber-format beam delivery in materials microprocessing, as well as the development of fiber endoscopes and fiber components for laser surgery, vessel photodisruption, laser ophthalmology, and optical histology.

Recent experiments (Fedotov et al. 2007) have demonstrated that the ionization-induced change in the refractive index of a gas can substantially blue-shift megawatt light pulses transmitted through hollow PCFs. Given the type of a fiber and the sort of gas filling the fiber core, the sign of the frequency shift of the laser field and its rate can thus be controlled by the input laser peak power, offering attractive solutions for the development of high-peak-power wavelength-tunable fiber-format sources of ultrashort pulses, as well as for the processing and regeneration of high-peak-power optical signals.

16.4 Nonlinear Raman microspectroscopy of silicon components

Along with bioimaging applications, CARS microspectroscopy offers much promise as an imaging technique for silicon electronic and silicon photonic chip components. Experiments reported by Mitrokhin et al. (2007) demonstrate CARS microscopy of silicon wafers using stretched pulses of a mode-locked Cr:forsterite laser as a pump and the frequency-shifted soliton output of a photonic-crystal fiber as a Stokes field. The laser system used in these experiments (Fig. 16.6(a)) consisted of a Cr^{4+}:forsterite master oscillator, a stretcher, an optical isolator, a regenerative amplifier, and a compressor. The master oscillator, pumped with an ytterbium fiber laser, generated 30–60-fs light pulses of radiation with a central wavelength of 1.25 μm at a repetition rate of 120 MHz. These pulses were transmitted through a stretcher and an isolator, to be amplified in a Nd:YLF-laser-pumped amplifier and recompressed to a variable pulse width of 100–800 fs with a maximum pulse energy up to 30 μJ at 1 kHz.

The amplified Cr:forsterite laser output was split into two beams (Fig. 16.6(a)). The first beam provided pump and probe fields for two-color CARS experiments (see Fig. 16.6(b)), while the second beam was launched in a silica PCF with a mode area of about 280 μm^2 (see Fig. 16.7(a)) in the region of anomalous dispersion to undergo a soliton self-frequency shift. The dispersion, nonlinearity, and length of the PCF were matched with input laser-pulse parameters in such a way as to deliver

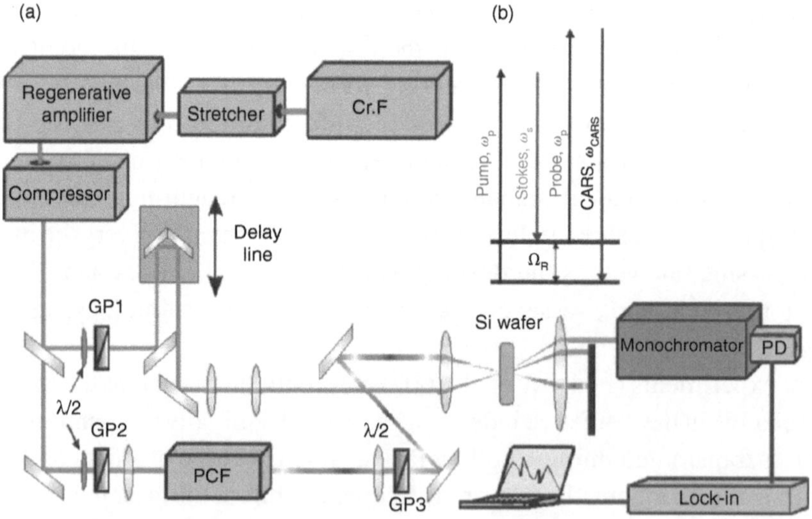

Fig. 16.6. (a) Diagram of the experimental setup: Cr:F, Cr:forsterite laser; PCF, photonic-crystal fiber; GP1–GP3, Glan prisms; $\lambda/2$, half-wave plates; PD, photodetector. (b) Diagram of a two-color CARS process with $\omega_{\text{CARS}} = 2\omega_p - \omega_s = \omega_p + \Omega_{\text{R}}$.

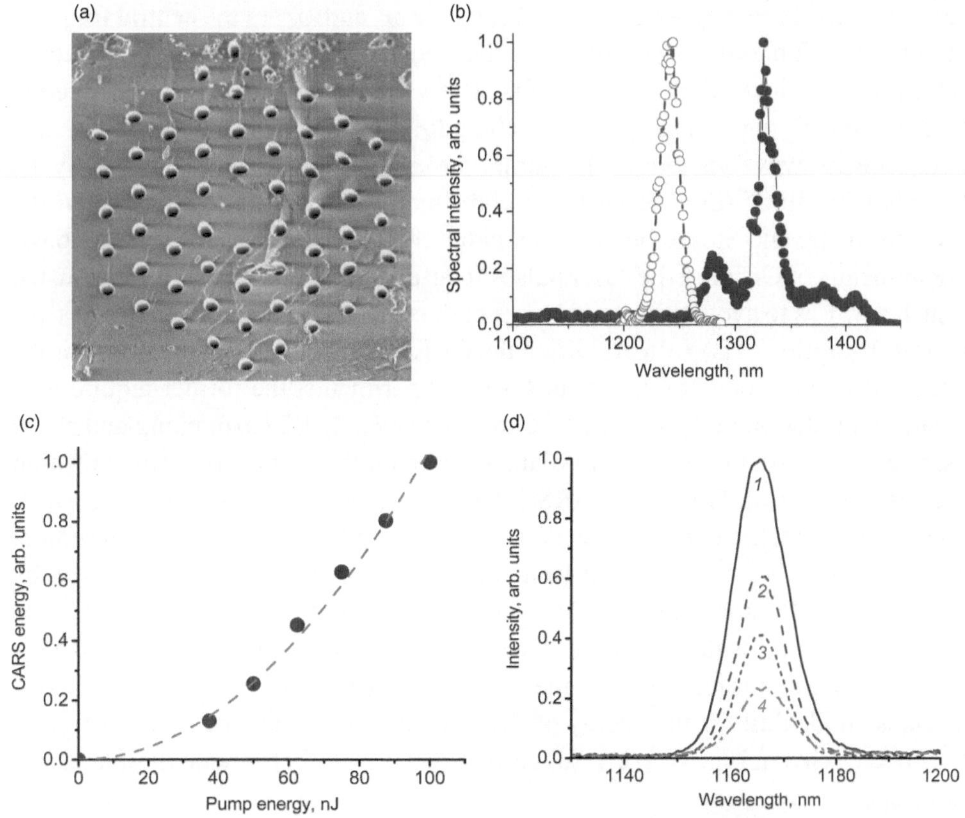

Fig. 16.7. (a) Cross-section image of the PCF. (b) Spectra of light field used for CARS experiments: (open circles) Cr:forsterite laser pulse and (filled circles) PCF output filtered by reflection off a diffraction grating. (c) Energy of the CARS signal measured as a function of the energy of Cr:forsterite laser pulses (filled circles) and the quadratic fit for this dependence (dashed curve). (d) Spectra of the CARS signal from a 1.5-mm-thick silicon wafer measured with Cr:forsterite-laser and PCF-output beams focused on the wafer surface (curve 1) and with a separation between the beam focus and the wafer surface set equal to 300 μm (2), 450 μm (3), 600 μm (4).

a soliton with a central wavelength λ_s (Fig. 16.7(b)) meeting the condition $\lambda_s^{-1} = \lambda_p^{-1} - \Omega_R$, necessary for a Raman-resonant excitation of a longitudinal optical (LO) phonon mode of silicon with the frequency $\Omega_R = 520\,\text{cm}^{-1}$ (Fig. 16.6(b)). The soliton PCF output at this wavelength had a pulse width of 300 fs and an energy of about 10 nJ. The efficiency of energy conversion from the input Cr:forsterite laser pulse to the frequency-shifted soliton was about 15% for the chosen PCF.

The Cr:forsterite-laser and PCF-output beams were focused into a spot with a diameter $d \approx 10\,\mu$m, giving rise to a CARS signal from a silicon wafer at the frequency $\omega_{CARS} = 2\omega_p - \omega_s = \omega_p + \Omega_R$, where ω_p and ω_s are the central frequencies of the Cr:forsterite laser pulses and the frequency-shifted soliton PCF output. In the folded-beam CARS geometry (Fig. 16.6), the CARS signal was generated in the direction of noncollinear phase matching $\mathbf{k}_{CARS} = 2\mathbf{k}_p - \mathbf{k}_s$, where $\mathbf{k}_p$, $\mathbf{k}_s$, and $\mathbf{k}_{CARS}$ are the wave vectors of the pump, Stokes, and CARS fields, respectively, thus allowing the CARS signal to be conveniently separated in space from the powerful pump and Stokes beams transmitted through a silicon wafer (Fig. 16.6).

Parameters of Cr:forsterite-laser pulses focused on a silicon wafer were adjusted in such a way as to avoid the laser damage of the silicon wafer and to minimize two-photon absorption (TPA) effects within the CARS beam-interaction region. For the 600-fs pulse duration chosen for our CARS experiments, the former requirement dictates a limitation of $E_p < 400$ nJ for the energy E_p of 1.24-μm pump and probe laser beams focused into a 10-μm-diameter spot on the surface of a silicon wafer. The requirement of TPA-free CARS interaction is expressed as $\beta Il < 1$, where β is the TPA coefficient, I is the field intensity, and l is the interaction length. With $\beta \approx 4.5 \cdot 10^{-10}$ cm/W for silicon and the CARS interaction length given by $l \approx d/\sin\varphi \approx 100\,\mu$m, where $\varphi \approx 5°$ is the phase-matching angle between the pump and Stokes beams inside the sample, we arrive at a limitation $E_p < 100$ nJ. For laser energies meeting this requirement, the scaling $\propto E_p^2$ (the dashed line in Fig. 16.7(c)) provides an ideal fit for the energy of the CARS signal measured as a function of E_p (filled circles in Fig. 16.7(c)), suggesting that TPA effects are negligible in this range of laser energies.

Figure 16.7(d) displays typical spectra of the CARS signal measured for different separations of the beam focus from the wafer surface. Because of a broadband probe used in our experiments, CARS spectra, as can be seen from Fig. 16.7(d), are substantially broader than the 520-cm^{-1} LO phonon Raman line, whose bandwidth is about 3.5 cm^{-1}. With the confocal parameter of focused beams estimated as 500 μm and the CARS beam interaction area estimated as 100 μm for the conditions of our experiments, a wafer displacement by 150 μm could be easily detected by measuring the CARS intensity as a function of the separation between the beam focus and the wafer surface (Fig. 16.7(d)).

For a CARS microspectrometry experiment, the CARS signal was measured in a transmission mode from a silicon wafer placed on a translation stage, allowing the incident light beams to be scanned over the surface of the sample. The shades produced by micron-size defects on silicon wafers are readily resolved in CARS scans. Uncertainties in the edges of the image result from the limited spatial resolution of the implemented CARS imaging scheme, controlled by the diameter of the focused light beams on the surface of the sample. With the spectral resolution of CARS improved to resolve the lineshape of the Raman response of the LO phonon mode in silicon, the CARS technique should allow an extension to on-line carrier generation detection in silicon structures, offering attractive options for on-line diagnostics of electronic and photonic silicon chip components and circuits in the process of operation.

16.5 Photonic-crystal fiber components for time-resolved studies of ultrafast molecular dynamics

In an experiment reported by Ivanov et al. (2006b), wavelength-tunable 100-fs pulses generated through the soliton self-frequency shift in a photonic-crystal fiber have been employed to visualize femtosecond coherence and population relaxation dynamics in molecular aggregates by means of time-resolved sum-frequency generation. The laser system used in those experiments consisted of a Cr^{4+}:forsterite master oscillator, a stretcher, an optical isolator, a regenerative amplifier, and a compressor. The master oscillator, pumped with an ytterbium fiber laser, generated 30-fs light pulses of 1.24-μm radiation at a repetition rate of 120 MHz. These pulses were amplified in a Nd:YLF-laser-pumped amplifier and recompressed to the 100-fs pulse duration with the maximum laser pulse energy up to 40 μJ at 1 kHz.

The Cr:forsterite laser system included four channels delivering pump and probe pulses for time-resolved nonlinear-optical measurements. The first channel generated amplified pulses of 1.24-μm Cr:forsterite laser radiation (curve 1 in Fig. 16.8(a)) with a pulse energy up to 30 μJ and a pulse width $\theta_1 \approx 100$ fs. In the second channel, 1.24-μm amplified 100-fs Cr:forsterite laser pulses were coupled

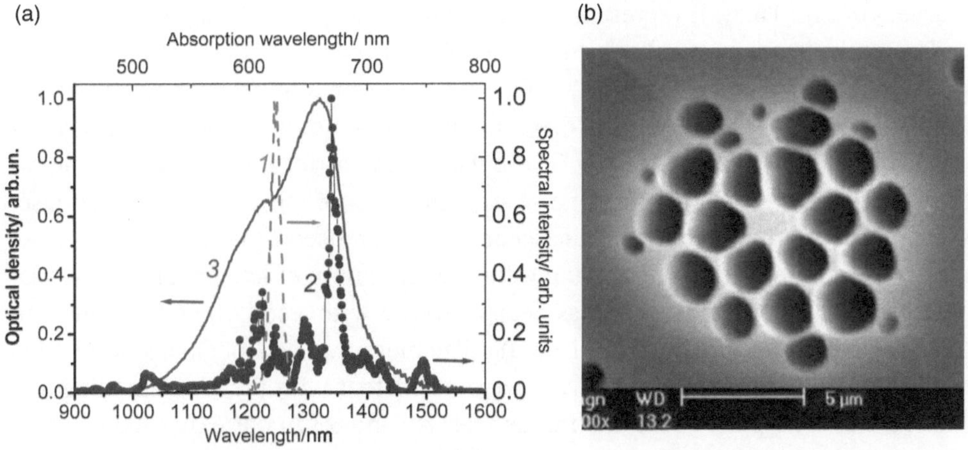

Fig. 16.8. (a) The spectrum of the amplified fundamental-wavelength output of the Cr: forsterite laser (1), the spectrum of the frequency-shifted soliton output of the first-type photonic-crystal fiber (2), and the absorption spectrum of a thin-film sample of molecular aggregates (3). (b) An SEM image of the PCF.

into a fused silica PCF with a core diameter of about 1.7 μm (see Fig. 16.8(b)), providing zero group-velocity dispersion at 730 nm. The laser pulses coupled into the PCF form solitons, which experience a continuous red shifting induced by the retarded fiber nonlinearity. This soliton frequency shift can be tuned by varying the input pulse energy and by changing the fiber length. A soliton with a pulse width $\theta_2 \approx 100$ fs and a central wavelength of 1.33 μm was generated using the PCF frequency shifter (curve 2 in Fig. 16.8(a)) to provide a two-photon-absorption-resonant excitation of molecules studied in our experiments. In the third channel, unamplified Cr:forsterite laser pulses with a pulse energy of 5 nJ and pulse width $\theta_p \approx 30$ fs passed through an adjustable optical delay line. Finally, in the fourth channel a 1-mm-thick BBO crystal was used to generate second-harmonic pulses with a central wavelength of 620 nm, pulse width of about 90 fs, and energy ranging from 10 to 80 nJ.

Experiments were performed with thin-film samples of J-type molecular aggregates of thiacarbocyanine dye. The samples were prepared by spin-coating a $5 \cdot 10^{-3}$ mol/l solution of this dye in a 2:2:1 mixture of acetonitrile, dichloroethane, and chloroform on a 1-μm-thick substrate. The thickness of the dye layer on the substrate was measured by the ellipsometric technique and was estimated as 30 nm. The absorption coefficient of the aggregate film at the wavelength of 633 nm was 10^5 cm^{-1}. Interactions between molecules in an aggregate give rise to collective electronic states delocalized over large chains of molecules, resulting in the formation of exciton energy bands (Kobayashi 1996). Broad absorption bands, typical of isolated dye molecules, display narrowing and spectral shifting in the aggregate state. A well-pronounced peak at 660 nm in the absorption spectrum of our samples (curve 3 in Fig. 16.8(a)) represents the excitonic absorption of J aggregates.

The method of time-resolved studies of coherent excitations in molecular aggregates adopted by Ivanov et al. (2006b) was based on sum-frequency generation (SFG) $\omega_{sf} = 2\omega_1 + \omega_2$ (see Fig. 16.9(a)) with a two-photon-absorption (TPA) resonance at the frequency $\omega_1 + \omega_2$. The pulses with carrier frequencies ω_1 and ω_2 have been delivered by the first and second channels of the above-described laser system. Two-photon-excited luminescence has been used to find the optimal frequency ω_2 of the solitonic PCF output for the maximum efficiency of the $\omega_1 + \omega_2$ TPA process. Such a maximum has been achieved with a soliton PCF output centered at 1.33 μm (curve 2 in Fig. 16.8(a)). The measured dependencies of the SFG signal on the excitation laser powers were consistent with the diagram shown in Fig. 16.9(a). The pulses with the frequencies ω_1 and ω_2 in our experiments could also reach the wing of the TPA spectrum of thiacarbocyanine monomers. However, due to the collective enhancement of optical nonlinearity in aggregates (Spano & Mukamel 1989), the SFG signal from aggregates in our experiments was at least an order of magnitude higher than the SFG signal from the monomers.

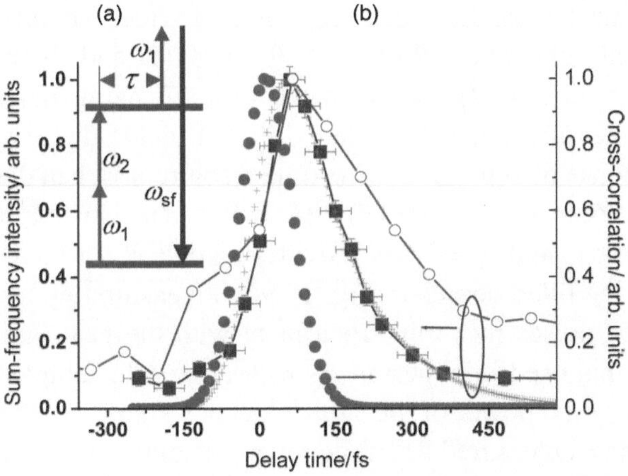

Fig. 16.9. Time-resolved SFG measurements with a solitonic output of the PCF. The filled circles show the power of the sum-frequency signal generated in a silica plate by amplified Cr:forsterite laser pulses, frequency-shifted pulses from the photonic-crystal fiber, and time-delayed unamplified pulses of the Cr:forsterite laser measured as a function of the delay time between the pulses (cross-correlation trace). Rectangles present the power P_{sf} of the sum-frequency signal from a molecular aggregate film measured as a function of the delay time τ between the probe pulse and the pulses providing a TPA-resonant coherent excitation of aggregates. Crosses show the fit of the experimental data with the use of expressions for P_{sf} and Q with the dephasing time $T_2 = 120\,\text{fs}$ and pulse widths $\theta_1 = \theta_2 = 100\,\text{fs}$. Open circles represent the inverted dependence of the sum-frequency signal from a molecular aggregate film on the delay time between the excitation second-harmonic pulse and the light pulses involved in SFG. The inset presents a diagram of SFG $\omega_{sf} = 2\omega_1 + \omega_2$ with a TPA resonance at $\omega_1 + \omega_2$ and a time-delayed probe pulse.

Time-delayed unamplified Cr:forsterite laser pulses with a pulse width $\theta_p \approx 30\,\text{fs}$, delivered by the third channel, served to probe coherent excitations driven by the first two pulses. The power of the sum-frequency signal P_{sf} measured as a function of the delay time τ between the probe pulse and the pulses driving the coherent excitation of aggregates is then given by $P_{sf}(\tau) \propto \gamma^2 L^2 \int_{-\infty}^{\infty} E_p^2(\xi - \tau) Q^2(\xi) d\xi$, where γ is the relevant nonlinear coefficient, L is the interaction length, $E_p(t)$ is the temporal envelope of the probe pulse, and Q is the amplitude of coherence induced in the medium by the TPA-resonant two-color field with central frequencies ω_1 and ω_2. When the TPA-resonant field consists of two pulses with temporal envelopes $E_1(t)$ and $E_2(t)$, the amplitude Q can be represented as $Q(t) = \int_0^{\infty} \exp(-\eta/T_2) E_1(t-\eta) E_2(t-\eta) d\eta$, where T_2 is the phase relaxation time. For two-photon-excited levels of aggregates accessed in our experiments, the main dephasing mechanism typically involves rapid intraband downward population relaxation towards the bottom of the exciton band.

In our experiments, the time-delayed pulse provides an ultrafast probe for coherently excited aggregates. With $\theta_1 \approx \theta_2 \approx 100\,\text{fs}$ and $\theta_3 \approx 30\,\text{fs}$, we have $(\theta_p/\theta_{1,2})^2 \approx 0.09$. In this regime, with the time T_2 being much less than the pulse width θ_p, measuring the dependence $P_{sf}(\tau)$ yields the temporal profile of the coherent response of aggregates. The time resolution is then determined by the cross-correlation $Y(\tau) = \int_{-\infty}^{\infty} E_1(\eta)E_2(\eta)E_p(\eta - \tau)d\eta$. This expression for $Y(\tau)$ with $\theta_1 = \theta_2 = 100\,\text{fs}$ and $\theta_p = 30\,\text{fs}$ provides a good fit for the cross-correlation trace (shown by the filled circles in Fig. 16.9(b)) measured by mixing the pump, Stokes, and probe pulses in a thin silica plate with the beam interaction geometry identical to that used in experiments with aggregate samples. Rectangles in Fig. 16.9(b) present the power of the sum-frequency signal P_{sf} from the film with molecular aggregates measured as a function of the delay time τ. A nearly perfect fit for the measured $P_{sf}(\tau)$ dependence is obtained by using the above expressions for P_{sf} and Q with $\theta_1 = \theta_2 = 100\,\text{fs}$ and the phase relaxation time $T_2 = 120\,\text{fs}$ (crosses in Fig. 16.9(b)). Notably, this phase relaxation time is much shorter than the typical population relaxation time T_{ex} of one-exciton states of aggregates. To compare T_2 and T_{ex} for our sample, we measured nonlinear absorption spectra of molecular aggregates excited by the second-harmonic output of the Cr:forsterite laser, providing a resonant population transfer from the ground state to the one-exciton band. The supercontinuum output of PCF was then applied with a variable delay time δt relative to the second-harmonic pulse. Differential absorption spectra measured in these experiments display well-pronounced minima at 665 nm, indicative of bleaching through pump-induced transitions between the ground state and the one-exciton band, and blue-shifted peaks at 640 nm, originating from induced absorption due to transitions between one- and two-exciton bands. The amplitudes of induced-absorption and bleaching peaks in nonlinear absorption spectra decrease with the increase in the delay time δt, allowing the relaxation time for one-exciton states of aggregates to be estimated as $T_{ex} \approx 770\,\text{fs}$.

To demonstrate an ultrafast switching of optical nonlinearity of molecular aggregates, we used the above-described SFG process $\omega_{sf} = 2\omega_1 + \omega_2$ with the delay time τ chosen in such a way as to provide the maximum SFG efficiency. The nonlinear-optical response of the aggregate sample was modified by applying a 90-fs pulse of 620-nm second-harmonic output of the Cr:forsterite laser, depopulating the ground state of aggregates. Open circles in Fig. 16.9(b) display the inverted dependence of the sum-frequency signal on the delay time Δt between the excitation second-harmonic pulse and the light pulses involved in SFG. The maximum in this trace corresponds to the minimum in the sum-frequency signal, and the background corresponding to large $|\Delta t|$ has been subtracted. For small Δt, as can be seen from this dependence, the resonant excitation of aggregates substantially reduces the sum-frequency signal. As Δt increases, the initial level of the sum-frequency signal (corresponding to large negative Δt) is recovered. A monoexponential time dependence fails to provide an adequate fit for the experimentally measured trace. However, the results presented in Fig. 16.9(b) demonstrate a 75% recovery of the sum-frequency signal within a characteristic delay time of about 450 fs.

16.6 Enhancement of guided-wave two-photon-excited luminescence response with a photonic-crystal fiber

Two-photon absorption (TPA) is the backbone of a broad variety of bioimaging techniques and biomedical strategies, including high-resolution microscopy (Denk et al. 1990; Zipfel et al. 2003), photodynamic therapy (Bhawalkar et al. 1997), as well as drug-delivery monitoring and drug activation (Fisher et al. 1997). Luminescence of TPA-excited molecules provides a measurable optical response that enables imaging of TPA-excited regions in two-photon microscopy and helps to visualize tiny objects and identify fine details of structure and morphology inside biological tissues (Denk et al. 1990; Zipfel et al. 2003). A combination of TPA approaches with capabilities of fiber-optic probes (Flusberg et al. 2005b) offers numerous advantages, suggesting a convenient format for beam delivery, facilitating manipulation of excitation radiation, and allowing this excitation to be applied locally and selectively. While many attractive fiber solutions have been developed to meet the needs of TPA-based technologies, collection of the two-photon-excited luminescence (TPL) response still leaves much room for optimization, with the main source of problems being related to a small aperture of fiber probes, making it difficult to efficiently collect the TPL signal, which is uniformly emitted in a 4π solid angle.

Photonic-crystal fibers (PCFs) can substantially improve the collection efficiency of the TPL response from TPA-excited molecules in waveguide modes, thus allowing the sensitivity of TPA-based techniques to be radically improved. In the earlier relevant work, PCFs of various types have been shown to offer numerous advantages as efficient optical sensors (Monro et al. 2001; Jensen et al. 2004). Dual-cladding PCFs have been employed to enhance the efficiency of two-photon biosensing (Myaing et al. 2003).

As shown in a recent work (Doronina et al. 2009a), with only a few nanoliters of TPA-excited molecules filling air holes in a specifically designed PCF, the guided-wave TPL signal can be enhanced by several orders of magnitude relative to the maximum TPL signal attainable from a cell with the same dye excited and collected by the same PCF. We will also discuss biophotonic implications of this waveguide TPL-response enhancement in a PCF, which include fiber-format solutions for online monitoring of drug delivery and drug activation, visualization and initiation of neural activity, biosensing, as well as locally controlled singlet oxygen generation in photodynamic therapy.

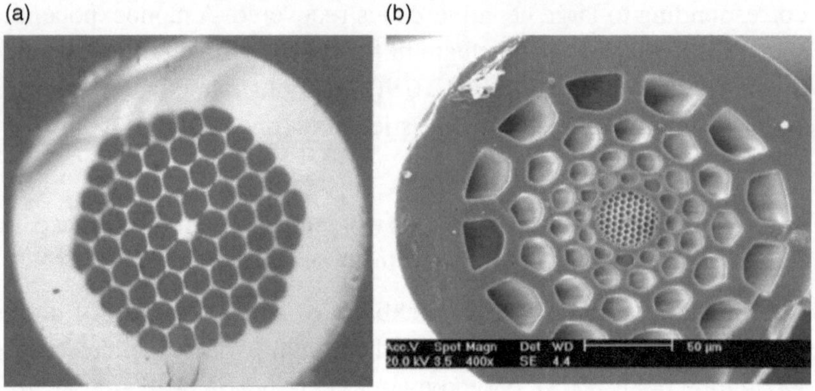

Fig. 16.10. SEM images of photonic-crystal fibers used in experiments on two-photon-excited luminescence.

In experiments, the TPL response is collected using a dual-cladding fiber probe. Two different PCF versions of such a fiber are shown in Fig. 16.10(a and b). The dual-cladding fiber includes a core and a two-section cladding. The fiber core is intended to deliver excitation light to the interaction region, while the inner part of the cladding is used to collect the luminescent response within a high numerical aperture, bounded by the outer, large-diameter part of the cladding. Such dual-cladding fibers have been earlier shown to offer attractive solutions for fiber-based imaging and sensing applications (Myaing et al. 2003).

We compare two arrangements for the collection of the TPL response with a dual-cladding fiber. In the first scheme, excitation light is delivered along the fiber core to an object adjacent to the fiber end. The TPL signal is then collected by the inner part of the fiber cladding. As shown by Myaing et al. (2003), this method of TPA excitation using dual-cladding PCF probes offers important advantages for TPA bioimaging. In the second approach, we let a tiny amount of the liquid- or gas-phase substance fill the air holes of the PCF. The TPL signal is then excited by the evanescent field of the mode guided along the fiber core and is captured into a waveguide mode inside the inner part of the fiber cladding.

A Ti:sapphire laser, delivering 40-fs, 0.5-W pulses of 805-nm radiation at a pulse repetition rate of 90 MHz, was employed as a source of light inducing two-photon luminescence in solutions of various dyes. Light pulses with an average power of approximately 50 mW were coupled into the fiber probe with a micro-objective. Because of the nonlinear nature of the TPA process, efficient TPL generation requires excitation of molecules by light pulses with a sufficiently high peak power. It is therefore critical for highly efficient TPL excitation to minimize temporal spreading of femtosecond light pulses transmitted through the fiber. To this end, the PCF was designed in such a way as to provide a low group-velocity dispersion at the central laser wavelength. In experiments, a laser pulse with an input pulse width of 40 fs transmitted through a 50-cm-long piece of PCF with a GVD parameter $\beta_2 \approx 5 \cdot 10^{-3}$ ps^2/m was stretched up to 75 fs. This pulse width was still short enough to provide a sufficient laser peak power for efficient TPA excitation of rhodamine 6G molecules.

In the first scheme of TPL detection, laser pulses were delivered to a solution of rhodamine 6G in a cell through the central core of the fiber with a diameter of about 2 μm. The inner fiber cladding with a diameter of 2.3 μm serves to collect the TPL response from the sample and to guide it in the backward direction, toward the input end of the fiber. Outside the fiber, the TPL signal is separated from the 800-nm light and is sent to a detection system by a dichroic mirror. In this regime, the TPL signal is reliably detectable, but is much weaker than the background signal in the spectral range of 690–810 nm (the dashed line in Fig. 16.11(a)), which includes back-reflected laser light centered at 805 nm and the nonlinear signal in the spectral region of 700–715 nm generated in the PCF through dispersive-wave emission by optical solitons (Herrmann et al. 2002). The ratio of the amplitude of the TPL peak in the spectrum of optical response shown by the dashed line in Fig. 16.11(a) to the amplitude of the peak corresponding to the back-reflected laser light at 805 nm is $\xi_1 \approx 0.04$.

In the second scheme of TPL detection, a PCF is brought into contact with a liquid-phase sample so that a small amount of liquid fills the air holes in the fiber

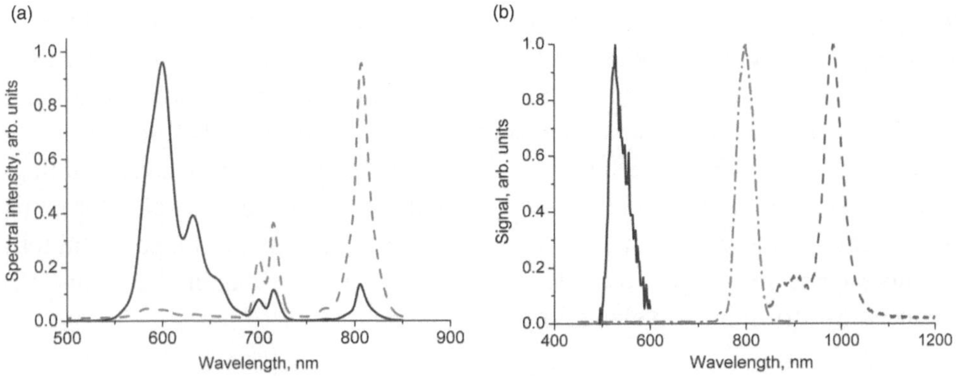

Fig. 16.11. TPA excitation and TPL response collection with a dual-cladding fiber probe: (a) excitation light is delivered through the fiber core to an object adjacent to the fiber end, while the TPL signal is collected by a high-numerical-aperture inner part of the fiber cladding; (b) an evanescent field of the mode guided along the fiber core generates the TPL signal through a TPA excitation of dye molecules in a solution filling the air holes in the PCF cladding, with the inner part of the fiber cladding capturing the TPL response into a waveguide mode and delivering this signal to a detection system.

cladding due to sample flows induced by surface tension. Dye molecules in the liquid filling the air holes in the fiber cladding are interrogated with the evanescent field of PCF waveguide modes. The TPL signal generated by dye molecules is detected in the backward direction using a dichroic mirror, a monochromator, and a photomultiplier. With such an experimental arrangement, the ratio of the amplitude of the TPL peak in the spectrum of an optical response from rhodamine 6G dye penetrating into the air holes inside the PCF over $L \approx 1.5\,\mathrm{cm}$ is $\xi_2 \approx 6.3$ times higher than the intensity of the back-reflected laser signal at 805 nm (the solid line in Fig. 16.11(a)). An overall volume of interrogated dye solution in this experiment is about 4 nanoliters. Using the back-reflected signal at 805 nm as a reference, we then find that the enhancement of the TPL response provided by the second regime of TPL fiber probing is $\eta = \xi_2/\xi_1 \approx 160$.

While a fiber probe where air holes are partially filled with a dye solution allows a substantial enhancement of the TPL response relative to the maximum TPL signal attainable from a cell with the same dye excited and collected by the same fiber probe, both regimes of fiber-based TPA excitation considered here offer significant advantages for bioimaging, including a high-precision positioning of the excitation laser beam on a sample and collection of the TPL signal within a high-numerical-aperture fiber cladding section, which helps to improve the detection sensitivity of the method.

In a related experiment, Doronina et al. (2009b) have employed a highly nonlinear PCF to shift the central wavelength of Ti:sapphire laser pulses (790 nm) to an output wavelength of 975 nm in order to provide the maximum efficiency for the TPA excitation of a microdroplet of an AlexaFluor® 488 dye (Fig. 16.11(b)). The PCF was designed in such a way as to keep the frequency-shifted soliton shorter than 80 fs, which helps to enhance the TPA process in a dye microdroplet. The inner part of the cladding of the same PCF is then used to deliver the TPL response to a detector. With AlexaFluor 488 dye widely used as tissue and cell labels, enabling, in particular, high-contrast, high-brightness brain imaging (Lewis et al. 2006), this experiment demonstrates that PCF soliton frequency shifters offer a powerful tool for fluorescence microspectroscopy, cell biology, and brain imaging based on TPL enhanced by ultrashort wavelength-tunable pulses.

16.7 Conclusion

We have shown in this chapter that PCFs offer an attractive platform for the development of the key components for nonlinear-optical microspectroscopy. Enhanced nonlinear-optical processes in solid- and hollow-core PCFs enable the generation of frequency-tunable radiation within the wavelength range stretching from the ultraviolet to the near-infrared with output pulse widths ranging from tens of femtoseconds to a few picoseconds and peak powers from a few watts to several megawatts, suggesting a convenient, efficient, and compact fiber format of short-pulse sources for time-resolved nonlinear spectroscopy and microscopy. Modulation-instability-type four-wave mixing in highly nonlinear PCFs can generate optimally pulse-shaped field waveforms for coherence-controlled nonlinear Raman scattering, including single-beam coherent anti-Stokes Raman scattering microspectroscopy. Hollow-core PCFs allow a substantial enhancement of CARS, providing an improved sensitivity of gas-phase sensing, spectroscopy and coherent control. PCFs can also substantially increase the guided-wave luminescent response from molecules excited through two-photon absorption by femtosecond near-infrared laser pulses. With only a few nanoliters of TPA-excited molecules filling air holes in a specifically designed PCF, the guided-wave two-photon-excited luminescence signal is enhanced by more than two orders of magnitude relative to the maximum TPL signal attainable from a cell with the same dye excited and collected by the same PCF. Biophotonic implications of this waveguide TPL-response enhancement include fiber-format solutions for on-line monitoring of drug delivery and drug activation, interrogation of neural activity, biosensing, endoscopy, as well as locally controlled singlet oxygen generation in photodynamic therapy. Photonic-crystal fiber soliton frequency shifters have been shown to offer a powerful tool for neural activity imaging in brain using TPL enhanced by ultrashort wavelength-tunable pulses.

References

Agrawal, GP 2001, *Nonlinear Fiber Optics* (Academic, San Diego).

Akimov, DA, Serebryannikov, EE, Zheltikov, AM, Schmitt, M, Maksimenka, R, Kiefer, W, Dukel'skii, KV, Shevandin, VS & Kondrat'ev, YuN 2003, Efficient anti-Stokes generation through phase-matched four-wave mixing in higher-order modes of a microstructure fiber, *Opt. Lett.* **28**, 1948.

Andresen, ER, Paulsen, HN, Birkedal, V, Thøgersen, J & Keiding, SR 2005, Broadband multiplex coherent anti-Stokes Raman scattering microscopy employing photonic-crystal fibers, *J. Opt. Soc. Am. B* **22**, 1934–1938.

Benabid, F, Knight, JC, Antonopoulos, G & Russell, PStJ 2002, Stimulated Raman scattering in hydrogen-filled hollow-core photonic crystal fiber, *Science* **298**, 399–402.

Bhawalkar, JD, Kumar, ND, Zhao, C-F & Prasad, PN 1997, Two-photon photodynamic therapy, *J. Clin. Lasers Med. Surg.* **15**, 201–204.

Cregan, RF, Mangan, BJ, Knight, JC, Birks, TA, Russell, PStJ, Roberts, PJ & Allan, DC 1999, Single-mode photonic band gap guidance of light in air, *Science* **285**, 1537–1539.

Denk, W, Strickler, JH & Webb, WW 1990, Two-photon laser scanning fluorescence microscopy, *Science* **248**, 73–76.

Doronina, LV, Fedotov, IV, Ivashkina, OI, Zots, MA, Anokhin, KV, Mikhailova, YuM, Lanin, AA, Fedotov, AB, Shneider, MN, Miles, RB, Sokolov, AV, Scully, MO & Zheltikov, AM 2009a, The second law of thermodynamics, Maxwell's demons, photonic-crystal fibers, and optimal waveguide solutions for biophotonics, in Proceedings of the 18th International Workshop on Laser Physics (Barcelona, 2009).

Doronina, LV, Fedotov, IV, Voronin, AA, Ivashkina, OI, Zots, MA, Anokhin, KV, Rostova, E, Fedotov, AB & Zheltikov, AM 2009b, Tailoring the soliton output of a photonic crystal fiber for enhanced two-photon excited luminescence response from fluorescent protein biomarkers and neuron activity reporters, *Opt. Lett.* **34**, 3373–3375.

Dudley, JM, Genty, G & Coen, S 2006, Supercontinuum generation in photonic crystal fiber, *Rev. Mod. Phys.* **78**, 1135–1176.

Dudovich, N, Oron, D & Silberberg, Y 2002, Single-pulse coherently-controlled nonlinear Raman spectroscopy and microscopy, *Nature* **418**, 512–514.

Eckbreth, AC 1988, *Laser Diagnostics for Combustion Temperature and Species* (Abacus, Cambridge, MA).

Eesley, GL 1981, *Coherent Raman Spectroscopy* (Pergamon, Oxford).

Evans, CL, Potma, EO, Puoris'haag, M, Côté, D, Lin, CP & Xie, XS 2005, Molecular imaging of tissue in vivo with video-rate coherent anti-Stokes Raman scattering (CARS) microscopy, *Proc. Natl. Acad. Sci.* **102**, 16807–16812.

Fedotov AB, Zheltikov AM, Tarasevitch, AP & von der Linde, D 2001, Enhanced spectral broadening of short laser pulses in high-numerical-aperture holey fibers, *Appl. Phys. B* **73**, 181–184.

Fedotov, AB, Serebryannikov, EE & Zheltikov, AM 2007, Ionization-induced blue shift of high-peak-power guided-wave ultrashort laser pulses in hollow-core photonic-crystal fibers, *Phys. Rev. A* **76**, 053811.

Fisher, WG, Partridge, Jr, WP, Dees, C & Wachter, EA 1997, Simultaneous two-photon activation of type-I photodynamic therapy agents, *Photochem. Photobiol.* **66**, 141–155.

Flusberg, B, Jung, J, Cocker, E, Anderson, E & Schnitzer, M 2005a, In vivo brain imaging using a portable 3.9 gram two-photon fluorescence microendoscope, *Opt. Lett.* **30**, 2272–2274.

Flusberg, BA, Cocker, ED, Piyawattanametha, W, Jung, JC, Cheung, ELM & Schnitzer, MJ 2005b, Fiber-optic fluorescence imaging, *Nature Methods* **2**, 941–950.

Hartl, I, Li, XD, Chudoba, C, Rhanta, RK, Ko, TH, Fujimoto, JG, Ranka, JK & Windeler, RS 2001, Ultrahigh-resolution optical coherence tomography using continuum generation in an air–silica microstructure optical fiber, *Opt. Lett.* **26**, 608–610.

Herrmann, J, Griebner, U, Zhavoronkov, N, Husakou, A, Nickel, D, Knight, JC, Wadsworth, WJ, Russell, PStJ & Korn, G. 2002, Experimental evidence for supercontinuum generation by fission of higher-order solitons in photonic fibers, *Phys. Rev. Lett.* **88**, 173901.

Ivanov, AA, Podshivalov, AA & Zheltikov, AM 2006a, Frequency-shifted megawatt soliton output of a hollow photonic-crystal fiber for time-resolved coherent anti-Stokes Raman scattering microspectroscopy, *Opt. Lett.* **31**, 3318–3320.

Ivanov, AA, Alfimov, MV & Zheltikov, AM 2006b, Wavelength-tunable ultrashort-pulse output of a photonic-crystal fiber designed to resolve ultrafast molecular dynamics, *Opt. Lett.* **31**, 3330–3332.

Jensen, JB, Pedersen, LH, Hoiby, PE, Nielsen, LB, Hansen, TP, Folkenberg, JR, Riishede, J, Noordegraaf, D, Nielsen, K, Carlsen, A & Bjarklev, A 2004, Photonic crystal fiber based evanescent-wave sensor for detection of biomolecules in aqueous solutions, *Opt. Lett.* **29**, 1974–1976.

Jones, DJ, Diddams, SA, Ranka, JK, Stentz, A, Windeler, RS, Hall, JL & Cundiff, ST 2000, Carrier-envelope phase control of femtosecond modelocked lasers and direct optical frequency synthesis, *Science*, **288**, 635–639.

Kano, H & Hamaguchi, H 2003, Characterization of a supercontinuum generated from a photonic crystal fiber and its application to coherent Raman spectroscopy, *Opt. Lett.* **28**, 2360–2362.

Kano, H & Hamaguchi, H 2005, Vibrationally resonant imaging of a single living cell by supercontinuum-based multiplex coherent anti-Stokes Raman scattering microspectroscopy, *Opt. Express* **13**, 1322–1327.

Knight, JC 2003, Photonic crystal fibers, *Nature* **424**, 847–851.

Kobayashi, T (ed.), 1996, *J-Aggregates* (World Scientific, Singapore).

Konorov, SO, Fedotov, AB & Zheltikov, AM 2003, Enhanced four-wave mixing in a hollow-core photonic-crystal fiber, *Opt. Lett.* **28**, 1448–1450.

Konorov, SO, Akimov, DA, Serebryannikov, EE, Ivanov, AA, Alfimov, MV & Zheltikov, AM 2004a, Cross-correlation FROG CARS with frequency-converting photonic-crystal fibers, *Phys. Rev. E* **70**, 057601.

Konorov, SO, Fedotov, AB, Mitrokhin, VP, Beloglazov, VI, Skibina, NB, Shcherbakov, AV, Wintner, E, Scalora, M & Zheltikov, AM 2004b, Laser ablation of dental tissues with picosecond pulses of 1.06 μm radiation transmitted through a hollow core photonic-crystal fiber, *Appl. Opt.* **43**, 2251.

Légaré, F, Evans, CL, Ganikhanov, F & Xie, XS 2006, Towards CARS endoscopy, *Opt. Express* **14**, 4427–4432.

Lewis, JD, Destito, G, Zijlstra A, Gonzalez, MJ, Quigley, JP, Manchester, M, & Stuhlmann, H 2006, Viral nanoparticles as tools for intravital vascular imaging, *Nature Medicine* **12**, 354–360.

Lim, H, Ilday, F & Wise, F 2002, Femtosecond ytterbium fiber laser with photonic crystal fiber for dispersion control, *Opt. Express* **10**, 1497–1502.

Limpert, J, Röser, F, Schreiber, T & Tünnermann, A. 2006, High-power ultrafast fiber laser systems, *IEEE J. Sel. Top. Quantum Electron* **12**, 233–244.

Mitrokhin, VP, Fedotov, AB, Ivanov, AA, Alfimov, MV & Zheltikov, AM 2007, Coherent anti-Stokes Raman scattering microspectroscopy of silicon components with a photonic-crystal fiber frequency shifter, *Opt. Lett.* **32**, 3471–3473.

Monro, TM, Belardi, W, Furusawa, K, Baggett, JC, Broderick, NGR & Richardson, DJ 2001, Sensing with microstructured optical fibres, *Meas. Sci. Technol.* **12**, 854–858.

Myaing, MT, Ye, JY, Norris, TB, Thomas, T, Baker, Jr, JR, Wadsworth, WJ, Bouwmans, G, Knight, JC & Russell, PStJ 2003, Enhanced two-photon biosensing with double-clad photonic crystal fibers, *Opt. Lett.* **28**, 1224–1226.

Ouzounov, DG, Ahmad, FR, Müller, D, Venkataraman, N, Gallagher, MT, Thomas, MG, Silcox, J, Koch, KW & Gaeta, AL 2003, Generation of megawatt optical solitons in hollow-core photonic band-gap fibers, *Science* **301**, 1702–1704.

Paulsen, HN, Hilligsøe, KM, Thøgersen, J, Keiding, SR & Larsen, JJ 2003, Coherent anti-Stokes Raman scattering microscopy with a photonic crystal fiber based light source, *Opt. Lett.* **28**, 1123–1125.

Ranka, JK, Windeler, RS & Stentz, AJ 2000, Visible continuum generation in air–silica microstructure optical fibers with anomalous dispersion at 800 nm, *Opt. Lett.* **25**, 25–27.

Rarity, JG, Fulconis, J, Duligall, J, Wadsworth, WJ & Russell, PStJ 2005, Photonic crystal fiber source of correlated photon pairs, *Opt. Express* **13**, 534–544.

Reeves, WH, Skryabin, DV, Biancalana, F, Knight, JC, Russell, PStJ, Omenetto, FG, Efimov, A & Taylor, AJ 2003, Transformation and control of ultra-short pulses in dispersion-engineered photonic crystal fibres, *Nature* **424**, 511–515.

Russell, PStJ 2003, Photonic crystal fibers, *Science* **299**, 358–362.

Russell, PStJ 2006, Photonic-Crystal Fibers, *J. Lightwave Technol.* **24**, 4729–4749.

Serebryannikov, EE & Zheltikov, AM 2007, Ionization-induced effects in the soliton dynamics of high-peak-power femtosecond pulses in hollow photonic-crystal fibers, *Phys. Rev. A* **76**, 013820.

Sidorov-Biryukov, DA, Serebryannikov, EE & Zheltikov, AM 2006, Time-resolved coherent anti-Stokes Raman scattering with a femtosecond soliton output of a photonic-crystal fiber, *Opt. Lett.* **31**, 2323–2325.

Smith, CM, Venkataraman, N, Gallagher, MT, Muller, D, West, JA, Borrelli, NF, Allan, DC & Koch, K 2003, Low-loss hollow-core silica/air photonic bandgap fibre, *Nature* **424**, 657–659.

Spano, FC & Mukamel, S 1989, Nonlinear susceptibilities of molecular aggregates: enhancement of $\chi^{(3)}$ by size, *Phys. Rev. A* **40**, 5783–5801.

Südmeyer, T, Brunner, F, Innerhofer, E, Paschotta, R, Furusawa, K, Baggett, JC, Monro, TM, Richardson, DJ & Keller, U 2003, Nonlinear femtosecond pulse compression at high average power levels by use of a large-mode-area holey fiber, *Opt. Lett.* **28**, 1951–1953.

Udem, Th, Holzwarth, R & Hänsch, TW 2002, Optical frequency metrology, *Nature* **416**, 233–237.

von Vacano, B, Wohlleben, W & Motzkus, M 2006, Actively shaped supercontinuum from a photonic crystal fiber for nonlinear coherent microspectroscopy, *Opt. Lett.* **31**, 413–415.

Zheltikov, AM 2006, Let there be white light: supercontinuum generation by ultrashort laser pulses, *Phys. Uspekhi* **49**, 605–628.

Zheltikov, AM 2007a, Photonic-crystal fibers for a new generation of light sources and frequency converters, *Phys. Uspekhi* **50**, 705–727.

Zheltikov, AM 2007b, Fiber-optic synthesizer of controlled sequences of ultrashort light pulses for single-beam coherent anti-Stokes Raman scattering microspectroscopy, *JETP Lett.* **85**, 539–543.

Zheltikov, AM & Koroteev, NI 1999, Coherent four-wave mixing in excited and ionized gas media: four-photon spectrochronography, ellipsometry, and nonlinear-optical imaging of atoms and ions, Phys. Uspekhi **42**, 321–351.

Zipfel, WR, Williams, RM & Webb, WW 2003, Nonlinear magic: multiphoton microscopy in biosciences, *Nature Biotechnol.* **21**, 1369–1377.

Zumbusch, A, Holtom, GR & Xie, XS 1999, Three-dimensional vibrational imaging by coherent anti-stokes raman scattering, *Phys. Rev. Lett.* **82**, 4142–4145.

Index